OXFORD CLASSICAL MONOGRAPHS

Published under the supervision of a Committee of the Faculty of Classics in the University of Oxford

The aim of the Oxford Classical Monograph series (which replaces the Oxford Classical and Philosophical Monographs) is to publish books based on the best theses on Greek and Latin literature, ancient history, and ancient philosophy examined by the Faculty Board of Classics.

Imperial Mines and Quarries in the Roman World

Organizational Aspects 27 BC – AD 235

ALFRED MICHAEL HIRT

OXFORD
UNIVERSITY PRESS

OXFORD
UNIVERSITY PRESS

Great Clarendon Street, Oxford OX2 6DP

Oxford University Press is a department of the University of Oxford.
It furthers the University's objective of excellence in research, scholarship,
and education by publishing worldwide in

Oxford New York

Auckland Cape Town Dar es Salaam Hong Kong Karachi
Kuala Lumpur Madrid Melbourne Mexico City Nairobi
New Delhi Shanghai Taipei Toronto

With offices in

Argentina Austria Brazil Chile Czech Republic France Greece
Guatemala Hungary Italy Japan Poland Portugal Singapore
South Korea Switzerland Thailand Turkey Ukraine Vietnam

Published in the United States
by Oxford University Press Inc., New York

First published 2010

British Library Cataloguing in Publication Data
Data available

Library of Congress Cataloging in Publication Data
Library of Congress Control Number: 2009938800

Typeset by SPI Publisher Services, Pondicherry, India
Printed in Great Britain
on acid-free paper by
CPI Antony Rowe, Chippenham, Wiltshire

ISBN 978–0–19–957287–8

1 3 5 7 9 10 8 6 4 2

Für Mueter u Vater, Gotte u Götti, für e Michael u dr Heinz

Preface

Roman mines and quarries have emerged in recent years as a topic of study by scholars interested in the Roman economy. Archaeologists in particular have directed their attention to specific mining sites like Las Médulas, Roşia Montană, and Dolaucothi, or offered detailed analysis of quarrying sites such as Gebel Dokhan, Gebel Fatirah, or Chemtou. These archaeological studies set aside, numerous finds of Latin and Greek inscriptions on stone monuments, quarried items, ingots, or bronze tablets, as well as writing tablets, ostraca, and papyri provide additional insights but have been published disparately. The main aim of this study is to unite the various snippets of information with respect to their archaeological and historical context in order to gain an overview of various aspects of the internal organization of mining and quarrying ventures under imperial control and their place within the administrative framework of the Roman empire.

An initial version of this study was submitted as a DPhil thesis at Oxford University in 2004. The thesis as a whole was reworked in the following years in order to include additional material and completed in late 2007. The study owes much to numerous discussions with Michael A. Speidel who freely shared his profound insights into the workings of the Roman empire; and to Heinz E. Herzig who at the University of Berne created the ideal academic environment and supported this endeavour from the beginning. At Oxford, my supervisor Andrew Wilson pointed the reluctant ancient historian towards the relevancy of the archaeological data and offered superb advice on many of its aspects. Many more at Oxford contributed (knowingly or not) to the genesis of this study. Alan Bowman and David Mattingly provided helpful comments on the thesis and encouraged publication of the work, Alan taking on the time-consuming task of correcting a draft version. Further thanks must go to J. J. Coulton and Nicholas Purcell, Bill Leadbetter, Adam Bülow-Jacobsen, Michael Crawford, Peter Herz, Stefanie Martin-Kilcher, for advice and bits and pieces of helpful information. Caillan Davenport, Caroline Brehaut, Charlotte Walden, Lesley Cousins, and, last but not least, Oskar Kaelin, read the texts at different stages and improved it by eradicating various remnants of my 'Helvetic' syntax. Galina Rusak and Colin Hughes helped me with the index and Kathleen Fearn and Abigail Coulson at OUP turned the submitted text into the typescript presented here. Any remaining mistakes are of course mine. Adam Bülow-Jacobsen's 'Mons Claudianus Ostraca graeca et latina IV' (2009) was published too late to be included in this survey.

My stay at Oxford was made possible by generous grants by the Swiss National Fund, the Janggen-Pöehn Stiftung in St Gallen, Switzerland, the Kármán-Stiftung of the University of Berne, Switzerland.

Finally, I would wish to thank my parents, the extended Hirt-Bolliger clan, and my friends who endured my *gravamina* and offered their encouragement and support.

A.M.H.

Contents

List of Illustrations

List of Abbreviations

AE	*Année Epigraphique.*
AIJ	V. Hoffiller and B. Saria, *Antike Inschriften aus Jugoslawien, 1. Noricum und Pannonia Superior*, Zagreb 1938.
CBFIR	E. Schallmayer et al (eds.), *Corpus der griechischen und lateinischen Beneficiarier-Inschriften des Römischen Reiches*, Stuttgart 1990.
CIG	*Corpus Inscriptionum Graecarum*, Berlin 1828–77.
CIL	*Corpus Inscriptionum Latinarum.*
CILA	*Corpus de Inscripciones Latinas de Andalucía*, Sevilla.
Corinth VIII/III	J. H. Kent, *The Inscriptions 1926–1950* (Corinth 8/3), Princeton 1966.
EAstorga	T. Mañanes Perez, *Epigrafia y numismática de Astorga romana y su entorno* (Acta Salmanticensia. Filosofia y Letras 134), Salamance 1982.
Eph. ep.	*Ephemeris epigraphica. Corpus inscriptionum Latinarum supplementum*, Berlin 1, 1872–9, 1903–13.
FIRA	*Fontes iuris Romani anteiustiniani*, Florentiae 1940–3.
HEp	*Hispania Epigraphica*
I. Akoris	É. Bernand, *Les Inscriptions Grecques et Latines d'Akôris*, Cairo 1988.
IDR	*Inscriptiones Dacicae Romanae*, Bucarest.
IG	*Inscriptiones Graecae*, Berlin.
IGBulg	G. Mihailov, *Inscriptiones Graecae in Bulgaria repertae*, Sofia 1958–97.
IGLS	*Inscriptions Grecques et Latines de la Syrie*, Beyrouth & Paris.
IGRR	R. Cagnat et al., *Inscriptiones Graecae ad Res Romanas pertinentes*, Paris 1906–27.
IGUR	L. Moretti, *Inscriptiones Graecae Urbis Romae*, Roma 1968–90.
IK	*Inschriften griechischer Städte aus Kleinasien*, Bonn.
I. Ko.Ko.	A. Bernand, *De Koptos à Kosseir*, Leyden 1972.
ILAfr	R. Cagnat, A. Merlin, and L. Chatelain, *Inscriptions Latines d'Afrique (Tripolitaine, Tunisie et Maroc)*, Paris 1923.
ILA Vell.	B. Rémy, *Vellaves* (Inscriptions latines d'Aquitaine 3), Bordeaux 1995.

ILER	J. Vives, *Inscripciones latinas de la España romana: antalogía de 6 800 textos*, Barcelona 1971–2.
ILJug	*Inscriptiones Latinae quae in Iugoslavia inter annos MCMII et MCMLXX repertae et editae sunt*, Ljubliana.
ILLPRON	*Inscriptionum lapidariarum Latinarum provinciae Noriciusque ad annum MCMLXXXIV repertarum indices*, Berlin.
ILS	H. Dessau, *Inscriptiones Latinae Selectae*, Berlin 1892–1916.
ILTun	A. Merlin, *Inscriptions latines de la Tunisie*, Paris 1944.
I. Memnon	A. Bernand and É. Bernand, *Les inscriptions grecques et latines du Colosse de Memnon*, Cairo 1960.
I. Metr.	É. Bernand, *Inscriptions Métriques de l'Égypte grécoromaine. Recherches sur la poésie épigrammatique des Grecs en Égypte*, Paris 1969.
IMS	*Inscriptions de la Mésie Supérieure*, Beograd.
Insc.It.	*Inscriptiones Italiae.*
Inscr. Aqu.	J. B. Brusin, *Inscriptiones Aquileiae*, Udine 1991–3.
I. Pan	A. Bernand, *Pan du Désert*, Leyden 1977.
I. Portes	A. Bernand, *Les portes du désert. Recueil des inscriptions grecques d'Antinooupolis, Tentyris, Koptos, Apollonopolis Parva et Apollonopolis Magna*, Paris 1984.
IRC I	G. Fabre et al., *Barcelone* (Inscriptions romaines de Catalogne 1), Paris & Bordeaux 1984.
IRLeón	F. Diego Santos, *Inscripciones romanas de la provincia de Léon*, Léon 1986.
IRPac	J. d'Encarnação, *Inscrições romanas do Conventus Pacensis: subsidios para o estudo da romanização*, Coimbra 1984.
IRPL	F. Arias Vilas, P. Le Roux, and A. Tranoy, *Inscriptions Romaines de la Province de Lugo*, Paris 1979.
IRT	J. M. Reynolds and J. B. Ward-Perkins, *The Inscriptions of Roman Tripolitania*, Rome 1952.
IScM	*Inscriptiones Scythiae Minoris graecae et latinae.*
I.Th.Sy.	A. Bernand, *De Thèbes à Syène*, Paris 1989.
LGPN	*A Lexicon of Greek Personal Names*, Oxford.
LMD	*lex metallis dicta*
LMV	*lex metalli Vipascensis*
MAMA	*Monumenta Asiae Minoris Antiqua*, Manchester.
O.Claud.	*Mons Claudianus. Ostraca Graeca et Latina*, Cairo.
OGIS	W. Dittenberger, *Orientis graeci inscriptiones selectae*, Leipzig 1903–5.

PIR	*Prosopographia Imperii Romani*, Berlin.
RIB	*Roman Inscriptions of Britain.*
SB	*Sammelbuch der griechischen Urkunden aus Aegypten.*
SEG	*Supplementum Epigraphicum Graecum.*
Supp.It.	*Supplementa Italica*, Roma.
Syll.^{3}	W. Dittenberger, *Sylloge inscriptionum graecarum*, Leipzig 1915–24.
TAM	*Tituli Asiae Minoris*, Vienna.
TIR	*Tabula Imperii Romani.*
TLL	*Thesaurus Linguae Latinae*, Leipzig.

For Papyri and Ostraca, cf. *Checklist of Editions of Greek, Latin, Demotic and Coptic Papyri, Ostraca and Tablets* (http://scriptorium.lib.duke.edu/papyrus/texts/clist_ostraca.html, 21 February 2009).

1

Introduction

Modern visitors to Chemtou in the Tunisian Medjerda Valley or Las Médulas in northwestern Spain are easily overawed by the grand scale of opencast quarrying or mining operations. The massive impact of Roman engineering on nature is still strikingly visible at these sites, even though much of their original contexts has withered away. These landscape monuments to Roman ingenuity reflect the purpose of their formation. The control over marble or metal resources was undoubtedly of high significance to Rome; the command of marble was not merely a luxury but central to the representation of imperial wealth and power, and uninhibited access to metal was vital for the economic and political survival of the Roman empire. The relevance of this study on Roman mines and quarries, however, lies not in the analysis of stimuli for the demand for these raw materials, but in the examination of the organizational measures Rome devised to ensure their continued extraction. In the wider context of the administration of the Roman empire this particular focus gains in substance: the study of administrative configurations and hierarchies, its control and command of human and material resources, and the organizational procedures set in place to ensure its continued existence may provide a partial insight into the sources of the strength and longevity of the Roman empire. The present study, thus, strives to analyse the organizational measures devised for the extraction of metals and marble. It is restricted to the examination of evidence from *imperial* mines and quarries. These are hereby defined as extractive operations where the presence of imperial officials i.e. members of the *familia Caesaris* or equestrians, and/or Roman army personnel, are documented in the epigraphic record of the mine/quarry. In consequence, imperial mines and quarries are part of a wider organization, the administration of the Roman empire under the ultimate control of the Roman emperor and his Palatine bureaux. The main epigraphic evidence for imperial mines and quarries limits this study to the period from the first to the third century AD.

Organizational aspects have not exactly been at the heart of the scholarly discussion on Roman extractive operations. The majority of papers and studies centre on the archaeological data and written evidence emerging from individual mines and quarries and rarely have a general conspectus in mind. Roman mining and quarrying management was usually dealt with summarily in extensive studies on the administration of the Roman empire, in particular Otto Hirschfeld's work on the 'Kaiserliche Verwaltung' or Michael Rostovtzeff's 'Geschichte der Staatspacht', whereas evidence on procuratorial posts and imperial personnel responsible for imperial mines and quarries is scattered throughout the works of Hans-Georg Pflaum and Gérard Boulvert.[1] In his concise chapter on the administration of Roman extractive operations—mines and quarries were subsumed under the same heading, 'Bergwerke'—Hirschfeld presented a detailed overview of the written evidence available at his time. The topics covered by Hirschfeld ranged from the question of ownership of mines, over the responsibilities of mining officials i.e. *procuratores*, their position within the provincial administration and the system of exploitation, to the fiscal connections between mining operations and Rome. Hirschfeld had identified mining ventures under imperial control as being owned by the Roman state, but did not exclude the possibility of private individuals or municipalities continuously owning mines throughout the Principate. In this opinion he was not followed by most scholars; in fact, the preponderant opinion sees Roman mines as the private property of the emperor (cf. 3.3.1).[2] According to Hirschfeld, the administration of a certain type of (gold, silver, or iron) mine in one, rarely in two provinces was dealt with by a freedman or equestrian procurator—a task which, he believed, required profound knowledge of mining technology.[3] As regards further responsibilities of procurators, Hirschfeld observed that the mining 'laws' for Vipasca/mod. Aljustrel in Portugal described judicial and other powers of the mining procurator in charge comparable to those of municipal authorities.[4] Most influential was his view on the system (or mode) of exploitation of mines by the central authorities: he assumed that during the Principate most mines were farmed out to small-scale lessees, 'Kleinpächter', in return for rent collected by *conductores* or large-scale lessees, 'Grosspächter'. By the later second century these *conductores* were replaced by imperial procurators who now dealt directly with the small-scale lessees. This view was followed by Michael Rostovtzeff, Ulf Täckholm, and others.[5] Täckholm, however, did point

[1] Marquardt 1884; Rostovtzeff 1904; Hirschfeld 1905; Pflaum 1960–1; Boulvert 1970; Pflaum 1982; Brunt 1990*b*; Brunt 1990*c*; Eck 1997*a*; Eich 2005.

[2] Hirschfeld 1905: 145–50, 158 f.

[3] Hirschfeld 1905: 174.

[4] Hirschfeld 1905: 160 f.

[5] Rostovtzeff 1904: 446; Hirschfeld 1905: 152; Täckholm 1937: 109. In the light of the abundant written evidence for convicts condemned to the mines Hirschfeld (1905: 162) does not exclude the possibility of a more direct control of exploitation by the Roman state.

out that the use of *conductores* may have been restricted to iron mines only and he acknowledged the possibility that modes of exploitation and administrative systems were perhaps not as uniform as Hirschfeld and Rostovtzeff suggested.[6] Peter Brunt set a different tone by offering an alternative view: he argued that *conductores* set in place either for the collection of rent or the exploitation of a mine were supervised by procurators with similar geographical circumscriptions and thus coexisted simultaneously.[7] With regard to mining procurators, Stanisław Mrozek analysed their titles throughout the empire and noted the difference in geographical extent of their jurisdiction which he believed to be dependent upon the social status of the freedmen or equestrian incumbents. He furthermore observed that the administration of silver mines both in Pannonia and Dalmatia was merged perhaps under Marcus Aurelius. Mrozek tried to assess the position of equestrian mining procuratorships within equestrian careers and outline their administrative responsibilities.[8] Brunt's astute analysis of procuratorial judicial powers in general added to the picture provided by the Vipasca mining regulations.[9] Both were combined by Danielle Capanelli in order to outline the judicial and other responsibilities of mining procurators.[10] As for the geographical extent of procuratorial jurisdiction, Géza Alföldy gave an original impetus by identifying a vast 'economic' district, the *patrimonium regni Norici*, at the heart of the province of Noricum. Based on Alföldy's method Slobodan Dušanić and Christoph Noeske wished to identify and outline further mining districts throughout the Danube provinces and Dacia.[11] The study by Christoph Noeske on Ampelum/mod. Zlatna and on Alburnus Maior/mod. Roșia Montană offered primarily a survey of both 'mining' communities, but also dealt with the freedmen and equestrian procurators responsible for the gold mines, and detailed the functions of their staff. Noeske believed that during the later second century the freedmen procurators at Ampelum were subordinate to their equestrian colleagues—an arrangement Hans-Georg Pflaum termed 'collégialité inégale'.[12] As for the tasks of these procurators, Noeske assumed that—besides the general responsibilities of a procurator—the Vipasca mining 'laws' applied to the Dacian gold mines as well.[13] The

[6] Täckholm 1937: 109–13.
[7] Andreau 1989: 99 f.; Brunt 1990*c*.
[8] Mrozek 1968. In addition, Jenö Fitz (1972) provided an improved chronological compilation of equestrian mining procuratorships for Pannonia and Dalmatia in support of Mrozek's view.
[9] Brunt 1990*b*.
[10] Capanelli 1989.
[11] Alföldy 1970; Dušanić 1977; Noeske 1977.
[12] Noeske 1977: 300 f.
[13] Noeske 1977: 301.

ubiquitous application of the Vipasca tablets, however, was questioned by Claude Domergue who pointed out the specific geological conditions to which they apply.[14] The unique early second-century AD mining 'laws' from the mines of Vipasca/mod. Aljustrel in Portugal have received special attention since their discovery in the late nineteenth century. These incomplete tablets provide a rare insight into the organization of a mining district and have spawned numerous papers and monographs discussing their content. The majority of scholars understood the Vipasca tablets to regulate the lease of mines to interested parties, small-scale lessees ('Kleinpächter'), by the Roman state in return for part of the extracted ore.[15] Stamped lead ingots from Spain, Britain, and, most recently, from the Augustan 'provincia Germania' were understood to reflect similar mining arrangements with the mining lessee or the owner of the mine being named on the ingot.[16] A recent study by Antonio Mateo, however, suggests that mining plots at Vipasca were sold to interested personnel.[17] Besides the 'régie indirecte', it was argued that a 'régie directe', the direct involvement of the Roman state in the exploitation of mines, occurred as well.[18] This is assumed to be the case with vast opencast mining sites such as Las Médulas in north-western Spain. Recent archaeological surveys of mining landscapes in the same region appear to corroborate the picture drawn by written sources. In this context it has been argued by Claude Domergue, Ines Sastre-Prats, Almudena Orejas, and others that the indigenous population was forcibly resettled in order to work in the mines.[19] Besides locals being coerced to mining work, written sources give information on the condemnation of convicts to imperial *metalla*, a field of study that has met renewed interest.[20] Free labour in an imperial mining context is documented at Roşia Montană in Romania and in the light of the new evidence from the quarries at Mons Claudianus in Egypt has also drawn fresh attention.[21] The role of the Roman army in mining administration was emphasized early on by Hirschfeld and others—an aspect which Pascal Le Roux dealt with in greater detail.[22]

[14] Domergue 1983: 146 f. Täckholm (1937: 103) likewise refuted the universal validity of the Vipasca regulations for all imperial mining operations albeit without providing any explanation.

[15] See Flach 1979 with older bibliography; Domergue 1983; Lazzarini 2001.

[16] Domergue 1990: 253–77; *RIB* II/1; Rothenhöfer 2003; Eck 2004*b*.

[17] Mateo 2001; Domergue 2004.

[18] Domergue 1990: 303.

[19] Domergue & Sillieres 1977; Domergue & Martin 1977; Domergue & Hérail 1978; Orejas 1994; Orejas 1996; Sastre Prats 1998; Orejas & Sastre Prats 1999; Orejas & Sanchez-Palencia 2002.

[20] Mommsen 1899: 47 f., 960–3; Geerlings 1983; Millar 1984; Gustafson 1994; Lassandro 1995; Peacock 1995; Gustafson 1997; Salerno 2003.

[21] Mrozek 1969; Mrozek 1977: 102–6; Noeske 1977: 336–45; Andreau 1990: 89 f.; Cuvigny 1996*a*.

[22] Hirschfeld 1905: 172 f.; Täckholm 1937: 138; LeRoux 1989.

A number of studies focused on Roman mining administration in a specific region or province.[23] Perhaps the most comprehensive analysis of a mining region to date is Claude Domergue's work on ancient mining in Spain and Portugal which deals in part with questions pertaining to the administration of mining operations. Although the study is limited to Roman Iberia, he identifies and discusses the relevant issues, i.e. the duties of provincial authorities (financial and patrimonial procurators) in mining administration; the varying functions of a procurator within a mining district in the contexts of a direct or indirect regime of exploitation; the position of these procurators within the imperial administration, in particular their linkage with the Palatine bureaux at Rome; the role of their staff; and the relevance of technological challenges in organizing mining operations.[24] The latter issue has been particularly emphasized in recent years, not least as a result of the archaeological (re)examination of various mining sites, such as Dolaucothi in Wales, Três minas and Las Médulas in north-western Spain, or Roșia Montană.[25] In a concise report on Roman mining Jean Andreau attempted to widen the ambit by discussing recent research on ownership of mines, the modes of exploitation, the workforce, and technologies employed in mining operations throughout the empire. Reflecting scholarly discourse, administrative aspects of mining operations *per se* were not discussed extensively: The question of 'collégialité inégale' in the procuratorship of Asturia and Callaecia set aside, Andreau briefly explored the relationship between imperial personnel and modes of exploitation as well as the role of the Roman army in mining operations. Nonetheless, Andreau did stress the role of the Roman state in deciding the modes of exploitation and the use of work force—without identifying the parties to the decision process.[26] Recent papers have again presented arguments for a centralized administrative bureau in Rome responsible for mining ventures.[27]

Given the scarcity of written evidence, insights into the organization of quarrying operations were provided predominantly by labels on quarried products emerging from Pietro Ercole Visconti's excavations of Rome's marble yards between 1868 and 1870 and published by Luigi Bruzza in 1870. Based on these quarry labels, Otto Hirschfeld, followed by Michael Rostovtzeff, Charles Dubois, and Ulf Täckholm, argued that quarries must have been contracted out to private entrepreneurs who had produced a certain amount

[23] Gaul: Sablayrolles 1989. Dacia: Wollmann 1996. Illyricum: Dušanić 1977; Ørsted 1985; Dušanić 1989; Škegro 2000; Dušanić 2004*b*. For a general overview, cf. Blázquez Martínez 1989.

[24] Domergue 1990: 279–303, esp. 294 f.

[25] Damian 2003; Burnham & Burnham 2004.

[26] Andreau 1989: esp. 96–100; Andreau 1990.

[27] Dušanić 1989; Cuvigny 1996*a*.

of marble in return for a payment.[28] Kurt Fitzler, however, believed the quarries in Roman Egypt to have been farmed out to interested lessees who were allowed to keep the produce in return for a rent. Moreover, Fitzler argued for direct exploitation of quarries by the Roman state through the use of convicts and prisoners.[29] As for the administration of imperial quarries, Hirschfeld and others assumed that imperial freedmen, or less frequently equestrians, ran these operations as procurators, supported by *probatores* and military officers as technical experts.[30] In recent years quarry labels were studied more intensively as the available data was dramatically expanded by new material from Bacakale, Simitthus, Mons Claudianus, and elsewhere, allowing for refined interpretations of the modes of exploitation (cf. 7.2).[31] Given the dissemination of Latin terminology in quarry labels, J. Clayton Fant argued that this suggested the quarries whence the inscribed marble originated were under imperial control. This observation and the evidence for a *statio marmorum* at Rome were seen by Fant to suggest a centralized administration of imperial quarrying affairs.[32] The internal organization of quarries was reconstructed predominantly on the basis of these labels (cf. 7.2).[33] The administrative personnel involved in quarrying organizations and the workforce present has in recent decades not received the attention it deserves—primarily a result of a scanty evidential basis. With the abundant written and archaeological material yielded by excavations at quarries in the Egyptian Eastern Desert in the late 1990s the organizational mechanisms and structures of a quarrying community have moved to the centre of interest. The wealth of ostraca has, as yet, not been examined and analysed in full within the context of imperial administration.[34]

In the light of the abundance of written material emerging particularly from quarrying sites under imperial control, the special focus on the internal organization and administration of mines and quarries is warranted. Studying the administration of mines and quarries jointly is not least called for by the application of the term *metallum* or *μέταλλον* to quarries as well as mines.

[28] Rostovtzeff 1904: 454; Hirschfeld 1905: 166; Dubois 1908: xxviii; Täckholm 1937: 118 f.

[29] Fitzler 1910: 116 f., 119 f.

[30] Hirschfeld 1905: 171. On the administration of quarries, cf. also: Dubois 1908: xxx–xxiv; Fitzler 1910: 125–35; Täckholm 1937: 117 f.

[31] The foundation for the research on quarry labels was basically laid by Luigi Bruzza in 1870 who published the labels inscribed on to blocks at the Marmorata in Rome. For Bacakale, cf. Fant 1989*a*; Christol & Drew-Bear 1986, 1987, 1991; Drew-Bear 1994; Drew-Bear & Eck 1976; for Simitthus, cf. Kraus 1993*a*; for Mons Claudianus, cf. Peacock & Maxfield 1997.

[32] Fant1993*a*: 157–60; cf. also Maischberger 1997: 19 f.; contra Hirschfeld 1905: 176; Dubois 1908: xxxviii.

[33] Dodge 1988; Dodge 1991; Ward-Perkins 1992*a*; Ward-Perkins 1992*b*; Fant 1989*b*; Fant 1993*a*; Fant 1993*b*; Maischberger 1997.

[34] Bingen et al. 1992; Bingen et al. 1997; Cuvigny 2000*a*.

The basis for comparison is predominantly provided by the similarity of these extractive operations, particularly the common nature of organizational constraints faced by the authorities set in charge of running these ventures. The issues addressed by earlier research on quarries or mines have remained much the same: the issue of ownership, the modes of exploitation, the technology and workforce employed, the relation between imperial authorities and extractive operations, the function of administrative and military personnel involved, and the position of mines and quarries within the provincial administration. Nonetheless, the comprehensive collection and analysis of these issues from a strictly organizational/administrative viewpoint, that is, from the perspective of the imperial authorities responsible for mines or quarries in its control has not been attempted since the works of Hirschfeld and Täckholm.

An appropriate starting point of a study on organizational aspects of imperial mines and quarries is the definition of the term 'organization': any organization can be defined as the totality of measures directed towards the achievement of goals and/or the fulfilment of a purpose. These measures structure any social collective resulting in a division of labour, and arrange the activities of its members, the use of assets, as well as the processing of information. As a result of these processes, behavioural expectations or rules are deliberately set for the members of the system: the interactions amongst individuals within an organization are formalized. These formalized rules are shaped in order to achieve certain objectives and form the basis of membership within the system.[35] Consequently, the members of an 'organization' can be reasonably well distinguished from non-members; internal forms (buildings, machines, information, types of behaviour) are different from external ones. The clear distinction of the organizational domain is closely tied to the definition of membership of an organization. Members are either employed by the organization, or are contributing towards the purpose of the organization.[36] *Structures* within an organization may hence be defined as the arrangement of members of the organized social collective and of the stable relationships among them. The order of functional activities and procedures of the organization's members or actors are described as *operative processes* (work procedures) and *directive processes* (planning and controlling of assets and goals, direction of human resources etc.).[37] Any organization is embedded in an environment or context. This organizational environment not only provides input to the organization in the form of human resources,

[35] Hill, Felbaum, & Ulrich 1994: 17, 24 f.
[36] Sorge 2002: 4, 8.
[37] Hill, Felbaum, & Ulrich 1994: 26.

raw materials, information, and technology, but also receives the output of an organization, such as its produce, goods, and experienced personnel. The organizational environment sets limits to an organization through *constraints* such as legal and socio-cultural restrictions, or the scarcity of raw material.[38]

This very generalized definition of the term 'organization' offers a rudimentary idea of what the analysis of Roman mining and quarrying organizations might entail. However, not all aspects of the organization of Roman extractive operations can be explored to their fullest extent. This is due to the fragmentary and unbalanced state of our literary and epigraphic material, which hinders any attempt at a full description of the imperial extractive industry in the kind of detail a study of a modern organization would permit. Given the ramifications of our evidential basis, the analysis of the organizational structure and content of Roman imperial mines and quarries is limited to a few areas. Yet, unlike other branches of the Roman administration, the physical remains of mines and quarries and their spatial arrangement in relation to technical, geological, topographical, and geographical constraints render a rather clear idea of potential organizational tasks at hand. Any study on organizational aspects of imperial extractive operations thus has to commence by detailing organizational measures innate to any quarrying or mining venture. Against this backdrop the measures taken by Roman authorities can be set into the appropriate context and the variety of solutions applied to substantial problems recognized to its full extent. Based on this foundation, we can attempt to define the responsibilities within the 'directive' and 'operative' processes (e.g. recruitment and pay of a work force, management of work procedures, etc.) and their allocation to numerous 'functions', i.e. incumbents of 'civilian' or military posts (procurators, centurions, civil engineers, etc.), documented in the epigraphic records of mines and quarries. In view of the evidential basis, any precise allocation of all administrative responsibilities to individual functions remains fraught with difficulties. The scope of most relevant written materials is rarely wide enough to allow for a glimpse beyond the structure of a quarrying or mining district. Most records of information exchange between mines/quarries and the upper echelons of the imperial administration or the provincial authorities have perished or have yet to be recovered. This basically rules out a clear-cut reconstruction of the decision-making process and the position of imperial mining and quarrying organizations within the framework of Roman provincial administration. Nevertheless, the connections between quarrying/mining administrations, provincial governors, and the emperor/Palatine bureaux in Rome have left traces in the written sources which shall be explored further below.

38 Hill, Felbaum, & Ulrich 1994: 20, 319–66 with fig. II/3–9; Sorge 2002: 5–7.

The present study thus revolves around the core elements of (*a*) determining the elementary constraints and challenges faced by the quarrying/mining organization, (*b*) identifying the key figures in organizing extractive operations under imperial control, (*c*) allocating organizational tasks and responsibilities to these key figures, (*d*) evaluating their solutions for the main organizational challenges at hand, and (*e*) delimiting their position within a wider administrative framework. The study is structured accordingly: the determination of technical, as well as geological, topographical, and geographical constraints and of the organizational challenges for Roman mining and quarrying ventures deriving from these constraints (Ch. 2), are followed by the definition of mining or quarrying districts and their status by legal and other sources (Ch. 3). This is followed by a study of hierarchical structures of imperial officials and their staff (Ch. 4); the role of the Roman army in mining and quarrying operations (Ch. 5); the assignment of administrative and organizational responsibilities to imperial officials and staff members (Ch. 6); and the use of a non-imperial workforce and civilian partners in exploiting extractive ventures (Ch. 7). Finally, the position of mining or quarrying organizations within the framework of the imperial administration and the connections between the emperor/Palatine bureaux and the mining and quarrying administrations will be explored (Ch. 8).

2

Geological Constraints and Organizational Implications

The geology and topography of marble outcrops and ore deposits dictate to a large degree the technology which can be used and the scale of extractive operations. Consequently, these restrictions significantly influence the organizational structures and, together with the geographical location, determine the organizational challenges faced by the heads of these extractive operations. Moreover, the geology and topography of a mining or quarrying district also determine the spatial arrangement of operative (work procedures) and directive processes (planning and controlling of assets and goals, direction of human resources, etc.). This is best exemplified by the topographical layout of building structures, transport facilities, and extractive locations within imperial mining and quarrying districts.

2.1. IMPERIAL QUARRIES

The archaeological remains of extractive activity such as roads, slipways, loading ramps, huts, forts, towers, wells, and so forth allow us to outline the immediate effects of geology, topography, and technology, as well as the geographical location of organizational structures and do reflect major differences amongst individual quarries. Owing to their remoteness and the unique conditions for preservation in the Egyptian Eastern Desert, the imperial quarries at Mons Claudianus, Mons Porphyrites, Mons Ophiates, and Tiberiane provide us with the most complete archaeological record. While the quarries and the administrative centre at Simitthus/Chemtou are similarly well documented, most other imperial quarries such as Karystos, Dokimeion, or Luna/Carrara have not yielded the same amount of information, mainly because some of these quarries have been in use over the centuries.

Figure 1. Principal mines and quarries in the Roman empire

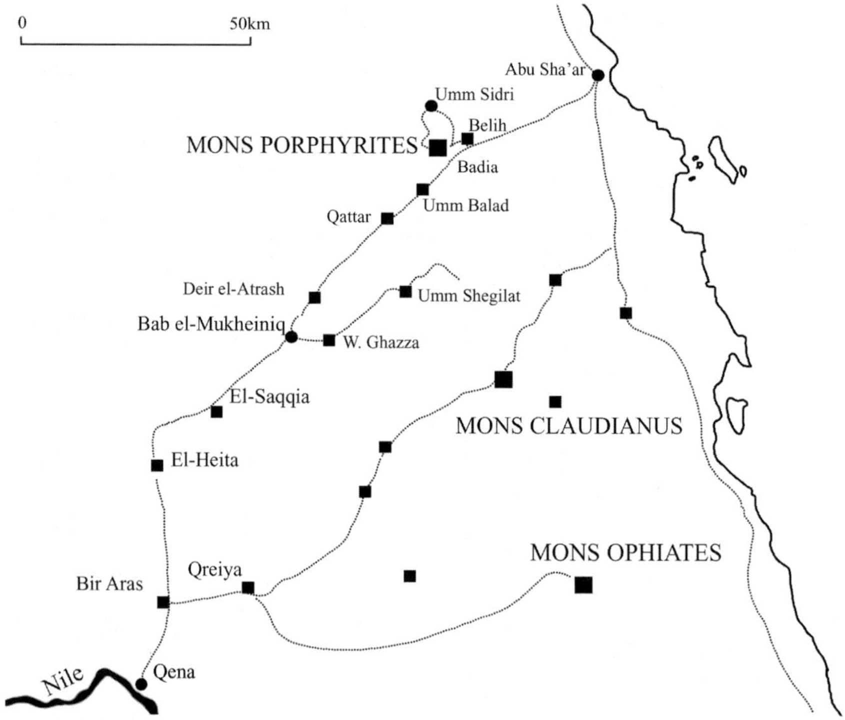

FIGURE 2. Imperial quarries in the Eastern Egyptian Desert (after Maxfield & Peacock 2001*b*: fig. 1.2)

2.1.1. Mons Claudianus/mod. Gebel Fatirah

As the name suggests, the quarries at Mons Claudianus/mod. Gebel Fatirah in Roman Egypt presumably supplied granodiorite stone from the reign of Claudius onwards. The earliest firm date for a permanent quarrying settlement at Mons Claudianus is given by an ostracon from the reign of Nero (AD 68). It was found at the 'Hydreuma', an early quarry settlement. The occupation of the site appears to end before the mid-third century AD. This date is provided by a dedication of a *cohors II Ituraeorum* to Zeus Helios Sarapis during the reign of Severus Alexander (AD 222–35).[1] The site saw its heyday during the first half of the second century AD, as most inscriptions and ostraca date to this period.[2]

[1] *O.Claud.*inv. 7363, cf. Peacock 1992: 9.
[2] For dating of ostraca, cf. Bingen 1996: 29–38.

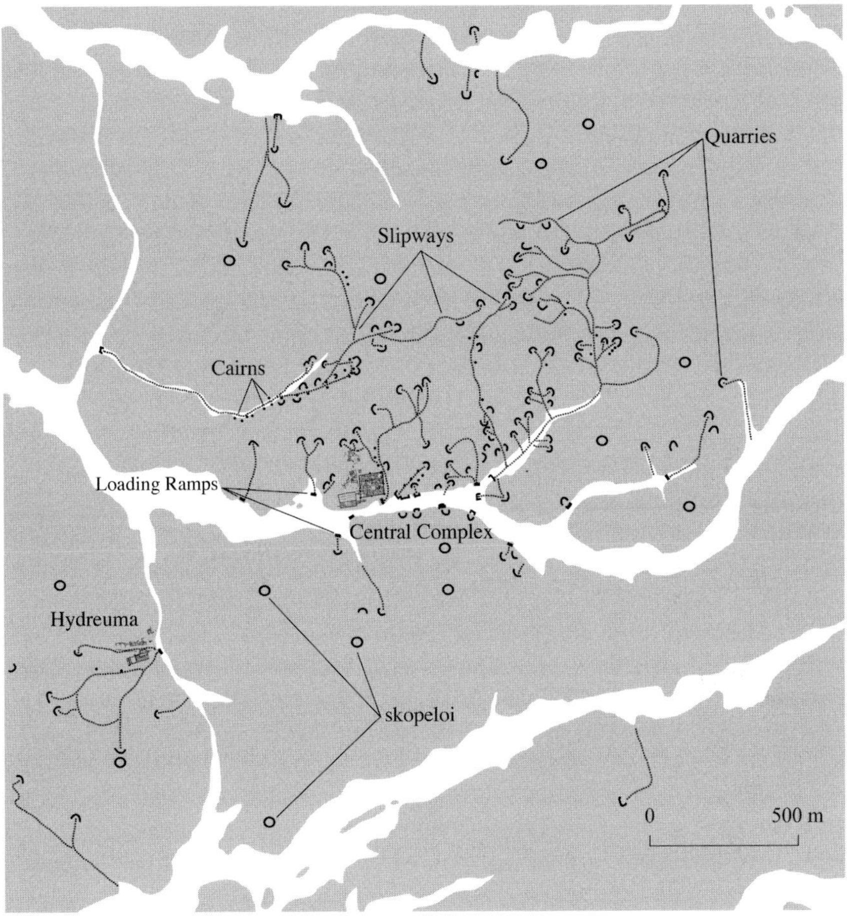

FIGURE 3. Mons Claudianus overall map (after Peacock & Maxfield 1997: fig. 1.2)

The central architectural features in Wadi Umm Hussein consist of a fort, a grain storage facility adjacent to animal lines, a 'private housing' area with a bath, and the temple for Zeus Helios Sarapis. The fort with its irregular layout and interior division is presumably not a genuine military structure, but is likely to reflect a mixed population consisting of civilians and military personnel.[3] Moreover, the examination of the rubbish dump (*sebakh*) directly south of the fort revealed numerous ostraca (*c.*4,000) and coins, suggesting that the majority of administrative activity within the quarry took place in this area.

[3] Peacock & Maxfield 1997: 50, 42 f., 84–6; Maxfield & Peacock 2001*b*: 17–19, 31, 59–76.

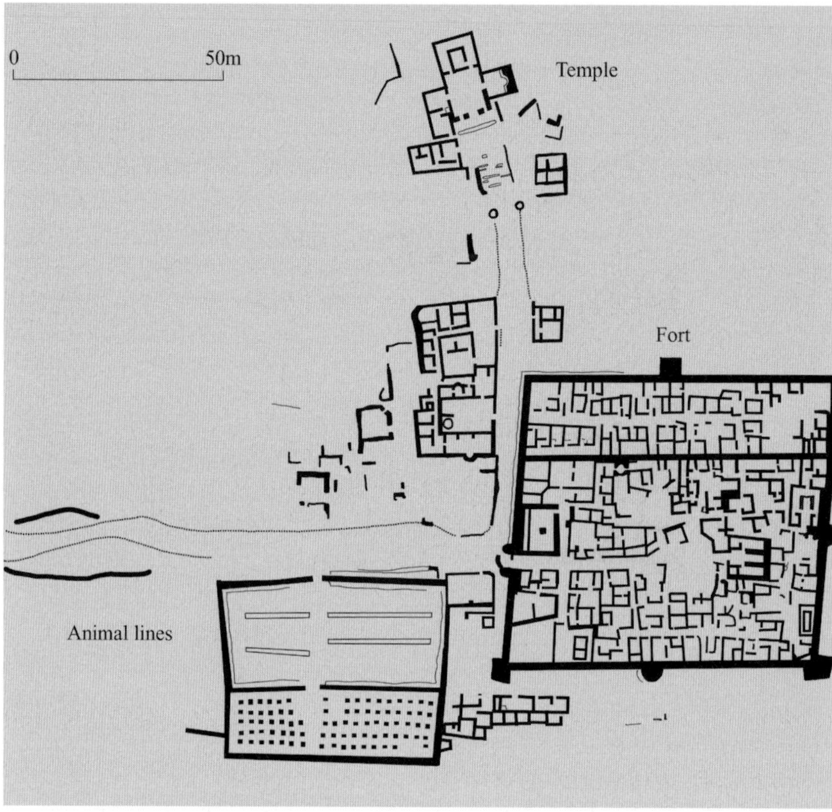

FIGURE 4. Mons Claudianus, fort (after Peacock & Maxfield 1997: fig. 2.14)

The adjacent animal lines, a feature common to most forts along the road from Kaine to Mons Claudianus or Mons Porphyrites, were enclosed by a low wall, housing camels, donkeys, and horses.[4] This structure should probably be identified with the stables mentioned in the ostraca.[5] A storage facility was built and a well sunk next to the animal lines.[6] From the Wadi floor a stairway led up to a temple structure for the god Sarapis. The temple was furnished with three inscriptions, which provide a date of construction around AD 118, with rooms added at a later date.[7] Next to the stairs up to the temple, a residential and bath-house complex was built, perhaps housing high officials of the quarrying

[4] Peacock & Maxfield 1997: 91 f., with fig. 2.56.
[5] *O.Claud.*inv. 2921, 4155.
[6] Peacock & Maxfield 1997: 86–101.
[7] *I.Pan* 37, 39, 42, cf. J.-M. Carrié, in Maxfield & Peacock 2001*b*: 142–6.

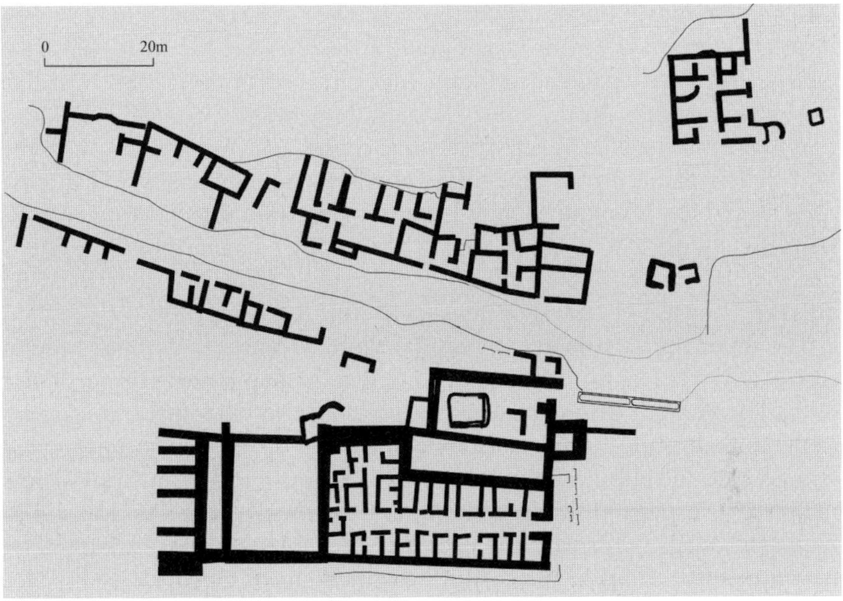

FIGURE 5. Mons Claudianus, hydreuma (after Peacock & Maxfield 1997: fig. 3.3)

administration. A graveyard or cemetery was also discovered but has yet to yield any information on the ancient inhabitants.[8]

South from the main complex in a tributary Wadi to Wadi Umm Hussein stands a building structure, the 'Hydreuma'. The site probably saw two phases of occupation during the reigns of Claudius(?) and Nero.[9] It consisted of a small fort with a water tank and stone troughs outside the enclosure to provide water for animals, as well as a house for the commanding officer and a workers' village. The ostraca discovered there, which date to the Trajanic period, document the same place name, Mons Claudianus, and an ostracon found below the rubbish dump (*sebakh*) in the workers' village dates to the reign of Nero (AD 68). This complex was probably the first structure housing the quarry administration and a garrison of roughly fifty soldiers. It presumably was replaced under Domitian by the fort at Wadi Umm Hussein. From the central complex at Wadi Umm Hussein, all 130 quarries at Mons Claudianus could be easily accessed from the Wadi floors or by short slipways, the longest being *c.*500 m. Stone cairns along the slipways provided material for the maintenance of the slipway surfaces. Blocks and columns were left lying

8 Peacock & Maxfield 1997: 118–34, 137 f.

9 J.-M. Carrié, in Maxfield & Peacock 2001*b*: 166 f.

on loading ramps at the end of the slipways. The ramps generally were 5–10 m in length and 0.6–0.9 m in height, and obviously were meant to allow the quarried produce to be loaded from sledge or rollers onto wagons. Near Bab el-Mukheniq tracks on the routes leading from Mons Claudianus or Mons Porphyrites to the Nile Valley usually had gauges of 2.7 m and 2.4 m. In the Naq' el Teir plain, a very large gauge of 3.5 m was measured.[10] What these carts looked like is unknown, but they must have been able to carry columns weighing up to 200 tons. Owing to the lack of bovine bones at Mons Claudianus, oxen have been ruled out as draught-animals. Valerie Maxfield argues for the use of camels or donkeys as work animals.[11]

The abundant outcrop of granodiorite allowed the extractive operations to spread out over *c.*9 km^2. Most quarries were to be found north or north-east of the fort in Wadi Umm Hussein. In comparison to other imperial quarries in the Roman empire, these quarries with a width of 20–30 m resemble small 'scoops' into the hillsides. Within some of these quarries columns and blocks in different shapes and sizes with engraved or painted inscriptions can still be found (cf. 7.2.3). Huts were discovered in fifty-three quarries, which might have functioned as resting places during work hours or perhaps were used for permanent housing. Apart from pottery, twenty-four huts yielded deposits of ash and iron slag. Four huts were thoroughly excavated, revealing not only water-tanks, but furnaces, ash, slag, pottery, and depressions for bellows.[12]

Throughout the Mons Claudianus quarrying territory, seventeen σκόπελοι or watchtowers were discovered. The location of these watchtowers within the quarrying 'district' excludes their construction for security purposes, as none of the towers overlooks the approaches to the site. Separated from each other by a distance of *c.*650 m, they allowed visual communication between the quarries and the fort at Wadi Umm Hussein. These factors suggest that the towers were either manned for the purpose of internal communication, or to guard convict labour (cf. 6.1.2.6).[13]

2.1.2. Mons Porphyrites/mod. Gebel Dokhan

The 'quarrying district' of Mons Porphyrites spreads out over an area of c. 9 km^2 and consists of several quarries and a central complex.[14] Geological

[10] Peacock & Maxfield 1997: 140–8, 177–253, 261–3, 287–314.

[11] Peacock & Maxfield 1997: 263 f.; Adams 2001: 183–8; Maxfield 2001: 159.

[12] Peacock & Maxfield 1997: 14 with fig. 1.2, 233–5, 246–8.

[13] Peacock 1992: 18.

[14] Kraus, Röder, & Müller-Wiener 1967; Sidebotham & Zitterkopf 1991: 575 f.; Maxfield 1994–5; Brown & Harrell 1995; Maxfield & Peacock 1996; Maxfield & Peacock 1997; Maxfield & Peacock 1998*a*; Maxfield & Peacock 1998*b*; Maxfield & Peacock 2001*a*; Maxfield 2001.

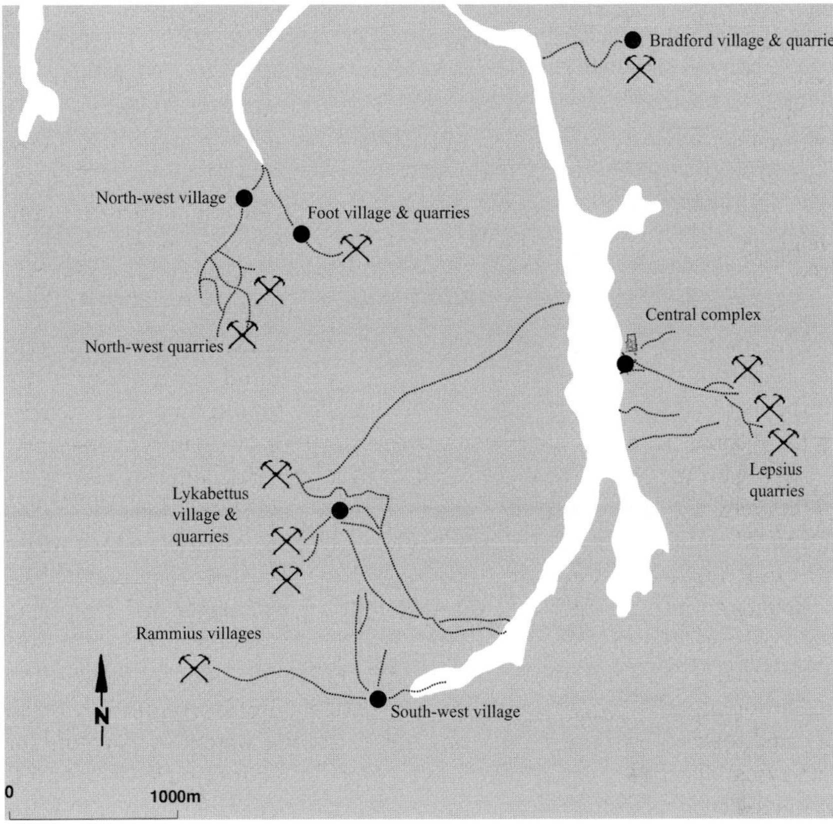

FIGURE 6. Mons Porphyrites, overall map (after Maxfield & Peacock 2001*a*: fig. 1.2)

constraints, however, complicated the pursuit of extractive procedures. The porphyry outcrops were not close to the Wadi floor: the stone had to be quarried some 500 m up the hillside and transported down on long slipways. The adverse topography determined the settlement pattern. Unlike Mons Claudianus, where settlements are clustered together around the administrative focal point at Wadi Umm Hussein, the workers' villages at Porphyrites tend to be dispersed along the routes leading from the central complex at Wadi Abu Ma'amel (630 m above sea-level) to the Lykabettus (1109 m), Lepsius (954 m), Rammius (1438 m), the North-West (1000 m) and the Bradford quarries (*c*.750 m).[15]

[15] Klemm & Klemm 1992: 82, Abb. 432.; Maxfield & Peacock 2001*a*: 2.

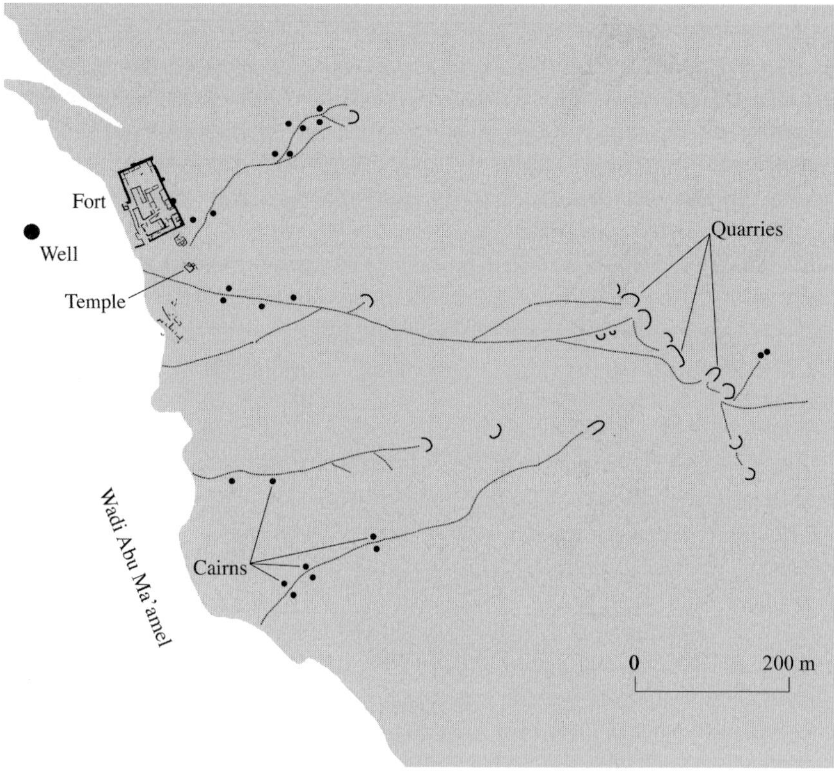

FIGURE 7. Mons Porphyrites, central complex (after Maxfield & Peacock 2001*a*: figs. 2.1, 2.4, 2.7)

The central complex of the Wadi Abu Ma'amel consists of a fort, a workers' village, two temples (for Sarapis and Isis), a bath-building and a cemetery.[16] The fort, hugging the hillside on a ridge above the Wadi floor, contained a large water tank for storage.[17] Excavations at the entrance of the fort yielded glass fragments, pottery, a substantial amount of woven textiles, as well as 165 ostraca. The range of documents included letters concerned with the purchase and transport of food or tools, lists of names and personnel (civilian and military), as well as *tituli picti.* One ostracon was dated to 13 August, AD 157, giving a rough date for the rubbish heap at the entrance. Another dump revealed 200 ostraca, mostly containing orders for bread. The interior layout of the fort is far less complex than at Mons Claudianus. Its main function was

[16] Maxfield & Peacock 2001*a*: 12–56.
[17] Maxfield & Peacock 2001*a*: 17 f.

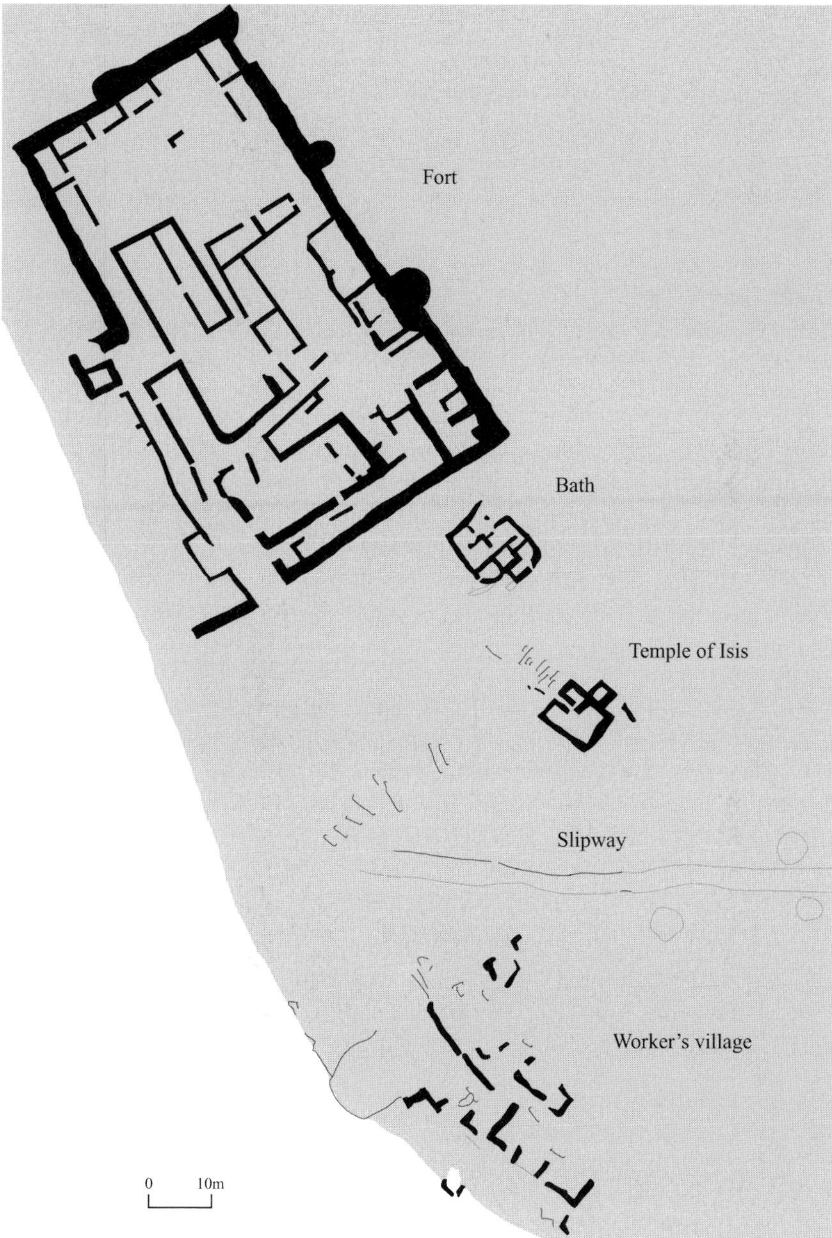

FIGURE 8. Mons Porphyrites, fort and temple of Isis (after Maxfield & Peacock 2001*a*: figs. 2.4, 2.7)

not only to provide shelter for the military garrison, but also to house administrative offices and officials.[18]

Close to the gate of the fort a building structure was discovered and—based on the large number of burnt tiles of a hypocaust found there—has been interpreted as a 'bath', or a house equipped with a heating system. From the Wadi floor a stairway led up to a simple and small rectangular building, the temple of Isis Megiste, dated by an inscription to the year AD 113.[19] A collection of huts follows to the south, which has been interpreted as a workers' village. Traces of smithing and particularly a layer of slag and charcoal, as well as random wedge holes cut for practice purposes, imply that quarrying tools were either produced or their quality tested there. At least two slipways lead up from the fort to the Lepsius quarries. North-west of the fort lies a necropolis and burials, consisting of simple mausolea (probably for more important individuals), cairn graves and pit graves.[20] South of the fort the once neatly sculpted tetrastyle temple of Sarapis, again dated by an inscription to the years AD 117–19, stands on a rocky knoll, accessible by a stairway.[21] A further rectangular temple was built across the Wadi on another hillside. According to an inscription found near the spot, the small temple might have been dedicated to Isis Myrionyma in AD 137/8.[22] Two wells guaranteed the supply of water, one north-east of the fort and one directly on the Wadi floor.[23]

Besides the central complex on the Wadi floor, further workers' villages were located close to the quarries and along the steep slipways. 'Bradford village' lies in the immediate vicinity of the black porphyry quarry of the same name, and consists of seven huts. An inscription naming a Cominius Leugas dates to the year AD 18, and probably makes the village one of the earliest focal points within the quarrying complex. The inscription itself was found in a small niched hut, probably a Paneion. The quarries and the village were connected by a slipway to the Wadi floor, the slipway ending in a flat pedestal and loading ramp littered with black porphyry chippings.[24] Leading up from a side Wadi of the Wadi Abu Ma'amel, a slipway reached the so-called 'Foot village' and its quarries, while another slipway led up from the same loading

[18] Kraus, Röder, & Müller-Wiener 1967: 170; Maxfield & Peacock 1996: 18; Maxfield & Peacock 1997; Maxfield & Peacock 1998*a*: 184; Maxfield & Peacock 2001*a*: 12–19.

[19] *I.Pan* 20; Maxfield & Peacock 2001*a*: 19–23.

[20] Maxfield & Peacock 2001*a*: 25 f., 135–51, 26–36.

[21] *I.Pan 21*, cf. Kraus, Röder, & Müller-Wiener 1967: 171–3; Maxfield & Peacock 2001*a*: 36–8.

[22] *I.Pan 22*, cf. Maxfield & Peacock 1998*a*: 190; Maxfield & Peacock 2001*a*: 39–42.

[23] Kraus, Röder, & Müller-Wiener 1967: 181 f.; Maxfield & Peacock 2001*a*: 42–55.

[24] Maxfield & Peacock 2001*a*: 58–62. 134.

ramp to the 'North-West' quarries. The North-West village was located on the slipway, closer to the Wadi floor than to the North-West quarries. The village consisted of five discernable houses or blocks and, based on the pottery and coinage finds, these buildings were probably occupied from the mid-first to the end of the second century AD. An inscription most likely originating from the North-West village dates to AD 27.[25] The North-West quarries might have seen two periods of extractive operations. While the small-scale quarries are understood to be of an earlier date, the larger quarries are late Roman. Quarries nos. 4 and 9 appear to be particularly vast: no. 4 reaches *c.*100 m in both height and width, whereas no. 9 displays the four faces of a monolithic cube of purple porphyry left standing.[26]

The 'Foot village' consists of sixteen huts, the pottery of which dates it to Tiberius' reign. While fifteen huts are grouped in a circle on one side of the slipway leading up to the adjacent quarries, one rectangular hut was built on the other side. The faces of the two quarries vary from 3 to 15 m (Foot 1) i.e. from 1.5 to 3 m (Foot 1a). The South-West village was situated at the lower end of a slipway running down from the Rammius quarries and terminating in the Wadi Abu Ma'amel. The assemblage of buildings includes a *contubernium*-like structure of seven uniform rooms, a rectangular building with an 'apsis', a long stretched L-shaped building with a large number of rooms, and an ensemble of five multi-roomed buildings. The L-shaped building was interpreted as providing accommodation for workers, a dining or social area, as well as spacious suites of rooms for officials/overseers and their families. The building was gradually extended. The pottery finds seem to suggest an occupation date of the late first to the mid-second century. Slipways led up to the Rammius quarries, the latter being surprisingly small and not exploited to the degree the size of the settlement at the foot of the slipway might lead one to expect.[27]

Two workers' villages were located along the slipway coming down from the Lykabettus quarries. One village, consisting of a small number of rectangular and circular rooms, was built besides the loading ramp, while the Lykabettus village lay close to the quarries.[28] The village seems to have comprised five housing complexes, lying on both sides of the slipway. Nearby, a cemetery yielded a late Roman/Byzantine tombstone, and it appears that the pottery finds and the coinage point towards a main occupation in the fourth century AD. However, there seems to be evidence to suggest earlier activity,

[25] *I.Pan* 18, cf. Maxfield & Peacock 2001*a*: 64–86.
[26] Maxfield & Peacock 2001*a*: 155–72.
[27] Maxfield & Peacock 2001*a*: 62–4, 86–104, 151–5.
[28] Maxfield & Peacock 2001*a*: 104–10.

perhaps under Nero.[29] The Lykabettus quarries are located on one of the highest mountains in the area. In one quarry (no. 4b) the inscription of a *centurio frumentarius* documents work going on there during the second or third century AD.[30] The quarry faces normally reach heights of up to 10 or 15 m, in one instance even 25 to 30 m (no. 4), and can be up to 70 m in width. The slipways leading down from these quarries and villages were long and steep; the slipway from the Lykabettus quarry, for example, reached 2.8 km, the one leading down from the Rammius quarries *c.*1.7 km, covering heights of 550 to 450 m. Numerous cairns were used as depots for the repair of the slipway surface. Loading ramps were positioned at the end of some of the slipways (e.g. from Lykabettus and Bradford quarries). Visual communication was possible via a central tower (*σκόπελος*) at the head of the Wadi Abu Ma'amel, allowing contact with the Lykabettus and Rammius quarries as well as the footpath leading up to the pass over to Badia, where a further tower stood. The footpath over the pass was guarded on the southern side by a small fortlet ('Footpath Station'). To the north of the central structure the Wadi opens into Wadi Umm Sidri. The walled structure at Umm Sidri once included animal lines and was probably built during the second century AD.[31] At the mouth of the Wadi Umm Sidri a main loading ramp was positioned, which was 1.75 m high and could accommodate vehicles of up to 22 m in length. A small settlement, consisting of circular and rectangular structures, was detected nearby.[32] The road then headed south to the Badia complex, consisting of a fort, animal lines, a fortified rock outcrop, a cemetery, and a well.[33] From there a footpath led over the pass into the Wadi Abu Ma'amel, while the main route continued south, joining the Nile—Abu Sha'ar road (cf. 2.1.5).

2.1.3. Tiberiane/mod. Wadi Barud

In comparison to the spacious imperial extractive operations at Mons Claudianus or Mons Porphyrites, the quarries and architectural remains at Tiberiane are modest in size. The fort at Tiberiane is a 'miniature clone' of the fort at Claudianus: the general layout, the interior dividing wall, the bastions, and

[29] *I.Pan* 29. Tile/veneer from Nero's Domus Transitoria has been determined on the basis of magnetic susceptibility to originate from one of the Lykabettus quarries, cf. Maxfield & Peacock 2001*a*: 110–29, 318.

[30] *I.Pan* 24.

[31] Maxfield & Peacock 2001*a*: 172–90, 195 f., 200–9.

[32] Kraus, Röder, & Müller-Wiener 1967: 199–201; Maxfield & Peacock 2001*a*: 209–14.

[33] Kraus, Röder, & Müller-Wiener 1967: 201; Maxfield 1994–5: 19 f.; Maxfield & Peacock 1996: 19; Maxfield & Peacock 2001*a*: 215–39.

the gate are all similar. A plaster-lined tank for water storage was found within the fort, and a survey yielded three quarries with a slipway, various huts, a tower (σκόπελος), and the remains of a cemetery.[34] The name of the site suggests a founding date under Tiberius; this is supported by the find of coarse quartz diorite (*granito bianco e nero*), originating from Wadi Barud, in the Domus Tiberiana. Pliny refers to a marble found in Egypt during the reign of Augustus and Tiberius called 'Tiberian', possibly the same as the one found at Tiberiane. 200 ostraca were unearthed at Tiberiane, of which two date to the year 10 of Antoninus Pius (AD 146/7).[35]

2.1.4. 'Mons Ophiates'/mod. Wadi Umm Wikala

In Wadi Umm Wikala a green/dark green stone with white speckles was quarried.[36] The stone probably was known to Pliny as *lapis ophyte* although he does not identify the origin of this 'snake-stone'. The ancient name of the area is revealed by an inscription as ἐν τῶι 'Οφιάτηι, from which the toponym 'Mons Ophiates' comes.[37] The numerous quarries and settlements are spread out over an area of *c*.2 km^2. Thirteen quarries as well as quarry prospections—small areas where weathered stone was removed—have been identified. The quarries appear to be located mostly on the hillsides or hill-tops (sometimes up to 100 m above the Wadi floor). From each quarry a slipway (made of rubble and of boulder revetments with a rubble fill) was cut into the bedrock and led down to the Wadi floor. Most slipways ended in platforms, variating in height between 0.6 and 1.7 m. Some of the platforms still have architectural elements stockpiled on them. Cylindrical stone cairns were discovered alongside a few slipways, presumably intended as reserve dumps of repair material. Crude huts were found in some of the quarries, but were only large enough to shelter one man. Isolated huts and towers (σκόπελοι) were distributed throughout the area. The towers were positioned at a regular distance from each other, allowing visual communication between the headquarters and the quarrying sites. The main settlement in the Wadi Umm Wikala consisted of a rectangular, bi-partite building structure (*principia*), a shrine, an assemblage of one-room-huts and, finally, two multi-roomed buildings. The *principia* comprised of twenty-eight rooms and a cistern, able to hold 8,800 litres, and was probably built as a living space for military officials and administrators.

[34] Peacock & Maxfield 1997: 275–83.

[35] Pliny, *NH* 36.55.

[36] Meredith 1953: 106 f.; Klemm & Klemm 1992: 408–10; Brown & Harrell 1995: 227; Sidebotham et al. 2001: 142–4.

[37] Pliny, *NH* 36.56; *I.Pan* 51; Cuvigny 1996*c*: 92 fn. 4.

The 'Hydreuma', a large fortified well, is situated at the entrance of the Wadi Umm Wikala, fortified by six or seven semicircular towers. Up to forty rooms can be distinguished in the interior of the fort, surrounding a circular well of unknown dimensions.[38]

2.1.5. Road System and Quarries in the Eastern Desert

The roads connecting Mons Claudianus, Tiberiane, Mons Porphyrites, and Wadi Umm Wikala with the Nile valley were built specifically for this purpose and protected by fortlets (see Fig. 2). Kaine/mod. Qena was initially constructed as a starting point for the roads leading into the desert and served as a river harbour for the reloading of pillars and other products from the quarries.[39] Remnants of the way-stations or forts are still preserved today. The route from Mons Porphyrites to Kaine was guarded by five forts (Qattar, Deir el-Atrash, El Saqiya, El-Heita, Bir Aras), while four forts guarded the vast quarrying zone of Mons Porphyrites (Badia, Umm Sidri, Abu Ma'amel, and Belia) and the small quarrying site at Umm Balad.[40] For the route from Mons Claudianus to Kaine at least three forts—Abu Zawal, Tal 'et el-Zerqa, Qreiya—are attested in addition to the forts at Mons Claudianus itself and at Tiberiane. The route from Wadi Umm Wikala to Kaine was guarded only by one fort. The forts, which were built near wells or enclosed them within their walls, reached sizes from 0.46 (Mons Claudianus) to 0.05 ha (El-Saqiya). Most of them—with the exception of the forts at Mons Porphyrites, Barud, Umm Balad, and Wadi Umm Wikala—were equipped with animal lines.[41] The large animal lines seem to be a special feature of the routes leading to and from these great imperial quarries, and apparently cannot be observed along the other roads through the Eastern Desert. The forts were roughly 25 km apart from each other, making it possible to reach the next way-station within a day's travel.[42] Besides being a shelter for travellers, the forts at Abu Zawal and Barud perhaps protected smaller quarries nearby.[43]

[38] Sidebotham et al. 2001: 146–55, 157 f.

[39] Cuvigny 1998: 90.

[40] Meredith 1952; Meredith 1953; Sidebotham & Zitterkopf 1991: 580–2; Brown & Harrell 1995: 223–6; Maxfield 1996; Maxfield 2001: 159–61.

[41] Maxfield 2001: table 7.4.

[42] Hamilton-Dyer 2001: 298; Maxfield 2001: 160.

[43] Meredith & Tregenza 1949; Klein 1988: 20.

2.1.6. Simitthus, Dokimeion, and other Imperial Quarries

2.1.6.1. *Simitthus*

The excavations of 1968–79 and 1992–8 at Simitthus/mod. Chemtou in Africa proconsularis, where a yellow-orange-pink stone was quarried, uncovered a tripartite 'work' camp. The camp lies *c.*550 m north-west of the quarries on the road from Simitthus to Tabarka. At the heart of this camp lay a stone building with six long and narrow rooms with latrines at the northern end, identified by its excavators as a penitentiary centre, an *ergastulum*. Coinage and pottery unearthed within this central structure provided a construction date of *c.*170 AD. A bath, built in the 190s, was added adjacent to the *ergastulum*, and one may expect 'military' barracks immediately to the east of the *ergastulum*.[44] This central prison area was abandoned by the imperial administration at the beginning of the third century AD and partly turned into a *fabrica*, where various refined products were made of the local *giallo antico* marble. With the transformation of the prison camp to a *fabrica*, the central complex received an addition to the east around AD 230, likely an *officina marmorum* and a *horreum*. The buildings to the west of the former *ergastulum* were added later and included a sanctuary with a treasury. The whole tripartite area was abandoned in the late third century.[45]

A further religious 'centre' appears on the hillflank south of the Numidian sanctuary on top of the 'Tempelberg'. A series of *c.*200 votive reliefs dedicated to Saturn found on the hillside indicates the popular veneration of this god.[46] The extraction history of the Simitthus quarries is rather complex and almost impossible to date. Simitthus is privileged in having suffered only a limited exploitation of *giallo antico* marble in recent years, a fact which facilitates a reconstruction of ancient quarrying procedures. The quarries can be found along the hill range immediately north-west of the city of Simitthus, the Gebel Chemtou. Three peaks mark the Gebel, which have been named 'Stadtberg', 'Gelber Berg', and 'Tempelberg'. The quarrying area covers no more than *c.*0.4 km^2. Compared to the extraction areas of other imperial quarries, the marble outcrop at Simitthus is significantly smaller. Most quarries at Simitthus appear to be clearly defined entities. Quarrying walls have been left standing, outlining the separate areas of extraction. Some quarry faces reach heights of up to *c.*23 m. These distinctions can be observed at the 'Gelber Berg' where the quarry walls divide the hillside into three quarrying zones.[47]

[44] Mackensen 2005: 88 f., 100.

[45] Mackensen 2000: 490, Abb. 2, 493, 502; Mackensen 2005: 87–101, 111–14.

[46] Kraus 1993*b*.

[47] Röder 1993: 17, 25, 27 f.

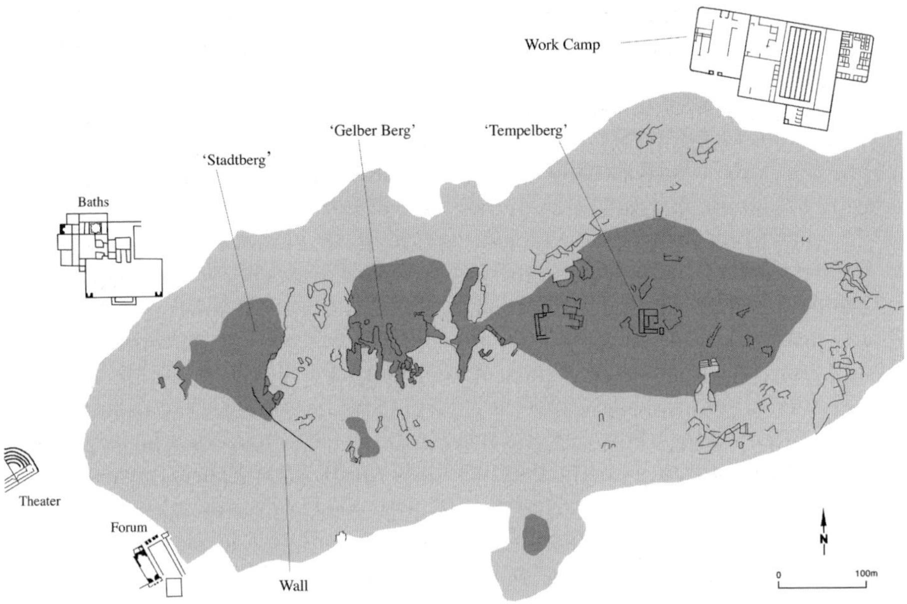

FIGURE 9. Simitthus, overall map (after Mackensen 2000: Abb. 1)

Continuous quarrying has removed part of these initial divisions. Between the 'Stadtberg' and the 'Gelber Berg', as well as between the 'Gelber Berg' and the 'Tempelberg', the quarries were dug through to the other side of the hill. Quarries were cut into the southern side of the saddle between 'Stadtberg' and 'Gelber Berg', while at a later stage extractive operations were begun on the northern flank. From the north, two further quarries cut through the hill flank between 'Gelber Berg' and 'Tempelberg', with only a thin wall left standing. Röder argues that the earliest quarries were extracting the marble horizontally ('Flachausbau'), but this was later replaced by vertical extraction ('Tiefausbau'). An exact chronology has not been provided by the German archaeological team, but it is probable that both techniques were used simultaneously. As a result of the small space available for manoeuvre, one major problem for the quarrying administration at Simitthus was the management of rubble deposits. Generally, half of the extracted marble was reduced to rubble during the extraction process and when the quarried rough blocks were dressed. The clearing out and disposal of rubble at previously designated dumps within the quarry necessitated a detailed extraction plan, particularly in the case of the 'Gelber Berg' where the most desired marble colour was available in abundant quantities.[48]

[48] Röder 1993: 25 f., 42–4.

It should be pointed out that the quarries at Simitthus were separated from the town of Simitthus by a wall. The wall itself presumably indicated the extent of the quarrying district, with the tripartite camp at its centre. Yet, not all extraction sites were enclosed by the wall. Some quarries at the 'Stadtberg' remained outside of the quarrying district.[49] The impression of an autonomous quarrying district disconnected from the town of Simitthus is further strengthened by Rakob's observation that no *giallo antico* was used in great quantities for the town's construction projects.[50] The imperial quarries of Simitthus lay in the immediate vicinity of the river Bagradas/mod. Medjerda, which seemingly provided an ideal mode of transport for the quarried *giallo antico* blocks. A broad stone plastered road was uncovered leading from the quarries to the Medjerda. Subject to the water level of the river, the blocks and columns may have been transported to Carthage by road.[51] An alternative land route was also in use. Lost column shafts were discovered by a French officer in 1895 along a Roman road from Simitthus to the harbour town of Tabarka. The road was 60 km long, had once been 6 m wide, and mastered gradients of 4 to 8.5 per cent while heading over a mountain range reaching heights of c. 800 m above sea-level. According to a series of milestones, the construction or restoration(?) of this road appears to have taken place in AD 129, perhaps mainly for the export of columns and blocks.[52]

2.1.6.2. *Dokimeion/Bacakale*

Similar problems beset the analysis of quarries near Dokimeion/mod. Iscehisar. There, little is left from the original Roman *pavonazetto* quarries near the town, owing to their reuse in the 1980s. Consequently, no architectural structures are associated with the ancient quarries, nor can much be said about the layout of the old extraction zones.

The Roman *pavonazetto* quarries appear to have covered an area of *c.*0.24 km^2 in the marmoriferous zone south-east of Iscehisar. Extraction efforts were concentrated at the Bacakale marble outcrop, leaving a gap of 200 m in length, 80–110 m in width and 40–5 m in depth, cut into the side of the marble ridge.[53] The quarries yielded a white marble and a *breccia* of this white marble with thin grey or violet veins (*pavonazetto*).[54] The Bacakale quarries may have been even larger than the size estimated by Joseph Röder, as

[49] Röder 1993: 21.
[50] Rakob 1997: 4.
[51] Röder 1993: 50–2; Mackensen 2005: 15 f.
[52] Winckler 1895: 38–47; Salama 1951: 61. On the road, cf. Röder 1993: 51; Hirt 2003: 118 f.
[53] Röder 1971: 265 f., Abb. 10, 11; Fant 1989*a*: 4 f.
[54] Röder 1971: 259.

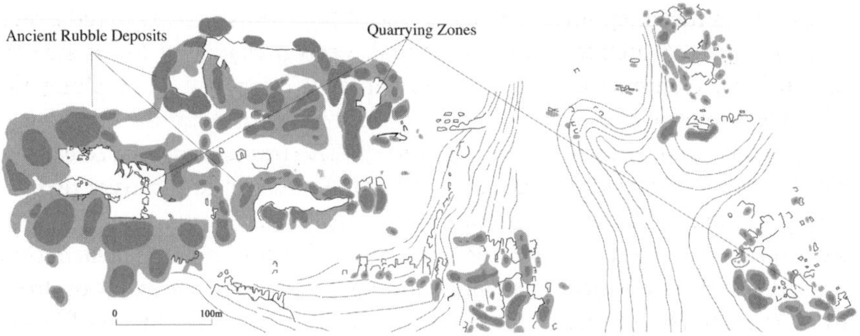

FIGURE 10. Dokimeion/Bacakale, overall map (after Röder 1971: Abb. 10)

modern quarrying operations removed vast amounts of the ancient rubble still lying within the quarries.[55] Once more a number of quarry walls suggest separate sections of extraction. Röder argued that the orderly fashion in which rubble was disposed of clearly indicates a deposit plan and, therefore, an overall design for extracting the precious marble. Surely, such plans were altered according to the demand for the 'Phrygian' marble and might not have been in place since the beginning of extraction.[56] Apart from the disposal of quarrying waste, a further logistical problem arose—the marble from the landlocked Dokimean quarries had to be transported over a considerable distance to the nearest harbour. Presumably, the *pavonazetto* was exported via Synnada, Apameia, and Laodiceia ad Lycum, down the Maeander to Tralles and then to Milet or Ephesos.[57]

2.1.6.3. *Karystos*

Other imperial quarries have not, as yet, revealed the remains of any substantial buildings, nor has their physical layout been properly studied so as to allow a comparison with either Simitthus or the Egyptian quarries. The *cipollino* quarries near Karystos on Euboea, for example, are numerous, yet their relationship to one another and their physical structure as a whole has not been analysed. A Belgian team surveyed a part of the quarries near Styra and Pyrgari and the associated road system. The quarries in the Haghios Nikolaos and Krio Nero area near Styra were not subject to an intensive

[55] Fant 1989*a*: 41–8.
[56] Röder 1971: 269, 271.
[57] Hirt 2003: 119 fn. 15, with further bibliography.

study—with the exception of the *Latomeia Claudiana*, which displayed a rock face of up to 16 metres in height. Other quarries appear to be of smaller dimensions. Columns and blocks still litter the quarries or lie along the slipways. Some of the quarried columns and slabs were used as pavement for the construction of roads leading from the quarries down to the coast. The roads show rut marks, indicating the width of ox-carts which were used to transport quarried blocks or columns.[58] At Pyrgari, nine large quarrying areas were discovered, and most of the floors were littered with ancient debris. A road system connected these quarries to a steep slipway, which ran down from an altitude *c.*320 m above sea level to the coast with a gradient of 20–40 per cent. Rectangular holes discovered along these steep slipways probably once held wooden posts. Around these, ropes would have been slung to work as a pulley on which the sledges transporting the columns or blocks could be lowered down safely.[59] Stone huts of unknown date have been closely examined, and interpreted as military buildings, positioned at strategic points of the quarries near Karystos. This hypothesis, however, has not been corroborated by any epigraphic or archaeological evidence.[60]

2.1.6.4. *Luna/Carrara*

The description of the quarrying infrastructure and its layout at the white marble quarries of Luna/mod. Carrara in Italy is similarly problematic. Since marble of Carrara has been quarried more or less continuously since Roman times, few archaeological finds have been made and most of the Roman quarries were destroyed by medieval and modern extraction work. However, at some points the remains of Roman quarrying can still be detected. A preliminary list of ancient quarries and the findspots of iron smithing furnaces, tools, blocks and columns was compiled and mapped by Luisa Banti in 1931. Quarrying sites were located at the Cava del Polvaccio, the Cava dei Fantiscritti, Cava della Carbonera, Canalgrande, Fossacava, Colonnata, Cava Gioia, Bedizzano, and Torano.[61] In the 1980s Enrico Dolci discovered two further sites, one at Scalochiello above the Valle Colonnata and another at Monte Strinato near Carrara.[62]

The physical layout of the quarries can no longer be determined. The ancient extraction area seems to cover *c.*4 km^2. An administrative

[58] Named after an inscription on a column discovered there, cf. no. 161. Papageorgakis 1964; Hankey 1965; Vanhove 1996: 16–22.

[59] Vanhove 1996: 22–33.

[60] Kozelj & Wurch-Kozelj 1995.

[61] For an overview, see map and list in Banti 1931 and Dolci 1981–2; Dolci 1983.

[62] Dolci 1997; Dolci 1998.

complex—if one existed—was possibly centered near the modern town of Carrara (as some architectural remains and the epigraphic records might suggest). If the funerary inscriptions are anything to go by, the graveyards were situated near the main quarrying sites up on the mountain ridges, indicating where the quarry workers most likely would have been accommodated. Furthermore, any traces of an associated road system have disappeared.

2.1.7. Imperial Quarries and Organizational Constraints and Challenges

The physical layout of the quarries, especially evident in the extensive remains at Simitthus and the Egyptian Eastern Desert, reveal the main constraints and challenges faced by the authorities responsible for their construction. The formation of work procedures and the technology applied differed little from one quarry to the next. The extraction of stone *per se* did not involve any major technological challenges, but it was a 'low tech' affair. After a suitable spot for a quarry was located (occasionally with the help of trial trenches) the withered rock surface was removed. Then broaching or channelling techniques were applied by which a line of holes was chiseled in a channel across the rock surface, perpendicular to the cleavages and joints of the underlying rock. Metal wedges were inserted into the stone and hammered until the stone split off. The stone blocks were then worked with chisels providing it with the rough form.

The organizational difficulties emerged not from the quarrying and fabrication procedure, but from problems inherent in managing such an undertaking. In order to quarry the stone the required labour force, preferably skilled and experienced, needed to be assembled at the quarry and equipped with tools. The workers had to be recruited and paid, or at least have their basic needs (food, water, shelter) covered. The use of tools warranted a constant supply of raw materials for fabrication or repair (iron, wood, charcoal). This in turn required the presence of specialists acquainted with the art of manufacturing these implements. The marble output then had to be removed from the quarry and forwarded to the 'client'. Draught animals and wagons and/or ships had to be organized so as to deliver the produce (i.e. mostly blocks and columns) to its destination. The topography of some quarrying sites, in particular Mons Porphyrites, Karystos, and Carrara, complicated the problem of exporting stone by adding a considerable difference in altitude into the equation. In order to master the heights slipways had to be put in place, sturdy wooden sleds built, rollers cut, and rope ordered in considerable quantities. The lifting of marble blocks or columns on to wagons warranted the use of more complex contraptions such as cranes. Moreover, the archaeological evidence (the location of 'workers' villages', cemeteries, and funerary inscriptions) creates the impression that the work force was housed as close as possible to the quarries, in order to shorten the daily 'commute' to the workplace.

Additional difficulties were caused by geological constraints such as the size of the stone outcrop. Whereas the numerous quarries of Mons Claudianus were spread out over the vast area of *c.*9 km^2, the quarries at Simitthus and Dokimeion were limited to small marble outcrops, their quarrying districts, if detectable, covering maybe 0.4 or 0.24 km^2. Unlike the spacious quarries of Mons Claudianus and Mons Porphyrites (and perhaps Luna and Karystos?), the disposal of rubble at Simitthus or Dokimeion was a significant problem. Owing to the constricted nature of these quarries, the waste had to be removed quickly so it did not hinder ongoing and future extractive operations. Consequently, the disposal of rubble had to be planned before commencing large extractive operations. Moreover, the small space limited the available locations for stocking extracted columns or blocks. While at Mons Claudianus, Mons Porphyrites, Karystos, and Luna extracted stones could be left lying in the quarry until they were dressed into the desired shape and shipped out to their respective destinations, the columns, blocks or sarcophagi at Simitthus and Dokimeion had to be moved from the extraction zone and were probably dressed and stocked at a location away from the quarrying section. Within the quarries, particularly in the *giallo antico* quarries at Simitthus, different quarrying sections are detectable, perhaps reflecting the extractive operations of different work parties. This and the limited space available implies that the quarrying authorities are likely to have coordinated quarrying work by different teams. Moreover, mechanisms had to be devised on processing incoming orders and converting this information into the desired output: the correct execution of these procedures and the produce had to be checked and various materials, tools, transport and lifting devices organized. The location of a quarry far away from the populated regions of the province had further repercussions for the organization of stone quarries. The extractive operations in the Eastern Desert, at a distance of *c.*120 km or more from the Nile valley, resulted in a wide array of logistical and infrastructural problems in addition to those already discussed. The workforce had to be brought in, provided with water and food, sheltered from the heat and the cold, and protected against nomadic tribes. Thus fortified housing and storage facilities had to be built. Moreover, supply routes had to be put into place and guarded, fortified structures had to be constructed and manned by soldiers, who in turn had to be supplied with shelter, water, and food. The constant demand for provisions required the construction of wells and storage facilities (cisterns, granaries) as well as a regular import of goods by transport animals. The use of camels and donkeys increased the need for further infrastructure and logistics such as animal lines, fodder, and water. The limited availability of water probably posed the most serious problem. This restricted the sustainable size of the population in the quarrying districts. Any migration to the quarries had to be carefully controlled. The infrastructure of the quarries at Mons Ophiates, Tiberiane, Mons Porphyrites, and Mons Claudianus appear to imply a fairly regular, if not constant presence of workteams, officials, and soldiers. Apart from (occasionally fortified) housing complexes, the existence of temples, baths, and cemeteries with the occasional gravestone of a Roman soldier strongly implies that these settlements were not makeshift lodgings but intended for permanent residence—at least for the foreseeable future.

The quarries at Simitthus or Dokimeion both lay in the immediate vicinity of sizeable towns and fertile plains which alleviated some of the ramifications faced by the quarrying authorities for the Egyptian Eastern Desert—the workforce could be brought in from the immediate surrounds and the supply of water and food did not pose a logistical challenge. The proximity to populated and fertile regions, however presented problems of a different kind. The alleged use of convicts at Simitthus, for example, resulted in additional infrastructural requirements, and a special compound had to be constructed north of the quarries for the captives or slaves. It is open to speculation whether the wall between the town of Simitthus and the quarries was erected to protect the townspeople or to prevent unauthorized quarrying.

Given the range of organizational tasks dealt with by the quarrying authorities, this must have left a considerable paper trail: information on orders, contracts, delieveries of various items, stocks, payments, rations, etc., had to be recorded and filed. This in turn required a functioning archival system, the assignment of clerks and the allocation of rooms or of a building to these tasks. This had further implications, such as shelter and supplies for the subaltern personnel, the construction of buildings, the supply of writing material etc. As we shall see further on, these activities were not necessarily centred at the quarry.

2.2. IMPERIAL MINES

Unlike imperial quarries, the archaeological remains of most mining sites or mining zones under imperial control are, as yet, not well documented outside Roman Spain or Britain. Consequently, a comparison of the layout of imperial mines throughout the empire can not be nearly as detailed as for imperial quarries. Most Roman(?) mining sites, especially in former Yugoslavia or in

Romania, are barely identifiable as Roman mines, mostly by heaps of slag and corroded tools, in some cases also by leats, water tanks, or mining tunnels. In particular in former Yugoslavia most observations on mining ventures were made in Bosnia-Herzegovina during the period of occupation by the Austro-Hungarian empire. Moreover, at a number of sites exploitation was continued through the medieval period into the eighteenth or nineteenth century. As most surveyors at the turn of the last century hardly can be described as trained archaeologists it remains doubtful whether or not they dated the remains of mining ventures precisely. More recent research on mining sites in Dalmatia or former Yugoslavia is barely available. Many of the known sites in Dalmatia or Romania have not been studied in depth and a map of the layout has been produced only in a few cases. There are some exceptions though: Alburnus Maior received broad attention in recent years, resulting in an exceptionally detailed study. The archaeological record of mines in Spain and the discoveries made at Dolaucothi in Britain provide a satisfying amount of physical evidence to attempt a comparison of mining techniques and physical layouts on a rudimentary level.

Unlike quarrying, mining techniques can be grouped in two distinctive categories—opencast mining and underground mining. Both are influenced by different geological constraints and different applications of extraction technologies with various implications for the organization of imperial mining operations.

2.2.1. Opencast Mining

2.2.1.1. *Secondary Gold Deposits and Hydraulic Technology in Northwestern Spain*

The specific geology of the gold deposits in the Douro basin west of Asturica Augusta allowed a large-scale approach to mining. Gold which had been eroded out of the mountainous regions and deposited as placers in the high-level alluvium or in the rivers below (known as secondary deposits) could be exploited far more easily than the vein ore from primary deposits. The higher content of free gold in secondary deposits made their treatment simpler.[63] Roman ingenuity in the use of waterpower allowed the easy removal of the gold-containing soil and the overburden in order to expose the gold deposit. The hydraulic mining techniques used, *hushing* and *ground sluicing*, necessitated a vast infrastructure for the supply of large quantities

[63] Wilson 2002: 17.

of water. The latter technique, ground sluicing, allowed the constant removal of gold-containing soil using the erosive force of water. The alluvial gangue was directed into sluice boxes consisting of a series of stepped troughs where the gold was filtered out. This technique is well documented in the Duerna valley, south-west of Asturica Augusta/mod. Astorga, a gold-mining zone where imperial control is documented during the late second century AD (cf. 3.2.5). The right bank tributary rivers of the Duerna provide ideal conditions for the use of waterpower.[64] Ten ground sluicing zones were discovered mostly on the southern banks of the Duerna valley, some of them extending for several hundred metres along the hillsides.[65] Not all of these ground sluicing zones were worked at the same time, and in some zones this hydraulic technique was perhaps used only as a means of prospection.[66] The other hydraulic mining technique, hushing, was particularily useful in removing large amounts of overburden to reveal the more promising strata of alluvial deposits. Water was collected in large reservoirs above the opencast. The gates of the reservoirs were opened from time to time, suddenly unleashing the collected water with erosive force and washing away the loosened overburden. The remnants of hushing are still detectable in the auriferous hillsides of the Duerna valley.[67] At Las Médulas a variant of the hushing technique has left colossal scars in the landscape. The rich alluvial gold deposits (Santalla facies: 20–100/ 60–300 mgr. Au/m^3) were covered in some places by other deposits which contained little gold (Las Médulas facies) of *c.*100 m in height.[68] The removal of this massive overburden called for exceptional measures. Archaeological surveys have revealed the gigantic hydraulic infrastructure which was used. Seven aqueducts of 2–3 m in width and with a length of up to *c.*50 km tapped the tributary rivers of the Duero or Sil, directing water to tanks with a capacity of up to 24,000 m^3.[69] Remains of tunnels and galleries are still visible at Las Médulas. Pliny the Elder describes how the system worked: a network of tunnels was cut into the overburden and then suddenly flooded bringing about the collapse of the rockface (*ruina montium*). The debris of the collapsed hill side could then be removed by hushing, until the gold-bearing strata were uncovered.[70] The debris was washed away in hush gullies, perhaps enabling the formation of the Lake of

[64] For a description of both techniques, cf. Rosumek 1982: 51–65; Woods 1987: 625 f.; Domergue 1990: 462–77; Wilson 2002: 17 f. with fn. 99.

[65] Domergue & Herail 1978: 53–246.

[66] Wilson 2002: 18.

[67] Domergue & Herail 1978: 263–78.

[68] Fernández-Posse & Sánchez-Palencia 1988: 243–8; Hérail & Pérez García 1989; Sánchez-Palencia Ramos, et al. 1996: 61–71, esp. 68.

[69] Lewis & Jones 1970: 174–8; Sánchez-Palencia Ramos et al. 1996: 74–84; Wilson 2002: 19.

[70] Pliny, *NH* 33.70–8. Sánchez-Palencia Ramos, et al. 1996: 81; Wilson 2002: 19.

Carucedo some 5 km north of the Las Médulas mines.[71] The remains of buildings and settlements discovered in the vicinity of Las Médulas do not provide us with a clear layout nor can their function in an organizational context be determined. The dearth of epigraphic evidence from Las Médulas forbids any conclusions on the function of buildings. However, recent surveys in Northwestern Spain have provided evidence for a significant shift in settlement patterns with the advent of Roman rule (cf. 6.1.3.2).

2.2.1.2. *Primary Gold Deposits and Opencast Hard-Rock Mining in Spain and Britain*

Far more promising in reconstructing the organizational layout of extractive operations under imperial control are the hard-rock opencast and underground mines at Trêsminas in north-western Spain and at Dolaucothi in Britain. Unlike secondary deposits, primary deposits of gold ore occur as veins in hard rocks of mountainous regions. The overburden, however, can still be removed by hushing, and the geology in some cases allowed opencast mining at the surface. At Braña de Folgueirosa, for example, the loosened topsoil was washed away by hushing, exposing the outcrop of an auriferous vein. Moreover, the water tanks could be used for periodic hushing in order to remove unwanted rock debris resulting from hacking out the gold ore.[72] In the case of the imperial gold mines at Trêsminas, water technology appears to have been put to use in a different manner. A find of anvil or mortar stones suggests that water-powered ore crushing machines were used to accelerate the crushing of the extracted ore to smaller stones, which then could be milled. The powder resulting from the milling was then washed in troughs in order to separate the sterile rock from the gold ore. This complicated procedure (beneficiation) minimized the dross of each smelting charge and therefore reduced the waste of fuel.[73] The Trêsminas hard rock mines consisted of two large opencasts: the Corta de Covas and the Corta da Ribeirinha. Remnants of double shafts in the rockface suggest earlier underground mining attempts before the hardrock had been sufficiently removed to allow opencast mining. From the opencasts, five galleries either functioned as adits draining water (as in the case of the Galeria do Pilar) or, in combination with

[71] Sánchez-Palencia Ramos et al 1996: 91–8.

[72] Bird 1972: 37–48. On hardrock mining, cf. Domergue 1990: 477–80; Rosumek 1982: 15–51.

[73] On the technology of ore-crushing machines and other findspots of anvil stones, cf. Wahl 1993: 139–42; Lewis 1997; Wilson 2002: 21 f. with fn. 110. See furthermore the 'bucking stone' or Carreg Angharad Fychan at Erglodd/Wales, in Schlee 2007: 8.

shafts, permitted the continued underground exploitation of gold lodes. The two opencasts cover an area of 430 × 140 m and 370 × 90 m. A third site of far smaller dimensions, the Corta dos Lagoinhos, consisted of a tunnel of *c.*100 m in length, connecting shafts sunk into the hardrock, whose ceiling collapsed at some point. Apart from a water reservoir and an aqueduct—both unrelated to the mining works—as well as a wooden amphitheatre, the preliminary excavations yielded hardly any building structures.[74] The epigraphic record from the cemetery at the site does document the presence of a military detachment and a number of immigrant workers, while pottery finds support an early date for the commencement of mining operations (probably under Tiberius). The oil lamps found in the mining tunnels point towards a peak in the extraction of ores during the first half of the second century AD.[75]

Hushing and ground sluicing were not only used on the Iberian Peninsula, but were also employed in the extraction of the gold deposits at Dolaucothi in Britain.[76] Though epigraphic evidence for the presence of members of the Roman administration or the Roman army is absent, the discovery of a Roman fort in the vicinity of the opencast and subterranean mining operations appears to suggest the involvement of the Roman state at Dolaucothi. A survey and excavations at this 'gold-mining' complex from 1987 to 1999 have added to the knowledge of the site.[77]

The mining operations at Dolaucothi consist of an opencast exploitation site combined with underground galleries and tunnels cut into the hardrock. A system of leats tabbed water from the Annell and Cothi conducting it to vast tanks situated above the opencast area. Jones and Lewis provided a developmental model explaining the sequence in which the hydraulic system set in place for hushing and ground-sluicing techniques facilitated prospection and exploitation by the Romans. The removal of the topsoil uncovered primary hardrock gold deposits which were then exploited by opencast mining and underground stoping. Pre-Roman exploitation was marginal, according to Jones and Lewis.[78] Based on an unpublished study by Cauuet, Burnham now proposes an alternative model: the Dolaucothi mining area included a primary deposit of hardrock which was exploited by tool-working and fire-setting. The secondary alluvial deposits were exploited with the hydraulic system. Cauuet suggests that the primary deposit appears to have been already extensively

[74] Wahl 1993: 126–39, 143–7.

[75] Wahl 1993: 146–50.

[76] Lewis & Jones 1969; Jones & Lewis 1971; Jones & Little 1973; Jones 1991; Stewart 2002: 233–8.

[77] For an general overview and detailed results from the campaigns of 1987–99, cf. Burnham & Burnham 2004.

[78] Jones & Lewis 1971: 297–300 with fig.9; Burnham & Burnham 2004: 6–8.

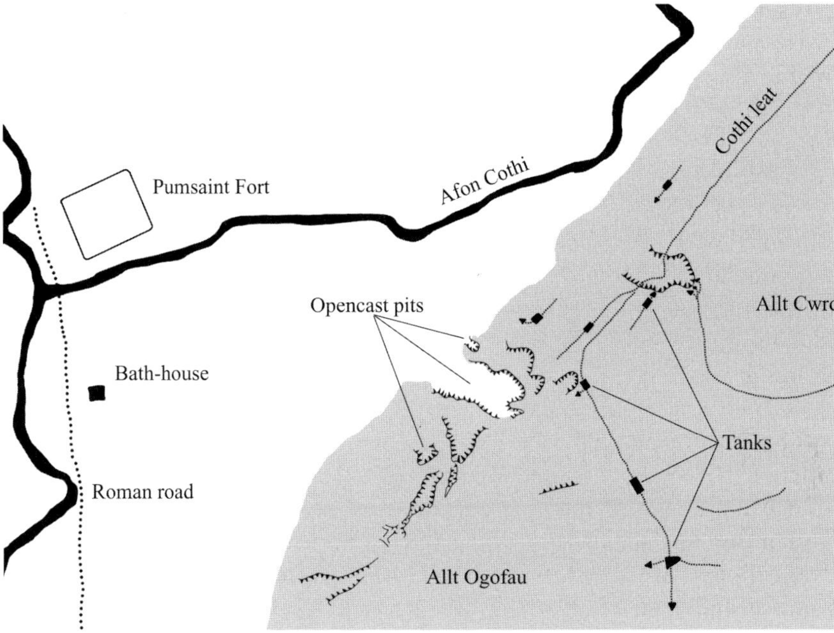

FIGURE 11. Dolaucothi, overall map (after Burnham & Burnham 2004: fig. 1.2)

exploited in the pre-Roman era. The Romans, arriving in the 70s AD found an opencast site which they sought to rework more thoroughly, in part by extending it further underground (fragments of a waterwheel attest such activity) and in part by working the secondary deposits with a large-scale hydraulic system.[79] As yet there is not sufficient evidence to resolve the early mining history of Dolaucothi. Even so, the discovery of a mortar stone, most likely belonging to a water-driven hammer mill of the Roman period and a conical spoil heap of processing wastes and rubble, offer additional insights into the mining operations. The discovery of an increasing proportion of pyritic shale strongly indicates the employment of roasting techniques in order to free the sulphide-locked gold.[80] Apart from the surviving traces of the mining operations, recent surveys have yielded settlement remains alongside a Roman road to Llandovery and in the vicinity of the Roman fort or fortlet. The military fort at Pumsaint was founded in the mid-70s AD on a level terrace above the confluence of the river Cothi and Twrch. The size of the fort probably was *c.*1.9 ha, enough for a *cohors quingenaria*. The interior of the fort/fortlet

[79] Burnham & Burnham 2004: 329 f.
[80] Burnham & Burnham 2004: 284–5, 328.

revealed a complex sequence with five phases of construction. The internal layout of the fort/fortlet, however, remains unclear. Around 100 AD the size of the fort was reduced and the new fortlet saw a refurbishment of its defences with a stone wall. Whether this alteration of the fort's size reflects a change in function and/or a reduction of troop level, as observed in other parts of Wales, cannot be confirmed. The fortlet continued to be occupied until the mid-120s AD and, like other military sites in south Wales, was abandoned during Hadrian's reign. A civilian settlement consisting of timber structures and at least one stone-built structure was discovered on the east side of the fort. Further settlement activity was located to its immediate south. With the abandonment of the fortlet the associated settlements were probably abandoned as well.[81] South of the Cothi river traces of a further settlement were discovered, sprawling alongside the Roman road to Llandovery and perhaps including a 'bath-house'. The finds appear to indicate a date range from the later first to the early second century AD and suggest that perhaps the settlement may have outlasted the Roman fortlet by one or two decades. However, no clear links to the gold-mine have been established.[82]

The settlement layout at Dolaucothi certainly draws the eye to the military fortlet around which the settlements appear to be grouped. These were most likely inhabited by miners and those supplying auxiliary services. In line with the layout of Roman quarrying ventures in the Egyptian Eastern Desert, it is tempting to see the Roman fortlet as the seat of the administration of the mining ventures. The lack of any written evidence, however, prevents the allocation of this function to the fort. Nevertheless, given the fact that gold was mined at Dolaucothi, the involvement of the Roman state is likely to have extended beyond the mere provision of security.

2.2.2. Underground Mining

2.2.2.1. *Primary Silver/Lead Deposits in South-Western Spain*

In contrast to the opencast extraction of primary and secondary deposits, the problems posed by underground mining were of a different nature. In a tedious process, the ore itself had to be extracted via shafts, galleries, and tunnels, creating technical problems such as ventilation and drainage, as well as the prevention of tunnel collapse. The mining procedures were severely limited by the water-table: mining below the water-table required the installation of Archimedean screws and treadwheel-operated waterwheels to

[81] Burnham & Burnham 2004: 318–20.
[82] Burnham & Burnham 2004: 320 f.

evacuate the water—a considerable investment. Mines also could be drained by a sloping channel or adit.[83] Moreover, the galleries and tunnels had to be secured. This was not only achieved by limiting their dimensions. Certain geometrical profiles such as trapezoidal shapes were better suited to deal with the strata pressures on the tunnel.[84] Additional security measures to prevent the collapse of tunnels were provided by timber frames and pillars.[85] Underground mining also needed lighting, with oil lamps satisfying basic lighting requirements.[86] The lack of space in the underground tunnel furthermore diminished the number of workers that could batter the rock with metal hammers after the rock-face had been softened by setting fire to it. The output in ore from an underground mine was thereby significantly limited and the removal of rubble posed a significant problem.[87] However, certain perfectly stratified primary deposits allowed the ore lode to be exploited simultaneously at several locations. Moreover, the appearance of slag heaps at sites like Sotiel Coronada points to the instant beneficiation and smelting of ore in furnaces at the site.[88] Such optimal geological conditions were encountered at argentiferous lead mines like Cabezo de los Silos/La Zarza, the copper mines at Sotiel Coronada and at Cabezo de los Silos/Cabezas de los Pastos in south-western Spain. At these mining sites the surface was literally perforated by hundreds of twin shafts or single shafts sunk to reach the vein ore. At La Zarza up to 800 pits or shafts, at Sotiel Coronada 100 twin shafts and around fifty single shafts, and at Cabezas de los Pastos 145 twin and again around fifty single shafts, have been discovered. This specific layout of mining operations is recorded in the *lex metallis dicta*, a mining regulation inscribed on a bronze tablet found at the imperial silver and copper mines of Vipasca/mod. Aljustrel.[89] The organizational layout of these mining operations can occasionally still be observed. Apart from the remains of mining operations and vast heaps of slag, Sotiel Coronada yielded remnants of a Roman settlement. The report of

[83] Craddock 1995: 76–81; Wilson 2002: 17, 20 f. Treadwheel-operated waterwheels have been discovered at Rio Tinto in Spain as well as Dolaucothi in Wales, while the mines at Posadas in Spain yielded batteries of four Archimedes screws, cf. Oleson 1984: 191, 200, 221, 249 f., 270–2. Pumping devices have beendiscovered at Sotiel Coronada and in the Sierra de Cartagena, cf. Domergue 1990, pl. XXIIa, b.

[84] Rosumek 1982: 28; Shepherd 1993: 26–30.

[85] Cf. Rosumek 1982: 31–42.

[86] Rosumek 1982: 42–6.

[87] On these tools and fire-setting, cf. Craddock 1995: 33–7, 47 f.; Rosumek 1982: 15–28.

[88] For slag at Sotiel Coronada, cf. Domergue 1987*a*: 221 f.

[89] Domergue 1983: 151–66, in part. 158 f.; Domergue 1987*a*: 217–23, 231–3, with older bibliography.

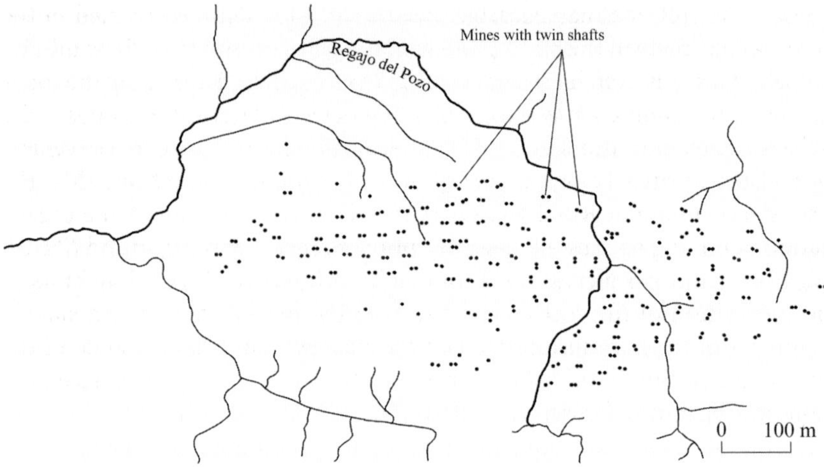

FIGURE 12. Cabezas de los Pastos, overall map (after Domergue 1990: fig. 18)

the Companhia Mineira of 1885 identified a fortified settlement, and minor settlements throughout the mining zone and a cemetery on the top of a hill close by were still visible at the time. The coin finds demonstrate the longevity of the settlement from the Republic to the first and second century AD. Some of the most notable finds were three Archimedean screws and a water pump. Moreover, an inscribed bronze tablet (100 × 80 cm) setting out the details of slaves being put to work, was discovered in the 1960s but subsequently lost.[90] A Roman mining settlement was also recorded at Rio Tinto, perhaps the ancient *Urium* known to Ptolemy.[91] A great number of buildings lay in the immediate vicinity of the mining operations. Most prominent is the vast settlement north of the Corta Dehesa, where in the eighteenth century walls, doors, columns, and pillars could still be seen. Apart from the settlement, a cemetery extended for 2 km along the former road to and from the mines. Excavations there yielded numerous lamps and Spanish *terra sigillata*. Most recorded inscriptions may originate from this locality. Moreover, the remains of two tower tombs might indicate the burial of members from the upper echelons of the mining community. Further cemeteries were discovered at Nerva, Planes, Bella Vista, and San Dionisio around the mining area. Furthermore, ancient slag deposits were found along the northern edge of the Corta Dehesa and Corta Lago[92]

[90] Domergue 1983: 222 f.
[91] Ptol. 2.4.10.
[92] Jones 1980: 154–7; Domergue 1987*a*: 241 f.; Willies 1997.

For both sites one may observe that the work force lived and died in the immediate vicinity of the mining site. Some wealthy members of these mining communities may even have been buried close by, presumably along the roads leading to the mining settlement. This is suggested by large tower tombs at the Dehesa cemetery at Rio Tinto. Modern exploitation of these sites prevents a more detailed archaeological survey and excavation. Apart from this, the inscriptions found at Rio Tinto contribute little information on the organization of mining operations. An inscription naming a *procurator* and *libertus Augusti* of emperor Nerva was found at Rio Tinto, which certainly hints at the supervision of the copper and silver mines there by the Roman state.[93] A number of funerary inscriptions do provide some insights on some of the inhabitants at Rio Tinto.[94] The inscriptions, however, do not inform us on the economic purpose of their presence.

2.2.2.2. *Goldmining in Alburnus Maior/Dacia*

Recent research on the underground gold-mining operations and mining settlements in the vicinity of Alburnus Maior/mod. Roşia Montană in Roman Dacia provides a limited insight into the spatial layout of this 'district'. The extensive exploitation of the gold lodes in modern times prohibits the complete reconstruction of the ancient mining topography. Even so, the recent surveys and excavations by Romanian and French teams have revealed a number of Roman settlements and remains of Dacian and Roman mining activities, spreading out over an area of *c.*5.5 km^2. These recent results add to previous archaeological work in the area of Roşia Montană.

Roman mining activity and settlements have been located at a number of sites on each side of the Roşia valley. Seven mining galleries discovered in 1983/4 on Carpeni hill display the use of timber to secure the galleries, niches cut especially for oil-lamps, and a ladder carved from a tree trunk. The galleries are cut in a trapezoidal form with a wide base (with heights between 1.5 and 1.9 m). Furthermore, excavations yielded remains of a settlement at 'Balea' as well as pottery and tools. Further building remains were uncovered at the 'Tomuş' and 'Bisericuţă' sites on Carpeni hill. One structure unique to Roşia Montană displays stone-mortar walls, a hypocaust and a vast number of tiles, some of which carry the stamp of the *legio XIII Gemina*.[95]

[93] *CIL* II 956.

[94] *CIL* II 957, 958; *AE* 1965: 298–300, 302; *AE* 1965: 301 = *HEp* 3, 1993: 221; *AE* 1991: 1000–2 = *HEp* 3, 1993: 215–17; *HEp* 3, 1993: 218–25.

[95] Wollmann 1996: 376–8; C. Crăciun and V. Moga, in Damian 2003: 35 f.; V. Rusu-Bolindeţ et al., in Damian 2003: 383–427; E. Bota et al., in Damian 2003: 429–42. O. Ţentea and V. Voişian, in Damian 2003: 443.

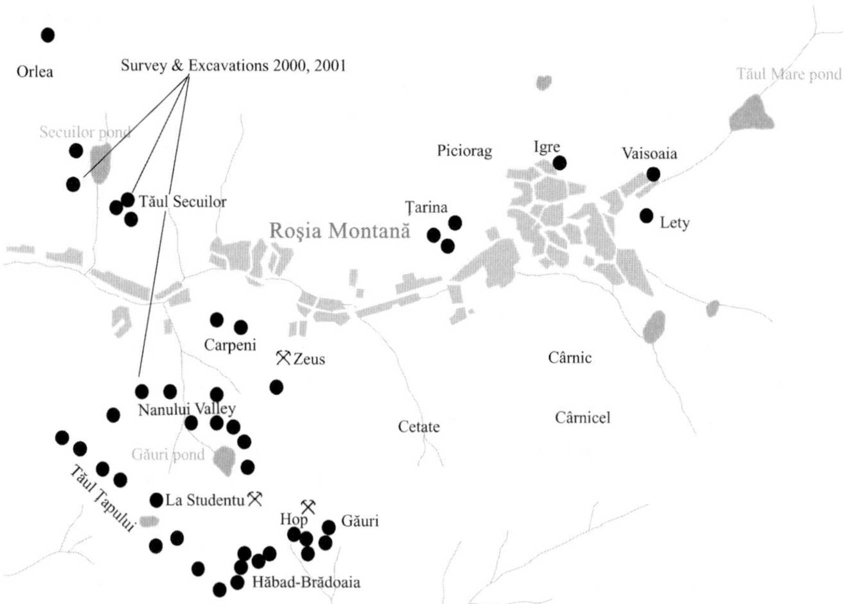

FIGURE 13. Roşia Montană, overall map (after Damian 2003: fig. 12 and maps; Wollmann 1996: plate LXXXVIII)

Further discoveries were made at the so-called 'Găuri', 'Hop-Botar', and 'Hăbad-Brădoaia' sites, as well as in the 'Nanului Valley'. At Găuri remains of a settlement have been detected, consisting of wooden and stone-built structures, in one of which a votive altar to Ianus was found. Particular finds ('household' pottery, grinders, oil lamps, needles, etc.) and the lack of mining tools have been used to identify the settlement in question as one of the civilian *kastella* referred to in the epigraphic evidence from the site.[96] Further building remains of Roman origin were yielded by excavations at Hop-Botar and nearby Tăul Ţapului, as well as at Hăbad-Brădoaia. The exact function of either of these edifices remains mostly unknown.[97] One complex at Hăbad-Brădoaia was identified as an 'area sacra', including 'cult' buildings and 35 inscribed altars. Amongst the numerous votive inscriptions the dedications of three *beneficiarii consulares* stand out, whereas four inscriptions appear to name a *k(astellum) Ansi/s* (perhaps to be equated with Hăbad-Brădoaia).[98] A cemetery near Hop-Botar yielded 169 incineration burials.[99] In the Nanului Valley the excavators unearthed three structures probably erected

[96] C. Crăciun and V. Moga, in Damian 2003: 34. V. Moga et al., in Damian 2003: 45–79.
[97] P. Damian et al., in Damian 2003: 81–120; A. Pescaru et al., in Damian 2003: 121–46.
[98] Wollmann 1996: 375; S. Coiciş et al., in Damian 2003: 147–89.
[99] V. Moga et al., in Damian 2003: 191–249.

for religious purposes. The 'cult edifice' at 'Szekely' included three inscribed votive altars.[100] Similary, in a structure at 'Drumuș' seven inscribed altars and a number of altar fragments were found, three naming a *collegium Sardiat(ensium)*.[101] Moreover, the 'Dalea' site also consists of a religious precinct and a number of inscribed, anepigraphic and fragmentary altars.[102]

Immediately to the east of these settlements on the Cetate mountain the archaeologists examined the traces of Roman underground mining operations. The 'Zeus' mining sector consists of mining tunnels, either for prospection or exploitation purposes, the latter following the lode in a vertical slant. As remaining niches suggest, the exploitation tunnels appear to have been secured with wooden beams. The examination of the organic and non-organic debris inside the 'Zeus' mine allowed a date for the beginnings of the exploitation during the first or second century AD and a continuation perhaps into the third and fourth century. Further works are found at the 'Găuri' sector which might have been exploited in Roman times.[103] East of the Cetate mountain and south of Roșia Montană a further hill, the Cârnic ridge, was mined for gold. Some of the underground mines at this site appear to be of Roman origin, consisting of a number of galleries and tunnels. The French excavators discovered traces of fire-setting, lamp niches typical of Roman mines and charcoal remains, the latter providing a Carbon 14 date for the mining works ranging from the second to the fourth century AD. Lamps and fragments of oil-lamps found in the rubble of the Cârnic mine date to the second century. Surprisingly, in some sections of the mine wooden remains provided a Carbon 14 date ranging between the mid-third century BC and the first century AD. Provided the Carbon 14 dating is trustworthy, mining operations at Alburnus Maior commenced in pre-Roman times.[104]

Similar data were gained from the analysis of organic material from a survey of the Țarina mining site on the northern bank of the Roșia Valley.[105] In the vicinity of the mine, remains of a cemetery were still visible in the late nineteenth century. Funerary stelae and a sarcophagus lid were interpreted as signs of wealth of the local mining community. Furthermore, the discovery of a votive altar suggests the presence of a sanctuary there.[106] Surveys on Orlea hill yielded two mining galleries with vast timber constructions, niches for oil

[100] C. Crăciun et al., in Damian 2003: 253–83.

[101] Perhaps a reference to the tribe of the *Σαρδύαται* or *Sardeates* in Dalmatia, cf. C. Crăciun and A. Sion, in Damian 2003: 285–333.

[102] V. V. Zirra et al., in Damian 2003: 335–80.

[103] B. Cauuet et al., in Damian 2003: 486–501.

[104] B. Cauuet et al., in Damian 2003: 467–526.

[105] B. Cauuet et al., in Damian 2003: 503.

[106] Wollmann 1996: 379 f.; C. Crăciun and V. Moga, in Damian 2003: 40 f.

lamps and water conduits still preserved. Moreover, seven incineration burials were excavated, and in the vicinity of Orlea at Tăul Secuilor a further cemetery was unearthed.[107] To the east of the Orlea–Ţarina mining area *c.*50 wooden writing tablets and mining tools were found in Roman mining tunnels at Igren, Văidoaia, and Lety during extractive operations in the eighteenth and nineteenth century.[108] Further tablets were discovered during mining operations at Carnicul Mare in the early nineteenth century.[109]

Although the intensity of modern mining and settlement activity destroyed much of the original Roman layout, the recovered traces allow for some conclusions to be made. The scattered settlements, cemeteries, and sanctuaries are located close to the mining operations on the hills surrounding the Roşia valley. Undoubtedly, the population involved in mining saw it necessary to settle in the immediate vicinity of their working place. This is reflected in the epigraphic evidence and the 'tabulae ceratae' as well: apart from a *vicus Pirustarum*, a *k(astellum) Baridust(arum)* and a *k(astellum) Ansi* or *Ansis* are named.[110] Whether or not there existed an administrative centre or a military fort at Alburnus Maior remains unknown. Tiles of the *legio XIII Gemina* might suggest the occasional presence of a military contingent and the discovery of inscriptions for *beneficiarii consulares* certainly implies some sort of administrative control, the 'bureau' perhaps located somewhere beneath the modern town of Roşia Montană.

2.2.3. Mining Techniques and their Organizational Implications

The exploitation of secondary gold deposits in the river valleys of northwestern Spain required a considerable investment in hydraulic infrastructure.[111] Huge aqueducts had to be constructed leading over difficult terrain, covering long distances and considerable differences in altitude. Large hushing tanks had to be built, channels had to be dug, sluice boxes constructed, tunnels cut into the overburden, and so forth. The installation and maintenance of this technological infrastructure as well as the mining procedure itself required a large and constantly available workforce and the presence of skilled engineers. The workforce needed to be supplied with pay, food, and tools, and security measures had to be implemented. The mining district required protection from brigands and the workforce

107 Wollmann 1996: 378 f.; C. Crăciun and V. Moga, in Damian 2003: 36 f.
108 Wollmann 1996: 380 f.
109 Wollmann 1996: 381.
110 Piso 2004: 291, 297 f.; see now Ciongradi et al. 2008.
111 Wilson 2002: 17.

(especially when coerced) had to be guarded. Military personnel were furthermore needed for the protection of convoys exporting the metal. The complexity of organizing a mining operation, the investment in a vast infrastructure, the command of human and material resources, and the huge scale of operations render a concerted extraction effort likely. The well-structured opencast gold mines at Trêsminas or Dolaucothi, in particular the galleries for water drainage from the opencast as well as those cut into the hardrock following the gold lode, imply a concerted effort under the control of one party. Unlike opencast mines used for the exploitation of secondary deposits, the removal of hardrock in the opencast extraction of primary deposits was far more tedious. In the case of Trêsminas, it is not quite clear whether or not hydraulic technologies could be applied to the removal of rock debris. This limited extraction to a smaller scale, and the technical process prior to the smelting of gold was lengthened. Besides the usual requirements in supplying and guarding or protecting the workforce, a hydraulic infrastructure had to be put in place to evacuate water, and to accommodate the mechanized crushing, milling and washing of gold ore.

The spatial restrictions of underground mining certainly reduced the number of workers extracting ore, decreasing the ore output of a mine. Moreover, the investments in technical infrastructure were considerable. Timber and wood were needed to set fire to the rockface or to prevent tunnels, shafts, and galleries from collapsing.[112] Complicated water-lifting devices had to be constructed in order to deal with rising water-tables. Metal tools were required and had to be repaired, in turn making the import of iron and charcoal necessary. Furnaces were constructed for the smelting of ore, while mills and ore crushing devices were needed for the beneficiation process. The vast quantities of rubble had to be removed from the tunnels and shafts. Besides the basic requirements of food and shelter for the workforce, the organizational challenges were significant, although perhaps not quite as colossal as the large opencast extractive operations in north-western Spain. Furthermore, in certain cases such as Sotiel Coronada or Vipasca the geological formation of the ore deposit allowed for a compartmentalization of mining operations. The imperial administration was thus in the fortunate position of being able to delegate the immediate burden of organizing the exploitation of the ore deposits (i.e. the operative process) to either different private contractors or different owners (cf. 7.1.1). The burden shouldered by imperial officials was limited to the construction of drainage adits.[113]

[112] This is suggested for the marked logs found at Mazarrón, cf. Rosumek 1982: 38.
[113] Cf. Vipasca/mod. Aljustrel Domergue 1983: 7–23, 159.

Given how remote some of these mines were from the more populated regions, the mining authorities were likely forced to 'import' the work force over considerable distances. The populace involved in hardrock extractive operations seems to have been settled in the immediate vicinity of the mines. The location of cemeteries close by and the construction of sanctuaries and baths at larger mining settlements appear to be the expression of a long-term commitment to operating the mines. In contrast to the settlements of the miners, the administrative mining centres were not necessarily to be found in the mining zones but—as at Dolaucothi and perhaps Roșia Montană—in the valleys near the main communication links i.e. roads or rivers (cf. 4.3).

2.3. SUMMARY

The comparison of the restrictive framework set by geological, technological, topographical, and geographical constraints reveals genuine organizational differences between imperial quarries and mines. Although the workforce lived in housing as close as possible to the extractive operations in both cases, and there are similar solutions to basic problems in sustaining and protecting the workforce in imperial mines and quarries in general, certain logistical or technological requirements remain unique to each category of extractive operation. Large-scale opencast mining required a complex hydraulic infrastructure consisting of aqueducts, canals, and tanks, while underground mining in some cases was in need of an elaborate scheme of water conduits and water-wheels in order to evacuate ground water. Moreover, ore crushing machines and mills could make a significant contribution towards speeding up the processing of ore. These infrastructural and technological possibilities, if fully exploited, undoubtedly increased the capital intensity of such ventures.

In contrast to imperial mining operations, there was little difference in extractive technology applied to imperial quarries and quarry operations in general. Stone was broken out of the rockface and split into the desired size by applying a series of wedgeholes for wooden or iron wedges, or using stone saws. The blocks were then dressed to the desired shape with chisels.[114] In comparison to opencast or underground mining procedures, the technology applied did not require a vast infrastructure. The main organizational problems emerged when the marble products had to be moved out of the quarry. The construction of special carts, slipways, and roads, as well as the supply of draught animals,

[114] Rosumek 1982: 122–31; Adam 1999: 22 f. On stone saws, cf. Rosumek 1982: 131–6.

etc., demanded a considerable logistical effort from the authorities in charge. The export of metal ingots, however, hardly necessitated preparations of a comparable scale. Once the technological infrastructure for the mines was in place, its maintenance perhaps was a lesser organizational burden than the constantly recurring strain on human, animal, and material resources imposed by the immense logistical problems connected with the movement of bulky marble items.

Therefore, the authorities running imperial quarries or imperial mines were faced with distinct organizational challenges. Even amongst imperial mines different geological and topographical constraints posed different problems (opencast or underground mining) and set limitations on the scale of the mining operations possible. Similarly, the size of the marble outcrop could determine the complexity of extractive procedures. Small marble outcrops necessitated well planned operative procedures in order to cope with the disposal of rubble and the quarrying activities of different parties. These fundamental differences certainly had an effect on directive and operative processes and in most cases demanded a customized organizational approach. In other words, the authorities in charge were forced by the geological and topographical constraints as well as the limitations of the available technology to find individual organizational solutions for each imperial quarrying or mining district.

3

Mining and Quarrying Districts

Mining and quarrying operations under imperial control appear to have taken place within a strictly defined territorial entity distinct from the colonial, municipal and other territories within a province. The enforcement of various organizational measures by those in charge of Roman mining and quarrying operations was likely focused on these delimited districts. The existence of such districts is highlighted most prominently in the Vipasca tablets. Some scholars have tried to identify further imperial mining and quarrying districts, their extent, and their setting within the ancient landscape of Roman provinces. Not all their conclusions can remain uncontested.

3.1. *METALLUM* AND *TERRITORIA METALLORUM*

The bronze tablets found at Vipasca/mod. Aljustrel (Portugal) in the early twentieth century clearly describe the mining area of Vipasca as an administrative district with its own boundaries. The passages in these texts stating this are numerous: according to the *lex metallis dicta* (*LMD*) slaves convicted of stealing ore were flogged and sold under the condition that they never return to *metalla* or *territoria metallorum* (ll. 27–8). In the case of a freeman caught stealing ore, the procurator confiscated his property and banned him from the mining territory, the *fines metallorum* (ll. 29). Similar rules applied to saboteurs of mining operations: slaves who had been punished or sold were to stay away from any *metalla*, while *liberi* were prohibited by the procurator from entering the *fines metallorum* ever again (ll. 32–5). The same terminology regarding the mining territory is repeated in the clauses concerning the punishment for breaching technical protocol of mining procedures (ll. 40–1). In the *lex metalli Vipascensis* (*LMV*) contractors were authorized to collect certain fees *intra fines metalli Vipascensis* (ll. 1–9 and 59–61). Moreover, the barbers and slag-collectors, as contractors of these services, were assured a monopoly of their trade within the confines of the *vicus metalli Vipascensis* or its territories (ll. 37–8 and 47–57).

The *lex metalli Vipascensis* and the *lex metallis dicta* distinguish between three territorial entities, 'vicus', 'metallum' and 'territoria metallorum' or 'fines metallorum'.[1] The enclitic particle *–ve* is used in both texts to distinguish the mining region and its territories: *LMV* ll. 37–8 *in vico metalli Vipascensis inve territoris eius*, 'in the *vicus* of the *metallum Vipascense* **or** its territories', and in *LMD* l. 27 *in ullis metallis territorisve metallorum*, 'in any *metalla* **or** *territoria metallorum*'. If we assume that the distinction is a deliberate one, the question arises as to what these terms actually mean. We learn from the *LMV* that the *metallum Vipascense* included a *vicus* with a bath, a school, barber shop, and, as the inscriptions found at Vipasca suggest, a cemetery.[2] In 1.2 of *LMV* one reads that contracts *intra fines metalli Vipascensis*, 'within the area of the *metallum Vipascense*', are subject to a 1 per cent tax. The text thus suggests that the *metallum Vipascense* consisted of a delimited mining zone, a *fines metalli*. In contrast *LMD* notes on several occasions *fines metallorum* or *territoria metallorum*.[3] The use of the singular in the *LMV* and the plural of *metallum* in the *LMD* most probably reflects the different purposes of these regulations. Whereas the *LMV* regulates affairs within the *metallum Vipascense* only, the *LMD* probably was enforced in a number of mining operations in Southern Spain.[4] A tentative conclusion may be proposed: a *metallum* could be an administrative district in its own right with its own territory or territories.[5]

Interestingly, the *lex metalli Vipascensis* refers to only one *metallum Vipascense*, whereas the title of the procurator at Vipasca, *procurator metallorum*, suggests that his responsibilities covered more than one *metallum*.[6] This is corroborated by an inscription found at Vipasca which was set up by the *coloni (...) metalli Vipascensis* for the procurator Beryllus, the *restitutor metallo[r]um*.[7] The singular 'metallum' identifies the basic territorial unit such as the mining area at Vipasca. Consequently, the *procurator metallorum* named in the Vipasca tablets and the inscription by the *coloni* would not only have supervised the *metallum Vipascense* but a number of other *metalla* as well.[8] At Pergamon in Asia Minor, an honorific inscription for a Saturninus,

1 Dušanić 2004*b*: 249.

2 *IRPac* 122–6, 128, 130–41.

3 *LMD* ll. 27–8, 34, 41–2.

4 Domergue 1983: 161.

5 The sentence 'in vico metalli Vipascensis inve territori⟨i⟩s eius'(*LMV* l. 37–8), suggests that the *metallum Vipascense*—rather than its *vicus*—consisted of multiple 'territoria', cf. Domergue 1983: 54, 88. The singular forms 'territorium metalli' or 'territorium metallorum' are not attested in the Vipasca tablets.

6 *LMV* l. 2: *(...)intra fines metalli Vipascensis (...) proc(urator) metallorum (...)*

7 *IRPac* 121.

8 Hirschfeld 1905: 160 with fn. 2.

listing his career, documents the office of an *ἐπίτροπος μετάλλων Οὐλπ[σ]κηνσίων* or *procurator metallorum Vipascensium*, evidently the plural of *metallum Vipascense*.[9] This does not necessarily contradict our previous assumption.[10] Vipasca was probably the administrative centre for the *metallum Vipascense* and perhaps for a number of other *metalla* in the area as well.[11]

The distinction between 'territoria/fines metallorum' and 'metallum', as observed in the Vipasca tablets, is not reflected in other legal sources. Moreover, Roman jurists hardly ever employ the term 'territorium' in a mining or quarrying context, while it is often used to define the territory of a community.[12] In fact, the writings of the Roman land surveyors demonstrate the precise use of the term 'territorium' for the lands of *coloniae*, *municipia*, *civitates* etc. An additional aspect, which only is hinted at marginally in the *Digest*, is the apparent understanding by Frontinus or Agennius Urbicus of a *territorium* as a jurisdictional unit, which perhaps even had limited autonomy.[13] Given the deliberate use of the term in the Vipasca tablets it could reflect the judicial powers bestowed on the *procurator metallorum* and might define the limits of his jurisdiction.[14] Even the term 'metallum' is defined rather vaguely in the Roman legal sources. For Ulpian both stone quarries and any other extractive operations can be identified as 'metalla'.[15] This seems to reflect the general understanding—observable in most literary sources—of *metallum* as a location where stones, metals, or other resources can be extracted.[16] However, in his ninth book on the duties of a proconsul, Ulpian explains that *metalla*, although numerous, do not exist in all provinces.

[9] Cf. Christol & Demougin 1990: 170.

[10] The argument that either the scribes of the bronze tablets or the Pergamene inscription mistakingly used the plural has to be ruled out. The legal content of the Vipasca tablets theoretically excludes an unsystematic use of terminology. The explanation that the Greek scribe in Pergamon misunderstood the legal niceties of the terminology does not quite work as the term *μέταλλα Οὐλπασκήνσια* seems to be a direct translation of *metalla Vipascensia*, albeit with a misinterpretation of the letter 'i' as 'λ', cf. Christol & Demougin 1990: 170 with fn. 60.

[11] Based on the recent excavations at Aljustrel the question whether or not the so called 'House of the Procurator' actually was an office for the imperial mining administration cannot be settled, cf. Domergue 1983: 31 f.

[12] For a definition, see Pomponius, *Dig.* 50.16.239.8: *Territorium est universitatis agrorum intra fines cuiusque civitatis (...)*. Other attestations: *Dig.* 2.1.20, 10.1.7, pr. 30.41.5, 47.12.3.4, 48.22.7.19, 50.4.18.25, 50.12.8, 50.15.4.2.

[13] See 'territorium' in Campbell 2000: index, glossary.

[14] Hirschfeld 1905: 160, compares the judicial and other powers of the mining procurator at Vipasca with those of the municipal authorities. On the responsibilities of the mining procurator at Vipasca, cf. 6.1.3.1.

[15] *Dig.* 7.1.9.2–3, see also *Dig.* 7.1.13.5, 27.9.3.6.

[16] *TLL* VIII M, 870–5.

Convicts condemned to the mines in a province without *metalla* have to be transferred to provinces with *metalla*.[17] In this context the word 'metallum' is not used with the usual ambivalence, but must describe one specific kind of extractive operation. Otherwise the explanation that not all provinces had *metalla* (although all sorts of extractive operations such as quarries, salt mines etc. are documented in every province) would not make much sense. Ulpian therefore must be using 'metalla' as reference to a specific category of extractive operation to which the governor of a province could condemn convicts.

3.2. MINING AND QUARRYING DISTRICTS

The existence of clearly defined territorial entities exclusively reserved for extractive operations under imperial supervision appears to be sufficiently documented by the evidence discussed above. Yet, when it comes to identifying 'metalla' and 'territoria metallorum' at places other than Vipasca, let alone defining the precise extent of the mining district at Vipasca itself, the epigraphic and archaeological material provides us with little evidence. Although Roman mining and quarrying ventures *per se* can be detected throughout the empire, one usually is left in the dark about the precise delimitation of the district. Only on rare occasions can the outline of a territory still be traced: the district of the Simitthus quarry, for example, was partially demarcated by a wall which provides a rough idea of the extent of the quarrying district. Moreover, the written material from the quarries in the Egyptian Eastern Desert demonstrates the existence of an administrative division consisting of numerous quarries.

3.2.1. Quarrying Territories in Roman Egypt

The archaeological remains at Mons Claudianus, Mons Porphyrites, Mons Ophiates, and Tiberiane display uniform characteristics in their assemblage of infrastructural elements (forts, slipways, cairns, etc.). These similarities seem to point towards the inclusion of these four quarries within the same organizational unit. Moreover, Peacock's discovery of red porphyry fragments at all four sites provides evidence for significant interaction amongst these quarries. This is confirmed by the written material which the excavations have yielded

[17] *Dig.* 48.19.8.4.

thus far. The management of the stone quarries at Mons Claudianus, Mons Porphyrites, and Mons Ophiates(?)/mod. Wadi Umm Wikala, fell partly under the responsibility of one *ἐπίτροπος τῶν μετάλλων*/*procurator metallorum*, as inscriptions naming the same procurators at different quarries appear to indicate. Marcus Ulpius Chresimus, for example, is documented in the epigraphic record at Mons Claudianus and at Mons Porphyrites, while one Ulpius Himerus is known at Mons Claudianus and Mons Ophiates. Furthermore, Epaphroditos Sigerianos, an imperial slave and *μισθωτὴς τῶν μετάλλων*, is documented at Mons Claudianus and at Mons Porphyrites (cf. 4.1.1).

Apart from the imperial officials in charge, further documentary evidence seems to corroborate the view of a single vast administrative division containing these four individual quarrying districts. The *φαμελιάριοι* (i.e. members of the *familia Caesaris*), are noted in the ostraca as belonging to the *ἀριθμός* of Mons Claudianus. Together with the *ἀριθμοί* (subdivisions) of the quarries at Tiberiane and Mons Porphyrites, the *ἀριθμὸς Κλαυδιανοῦ* formed a subdivision of the *νούμερος Πορφυρείτου*.[18] An ostracon from Maximianon/mod. Al-Zarqa' on the road between Koptos and Myos Hormos mentions a man *ἐκ φαμ[ιλίας---] Παρφιρίτου*(!), suggesting that other areas beyond the known quarries might have been included in this vast administrative division.[19] Further evidence is provided by an inscribed limestone block at El-Ashmunein. The building inscription attests the existence of a 'hosp(itium) tabular(iorum) Porphyr(itis) et aliorum metallorum', quarters for *tabularii* of the Mons Porphyrites quarries, or perhaps even of the *numerus* of Mons Porphyrites (provided that *aliorum metallorum* means the quarries of Mons Claudianus, Tiberiane, and Mons Ophiates).[20] These imperial quarries undoubtedly were part of a vast—though not necessarily continuous—quarrying district in the Eastern Desert. There are indications in the documentary evidence of at least one other quarrying *territorium* in Roman Egypt: apart from the *numerus* of Porphyrites, a *numerus* of Alabastron is mentioned twice in the published Claudianus ostraca.[21] Other hitherto unpublished ostraca refer to a *σκλ[ηρουργῷ] | Ἀλαβαστρ.[-]* and a *μέταλλον ἀλαβαστρίνης*.[22] Moreover, two graffiti in Wadi al-Hammamat appear to

[18] What status the *familia* of Mons Ophiates had, whether it belonged to the *numerus* of Mons Porphyrites or to another *numerus*, cannot be determined at present. Cuvigny 1996*c*: 97 fn. 16; Cuvigny 2000*a*: 36 f., cf. also index 315, 317.

[19] M 997, cf. Cuvigny 2000*a*: 37; Cuvigny 2003*c*: 373.

[20] Cockle 1996: 23–8; Maxfield 2001: 153. On the term 'hospitium', cf. Cuvigny 2000*a*: 37, fn. 128.

[21] *O.Claud.* 528, 587; for the location of Alabastron, cf. Cuvigny 2000*a*: 37 f.

[22] *O.Claud.*inv. 3285, 7149, cf. Cuvigny 2000*a*: 38.

refer to Alabastrine.[23] The territory covered by this *numerus* probably included the alabaster quarries and other quarries in the area of al-Miniya and Asyut near Alabastrine/mod. Qum el-Akhmar.[24]

3.2.2. *Patrimonium Regni Norici*

The inscriptional and archaeological evidence for such districts outside Roman Egypt is far from abundant. Nevertheless, in his study on the *patrimonium regni Norici* Geza Alföldy synthesized the evidence to produce an outline of a vast mining area at the heart of Noricum. The Alpine area contained the majority of the province's economic resources: Roman iron mines were located at Gurina, Hüttenberg, Lölling, Semlach, and Wiesenau in Carinthia.[25] The iron mines in the vicinity of Neumarkt and Friesach as well as in Eisenerz in Upper Styria also show signs of Roman exploitation.[26] Moreover, Strabo reports gold being gained from rivers in the vicinity of Noreia in Noricum.[27] Possible sources of gold are the gold-bearing rivers such as the Weissenbach (Tragin) or the Klieningbach in the Upper Lavant valley or the hardrock gold mines in the Hohen Tauern mountains, at Tragin and Siftlitz.[28] The presence of imperial slaves, soldiers, or *conductores* in the area is recorded by inscriptions, strongly suggesting imperial management of these resources.[29] However, it is not easy to determine whether these imperial officials controlled numerous and territorially dispersed mining and other districts (such as the salt mines at Hallstatt, the quarries of Kraig near St. Veit an der Glan and Schaidberg, and perhaps forests and pasture lands),

[23] *I.Ko.Ko.* 107, 108; Cuvigny 2000*a*: 38. To what quarrying territory, if at all, the *bekhen*-quarries at Persou/mod. Wadi al-Hammamat were attached is not clarified by the written material found there. The settlement at Hammamat appears to date to the 1st cent. AD. and, apart from the inscription by P. Iuventius Agathopus, προνοητὴς πάντων to the μεταλλάρχης in Egypt (*OGIS* II 660 = *SB* 8580 = *I.Ko.Ko.* 41; AD 18), there appears to be no further evidence for imperial control of the *bekhen*-quarries after the early Tiberian period, cf. Brun 2003*a*; Cuvigny 2003*a*: 280–4.

[24] Drew-Bear 1979: 59–61; Cuvigny 2000*a*: 37 f. Perhaps it was to these quarries that the *praefectus Aegypti* sent convicts, cf. Hagedorn & Shelton 1975; Bastianini 1986.

[25] For map, cf. Piccottini 1989: nos. 20, 123, 154, 225, 432, 508 with further bibliography.

[26] Alföldy 1970: 167. Roman involvement in the trade of iron products is attested for the early 1st cent. BC on the Magdalensberg, where lists painted on to the plastered walls by Italian merchants name a considerable number of different iron tools being traded, cf. Egger 1961; Alföldy 1974: 73.

[27] Strabo 5.1.8.

[28] Piccottini 1994*a*: 471–4.

[29] Tiffen: *CIL* III 4788 = *ILS* 1466; *AE* 1957: 108 = *AE* 1969/70: 454; *CIL* III 4787, 4822 = 11505, cf. Alföldy 1969*b*: 25, no. 34. Hohenstein: *CIL* III 4807, 4808, 4809 = *ILS* 1467; *CIL* III 14362 (p. 2328, 197) + 14363 = *AE* 1968: 408. Feldkirchen: *CIL* III 4861. Wieting: *AE* 1995: 1195.

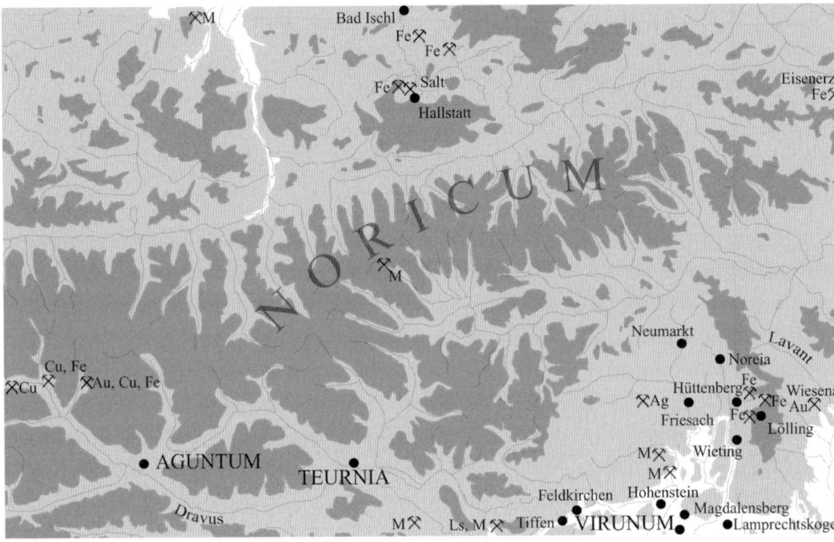

Figure 14. Mines in Noricum

or, alternatively, one continuous and vast territory embracing all these areas of economic interest. Alföldy examined the epigraphic evidence for Norican soldiers and the distribution of inscriptions erected for and by members of municipal authorities of Norican towns. A number of soldiers (mostly praetorian guards and *equites singulares*) did not name a town as their place of origin but simply claimed to be of the 'Norican nation'. Alföldy thus argued that these soldiers most likely came from a rural area which did not belong to the known municipal territories in Noricum.[30] More importantly, the distribution of inscriptions by members of municipal authorities and especially the dearth of such inscriptions in the Alpine area of Noricum, together with and the concentration of important economic resources within this central zone, led Alföldy to conclude that the heart of Noricum formed a coherent territorial unit under imperial control.[31] Judging from the epigraphic record, the centre of the mining administration appears to have been based at Virunum, the provincial capital until the reign of Marcus Aurelius, and later the seat of the financial procurator. There, inscriptions display the letters PRN following

[30] Alföldy 1970: 165 f. M. P. Speidel (1981–2) pointed out that the term 'natione Noricus' could also refer to soldiers of municipal origin. Thus, details on the origin of soldiers—as Alföldy (1989: 52) conceded later—can not be used to argue for non-municipal territories.

[31] Alföldy 1970: 167; Alföldy 1974: 100 f. For a discussion of Alföldy's views, cf. Ørsted 1985: 201 f. with further bibliography.

the functional grade of members of the *familia Caesaris*, e.g. *disp(ensator)* PRN. Alföldy expanded the abbreviation to *p(atrimonium) r(egni) N(orici)*, a reading he based on the assumption that the iron mines, quarries and pastures at the heart of Noricum were the private property of the emperor which Augustus probably had 'inherited' from the Norican kings.[32] The term *regnum Noricum* itself is attested in other inscriptions, which partly corroborates this reading of PRN.[33] Alföldy's reading *p(atrimonium) r(egni) N(orici)* appears to be the most plausible solution so far. Furthermore, Alföldy's hypothesis of a vast territorial unit at the Alpine heart of Noricum is partly confirmed by the presence of two *stationes* for the exaction of the *portorium Illyrici.* These toll stations are not found at the provincial border, but within Noricum itself at Bad Ischl and in the area of Lambrechtskogel east of Virunum, on the territorial borders of an imperial domain, as Alföldy argues.[34] Moreover, in Friesach a *conductor ferrariarum Noricarum* set up a votive inscription for *Termunes*, the gods of the 'dominial(?)' border.[35] Alföldy concluded that for the export of iron ore taxes had to be paid at the limits of the mining territory.[36] He thus offered the compelling hypothesis of a vast extra-municipal territory in central Noricum named *patrimonium regni Norici*, containing most natural resources of economic interest to the Roman state.[37] Yet, apart from the peculiarly positioned toll

[32] Alföldy 1970: 172. Virunum: *CIL* III 4828; *ILLPRON* 176; *CIL* III 4800 = *ILS* 4198; *CIL* III 4797 = *ILS* 1506. As an alternative to Alföldy's reading of the abbreviation PRN Winkler had offered *p(rocurator) r(egni) N(orici)*. Alföldy, however, pointed out correctly that subaltern officials of the imperial administration normally do not identify themselves as subordinates to a function like procurator but as members of an administrative branch. This and the rare attestations of the single letter P as an abbreviation for *p(rocurator)* led Alföldy to reject Winkler's reading, cf. Winkler 1969: 141–3; Alföldy 1970: 171; Ørsted 1985: 216–18.

[33] *arcar(ius) regn(i) Noric(i)*, cf. *CIL* III 4797 = *ILS* 1506, cf. Alföldy 1970: 171; *procurator Augusti regni Norici*, cf. Alföldy 1974: 79 and 242–7 (Appendix V).

[34] Alföldy 1970: 170 f.; Winkler 1977: 219 f.

[35] *CIL* III 5036, cf. Alföldy 1970: 169, 177.

[36] Alföldy 1970: 171. Alföldy's hypothesis has encountered opposition from Vetters and other scholars. The massive extent of the proposed imperial domain appears to have particularly troubled Vetters (1977: 307–11; 1980: 42 f.). His main argument was that the borders of Alföldy's imperial domain did not reflect any of the topographical realities. The location of inscription *CIL* III 5620, naming a toll station at Bad Ischl was questioned by Scherrer (1985). He claimed that, according to a manuscript by Petrus Apianus of 1534, this *statio Esc(ensis)* probably was originally located in the Chiemsee area in Bavaria and therefore on the border between Raetia and Noricum. Weber (1995), however, provided convincing arguments for relocating the station at Bad Ischl. Even so, the arguments hitherto put forward alter the picture only marginally.

[37] The inscriptions naming subaltern administrative personnel responsible for the *p(atrimonium) r(egni) N(orici)* are dated to the 3rd cent. AD, yet the term *patrimonium regni Norici* is an archaism referring to the early days of Roman rule in Noricum, when the properties of the indigenous king(s) came under imperial control. The nature of the evidence forbids us to trace possible alterations of territorial borders or the unity of the territory in question from the heyday of Roman rule in Noricum to the early 3rd cent., cf. Alföldy 1970: 171 f.; Vetters 1977: 310.

stations, there hitherto is no positive evidence such as boundary stones delimiting this estate or written documents describing such an extensive entity within the province. One therefore might also be looking at a patchwork of private or municipal lands and public mines, quarries, woods and pastures not amalgamated in one homogeneous imperial estate.[38]

3.2.3. Mining districts and toll stations in Pannonia, Dalmatia, Moesia Superior, and Thracia

Alföldy's hypothesis for the mining areas of central Noricum has been used as a model by Slobodan Dušanić to pinpoint other Roman mining districts throughout the Danube provinces. Toll stations of the *publicum portorii Illyrici* in the interior of the customs district and in the vicinity of Roman mining ventures have also been located in this region. Like their Norican counterparts, these are believed to have marked the boundaries of vast mining districts:[39]

- The inscription on a fragmentary altar by a *vil(icus) Moes(iae) r(ipae?) s(uperioris?)* found at Gorička in Pannonia Superior supposedly refers to a toll station associated with the Roman iron fields around Topusko, west of Siscia/mod. Sisak.[40] Dušanić argued that the iron ore deposits near Topusko together with those in the vicinity of Ljubija were part of a larger imperial domain within Pannonia, perhaps including remnants of iron-mining in the area around Briševo and Stari Majdan, both near Sanski Most.[41] The epigraphic evidence documents an administrative centre of

[38] Hadrianic coins with the legend MET. NOR., usually extended to *met(alla) Nor(ica)*, would refer to the existence of individual mining territories (?) in the early 2nd cent. AD—provided the abbreviation is read correctly and the term 'metallum' is used in the sense offered by the Vipasca tablets as a clearly defined territorial entity, cf. Strack 1933: no. 432a; Alföldy 1974: 114 with 318 fn. 72.

[39] Dušanić 1977: 63; Dušanić 2004*b*: 249 f. fn. 10.

[40] *CIL* III 3937 = 10821 = *Eph. ep.* II p. 413, no. 823 = *AIJ* 524. On the mines, cf. *TIR* L 33, p. 37.

[41] Along the banks of the Japra and Sana river large deposits of iron slag are still visible today. According to one estimate, the quantity of slag might amount to nearly 2,000,000 tons. A number of smelting furnaces and so-called 'administrative' buildings have been excavated, and amongst the numerous finds are mining tools, nails and iron ingots of 6–10 kg in weight. Moreover, ancient galleries have been documented near Liubija, although these have not been precisely dated. In one tunnel of 300 m length, Roman pottery, tools and a human skeleton have been discovered. Archaeological remains at Stari Majdan demonstrate the existence of a smelting furnace connected with a water adit running for 2 km, cf. Davies 1935: 184–6; Pašalić 1967: 127 f.; Dušanić 1977: 65, 69; Bojanovski 1982: 106–12; Dušanić 1989: 148; Škegro 2000: 118–23, with further bibliography; Dušanić 2004*b*: 251.

some sort at or near Ljubija.[42] According to Dušanić, the boundaries of this mining district might have been marked out by two further 'toll' stations, provided the two *mansiones* named 'Ad fines' in the Tabula Peutingeriana may be identified as such.[43] Whether or not these *mansiones* north of Ljubija marked provincial boundaries or those of a mining district cannot be ascertained.[44]

- A further district in Moesia Superior is claimed to have contained the silver/lead mining area in the vicinity of Kuršumlija in the Toplica valley and the mining areas of Janjevo, Novo Brdo, and Lece.[45] A mould inscription on a lead ingot discovered in Žuč northwest of Kuršumlija reads *metallo Imp(eratoris) Aug(usti)* and may originate from these mines.[46] Based on coinage with the caption *metalli Ulpiani*, András Mócsy claimed that the *municipium Ulpianum* or *Ulpiana*/mod. Gračanica (which reached municipal status under Hadrian) was founded within this mining district.[47] Although there is no evidence that this was the case, the municipality probably did owe its existence to the neighbouring mining ventures.[48] The vast district is understood to have been marked by toll stations recorded in the area: an altar from Kuršumlijska Banja, immediately south of Kuršumlija, names a *vilicus stat(ionis) Aquar(um) Bas(–)*; the inscription of a *vilicus stat(ionis) Ulp(ianensis)* was discovered at Laplje Selo immediately east of Ulpiana; a further votive altar of a *vilicus stationis Petobiensis et Moes(iae) r(–) aq(–)* originates from the village of Srbica. All four stations are situated in the province of Moesia Superior.[49]

[42] A number of *procuratores*, imperial freedmen as well as equestrians, and *vilici* set up dedications mostly to Terra Mater during the first half of the 3rd cent. AD, cf. *AE* 1958: 63 = *ILJug* 157; *AE* 1958: 64 = *ILJug* 158 (AD 228); *CIL* III 13240 = *ILJug* 161 (AD 247/8); *CIL* III 13329 = *ILJug* 162; *AE* 1973: 412 = *ILJug* 778 (AD 211–17); *AE* 1973: 411 = *ILJug* 779 (AD 201); *AE* 1973: 413 = *ILJug* 780 (AD 223); *AE* 1979: 414 = *ILJug* 781 (AD 229).

[43] One was located at Laktaši on the route Salona–Servitium, the other some 20 km southwest of Siscia on the road Emona–Siscia. Dušanić opposes the *opinio communis* that both *mansiones* stood on the provincial border between Dalmatia and Pannonia. He argues that they perhaps demarcated the northern boundary of the imperial mining domain, cf. *TIR* L 33, p. 19; Dušanić 1977: 65, with fn. 58. His view is partly confirmed by the discovery of an inscription at Mursa, set up by a *conductor ferrariarum* who is also recorded at Ljubija. The texts (Bulat 1989: 36; *AE* 1973: 411 = *ILJug* 779) clearly state that the *conductor* had contracted out *ferrariae Pannoniarum*.

[44] Fitz 1993–5: 740 f.

[45] On the mines, cf. Dušanić 1977: 72, with fns. 112, 113; Dušanić 2004*b*: 51.

[46] *AE* 1994: 1512 = *IMS* IV 136 = *AE* 1978: 705, cf. Dušanić 1994–5.

[47] Mócsy 1970: 77.

[48] Dušanić 1977: 72 f.; Dušanić 2004*b*: 257 fn. 51. The fragmentary inscription of a *beneficiarius consularis* discovered at *Ulpiana* (*AE* 1981: 735) certainly registers the administrative interest of the provincial governor in the area. Yet, the epigraphic record of the area has hitherto not provided any indication of the presence of imperial mining officials there.

[49] Kuršumlijska Banja: *AE* 1952: 192 = *AE* 1955: 221 = *IMS* IV 104 (AD 206), cf. Dušanić 1989: 152; Mócsy 1970: 25. On the Kuršumlija mining area, cf. *IMS* IV, p. 62; Laplje Selo: *AE*

- Imperial mining officials and military personnel are known from a further mining zone north of *Ulpiana* in the Sitnica valley. The epigraphic record of the *municipium D(ar)d(anorum)*/mod. Sočanica might document a *proc(urator) mm(etallorum) DD(ardanicorum)* and names *coloni arg[entariarum Dardanicarum].*[50] The find of a mine coin at Trepča, south of Sočanica, with the inscription *(metalli ?) Dardanici*, has been understood to record the name of the mining district.[51] A number of *stationes of beneficiarii consulares*, not to be confused with toll stations, can be observed in the Sitnica valley at Vučitrn (perhaps Vicianum?), at Novopazarska Banja, at Slatina, and at Kosovska Mitrovica, and might have been connected with the administration of the gold and silver mines in the vicinity of the *municipium D(ar)d(anorum).*[52]
- Dušanić detected a further possible district in Moesia Superior on the Kumanovo plain east of Scupi/mod. Skopje. Apart from the archaeological evidence for mining, the epigraphic record does document the presence of provincial officials, albeit not the usual mining administrators, as well as soldiers in the Kumanovo plain and in the vicinity of the mines.[53] A *statio Lamu(–)* or *Lamud(–)* of the *vect(igal) Illyr(ici)* near mod. Lopate, and *Vizianus*, probably Klečovce east of Kumanovo lay in the vicinity of the Lojane and Konjuh silver/lead mines.[54] Whether a further *statio* at Lomnica was perhaps related to the gold and iron mines in the Vlasina valley (Božica, Dolno Kobile, Gorno Kobile, Gorno Ujno, and Dolno Ujno) or constituted a border station between Thracia and Moesia Superior is currently uncertain.[55]

1903: 286 = *ILJug* 1413 (AD 227); Turičevac/Srbica: *AE* 1981: 724 = *AE* 1982: 841 (AD 225), cf. Šašel 1982. Vicianum: *TIR* K 34, p. 131.

[50] *ILJug* 501, 503: cf. Dušanić 1977: 72. 87 f. with fn. 223.

[51] Dušanić 1977: 72.

[52] Novopazarska Banja: *ILJug* 1404, cf. *TIR* K 34, p. 131; *AE* 1972: 514, 515. Statina: *AE* 1952: 193 = *ILJug* 68: the inscription is dedicated to a *genius stationis M(-) Dard(anorum).* Kosovska Mitrovica: *ILJug* 1392–5. On *stationes beneficiarii* and their functions, cf. Rankov 1983: 48–51; Nelis-Clément 2000: 259 f.

[53] Lopate: *IMS* VI 227. Lojane: *IMS* VI 241 = *AE* 1984: 790 (early 3rd cent. AD). Konjuh: *IMS* VI 236(?). On the mines, cf. *TIR* K 34, pp. 72 f. 'Konjuh', p. 80 'Lojane'; Dušanić 2004*b*: 258 fn. 52 with further bibliography.

[54] *IMS* VI 209 (Lopate), 212 (Klečovce), and pp. 44 f. (AD 211).

[55] Cf. *CIL* III 8256; Dušanić 1989: 152 with fn. 49. For the mines, cf. *TIR* K 34, pp. 46 ('Dolno Kobile', 'Dolno Ujno'), 58 ('Gorno Kobile', 'Gorno Ujno'); for an imperial procurator at nearby Pautalia/mod. Kjustendil, cf. *IG Bulg* 2052. An inscription recently discovered at Pautalia (*AE* 2004: 1313) names the construction of aqueducts for the baths and the city under the partial supervision of a *μεταλλικός*, a mining engineer(?).

- The Mt Kosmaj/Stojnik region in Moesia Superior, together with the mines at Avala, Železnik, and Rudnik were identified by Dušanić as belonging to one vast imperial domain (Šumadjia).[56] The argentiferous lead mining ventures of Avala, some 15 km south of Singidunum are suspected to have come under imperial control in the late third century.[57] The material remains of Kosmaj show that silver mining activity took place on a massive scale under imperial control.[58] Moreover, the second-century copper coinage discovered at Kosmaj displayed the caption *metallum* or *metalla* on the reverse of ten coins.[59] Dušanić at first assumed that four coins from Trajan's reign with the caption 'metalli Ulpiani' referred to the Kosmaj area and thus argued that this mining district was named *metallum Ulpianum*, a name which perhaps was changed later, in accordance with the caption 'metall(is) Aurelianis' on one further coin.[60] In the immediate vicinity of the Mt Kosmaj/Stojnik mining region, at Guberevac, a *vilicus vectigalis Illyrici* set up a votive altar to Mithras. Based on this inscription Dušanić postulated a toll station closeby, which delimited the mining territory.[61] A further toll station might be recorded at Kamenica *c.*10 km southwest of the Roman mining area at Rudnik in Moesia Superior.[62] Inscriptions attesting the presence of a *procurator (metallorum)* and *coloni* at Rudnik

[56] Dušanić 1977: 77; Dušanić 2004*b*: 259 with fn. 57.

[57] A *vir egregius* is recorded there under whose *cura* an inscribed monument to the Dea Orcia was set by the *duoviri* of the *colonia Singidunensium* in AD 287 (*IMS* I 20 = *CIL* III 1660 = 8151, p. 1022). The further attestation of a votive monument to Volcanus (*IMS* I 46 = *CIL* III 1661; AD 272) documents ties between Avala and Singidunum. The assertion by Dušanić (1990: 588f.) that a *vet(eranus) leg(ionis) IIII Fl(aviae) ex sig(nifero) p(raefectus?) k(anabarum) q(uin)q(uennalis) Sing (iduni), dec(urio) col(oniae) Sirmens(ium)* (*IMS* I 16 = *AE* 1910: 172 = *AE* 1911: 165; late 2nd cent. AD), controlled a *territorium* including the mines at Avala, overstretches the available evidence considerably. This inscribed monument to Terra Mater was found some 4–5 km south of the legionary camp at Singidunum near the road to Avala and not in the mines itself. On the mines at Avala, cf. Davies 1935: 215 f.

[58] The ancient slag deposits amounting to 1,000,000 t set aside, '5,000' (?) mining pits (said to be probably Roman) as well as furnaces are recorded, cf. Tomović 1995; Škegro 2000: 94–6. The epigraphic record of the Kosmaj area documents the presence of imperial officials and auxiliary units (*IMS* I 97, 98, 103, 116, 117, 118, 119), cf. Dušanić 1976: 101–7; Dušanić 2004*b*: 259 fn. 57.

[59] See *IMS* I 160–5 with Dušanić's commentary. On the mines, cf. Davies 1935: 214 f.; Täckholm 1937: 159–77; Dušanić 1976: 98 f.; Dušanić 1977: 78, 111 f.

[60] Apart from the copper coins, thirteen stamped lead ingots with barely readable abbreviations were also discovered, cf. Dušanić 1971*b*: 554; Dušanić 1976: 99. Dušanić 2004*b*: 257 fn. 51, absconds from the above position. For the coinage, cf. Simić & Vasić 1977 and below.

[61] *IMS* I 105 with commentary by S. Dušanić, cf. also Dušanić 1976: 102; Dušanić 1977: 151; Dušanić 1989: 151 f. On the Kosmaj region as part of Moesia Superior, cf. Dušanić 1976: 95–7.

[62] *IMS* I 167 and p. 115 n. 39; Dušanić 1989: 152 with fn. 49. The ancient silver, copper, and lead mining pits there had probably already been exploited by the Romans, although there is no hard evidence to support this assumption, cf. Dušanić 1976: 113 with fn. 31.

do suggest imperial control and consequently Roman mining activity at this site.[63] As for the inscription from Kamenica, its fragmentary state does not allow it to be linked convincingly with a *vilicus* of the *vectigal Illyrici*. In the absence of any secure evidence on the toll station, the extent of the mining district at Rudnik cannot be ascertained.

The hypothesis that the *stationes* mentioned above marked the extent of vast *territoria metallorum* rests solely on the analogy with the toll stations located near the alleged vast imperial domain in central Noricum. Unlike Noricum, no additional sources have been marshalled to corroborate this hypothesis: the epigraphic material, or rather the lack of it, prevents any attempts to define the extent of mining territories in Pannonia, Dalmatia, and Moesia. The few toll stations can not necessarily be synthesised to define the extent of mining territories, as the sites of stations within the district of the *portorium Illyrici* appear not to have been determined by the mining ventures alone. Particular stations may have been erected on *viae publicae*: the toll stations near Ulpianum/mod. Gračanica, Kuršumlijska Banja, and Laplje Selo are positioned on the road Naissus-Ulpianum-Scupi, whereas the station at Srbica perhaps lay on one of the approaches to the main road.[64] Similar observations can be made at some of the other toll stations: The station at *Lamu(d)(–)*/Lopate was positioned on the road Scupi—Anausaro, whereas Vizianus might have guarded one of the approaches from the Konjuh mines to the *via publica*.[65]

This raises the question whether these internal toll stations functioned at all as boundary markers outlining territorial mining 'enclaves' within the customs district of the *publicum portorii Illyrici*. Jérôme France pointed out that most stations of the *quadragesima Galliarum* 'sont quasiment toutes, [...], placée sur des lieux de passage et de transit, ports, carrefours routiers et fluviaux, points de rupture de charge. Dans le secteur alpin, on remarquera leur présence exclusive dans les basses vallées et sur les sites de piémont, à l'entré des axes de circulation, mais jamais à l'intérieur du massif montagneux lui-même'.[66] Furthermore, the lessees of the *portorium Asiae* were required to run stations within eight stadia (= 1 Roman mile?) from the provincial border, the border of the customs district, but not at every conceivable point of entry. The merchants and traders were obliged to report to the nearest station upon entry into Asia or, if the station was not occupied, to travel to the next town with a customs office. If there were two stations at a

[63] *IMS* I 168.
[64] On the road network, cf. Petrović 1979: 26–8.
[65] *TIR* K 34, p. 80.
[66] France 2001: 332; cf. also Vittinghoff 1953: 376 f.

FIGURE 15. Mines in Dalmatia, Pannonia, and Moesia Superior

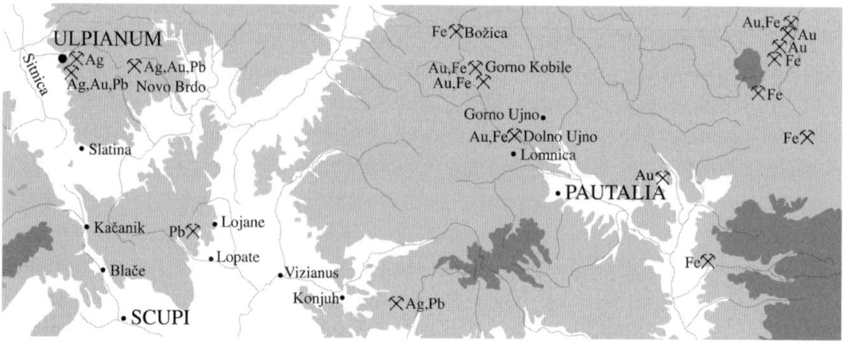

FIGURE 16. Mines in Moesia Superior

similar distance from the point of entry the traveller/trader was free to choose where to deposit his declaration.[67] Provided these findings can be generalized, the stations did not mark the borders of customs districts precisely, but were required to be located within a certain distance from them. There is no indication whatsoever that similar requirements constrained the placement of stations in the interior of a customs district. Although the proximity of toll stations to mining areas does suggest a strong link between mining and the exaction of a toll or tax, these stations were not necessarily placed at points of exit or entry into a vast mining district. Given the position of some of these toll stations on *viae publicae*, one might imagine the yield of different mining districts passing through one and the same station in order to calculate and pay the toll—provided this was its purpose. Siegfried de Laet and Friedrich Vittinghoff, in particular, assumed that stations in the interior of a customs district, usually positioned at important river crossings, crossroads or near mountain passes, merely exacted tolls for the use of these roads and bridges—the full customs duty was paid upon entering the customs district.[68] Even though Alföldy's study demonstrated a connection between these stations and mining activity, the conjecture that not all commodities passing through stations in the interior were subject to a toll remains plausible. Hence, it is possible that only ingots and metals emanating from the mines were taxed, provided they were not owned by the Roman state, the emperor or the Roman army.[69] Unlike

[67] The *lex portorii Asiae* provides a list of ports and other entry points where a station must be positioned, cf. Engelmann & Knibbe 1989: 166 and §§ 9 f., 12–17, 20 f., 28–30, 51. For further stations of *portoria* cf. Vittinghoff 1953: 351–75.

[68] De Laet 1949: 168 f., 368; Vittinghoff 1953: 378.

[69] Written evidence for the taxation of ore (?) or metal is offered by the *lex portorii Asiae* (§34) which notes that exports from the province of Asia to Rome were taxed according to a

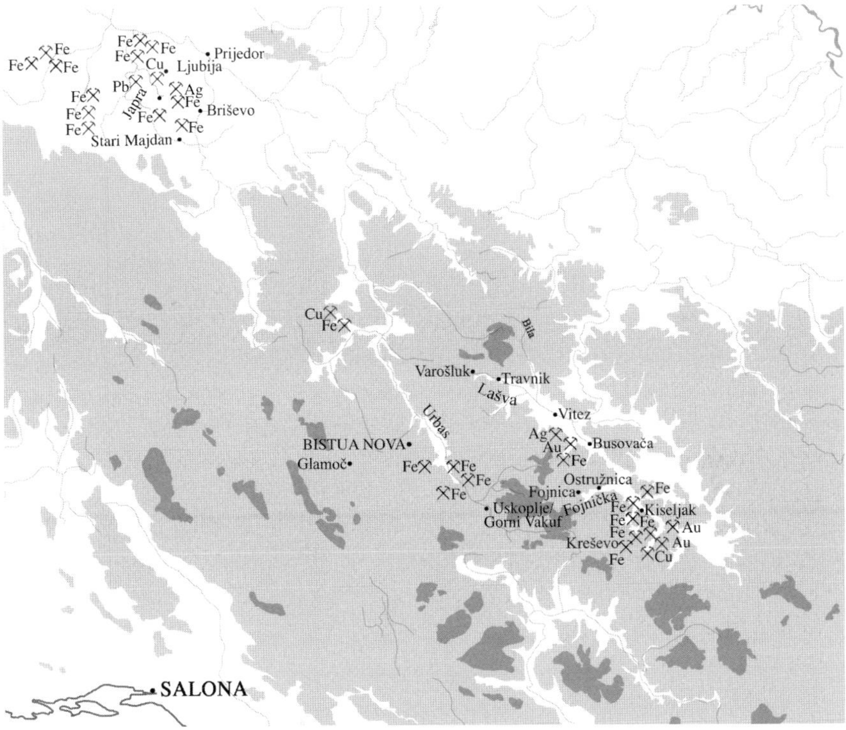

FIGURE 17. Mines in Dalmatia

customs stations at town gates in Roman Egypt, the *lex portorii Asiae* gives the impression that toll stations did not primarily function as points of control covering all traffic in and out of a customs district. They appear merely to have offered a place where merchants and travellers could submit a declaration on the value and quantity of imported/exported goods and pay the customs duty calculated on the basis thereof.[70] The contractors of the *portorium* were allowed to inspect the transported goods in order to verify the submitted toll declarations, confiscate goods or double the toll to punish traders in conflict with the regulations.[71] The marking of merchandise and

νόμος γεωρυχικός, a mining law. Moreover, § 27 may indicate that the export of mining samples may have been tax free, cf. Engelmann & Knibbe 1989: 167 and §§ 25 f. *Dig.* 39.4.9.7–8; De Laet 1949: 121; Vittinghoff 1953: 395; Dušanić 2004*b*: 249 f., fn. 10.

[70] In a first step, a declaration was provided orally. If required, a written declaration giving number, weight and value of the imported goods followed suit, cf. Engelmann & Knibbe 1989: 166 f. and §§ 2, 4, 6, 9, 11, 18, 20 f., 38, 51. On Roman Egypt, cf. Sijpestein 1987.

[71] Engelmann & Knibbe 1989: 167 and §§ 5, 8, 18, 19, 22, 23, 37, 38, 49; cf. also Vittinghoff 1953: 396 f.

the provision of documents and receipts to traders after payment of the toll indicates that trade was subjected to further controls within the customs district. The authority of the contractors to enforce compliance with toll regulations was probably limited to checking the accuracy of submitted declarations at the customs stations. Even though the *lex portorii Asiae* awards contractors the commodities confiscated from smugglers, there is no information on contractors running additional checkpoints in the hinterland or policing the borders of the customs district in order to undercut illegal smuggling activities. Given the interest of the Roman state in revenue, these tasks, amongst others, are likely to have been performed by military personnel, i.e. by *beneficiarii consulares* at local road stations on behalf of the provincial governor.[72]

Toll stations in the vicinity of mining zones imply a certain 'extraterritorial' status of nearby mining districts in relation to the customs district. The distinctive copper and brass *semisses* and *quadrantes*, the so-called 'nummi metallorum', found within some of the Roman mining areas of the Danube provinces were believed by some scholars to confirm this notion.[73] Based on the finds of this coinage in the Kosmaj mining region, the argument was put forward that it circulated exclusively in the mining districts. In analogy with the medieval and early modern 'Bergwerksmarken' it was assumed that the mining districts—owing to their economic 'autonomy'—required their own currency. Following the results of the chemical analysis of these coins Dušanić maintained that this coinage was probably minted for the mining districts in Pannonia, Dalmatia, and Moesia Superior.[74] However, Bernhard Woytek's recent study of these 'nummi metallorum' has altered the perception of their usage.[75] The coinage consists of *semisses* (half of an *as*) and *quadrantes* (quarter of an *as*), the former usually made of brass, the latter of copper and distinguishable from other coins by the captions on the reverse. These captions do appear to refer to mines or mining districts in the Danube provinces—without, however, providing any specifics on their location.[76] The earliest series of *metalla*-coins were minted during the reign of Trajan, displaying the emperor's bust with his title on the obverse, and on the reverse a depiction (i) of a female figurine (a personification of the *metallum*

[72] Engelmann & Knibbe 1989: 167 and §§ 22 f., 51. On lead seals as markings, cf. France 2000: 60–4, with further bibliography. On *beneficiarii* and toll stations, cf. Vittinghoff 1953: 396 f.; Nelis-Clément 2000: 252–4.

[73] Dušanić 1977: 62.

[74] Dušanić 1977: 56 f. fn. 4, 72 fn. 107.

[75] Woytek 2004*a*; Woytek 2004*b*.

[76] Woytek 2004*a*: 43 f. This is corroborated by mould marks on lead ingots recovered from Caesarea Maritima, cf. *AE* 1999, 1683: *met(alli) Dard(anici)*, and from Sarmizegetusa, cf. Dušanić 2004*b*: 257 fn. 51 (unpubl.): *me(tallo) Ulp(iano)*.

Dardanicum?) with the caption DARDANICI, (ii) of Aequitas with the caption METALLI ULPIANI, (iii) of Aequitas with the captions METALLI ULPIAN DELM or METALLI ULPIANI PAN, or (iv) of Aequitas with METALLI ULPIANI SC. Under Hadrian two groups of *semisses* and *quadrantes* are known, both showing the emperor's bust on the obverse and the letters MET NOR or AELIANA PICENSIA in a wreath on the reverse. Further small coinage was produced during the reign of Antoninus Pius, one series depicting the emperor's bust and his titulature on the obverse and a female figure with the caption DARDANICI on the reverse. Other *quadrantes* and *semisses* appear to belong to the reign of Antoninus Pius as well, although they do not show his portrait or bear his name. We find either the bust of Roma on the obverse and a female figurine with the caption DARDANICI on the reverse, or a number of coins portraying the busts of Apollo/Sol, Diana, or Mars on the obverse, and the letters METAL AURELIANIS, METAL PANNONICIS, or METAL DELM on the reverse. The widely accepted view is that the depiction of Apollo/Sol, Diana, and Mars refers to the three metals gold, silver, and iron.[77] The captions on the reverse either name *metalla* directly or provide toponyms most likely to be associated with a mining area. Apart from the legends with a rather broad geographical reach such as *met(alla?) Nor(ica?), metal(lis) Pannonicis, metal(li?) Delm(atici?), metalli Ulpiani Delm(atici)* or *metalli Ulpiani Pan(nonici)*, the attempts by Dušanić and Mócsy in locating a *metallum Dardanicum, metallum Ulpianum* or *metalla Aureliana* have not been convincing so far. Only the legend *(metalla) Aeliana Picensia* allows for the tentative localisation of these mines in the Pincus/mod. Pek river valley in Moesia Superior.[78]

The purpose of the 'nummi metallorum' is a contentious issue. The idea proposed by Dušanić and others of a coinage circulating exclusively in mining districts—similar to the medieval 'Bergwerksmarken' or token money—can not be upheld in the light of the coin distribution. Although the number of known specimens is small, the confirmed finds of such coins in secure stratigraphic contexts suggests that these *quadrantes* and *semisses* were issued not only for the Danubian provinces, but most likely circulated throughout the empire as well. According to Woytek, roughly half of the known coins were found in Rome and Italy, as well as Cyprus and Raetia.[79] The circulation of the *nummi metallorum* outside of mining areas, in fact, even outside the Danubian provinces, once again raises the question of the purpose of these coin issues. There is a scholarly consensus on the origin of the coinage which

[77] Woytek 2004*a*: 36–9, 40–2, 44 with further bibliography.
[78] Woytek 2004*a*: 47 f. For grammatical matters on coin captions, see Woytek 2004*a*: 44.
[79] Woytek 2004*a*: 52–4.

appears to have been minted at a central location, probably Rome.[80] However, the reason why low denomination coins with mining imagery were minted and circulated remains obscure. It has been interpreted as either marking a perceived change from an indirect to a direct exploitation regime of the mines, or—in analogy with the early modern 'Ausbeutegeld'—simply commemorating the opening of new mines and the influx of metals.[81] Woytek argued that during the preparations of Trajan for the Dacian wars the increased demand for metals to cover war costs or for the production of weapons and armour highlighted the importance of mining areas such as Dardania in Moesia Superior.[82]

A different interpretation may be put forward. Although the coinage did not circulate exclusively in mining districts one might argue that the demand for copper coinage of small denominations was specifically high in these areas owing to increased economic activity. The hiring of labour, the supply of mining works with wood and tools, the compliance with fiscal requirements (fees) or the payment for services (baths, hairdressers, shoe makers) within the mining district resulted in a high circulation of coinage, especially in coins of small denominations. Perhaps initially intended to remedy the situation in the mining districts and in order to reduce counterfeiting, a series of low-denomination coinage was introduced. A passage of the *Digest* concerning the punishments for counterfeiting in the vicinity of mining areas may indicate that mining districts were likely to receive new coinage.[83]

The location of toll stations in the vicinity of extractive operations appears to be a phenomenon limited to the Danubian provinces. In other provinces no such stations were installed near imperial mines or quarries. The *lex portorii provinciae Asiae* and other inscriptions indicate that they were mostly positioned in towns on the coast or near the provincial border of Asia. Similarly, the *quadragesima Galliarum* was also exacted in the border areas of this customs district.[84] Furthermore, no toll stations have been observed near the mining areas of the Iberian peninsula or Britain.[85] Although one may not necessarily conclude that the metals mined within these customs districts were not subject to a toll or tax, the exaction of possible tolls was apparently not carried out by establishing stations close to these mining zones.

The inscriptions naming either *stationes* and/or officials of the *vectigal Illyrici* at Guberevac, Goričko, Laplje Selo, Kuršumlijska Banja, Turičevac/

[80] Dušanić 1971*b*: 553; Dušanić 1977: 57 f.; Woytek 2004*a*: 48–50 with further bibliography.
[81] For discussion of earlier research on the issue cf. Woytek 2004*a*: 54–8.
[82] Woytek 2004*a*: 60 f.
[83] *Dig*. 48.19.16.9–10.
[84] Nicolet 1993: 931, map; France 2001: 333, fig.18.
[85] Vittinghoff 1953: 370 f. (Spain), 375 (Britain).

Srbica, Lopate, and Klečovce mostly date to the end of the second century AD or to the beginning of the third century AD. This coincides with De Laet's observation that during the reigns of Marcus Aurelius and/or Commodus the practice of farming out the collection of the *vectigal Illyrici* was replaced by direct imperial control.[86] The lack of any evidence for *stationes* in the vicinity of mining ventures prior to late second/early third century AD does not allow one to conclude that they were not present during the first and second centuries. Even so, the similar date of most attested *stationes* in the Danubian provinces may reflect an imperial effort in intensifying the exaction of tolls on metals near their point of origin. This is confirmed in part by a series of inscribed monuments identifying *stationes* of *beneficiarii consulares* located near mining ventures of the Ibar and Sitnica valley, around Scupi/mod. Skopje and the Kumanovo plain, Domavium/mod. Srebreniča, and near Stojnik, mostly dating to the late second / early third century AD.[87]

One can thus argue that toll stations alone do not provide a sound enough basis for demarcating the territorial extent of vast mining districts. Moreover, the so-called 'nummi metallorum' can not necessarily be synthesised to prove an 'extraterritorial' status of mining districts. Even though their existence can not be positively proven, the notion of vast mining districts ought not to be disgarded completely: the existence of imperial domains of considerable size is documented in the Lebanon (Hadrian's forest), in North Africa, and in Phrygia.[88] Yet, one might also have to consider the possibility of numerous smaller entities or *metalla* dispersed throughout the Danubian provinces, perhaps existing alongside larger territories, to be 'unified' solely in an administrative sense under the control of a *procurator metallorum*. The captions on the 'nummi metallorum' and the mould-marks on lead ingots may perhaps reflect generic administrative entities consisting of several distinct 'metalla'. However, no information is available on the geographical location, extent and coherence of these hypothetical units.

[86] De Laet 1949: 404.

[87] Ibar Valley: *CBFIR* 569 (Novi Pazar), 580 (Gračanica), 595 (Novopazarska Banja), 596 (Petova Crvka near Novi Pazar), 599 (Prokuplje), 610 (Vučitrn), 605–8 (Kosovska Mitroviča). Scupi and Kumanovo plain: *CBFIR* 575 (Blače), 581 (Kačanik), 597 f. (Prizren). Domavium: *CBFIR* 431 (Bajna Bašta), 455–8 (Skelani), 460 (Pljevlja). Stojnik-Mt Kosmaj: CBFIR 604, cf. Nelis-Clément 2000: 259 f., Alföldy 2003.

[88] *IGLS* VIII/3; Flach 1990: 82–122; Mitchell 1993: 80–98; Mitchell 1999: 37–46.

3.2.4. Mining Districts under Military Control(?) in Moesia Superior/Inferior

3.2.4.1. *Timacum Minus*

The indications of territorial entities under military control which might have included mining ventures are similarly problematic. This is the case with a funerary stela from the auxiliary fort of Timacum Minus(?)/mod. Ravna displaying the inscription of an Ulpius Aquilinus, soldier of the *legio VII Claudia* and *librarius offici(i) praef(ecti) ter{r}it(orii)*, a clerk to the office of the *praefectus territorii.*[89] The extent and composition, not to mention the legal status, of this *territorium* is difficult to establish. A number of inscriptions from the vicinity of auxiliary camps combine *territorium* with a toponym of a fort such as *territorium Capidavense* or *territorium Sucidavense* for the auxiliary forts Capidava in Moesia Superior and Sucidava in Dacia, or *territorium Matricensium* and *territorium(?) Vetussaliensium* for the auxiliary camps Matrica and Vetus Salina in Pannonia Inferior.[90] Likewise the *ter(ritorium) Abri(tanorum)* as well as the *territorium Troesmensium*(?) gained their toponyms from legionary camps in Moesia Inferior.[91] However, the content of the inscriptions naming these *territoria* does not suggest that there was any military involvement in their administration. They appear to have been under the control of the *vici* or *canabae* in the vicinity of auxiliary and/or legionary camps whose inhabitants not only copied municipal political structures but apparently held sway over their own *territorium.*[92]

There is no evidence that this was the case at Ravna.[93] The inscription clearly names an equestrian officer, a *praefectus*, in charge of an undefined *territorium* with his own administrative staff (*officium*) consisiting of legionary soldiers of *legio VII Claudia.* Besides this military bureau, the camp at Ravna was occupied by auxiliary units, the *cohors I Thracum Syriaca* from AD 70 to the early second century AD and a *cohors II Aurelia Dardanorum*

[89] *IMS* III/2, 31. Dušanić 1990: 589 f., with fn. 41; Petrović 1995*a*; Dušanić 2000: 354 f.

[90] *AE* 1980: 712 (Matrica/Szazhalombatta). *CIL* III 10305 (Vetus Salina). 12491 = *IScM* V, 77 (Capidava). *IDR* II 190 (Sucidava/Celei), cf. Bérard 1992: 90 f., 104 f., nos. 12–15.

[91] *IScM* V 135 (Troesmis, AD 163). *AE* 1985: 765 (Hisarlyka near Razgrad/Abritus, AD 244–7), cf. Bérard 1992: 89 f., 91, 104 f., nos. 11, 16. The inscription (*CIL* III 14370/10 = *ILS* 7111 = *AE* 1986: 534 = *AE* 1987: 792) from the legionary camp at Castra Regina/mod. Regensburg naming an *aedil territor(ii) contr(ibuti?) et k(anabarum?) R(eginensium?)* remains a conundrum, cf. Lieb 1998: 65 fn. 21.

[92] For a discussion, cf. Mócsy 1972: 157 f.; Vittinghoff 1974: 111; Mócsy 1980: 371 f.; Bérard 1992: 91–9.

[93] To date, there is no epigraphic evidence for a *vicus* in the vicinity of the auxiliary camp at Ravna which might have exerted quasi-municipal control over the territory, cf. Petrović 1995*b*: 42.

after AD 169.[94] The auxiliary unit was posted there primarily to control one of the major routes running from the Danube to Naissus.[95] Quite obviously, the auxiliary garrison at Timacum Minus was provided with administrative specialists from the legion stationed at Viminacium with the additional task of monitoring a *territorium*.

The explicit use of the term 'praefectus territorii' is reminiscent of the supervisory and juridicial function of *praefecti* over a *civitas*, *gens* or *natio*, often connected with the command over an auxiliary unit.[96] The office is attested predominantly in the first century whereas the inscription at Ravna dates to the mid-second century AD or later.[97] Furthermore, the title of *praefectus territorii* does not directly indicate that his responsibilities included the surveillance of a local tribe. However, one might also expect the readers of the funerary inscription to have been spared the full title of the *praefectus* well known to the locals. The inconclusive text raises the question what the function of this office and the content of the territory was. On the premise that 'territorium' is a technical term for a topographically continuous and clearly delimited area under administration (as appears to be its usage in the legal sources and the writings of the land surveyors), the *praefectus territorii* held control over a coherent territory and had judicial powers over its inhabitants.

The extent of the territory is difficult to trace as signs of any delimitation are absent. Dušanić argued that the *territorium* was most likely a mining territory centred on Timacum Minus, which might have included the silver mines at Lukovo-Valakonje, perhaps the gold, silver, iron, copper, and lead mines at Slot, Bor, and Rusman, and the mining areas of Kalna, Rgošte, Gradište, and Orešac, 10–15 km south of Ravna.[98] Moreover, a lead ingot with the mould-mark *metallo Caesaris Aug(usti)* found at Jasenovik south-west of Timacum Minus does suggest imperial control of some of the mines.[99] The fragmentary evidence for a possible toll station at Ravna might indicate the existence of mining districts closeby.[100] The assertion, however, that a *metallum Caesaris Augusti* or any other mining zone formed part of the

[94] On the garrison, cf. *IMS* III/2, nos. 13, 37–40 (*coh. I Thracum Syriaca*), nos. 7(?), 8, 9, 18–20, 22, 29, 43–52 (*coh. II Aurelia Dardanorum*).

[95] Dušanić 2000: 350 fig. 1.

[96] Zwicky 1944: 11–14; Dušanić 2000: 354.

[97] Zwicky 1944: 16; Dušanić 2000: 354; Eck 2001*b*.

[98] Dušanić 1977: 75 f. with fns. 147–50; Dušanić 1990: 589 f.; Petrović 1995*a*; Petrović 1995*b*: 37–43, 58 f.

[99] Petrović 1995*a*: 199.

[100] *IMS* III/2 84, cf. Petrović 1995*b*: 119 f.

territorium under the command of a prefect and his staff at Timacum Minus remains conjectural.

3.2.4.2. *Montana*

A further military *territorium*, perhaps including mines, might have existed in the area of Montana/mod. Mihailovgrad in Moesia inferior. A local votive inscription to the goddess Diana mentions a *beneficiarius consularis* of the *legio I Italica* and *ag(ens) t(erritorio) M(ontanensium)*.[101] The term *territorium Montanensium* is not otherwise documented in the epigraphic record—the term ***regio*** *Montanensium*, however, is well known.[102] Besides the documentation of *regionarii* at Montana, an *optio leg(ionis) XI Cl(audiae) agens r(egione) Mont[an(ensium)]* set up an inscription at Almus/mod. Lom on the Danube some 50 km north of Montana.[103] Two further inscriptions commissioned by *centuriones r(egionarii)* in AD 156–8 in Almus on the Danube and in AD 158–60 in Montana supposedly indicate the vast extent of the territory.[104] According to M. P. Speidel an additional inscription found north-west of Montana in Gromšin and dating to AD 161–9 may name further *regionarii Montanensium*.[105] In the light of the evidence for a *regio Montanensium* the expansion *ag(ens) t(erritorio) M(ontanensium)* is disputable. To my knowledge, there is no indication that *territorium* and *regio* were used alternately. Provided that a *territorium Montanensium* existed, one would have to assume a separate, contemporary, or perhaps subsequent territorial entity besides the *regio Montanensium*.[106] It appears likelier that the engraver made a mistake. Provided he copied the text from a wooden tablet written in a Latin cursive, the minuscule 'r' may have been misread as a 't'.[107] The inscription thus should read *bf. cos. ... ag(ens) r(egione) M(ontanensium)*, in accordance with the *optio ... agens r(egione) Mont[an(ensium)]* known at Almus.[108] Given the distribution of inscriptions naming a soldier *agens*

[101] *CBFIR* 647 = *AE* 1987: 881, cf. Bérard 1992: 91 f., 105, no. 17.

[102] Rankov 1983: 46 f.; Speidel 1984: 140.

[103] Montana: *centuriones regionarii*, cf. *AE* 1975: 745; *CIL* III 12371, 12380; others, cf. *AE* 1957: 341 *([ag(ens?)] reg(ione) Mont(anensium)*. Almus: *AE* 1969/70: 577, cf. Rankov 1983: 54.

[104] *CIL* III 7420 (Almus). *CIL* III 12371 (Montana); with Speidel 1984.

[105] Speidel 1984: 187.

[106] Cf. Vulpe 1976: 294–6. Bérard (1992: 92) argued that Montana was a 'communauté de type pseudo-municipale' consisting of *decuriones* as well as other magistrates, and equipped with its own *territorium*.

[107] Speidel 1996: 31–4.

[108] The epigraphic record on *beneficiarii* does not document a further *bf. cos. agens* ***territorio***; on the contrary, *bf. cos. agens* ***regione*** are attested in *CBFIR* 530, 531.

regione Montanensium or a *centurio regionarius*, the region may have reached north of Montana as far as Almus.[109]

A military force of considerable size was based at Montana. A *cohors I Sugambrum equitata* is attested there and was moved east during Hadrian's reign. It was replaced by a smaller legionary detachment of the *legio XI Claudia*, under the command of a *centurio*.[110] The garrison at Montana probably guarded the approaches to the Petrohan pass into the interior of Thrace and to the Sveti Nikola pass into Moesia Superior. Alternatively, this sizeable garrison in the hinterland of the Danube frontier may have guarded the silver and gold mines of the Ogosta and Zlatitsa valleys southwest of Montana throughout much of the second century.[111] Inscribed monuments naming a *primiscrinius officinae* and a *vikarius dispensatoris* have been taken as evidence for a mining administration based at Montana. Moreover, the presence of *beneficiarii consulares* at Montana has been explained by its 'proximity' to the mining area, owing to the number of other *stationes of beneficiarii* attested near mining areas throughout the Danubian provinces. Boris Rankov implied that the *regio Montanensium* may also have included the mining areas.[112] The evidence, however, remains circumstantial. There is no explicit documentation whatsoever for a mining procurator or mining bureau at Montana, nor for an incorporation of the mining ventures described above in a *regio Montanensium*.

3.2.5. Further Mining Districts under Imperial Control

For most other mining areas throughout the Roman empire, the existence of mining districts under imperial control is indicated only by inscriptions documenting the presence of military personnel and imperial officials in the vicinity of archaeologically attested mining ventures. The extent and coherence of the following districts is not ascertainable on current evidence:

- The argentiferous lead mines in the vicinity of Domavium/mod. Srebrenića in Dalmatia have yet to be the subject of a detailed archaeological survey,

[109] Speidel 1984: 141.

[110] Rankov 1983: 53.

[111] Mining galleries discovered near Gaganitsa, Elovitsa, Karilovtsa, and Diva Slatina might be Roman. Stone troughs (?) for washing alluvial deposits of ore were discovered at Govezhda and water channels at Kotenovtsi. The by-product of the extraction of gold-bearing sand, piles of stones, were found near Galanitsa, Govezhda, and Dulgi Del. Late Roman pottery provides the only basis for datation, cf. Rankov 1983: 41, 45 with fn. 29.

[112] *CIL* III 12379 =14207/19 (*vikarius dispensatoris*); 14209 (*primiscrinius officinae*), cf. Rankov 1983: 46–58.

but Roman remains of mining ventures (slag heaps and lamps) can be observed along the lower Drina and between the rivers Jadar and Drina near Srebreniča.[113] The inscription *metal(lis) Pannonicis* on the reverse of copper coins found at Kosmaj is taken to refer to these mining ventures.[114] Latin inscriptions attest imperial mining officials and Roman army personnel at Domavium (cf. 4.1.10). Perhaps the *stationes* of *beneficiarii consulares* recorded at Skelani and Bajna Bašta are related to the mining ventures some 25 km to the north-west.[115] Owing to its importance as a mining centre and as a migrational focal point, Domavium reached municipal status, as a series of inscriptions by the *ordo municipii Domavianorum* in the first half of the third century documents. Perhaps in the mid-third century the town became *col(onia) m(etalla?) D(omaviana).*[116]

- Three inscribed altars to Terminus and Terminus-Liber discovered at Ustikolina near Foča and near Goražde might (in analogy to the Terminus altar discovered at Friesach in Noricum) provide evidence of one or more mining districts in the Upper Drina Valley in Dalmatia. These perhaps included the silver mines around Ustikolina, Foča/Potpec, and near Pljevlja in the area of which the remains of 'Roman' mining activites were recorded.[117] A *statio beneficiarii* is documented at Pljevlja.[118]
- A further argentiferous lead mining area might have been located in the Lim Valley in the vicinity of Prijepolje, at Čadinj, where Roman excavations are well documented.[119]

It has been argued that the argentiferous lead mines listed above, including those west of Loznica on the Lower Drina and Brskovo, formed a vast mining district called *argentariae Delmaticae et Pannonicae,* in accordance with the

[113] Davies 1935: 194 f.; Wilkes 1969: 277–80; Dušanić 1977: 66; Škegro 2000: 84–7; Dušanić 2004*b*: 254 with fn. 38.

[114] Dušanić 1977: 66, 86.

[115] *CBFIR* 431 (Bajna Bašta), 455–8 (Skelani), 460 (Pljevlja), cf. Nelis-Clément 2000: 259 f.

[116] *CIL* III 12727 (AD 213–17), 12733 = 8363 (AD 217–18), 12720 = 8359, 8360 (AD 229–35), 12728, 12729 (AD 251–3), cf. Wilkes 1969: 280; Zaninović 1977: 797 f.; Bojanovski 1982: 94, 99–106, 19.

[117] Besides slag heaps and furnaces, ancient mining tunnels, and cast moulds were discovered along the Čeotina river valley between Foča and Pljevlja and in the area of Pljevlja, cf. Škegro 2000: 87–9; Dušanić 1977: 68. On Terminus/Terminus-Liber, cf. *AE* 1939: 301; *ILJug* 1572, 1573; *CIL* III 8371.

[118] *CBFIR* 460.

[119] An *argenti actor* and a *vilicus* are attested at Kolovorat near Prijepolje (*ILJug* 1685, 1690; Škegro 2000: 92 with fn. 259; Mirković 1975: 95–108, nos. 1 and 7). There is no indication that they were members of the *familia Caesaris.* Roman mining tools, slag heaps and remains of a smelting furnace supposedly were discovered in the area, cf. Škegro 2000: 89 f. with further bibliography.

procuratorial titles recorded at Domavium and elsewhere. Dušanić suggested that the provincial border of Pannonia and Dalmatia cut through the mining district—hence the name *argentariae Delmaticae et Pannonicae.*[120] There is little to support the view of a territorial district of this size, yet on the administrative level the subordination of mines from different provinces under one procurator is well attested.[121]

A further district may be recorded in Moesia Superior:

- The copper coins minted unter Hadrian with the legend *(metalla) Aeliana Pincensia* are understood to refer to the mining ventures in the Pek (anc. Pincus) and Mlava river valleys, for instance the copper, iron and gold mines at Majdanpek, the gold/copper/silver mines near Kučajna, and the lead/silver/iron mines at Vitovnica.[122] Further mines have been located along the Pincus river, but so far no administrative centre has been observed.[123] To date, there is no direct epigraphic documentation for imperial control of the mining areas in the Pincus Valley.

In the immediate aftermath of the Pannonian wars—according to Florus (2.25.12)—the Roman governor C. Vibius Postumus put the local Dalmatian population to work in the gold mines. Both Statius and Martial refer to goldmining in Dalmatia without providing any names for specific localities.[124] The archaeological evidence suggests that gold-mining mainly took place in Central Bosnia:

- Austrian surveyors operating mostly during the late nineteenth century identified a number of 'ancient' mining sites in the immediate vicinity of Uskoplje/Gornji Vakuf in the Vrbas valley. Based on the remaining traces of sluicing, the construction of leats and the further evidence for a Roman presence at Uskoplje, it was argued that the remnants were Roman

[120] On Loznica, see *TIR* L 34, p. 74. Davies 1935: 189–191; Dušanić 1977: 67 f.

[121] Alföldy 2003: 226.

[122] Davies 1935: 217–21; Dušanić 1977: 57, 74, 76; Dušanić 2004*b*: 259 f. with fn. 58; Woytek 2004*a*: 47 f. The recent excavations from 1971 to 1990 at the iron, copper and gold mining/smelting site of Kraku lu Jordan near Brodica yielded not only a peculiar fortified architectural structure with Roman pottery, tools and coinage, which—judging from the slags—was apparently erected for the sole purpose of smelting copper and iron ores and melting the gold gained from placer deposits in the Pek river into ingots. Fetters found in the nearby village of Železnik have led Miodrag Tomović (2000) to see slaves (or perhaps convicts?) as the major source of labour on the site.

[123] Davies 1935: 217, believed the administrative centre to be based at Viminacium, while Dušanić 1977: 76, locates the mining administration at Pincum/mod. Veliko Gradište, cf. also *TIR* L 34 Budapest, p. 90.

[124] Stat. *silv.* 3.3.89 f.; Mart. *epigr.* 10.78.1.5; Wilkes 1969: 272; Zaninović 1977: 796; Kienast 1999: 403.

gold-mining ventures. Moreover, iron was also mined in the vicinity of Uskoplje.[125]

- Traces of gold and iron-mining were observed in the Lašva and Bila river valleys near Travnik. Remnants of sluicing (?) were detected between Varošluk and Travnik and near Vitez on both sides of the Lašva valley, whereas mining tunnels were discovered at Gornji Slimeni near Travnik.[126]
- Evidence for gold and iron-mining (mostly sluicing) originates from the areas around Fojnica, near Višnjika, and Ostružnika, near Kiseljak, in the Fojnička valley, and near Kreševo, south of Kiseljak. Gold-mining appears to be attested in the area at a few sites west of Busovača.[127]
- The Central Bosnian gold-mining area appears to have been linked by a road from Salona to Haedum castellum Daesitiatium/mod. Breza in 19/20 AD. Scholarly opinion has Ad Matricem (mod. Bugojno?) as the possible headquarters of the mining administration.[128] However, there is no direct epigraphic evidence for imperial officials or military units being present in the area. An inscribed funerary monument of a *commentariesis* (sic) *aurariarum Delmatarum* and a *dispesator* (sic) from Salona (perhaps first century AD ?), as well as an undated inscription perhaps of a *[proc (urator)* (?) *m]etalloru[m* from Glamoč might suggest imperial control in the hinterland of Salona/mod. Split.[129] This assumption might be supported by the presence of a *beneficiarius consularis* in the third century at Glamoč.[130]

Similar difficulties in outlining mining districts and territories emerge in the case of Roman Dacia. Apart from epigraphic evidence for the mining administration of *aurariae Dacicae* being based at Ampelum/mod. Zlatna, there are no available details on the territorial configuration of the gold-mining areas under imperial control. Nevertheless, Noeske argued for a vast Dacian gold-mining district delimited in the west by the provincial boundary running south from the Meseş mountains to Baia de Criş, the eastern border perhaps being the Mureşul. The southern borders might have been created by the

[125] On the Vranica mountain east of Uskoplje traces of ground sluicing(?), as well as two leats tapping water from Suhodolpotok brook and a water tank have been discovered. At Zlatno Guvno water leats stretching for 9 km, and at several sites near Uskoplje in the Vrbas Valley hill slopes scarred by sluicing were reported. Škegro 2000: 74 f. (gold), 132 f. (iron).

[126] Škegro 2000: 75 f. (gold), 133 f. (iron).

[127] Škegro 2000: 76 f. (gold), 134 f. (iron), with further bibliography; Davies 1935: 184; Dušanić 1977: 69.

[128] Dušanić 1977: 68; Škegro 2000: 78 f.

[129] *CIL* III 1997 (Salona, 1st cent. AD ?). *ILJug* 1655 (Glamoč) with D. Sergejevski, *GZM* 39, 1927: 260 n. 9.

[130] *ILJug* 1463 (Halapič/Glamoč, AD 261).

FIGURE 18. Mines in Dacia

territorial boundaries of Micia and Apulum, home to the *legio XIII Gemina*. Noeske, however, was unable to trace the northern borders of his district.[131] To date, no evidence has come to light confirming Noeske's concept of a vast

[131] Noeske 1977: 274 f. Micia perhaps was a *statio* of a *beneficiarius consularis* connected with the mining zone, cf. *IDR* III/3, 86 = *AE* 1930: 11 = *AE* 1931: 119 = *AE* 1933: 9. As Rankov (1983: 49 with fn. 50, following Wagner 1938: 49 f., 53, 124–6, 208 f.; and Gudea 1977: 872, no. 19), argues, the auxiliary garrison at Micia did guard the approaches to the Dacian gold mining

territory. At the most, the disparate *metalla* or *territoria metallorum* were partly unified as *aurariae Dacicae* on an administrative level.[132] Moreover, control by imperial staff can only be verified in a few gold-mining areas in Dacia. The most compelling evidence stems from Alburnus Maior/mod. Roşia Montană.[133] Outside Alburnus Maior written evidence for imperial control begins to wear thin. Wollmann argued that a votive inscription found at Apulum by a *subprocurator aurariarum* might have originated in Baia de Criş where remains of Roman gold-mining have been located.[134] The stamped tiles of *legio XIII Gemina* found near the alluvial gold deposits of Pianul de Jos in the Arieşu Valley and in the valley of the Ogaşul Băieşului, or of *legio IIII Flavia Firma* found at gold-mining shafts near Bocşa-Vasiova, are an inadequate basis from which to argue for imperial control of these sites.[135]

Despite the wealth of archaeological material for extractive operations in Roman Spain, the sparsity of written documents from mining zones makes it difficult to identify or to delimit the extent of imperial mining districts. In north-western Spain the inscriptions set up by vexillations of the *legio VII Gemina* and auxiliary units in Luyego and Villalís, which name the (mining) procurator in charge, provide significant evidence for imperial control of the alluvial gold extraction in the Duerna Valley south-west of Asturica Augusta/mod. Astorga in the second half of the second century (cf. 4.1.6, 5.2). Furthermore, the presence of Roman military personnel at Trêsminas suggests that these opencast hardrock gold mines were also under imperial supervision.[136]

Turning to southern Spain, epigraphic material documenting imperial mining districts is likewise sparse. Perhaps a number of the silver/lead mines in the vicinity of Castulo were supervised by imperial officials. An inscription from the first half of the second century AD records a *libertus Augusti* at El Centenillo, where opencast as well as underground mining

district(s). For *beneficiarii consulares* and their *stationes* near salt mines in Dacia, cf. Rankov 1983: 49 with fn. 51.

132 On mining remains at, cf. Wollmann 1996 with further bibliography.

133 Noeske 1977.

134 *CIL* III 1088, p. 1390 = *IDR* III/3, 228, cf. Noeske 1977: 348, AMP 3. The gold mining of secondary deposits in the Crişul Alb valley south of Baia de Criş can be dated by a number of coins to the second half of the second century AD. cf. Wollmann 1996: 360 f.

135 Wollmann 1996: 390, 394, 397.

136 Although there is no direct epigraphic evidence for imperial officials, the presence of army units suggests a mine run by the Roman state, cf. Wahl 1993: 142. *CIL* II 2389 = *AE* 1907: 150. *AE* 1980: 582. *AE* 1907: 151.

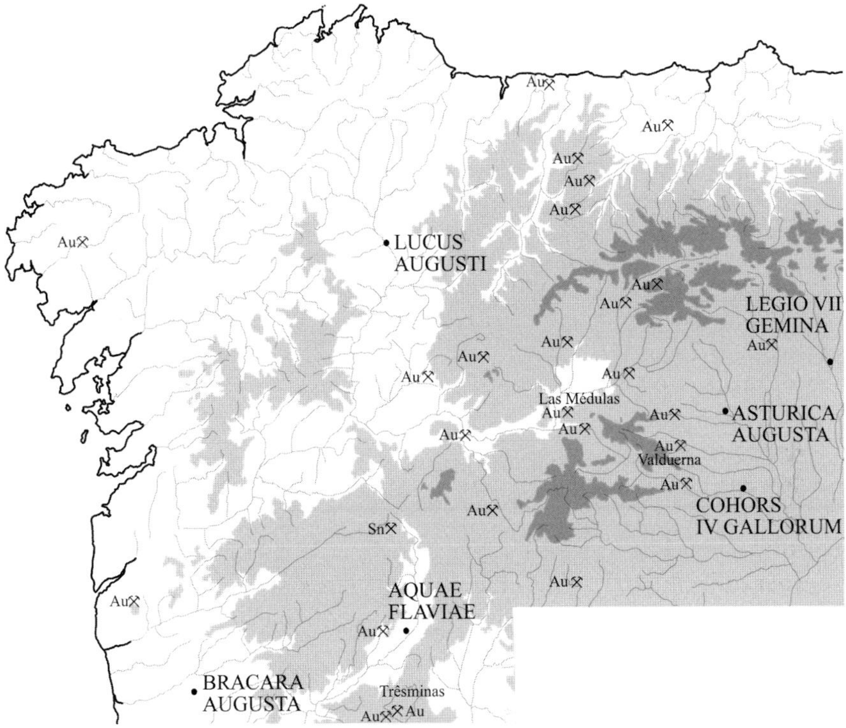

FIGURE 19. Mines in northwestern Spain

operations are attested.[137] Further evidence for the involvement of imperial officials in mining operations is provided by the epigraphic documents of Baetica. A *procurator montis Mariani* is attested in Hispalis at the end of the first century AD in charge of Mons Marianus, a gold mine or mining area close to Corduba.[138] The gravestone of an imperial *dispensator* in Regina/Los Paredones near Casa de Reina (BA), *c.*20 km west from the Azuaga mining area is the only indicator of imperial interest in the extractive operations on the northern slope of the Sierra Morena.[139] The connection of our *dispensator*

[137] *AE* 1922: 9, cf. Domergue 1987*a*: pl. XIXa and fig. 25. The discovery of five Archimedean screws for the drainage of ground-water, tools, oil lamps, smelting furnaces, and a rudimentary aqueduct, as well as coinage, is suggestive of a continuous exploitation since the 2nd cent. BC, and evidences a high level of investment and organizational effort, cf. Domergue 1987*a*: 267–72.

[138] *CILA* II 25. The location of Mons Marianus is problematic. While R. C. Knapp believes all of the Sierra Morena in the hands of Sextus Marius, others would like to place Mons Marianus in the mines at Cerro Muriano / Cerros Marianos, cf. Davies 1935: 35–9, 132–5; Knapp 1983: 39 f.; Domergue 1987*a*: 116–24. The epigraphic evidence seems to tip the balance more towards the latter location.

[139] *CIL* II^{2}/7 981. Near Azuaga, a vast number of 'rafas', small opencast extraction ditches, scar the landscape, cf. Domergue 1987*a*: 17–25.

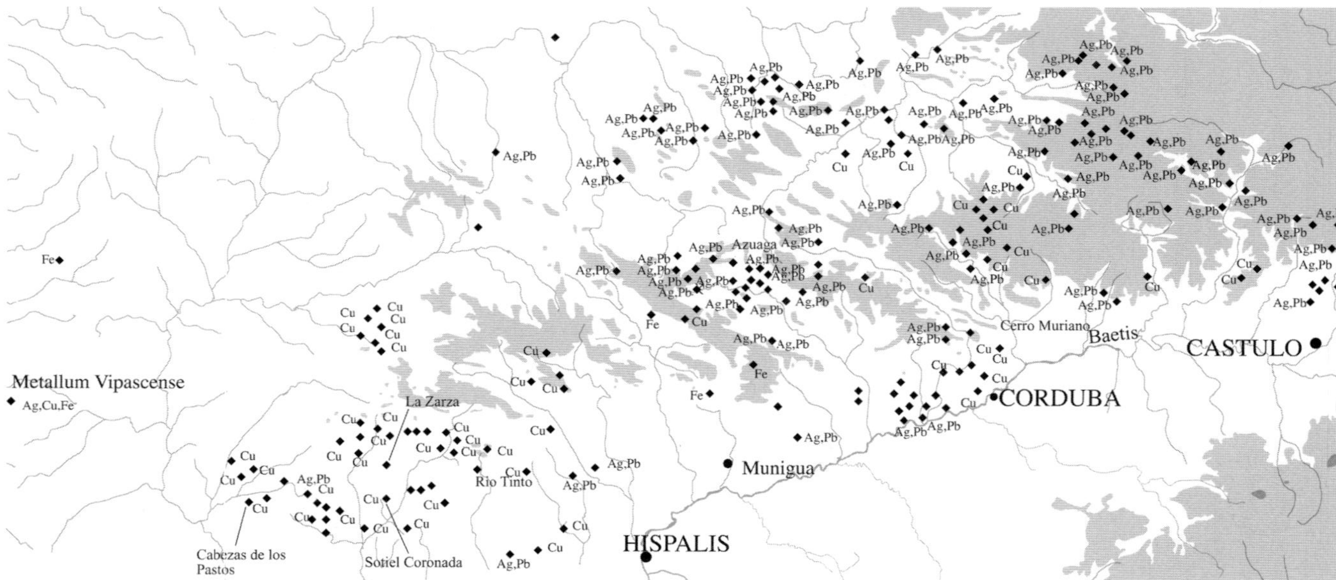

FIGURE 20. Mines in Baetica and Lusitania

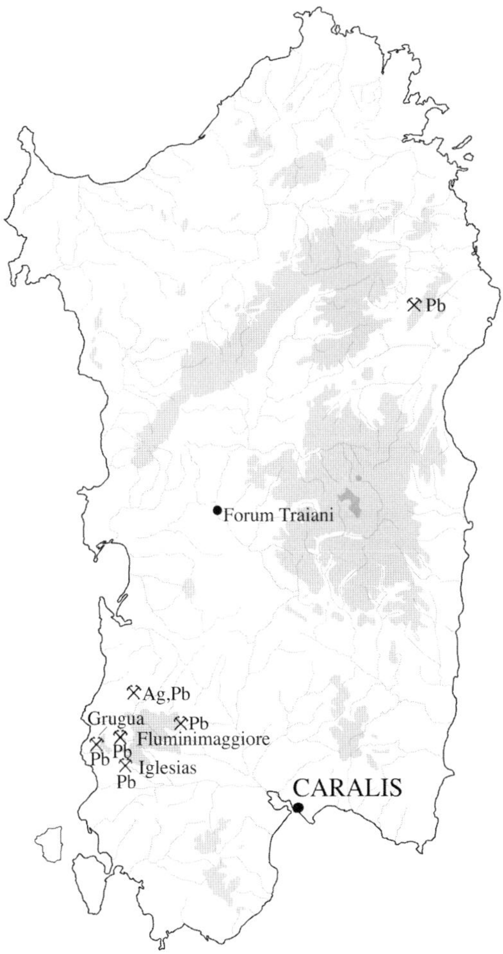

FIGURE 21. Mines in Sardinia

with these local mining operations appears likely, as there is no evidence for other activities (i.e. landed estates, military camps) in this area.[140] At the Rio Tinto mines a freedman *procurator* is attested during the reign of Nerva where he set up an honorific inscription for his emperor.[141]

140 Boulvert 1970: 429–33; Weaver 1972: 202.

141 *CIL* II 956. Latter-day silver/copper-mining operations have resulted in the destruction of the original topography of this Roman site. However, Domergue and Davies assume that the extractive operations resembled those at La Zarza, Sotiel Coronada, Cabezas de los Pastos, and Herrerías. Some mining galleries even yielded water-lifting devices. At least sixteen pairs of treadwheel-operated wheels have been recovered from the mining site at Rio Tinto, suggesting that there was underground mining below the water-table, cf. Davies 1935: 126–31; Jones 1980: 146; Domergue 1987*a*: 235–8, 244.

FIGURE 22. Mines in Britain (after Jones & Mattingly 1990: map 6.2)

Further epigraphic evidence for the imperial control of mining areas was recently discovered in Sardinia. The inscription mentioning an imperial *procurator metallorum et praediorum* at Forum Traiani/mod. Caddas has for the first time confirmed the notion, gained from other sources, that some mines were under imperial control.[142] The heartland of Roman mining on Sardinia seems to have been the Iglesiente. In this argentiferous region slag heaps and mining galleries yielding tools and oil lamps have

[142] *AE* 1998: 671, cf. Benito Serra & Bacco 1998: 1244 f.; Bruun 2001.

been found at Monteponi, Malacalzetta, at S. Giovanni, and at S. Giorgo near Gonnesa. Further Roman silver mines are recorded at Argentiera near Nurra.[143] Yann Le Bohec was able to show that a series of military inscriptions in the Iglesiente probably records the imperial control of the mines there.[144] Moreover, the mould marks on lead ingots discovered at Fluminimaggiore and at San Nicolao in the Iglesiente display the name of the emperor.[145] Similar labels naming Hadrian were found on nine ingots from Carcinadas and on ingots from the shipwreck of Pistis.[146] While the inscriptions on the ingots do not necessarily prove the direct management of mines by imperial officials on Sardinia during the first and second century AD, we certainly can suppose that part of the lead produced came into imperial ownership.[147]

Further evidence for a mining district under imperial control comes from Gaul. Apart from *tabularii (rationis) ferrariarum* at Lugdunum, members of the *familia Caesaris* are present in the mining areas at Villefranche-de-Rouerge/Aveyron. The inscribed slab for a *vilicus* Zmaragdus names the *familia Caesaris* of emperor Tiberius 'quae est in metallis'.[148] Although the presence of military personnel is attested archaeologically in Britain at the gold mines of Dolaucothi for the Flavian to Hadrianic period, the silver/lead mines at Charterhouse-on-Mendip, and perhaps in Weston-under-Penyard(?), there is to date no record of imperial administrative personnel for mines in Britain.[149]

3.2.6. Summary

Leaving aside the evidence for mining areas under imperial control in the Danubian provinces, the information on such districts in the western provinces in general is surprisingly scarce, in spite of the rich archaeological finds and remains of mines in Spain, Britain, or Gaul. This is not necessarily a result of the sparsity of epigraphic record, but might reflect more fundamental differences in administration in these areas in comparison to the Danube provinces. Moreover, differences between the Danubian provinces and Spain also appear in the chronological documentation of such districts. While

143 Meloni 1990: 179 f., 475 f. with further bibliography.
144 *CIL* X 8321 = Sotgiu 1988: C113. *AE* 1985: 485a = Sotgiu 1988: B115. *CIL* X 7535, 7537, cf. Le Bohec 1992*a*: 260–2.
145 *CIL* X 8073/1, p. 1002, 8073/2.
146 Zucca 1990: 803–9.
147 Le Bohec 1992*a*: 258 f.
148 *AE* 1892: 23 = *CIL* XIII 1550, cf. Domergue & Leroy 2000: 7 with further bibliography.
149 On the forts, cf. Wilson 1971; Jones & Little 1973: 13 f.; Stewart 2002: 134.

mining districts under imperial control are documented on the Iberian Peninsula from the beginning of the first to the end of the second century AD, most of the epigraphic evidence from Dalmatia and Moesia Superior might point to the creation of mining districts or territories during the second and beginning of the third century AD.

The existence of a vast mining district including massive tracts of land under imperial control is hitherto only documented in Noricum, albeit only by circumstantial evidence. In the Danubian provinces, the presence of toll stations in the vicinity of mining ventures may imply the exaction of a toll or tax on mined goods. Nevertheless, these toll stations did probably not delimit mining districts precisely. Moreover, the *nummi metallorum* do not demonstrate an extraterritorial status of mining ventures in the Danube provinces. Thus, the precise delimitation of *metalla* and *territoria metallorum*, as implied by the Vipasca tablets, is rarely traceable in the present day landscape.

3.3. THE LEGAL NATURE OF *METALLA* AND *TERRITORIA METALLORUM*

Given the clear demarcation of some of these imperial mines and quarries and the presence of imperial officials and elements of the Roman army in these districts, the question arises whose property these mining and quarrying zones were. The scholarly answers are contentious: Dominic Rathbone summed up the prevailing view when he wrote 'the Roman state also had fixed assets consisting principally of land, urban properties and mines.... Under Augustus and Tiberius almost all mines not already run by the state came into the hands of the patrimonium, and often if not normally were put under military supervision, and new mines, like those in Britain, followed suit'.[150] This echoes the assessment made in 1876 by Joachim Marquardt. He argued that during the Republic, most mines had been owned by private individuals, while in the Principate the most important mines were acquired by the emperors, partly 'für den Fiscus' (for the public treasury), and partly 'für das kaiserliche Privatvermögen', for the emperor's private property.[151] In the wake of Marquardt's theory, scholars such as Otto Hirschfeld, Michael Rostovtzeff, and Ulf Täckholm viewed public and patrimonial mines as coexisting with privately owned mines.[152] Hirschfeld added to Marquardt's view that the difference between public and patrimonial

[150] Rathbone 1996: 315 f. [151] Marquardt 1884: 259 f.
[152] Rostovtzeff 1904: 445; Täckholm 1937: 97 f.

mines was a result of how they had been acquired—as spoils of war, through confiscation or inheritance.[153] The view, however, that mines were exclusively part of the imperial *patrimonium* has continued. Ferdinand Orth saw most mines falling into the hands of the emperor, whom he described as 'eigentlichen Besitzer der Gruben', the true proprietor.[154] In his influential work on the Norican iron mines, Geza Alföldy simply noted that the mines there were the private property of the emperor (without providing a discussion of the known material).[155] William Harris also observed: 'The principal metals in question were gold, silver, copper, tin, bronze . . . lead and iron. Most of the ultimate sources of supply were the property of the emperor.'[156] Taking a more moderate position, Claude Domergue pointed out that it is impossible to maintain a distinction between state/public mines and patrimonial mines in the Roman empire:

> En fin de compte, il est, croyons-nous, plus simple et sans doute aussi plus conforme à la réalité de considérer ces mines administrées par les services financiers du Prince globalement et sans autre nuance comme des 'biens impériaux', donc des 'mines impériales', et il semble bien qu'à partir du IIe siècle ces dernières vont absorber celles qui, dans les provinces, dépendaient jusque là de l'aerarium.[157]

In the case of imperial quarries, the view voiced in 1884 by Joachim Marquardt, that the quarries under imperial control belonged to the *patrimonium*, seems to have prevailed.[158] Otto Hirschfeld, Charles Dubois, and Ulf Täckholm followed Marquardt's opinion. Recent scholarship has failed to directly address the question. John Ward-Perkins argued that 'certainly by the mid-first century AD the major quarries were apparently in imperial or state control'. For J. Clayton Fant, however, most imported marble in Rome came from quarries owned by the Roman emperor.[159] The only voice of dissent appears to be that of Kurt Fitzler, who argued that the quarries in Roman Egypt were public property and were not of patrimonial status.[160]

[153] 'Je nachdem die Art der Erwerbung war, werden dieselben [mines] entweder dem Fiskus (so bei der Eroberung des Landes[. . .]) oder dem *patrimonium* zugefallen sein, so besonders die *bona damnatorum*. Durch Erbschaft sind ohne Zweifel ebenfalls zahlreiche Bergwerke in kaiserlichen Besitz gekommen [. . .]', cf. Hirschfeld 1905: 146 f. with p.147 fn. 1.

[154] Orth 1924: 152 f.

[155] Alföldy 1970: 167; Alföldy 1974: 100. On ownership of mines in Noricum, cf. Ørsted 1985: 217 f.

[156] Greene 2000: 749; Harris 2000: 722.

[157] Domergue 1990: 239 f.; cf. also Andreau 1990: 86.

[158] Marquardt 1884: 263 with fn. 3; Ward-Perkins 1992*b*: 27, fig. 14.

[159] Hirschfeld 1905: 175 fn. 3; Dubois 1908: ix f.; Täckholm 1937: 113; Dodge 1991: 34; Ward-Perkins 1992*b*: 24; Fant 1993*a*: 145.

[160] Fitzler 1910: 112.

In the light of these conflicting and vague positions on the legal nature of mines and quarries under imperial control in the Roman empire, a reassessment of the literary, legal, and epigraphic sources is appropriate.

3.3.1. Private and Public Metalla

Suetonius notes in his 'De vita Caesarum' that the emperor Tiberius—besides confiscating the property of 'principes' in Gaul, Spain, Syria, and Greece on preposterous pretences—had rid most *civitates* and private individuals of old privileges (*immunitates*) and the *ius metallorum ac vectigalium.*[161] This passage has been interpreted by a number of scholars as evidence for the appropriation of *all* relevant mines and quarries by Tiberius, either in the name of the Roman state or for the emperor's *patrimonium*—without, however, creating an imperial monopoly on extractive operations.[162] In addition, Tacitus reports that the silver and gold mines of Sextus Marius, described as the richest man in Spain, were seized by Tiberius in AD 33 after Marius had been convicted of having an incestuous relationship with his daughter. Tacitus has been thought to accuse Tiberius of keeping the mines for himself, instead of transferring Marius' *aerariae* and *aurariae* into public property.[163] This quote from the *Annales* is usually taken to corroborate the above passage from Suetonius.

The texts of Tacitus and Suetonius, however, are in disagreement. Unlike Suetonius, the infringement of the *ius metallorum* or *ius vectigalium* of cities or private individuals by Tiberius is not mentioned in the surviving parts of the Tacitean text. Moreover, Tacitus underlines the abnormality of Tiberius' procedure regarding Marius' mines: it was unusual, perhaps even unlawful, that the property or *bona damnatorum* intended for the state was diverted to assets under his control.[164] Michael Alpers and Peter Brunt have provided additional insights on this passage. Both argued that the mines of Sex. Marius were not necessarily incorporated into Tiberius' private property, but that Marius' *bona*, including his mines, were made public property.

[161] Suetonius, *Tib.* 49.2: *plurimis etiam civitatibus et privatis veteres immunitates et ius metallorum ac vectigalium.*

[162] Hirschfeld 1905: 146 f.; Schönbauer 1929: 135 f.; Davies 1935: 3; Täckholm 1937: 97 f.; Domergue 1990: 236 f.; France 2001: 278–83; For the quarries, cf. Dodge 1991: 32; Ward-Perkins 1992*b*: 24 with fn. 13 (commentary by Hazel Dodge).

[163] Tacitus, *Ann.* 6.19: *Post quos Sextus Marius Hispaniarum ditissimus defertur incestasse filiam et saxo Tarpeio deicitur. Ac ne dubium haberetur magnitudinem pecuniae malo vertisse, aerarias aurariasque eius quamquam publicarentur, sibimet Tiberius seposuit*; cf. Hirschfeld 1905: 147 fn. 1; Täckholm 1937: 97; Millar 1992: 176; Levick 1999*a*: 101.

[164] On *bona caduca* and *bona damnatorum* going to the *fiscus* i.e. the public or state treasury, cf. Brunt 1990*a*: 141–6.

The distinction is made evident by Tacitus' (usually precise) use of the term 'publicare' instead of 'confiscare'. As stated by Brunt, Tiberius' transgression lay merely in having 'simply decided to manage and exploit [Marius' mines] through his own agents' instead of involving the senate.[165] Tiberius strictly adhered to the division of *fiscus Caesaris/patrimonium* and the *aerarium Saturni*, making it unlikely that the emperor had appropriated Marius' mines for his own *fiscus Caesaris*.[166]

Suetonius' claim that individuals and cities were losing their *ius metallorum* or *ius vectigalium* is not confirmed but contradicted by other sources. In the case of the *ius vectigalium*, there is no evidence of Tiberius' alleged measure in the epigraphic record of municipalities—towns like Veii in Italy, Stratonicea as well as Aezani in Asia, Irni, Malaga, and Munigua in Baetica, and others, continued to exact *vectigalia* throughout the first century and later.[167] Moreover, Gaius, a jurist writing sometime during the second century, refers in his Provincial Edict to *publicani* renting out *vectigalia* from the public authorities of municipalities.[168] As regards the *vectigalia*, Suetonius' statement is an untenable exaggeration. By analogy, one may assume that private individuals or municipalities kept control of mines and quarries. Evidence for this assumption is mainly provided by legal sources. In general, the rights of ownership over landed property included not only a delimited surface area, but were extended to the sky above and soil below.[169] In the case that stone quarries were discovered on private property, no one was to quarry the stone either in a private capacity or in the name of the state without the permission of the proprietor.[170] Moreover, the usufructuary of a private estate was allowed to make use of stone quarries or even open quarries and gold, silver, copper, iron, sulphur, and other *metalla* after the legacy of the usufruct had been made.[171] Javolenus and Ulpian discussed the consequences of a divorce on the ownership of marble quarries opened by the husband on land that formed part of his wife's dowry. Ulpian believed the husband to be entitled to keep the quarried marble still lying in the quarry, yet could not hope to be reimbursed for the expenses in opening and running the quarries, despite the rise in value of his divorcee's dotal land. Ulpian and Javolenus both saw

[165] Lang 1911: 77; Brunt 1990*a*: 145; Alpers 1995: 90 fn. 282.

[166] Alpers 1995: 59–95.

[167] *ILS* 6579 (Veii). Syll.^{3} 837, 9–10 (Stratonicea); *OGIS* 502 (Aezani); *AE* 1962: 288 (Munigua); *AE* 1986: 333 (Irni); *CIL* II 1964 (p. xliii 876. 704) (Malaga), cf. Lintott 1993: 84 with fn. 89; Galsterer 1988: 86 with fn. 40; González 1986: 167, ch. 63; LeRoux 1999: 157–63; Nonnis & Ricci 1999: 54–9 (Appendix).

[168] *Dig.* 39.4.13.1: *Praeterea et si quis vectigal conductum a re publica cuiusdam municipii habet, hoc edictum locum habet.*

[169] *Dig.* 8.4.13.1, 39.2.26; Kaser & Knütel 2005: 114.

[170] *Dig.* 8.4.13.1, 18.1.77.pr.

[171] *Dig.* 7.1.13.5, cf. 7.1.9.pr.-7.

marble quarries as a non-renewable resource (unlike chalk pits, silver and gold mines, or other mining ventures, and certain marble quarries in Gaul and Asia). The extracted marble therefore could not be regarded as fruits of the land.[172] In the thirty-fifth book of his 'Edict', Ulpian discusses a speech given by the emperor Septimius Severus on the Ides of June, AD 195, on the property of those in tutelage or care and the protection of their property from alienation (e.g. sale) by the tutor. Amongst other relevant issues, Ulpian argues that *lapidicinae* or other *metalla* owned by a person in tutelage (*pupillus*) were protected under Severus' ruling and thus were inalienable.[173] Although the question of ownership of mines and quarries is never directly addressed, Roman law did not prohibit the owner or the usufructuary of land to open and run quarries or mines or to profit from any other 'fruits' of the land. Thus, extractive operations could be owned privately.

There were, however, some restrictions: to Ulpian's discussion of Severus' speach the sixth-century compilers of the *Digest* added a short passage from Paulus' commentaries. It was inserted as a caveat following Ulpian's list of extractive operations which could be owned by a person in tutelage. The caveat implies that private ownership rights might not have included all mines or quarries (*Dig.* 27.9.4: *quod tamen privatis licet possidere*).[174] In addition, Marcianus, a jurist from the Severan era, records a rescript of Caracalla exempting any *dominus praedii*, landowner, from penalty whose *colonus* or slave produced iron (*ferrum facere*) on his estate without his knowledge and not in compliance with the law (*illicite*).[175] This restriction on iron production may have been in accord with a

[172] *Dig.* 23.5.18.pr.-1. (Javolenus), 24.3.7.13–4 (Ulpian).

[173] *Dig.* 27.9.3.6–27.9.5.pr. (Ulpian): *Si lapidicinas vel quae alia metalla pupillus habuit stypteriae vel cuius alterius materiae, vel si cretifodinas argentifodinas vel quid aliud huic simile,* [Paulus: *quod tamen privatis licet possidere*] *magis puto ex sententia orationis impediri alienationem.*

[174] The original context of Paulus' caveat can not be reconstructed. The use of 'licet possidere' evokes the formula 'uti frui habere possidereque licere' describing unimpeded rights of possession and usage over an object. Roman jurists distinguished property (*proprietas, dominium*) from mere possession (*possessio*). Real property, *dominium ex iure Quiritum*, was limited to Roman citizens and only applied to moveable goods or landed estates in Italy. With exemption of areas under *ius Italicum*, the owner of provincial land could theoretically never gain full property rights. Provincial land was seen as property of the Roman state, since a land rent (*tributum, stipendium*) had to be paid. The owner of provincial land, however, had rights of possession and usage (*uti frui habere possidereque licere*), basically equalling regular *dominium.* A Roman landowner in the province would hardly have noticed the legal differences, yet amongst Roman jurists such issues appear to have been discussed at length, cf. Gaius, *Inst.* 2.7, 2.21; *Inst.* 2.1.40; Kaser & Knüttel 2005: 111 f.

[175] *Dig.* 39.4.16.11: *Magnus Antonius rescripsit, si colonus vel servi domini praedii ferrum illicite in praedio fecerint ignorante domino, nulla poena dominum teneri.* The term 'ferrum facere' may imply the mining for and/or smelting of iron ore, cf. Herz 2005: 28.

lex ferrariarum documented in the *lex metalli Vipascensis*: the contractor of shoe-making within the mining district was to sell nails in line with the *lex ferrariarum*.[176] Moreover, a tax on nails and the use of iron are at issue in a rescript by Hadrian to the Aphrodisians.[177] Paulus (*Dig.* 39.4.11.pr.) relates that the sale of iron to barbarians is strictly forbidden and subject to capital punishment. Given the circumstantial evidence for legal and fiscal restrictions on the mining and trade of iron, the general exclusion of this metal from private ownership is plausible. There is, however, no direct documentation of a particular law to that effect.[178]

Set aside the legal codes, the private exploitation of mines and quarries is demonstrated by epigraphic and literary sources. Pliny, while discussing copper of different origins, points out that the *aes Sallustianum* in the 'Ceutronian Alps' (Tarentaise/F) and the *aes Livianum* in Gaul were named after the *domini* of the *metalla*, one a close friend of Augustus, the other Augustus' wife.[179] He identifies C. Sallustius Crispus and Livia as former owners (*domini*) of the copper mines. However, it is uncertain whether Pliny was aware of the linguistic niceties of Roman law and described the entitlement of Sallustius and Livia to these copper mines in Gaul with any accuracy. Papyri from Roman Egypt, however, provide interesting parallels. There, large imperial estates were named after the initial 'owner'. Apart from estates carrying the name of Livia, the estates of Maecenas, the *Μαικηνατιαναὶ οὐσίαι*, appear in papyri as part of Claudius' and later Nero's *patrimonium*.[180] The answer to the question as to how Maecenas or Livia acquired these estates has been controversial. George M. Parassoglou argued that with the Roman occupation of Egypt, the confiscated *δωρεαί* (lands granted by the Ptolemaic kings to members of the royal family and trusted government officials), and *κλῆροι* (allotments given to soldiers or local guards) 'were offered for sale at public auctions'.[181] As some of the reviewers of Parassoglou's study have pointed out, there is no evidence either to confirm or contradict his view.[182] Perhaps the distribution of patrimonial land to members of the imperial family or close friends was a viable option as well. The *senatus consultum de Cn. Pisone patre* provides some insight into this matter. Aside from additional punitive sentences, the Roman senate ordered the confiscation

[176] *LMV* ll. 34.

[177] Reynolds 1982: 115–18, no. 15, ll. 8–9, cf. Oliver 1989: 166, no. 69.

[178] The content of the *lex ferrariarum* in *LMV* or the *νόμος γεωρυχικός* recorded in the *lex portorii Asiae* (Engelmann & Knibbe 1989: §34, l. 78) remain unknown.

[179] Pliny, *NH* 34.3–4.

[180] Parassoglou 1978: 16 f.; Capponi 2005: 104–6.

[181] Parassoglou 1978: 4, 8.

[182] Crawford 1980: 252 f.; Hengstl 1981: 37; Rathbone 1993: 102 f.

of Cn. Calpurnius Piso's property for the state, with the exception of a *saltus* in Illyricum. This estate had been a gift from Augustus to Piso, and now was to be returned to Tiberius' private property.[183] By analogy with this passage, certain copper mines in the Alpes Ceutronum and in Gaul were perhaps given as a present to C. Sallustius Crispus and Livia by the emperor Augustus, and thus came into into private ownership.[184]

Outside Augustus' family and inner circle, wealthy individuals (besides Sextus Marius) are documented owning mines. Pliny the Elder names two Baetican lead mines, a *metallum Sama/lutariense* and a *metallum Antonianum* that were contracted out by the Roman state.[185] The *metallum Antonianum* is most certainly named after an 'Antonius' and hence is likely to have been privately owned before coming under imperial administration.[186] Nor was the ownership of mines limited only to men. An inscribed monument from Lugdunum/mod. Lyon set up in AD 226 names the *ferrariae* of Memmia Sosandris, a *c(larissima) f(emina)*, a woman of senatorial rank.[187] The monument was commissioned by the contractors of a *vectigal* on the *massa ferrariarum* (probably an estate including iron mines), of a Memmia Sosandris.[188] The use of the term 'vectigal' (mostly understood as state revenue or tax) demonstrates the public ownership of the mines in question. Thus, René Sablayrolles argued that the iron mines were not the private property of Memmia Sosandris. Yet, 'vectigal' can also describe revenues from private estates; perhaps the small company (*socii*) was awarded the contract for collecting revenue from the miners of these *ferrariae*.[189] As one might expect, the extant sources barely provide any insights on the private ownership of mines outside the upper echelons of Roman society. Even though Strabo

[183] Eck, Caballos, & Fernandez 1996: 45, ll. 84–6. For other estates given as *donativa* to trusted generals and then returned to the emperor when they had fallen out of favor, cf. Cornelius Gallus (Dio 53.23.7), C. Silius (Tacitus, *Ann.* 4.20.1). Eck, Caballos, & Fernandez 1996: 202–7, with further bibliography.

[184] Kaser & Knüttel 2005: 238 f.

[185] Pliny, *NH* 34.165.

[186] Domergue 1990: 235 with fn. 39, 270, 274 with fn. 52, 276; a hypothetical candidate is Marc Anthony, given that a further influential politician of the Late Republic, M. Licinius Crassus, reportedly owned silver mines, cf. Plutarch, *Crass.* 2.7.

[187] *CIL* XIII 1811 = *ILS* 8641 (Lugdunum), cf. Hirschfeld 1905: 158 fn. 4; Sablayrolles 1989: 159 fn. 24. Memmia Sosandris has been claimed to be the sister of Memmia, the wife(?) of Severus Alexander. On the name, cf. *PIR*[2] M 489. Syme 1971: 6, 8, understood the latter Memmia to be a fabrication by the author of *HA Alex.* 20.3.

[188] Hirschfeld 1905: 78 with fn. 2, argued that 'massa ferrariarum' must be understood as ore from the iron mines, cf. also *TLL* VIII 429/39 f., 1a. Brunt (1990*c*: 398 fn. 143) believed 'massa ferrariarum' to mean a territory or estate of iron mines, cf. *TLL* VIII, pp. 430 f., 2b.

[189] Sablayrolles 1989: 160 with fns. 28, 29; Domergue 1990: 240, esp. 361 with fn. 70. On C. Sallustius Crispus, cf. *PIR*[1] S 61. On 'vectigal', cf. Mommsen 1887: 427–34; Pékary 1979; Burton 1999; Mateo 1999: 90–3.

records private ownership for the mines near Carthago Nova (except for the gold mines) in the early days of the Principate, information on the owner's status is not provided.[190]

Besides private individuals, municipalities may also have profited from mining ventures. Suetonius' claim that Tiberius had rid most *civitates* of *ius metallorum* as well certainly implies some municipal ownership of *metalla*.[191] By analogy with the municipal *vectigalia* one may argue that municipalities retained control of mines after Tiberius. However, there is little written evidence to corroborate this hypothesis.[192] Furthermore, the issue whether or not provincial councils were entitled to revenues from mining ventures is a contentious one.[193]

The majority of archaeologically documented quarries undoubtedly remained in private or municipal hands during the Principate, and only a limited number of marble or granite quarries stood under direct imperial control.[194] The town of Luni may have held rights of ownership on the quarries at Carrara, whereas Herodes Atticus perhaps counted the quarries of the Pentelicon amongst his properties.[195] The small number of papyri refering to stone extraction do not directly touch on the issue of ownership:

[190] Strabo 3.2.10.

[191] Suet. *Tib.* 49.2.

[192] Domergue 1990: 235 f., had pointed out that lead ingots carrying the stamp Carthago Nova or Augusta Firma might originate from mines of the town's territory. A missive by Tiberius to the town elders of the *municipium Muniguense* refers to a quarrel over oustanding payments to the collector of the municipal *vectigalia* (*AE* 1962: 288 = *AE* 1972: 257 = *CILA* II 1052). Given the numerous iron mines recorded on the territory of Munigua in Roman Baetica (cf. Griepentrog 1995), the municipial *vectigalia* possibly could have included the returns from the known mining ventures.

[193] The Marble of Thorigny mentions a *iudex arcae ferrar(iarum)*, an office which T. Sennius Sollemnis appears to have held before AD 238. By analogy with the *arca Galliarum*, the *arca ferrariarum* has been understood to be independent from the imperial administration and under the supervision of the Council of the Three Gauls. It has been argued that part of the revenues (?) gained from iron-mining in Gaul belonged to the Council, cf. Hirschfeld 1905: 158; Pflaum 1948: 18; Sablayrolles 1989: 160 with fn. 31; Rémy 1995: 77 f. For a different view cf. Deininger 1965: 104 fn. 4. Two *adlectores(?) ferrariarum* are documented in inscriptions at Anicium/mod. Le Puy in an iron-mining area and were perhaps subordinate to the *iudex arcae ferrariarum* (*CIL* XIII 1576, 1577).

[194] Ward-Perkins 1992*b*: 24 with fn. 14.

[195] On Carrara, cf. 7.2.3. The marble quarries at Tainaron, opened in response to heightened Roman demand for marble, may have been owned privately, cf. Strabo 8.5.7; Pliny 36.135, 158. Herodes Atticus' ownership of the Pentelian quarries is indicated by an inscription on a block of Pentelic marble found at Rome (no. 1260, AD 166), which reads *caesura Cla(udii) Hier(odis) Attici et Apolloni Lupi (...)*. Based on this reading, Dubois believed the quarries to be the property of Herodes Atticus, who in the 2nd cent. AD used Pentelic marble for different building projects, cf. Pausanias 1.19.6, 6.21.2, 10.32.1; Philostratus, *Vit.Soph.* 2.550, cf. Ameling 1983*a*: 84–94; Korres 1995. It therefore is possible that Herodes Atticus and his partner(?), Apollonius Lupus, had leased or owned the quarries. Walter Ameling, however, believes the named Claudius Hierodes Atticus to be his freedman, cf. Ameling 1983*b*: 216; Fant 1993*a*: 167. Atticus himself was certainly not involved in the everyday work processes at the quarry.

these are usually work contracts in which stone-cutters or masons come to terms on the provision and transport of stone blocks and construction elements (bases, capitals).[196] Whilst the quarries from which the stone came are occasionally named, the papyri do not assist in deciding whether the named masons owned the stone quarries, were merely lessees of municipial or private quarries, or, on the basis of a *locatio-conductio operis*, were paid to run quarries and workshops of private owners.[197]

Other than private individuals and, perhaps, municipalities, the Roman state emerges as owner of mines and quarries. As a consequence of its expansion during the middle Republic, the Roman state acquired important mining districts in newly conquered territories. These were regularly farmed out by the state to private contractors.[198] The literary sources, however, rarely provide any particulars on public mining ventures. Limited insights are available for southern Spain. With the fall of Carthago Nova/mod. Cartagena in 205 BC Rome not only gained possession of considerable booty, but the silver mines near the former bastion of Barcid Spain came under control of the Roman state as well.[199] In 195 BC the consul M. Porcius Cato, during his Spanish campaign, ordered the exaction of *vectigalia* on iron and silver mines, including the extractive operations near Carthago Nova.[200] Strabo, quoting Polybios, claims that 40,000 miners worked in the silver mines within twenty stadia from Carthago Nova. The silver mines supposedly extracted silver worth 27,000 drachmas per day, which went to the Roman state (ὁ δῆμος τῶν Ῥωμαίων).[201] The territories near Carthago Nova (and, thus, the mines as well) remained

[196] A 2nd-cent. contract between Antonia Asclepias from Oxyrhynchos and stone-cutters revolves around the provision of different sizes and types of blocks, some of which were transported by camel from the 'northern quarry' (*P.Oxy.* III 498). Two petitioners, Pamunis and Pausiris, known from a papyrus found in El-Hibeh and dated to AD 177–80 (*P.Hibeh* II 273 + 217), ask to be paid for the transport of nine columns including the bases and capitals; as Dieter Hagedorn (1993: 97 f.) suggests, they had probably produced the items in question as well. An account of construction work on a temple from the Oxyrhynchite nome, dating to AD 117 (*SB* 11958) attests to the transport of 100 columns from the Ankyronon quarries to the port of Artemis.

[197] The papyrological evidence for lease and work contracts between potters and owners of kilns might provide a useful comparison: cf. *P.Oxy.* L 3595–7 with Strobel 1987: 92–7. Further written evidence for quarries worked by stone-cutters comes from Saint-Boil in Gaul, cf. Monthel & Lambert 2002: 111 f., fig. 86. An inscribed monument at Marignac near St-Béat notes two men *qui primi hinc columnas vicenarias c(a)elaverunt et exportaverunt* (*CIL* XIII 38 = *ILS* 3579 = *AE* 2000: 923) which Hirschfeld 1905: 147 fn. 5, assumes were perhaps also private owners of quarries. A series of inscriptions from Phrygia documents the wealth of individual stonemasons, cf. Reinach 1890: 55, no. 9; Mendel 1909; Waelkens 1977: 288; Waelkens 1986: nos. 417, 471, 486, 501, 502; Strubbe 1997: no. 256; Merkelbach & Stauber 2001: 200, no. 16/22/05.

[198] Polybios 6.17.

[199] Livy 26.47.

[200] Livy 34.21.7, cf. Domergue 1990: 242 fn. 12; Lintott 1993: 73; Mateo 1999: 127.

[201] Strabo 3.2.10 = Polybios 34.9.8–11.

property of the Roman state until the late Republic. In the year 63 BC Cicero—providing his audience with a list of *agri publici*—mentions the *agri propter Carthaginem Novam* being public land since 209 BC.[202] A passage from Livy supports the view that during the Republic, mines owned previously by local political entities or rulers fell to the Roman state. Livy reports on the senate's decree in 167 BC terminating the lease of Macedonian mines formerly owned by the Macedonian king Perseus. With the division of Macedonia, L. Aemilius Paulus allowed for the iron and copper mines to be reopened, whereas gold and silver mining was prohibited.[203] Strabo identifies further state-owned mines at Mt Sandaracurgium near Pompeioupolis in Pontus which were contracted to *publicani*, while Pliny alludes to a Republican *lex censoria* for the *aurifodinae Victumularum* (mod. Bessa) in the territory of Vercellae in northern Italy, which prohibited the *publicani* there from employing more than 5,000 workmen.[204] The gold-mines of the Salassi in the Aosta valley, too, had been farmed out to *publicani* prior to the conquest in 25 BC.[205] Alfenus Varus furthermore records whetstone-quarries on Crete, which were probably publicly owned, farmed out (*locare*) by Caesar to *redemptores*.[206] With the submission of Numidia to Roman rule, the quarries of Simitthus, formerly owned by the Numidian kings, passed into the hands of the Roman state and are likely to have been contracted out as well.[207]

Public mines continued to generate considerable returns for the Roman state during the Principate. Pliny's report on the contracting out of the *metallum Samariense* and the *metallum Antonianum* aside, he also names the *minium* mines of Sisapo as providing *vectigal* for the *populus Romanus*.[208] The legal texts once more provide the main bulk of evidence for state ownership of mines. In the third book of his study on *The Provincial Edict*, Gaius states that partners in collecting public *vectigalia* and running gold, silver and salt mines are exempted from the prohibition to form corporations.[209] Moreover, he declares in the thirteenth book of the same study that

[202] Cicero, *Leg. agr.* 1.5, cf. Domergue 1990: 229, 233.

[203] Livy 39.24.2, 45.18.3–5, 45.29.11; Cassiodor, *Chron.* ad 596 a.U.c.; Strabo 7.7.4; Domergue 1990: 243–4, 248.

[204] Strabo 12.3.40. Pliny, *NH* 33.78. For the localization of Vercellae, cf. Cavalieri Manasse et al. 1982: 76–7, also Strabo 5.1.12; *CIL* V, p. 715. Strabo 6.2.10 and Diod. 5.10, name an 'alum' mine on Lipara generating revenues for 'Rome'.

[205] Strabo 4.6.7; Dio 53.25.

[206] *Dig.* 39.4.15.

[207] Initial ownership by the Numidian kings is implied by the existence of an early temple at Simitthus. The continuous and extensive exploitation of the marble during the Principate and later has destroyed any older traces, cf. Rakob 1997: 2 f. On the history of the quarries near Dokimeion, cf. Fant 1989*a*: 6–11.

[208] Pliny, *NH* 34.118.

[209] *Dig.* 3.4.1.pr.

those who run salt mines, chalk-pits, or *metalla* are to be defined as *publicani* as well.[210] Provided the text of this second-century law scholar is genuine, public mines were still in existence in his time.[211] Ulpian in the tenth book of his 'Ad Edictum' defines public *vectigalia* as taxes from which the *fiscus*, i.e. the *aerarium Saturni*, derives revenues. Amongst these are to be counted the *vectigal metallorum*.[212] In this specific context 'vectigal' is to be understood as rent on state or public property, i.e. on state-owned mines, which suggests that at the beginning of the third century mines or quarries could still be owned by the Roman state.

Epigraphic testimonies for state ownership of mines and quarries are few. The Vipasca tablets, too, refer to the *fiscus* on several occasions, but they do not specify whether the term refers to the *fiscus Caesaris*, i.e the private funds of the emperor, or the *fiscus provinciae*, the provincial sub-chest of the *aerarium Saturni* in Rome containing public monies.[213] The *lex metallis dicta* (*LMD*), however, provides only vague clues on this matter. The text was sent by letter to an Ulpius Aelianus, most likely the *procurator metallorum* in charge of Vipasca. Although the author of the missive is not known—Flach surmises it perhaps originated with the financial procurator of Lusitania[214]—we can certainly exclude the emperor: the *LMD* does refer to emperor Hadrian, yet he is never noted in the first person singular. In fact the emperor's name is attested only once and in the third person singular. Moreover, the pronoun 'noster' for 'our Caesar', often used by procurators concerned with the emperor's private domains and by *coloni* working his estates, is omitted.[215] In itself these observations do not provide enough evidence to exclude the emperor as owner of the Vipasca mines. Additional evidence, however,

[210] *Dig.* 39.4.13.pr.

[211] For textual criticism, cf. Schönbauer 1929: 138 f.

[212] *Dig.* 50.16.17.1.

[213] *LMD* ll. 3, 6, 11, 21, 24, 26, 34; *LMV* ll. 13, 29. Current scholarly opinion is divided on this issue, cf. Brunt 1990*c*: 398; Mateo 1999: 124 f. (public operation); Flach 1979: 440–6; Domergue 1983: 171 f.; Flach 1990: 117 f. (emperor's estate). On dates of both, cf. Flach 1979: 400, 413. On the term *fiscus*: Jones 1960; Brunt 1990*a*: 134; Alpers 1995: 1–20 with further bibliography 308–11. For a different view, cf. Millar 1963*a*.

[214] Flach 1979: 413 f.

[215] Hadrian is still in power at the time the letter was written (the epithet *divi* is missing), cf. *LMD* ll. 5–6: *quorum pretia secundum liberalitatem sacratissimi Imp(eratoris) Hadriani Aug(usti) obser|vabantur* (...); on the date of the *LMD*, cf. Flach 1979: 440. Unlike the inscriptions from imperial estates in North Africa, where the deliberate usage of 'noster' mirrors the 'close' relationship of the procurator with the emperor, this is not highlighted in the Vipasca tablets by the words 'our Caesar', cf. *CIL* VIII 25943+26416 = *FIRA*^{2}, no. 101+102 (Ain el-Djemala), col. II, ll. 3–4: *Sermo procuratorum [im]p(eratoris) ⟨C⟩aes(aris) Hadriani Aug(usti): Quia Caesar n(oster)*...; (Ain Wassel), col. I, ll. 8–9: *Sermo procuratorum: Qui⟨a⟩ Caesar n(oster)*...

corroborates this notion: as noted by the *lex metallis dicta,* free men convicted of destroying mining infrastructure forfeited their *bona* to the *fiscus* and were banned from the mining district.[216] This procedure seems to go beyond regular liabilities of private law, as the usual penalty for the destruction of property was the payment to the owner of damages equivalent to the loss.[217] Furthermore, the property of the offender is collected for the *fiscus* by the procurator at Vipasca. If one understands the *fiscus* to be the *fiscus Caesaris,* the private treasury of the emperor, the appropriation of *bona* by the *procurator metallorum* would be highly irregular. However, Alpers and Brunt have clearly shown that *bona damnatorum* were generally appropriated to the *aerarium Saturni,* not to the *fiscus Caesaris.*[218] Consequently, the property of convicts at Vipasca was forfeited not to the *fiscus Caesaris,* but to the *aerarium Saturni,* that is, the *fiscus provinciae* as the provincial sub-chest of the *aerarium Saturni.* The *LMD* does not make any distinction between the *fiscus* receiving the bona of the offender or the *fiscus* receiving the *pretium* for half of the extracted ore. This leads us to the tentative conclusion that the Vipascan mines were public property. Further, it is in line with our comprehension of internal arrangements at Vipasca: the 'sale' of *proprietas partis dimidiae ad fiscum pertinentis* to *occupatores* appears to stand in the tradition of public land apportioned to *occupatores* in return for a *vectigal* (cf. 7.1.1). The regular transfer of property, albeit in a very restricted sense, does not fit the idea of a privately owned estate.

3.3.2. *Metalla* and the Roman Army

Further indications towards the legal status may be provided by the personnel present at extractive operations. The *lex metalli Vipascensis* (*LMV*) notes imperial freedmen and slaves serving in the *officia* of the procurator, those receiving *commoda* (salaries, pay), children and soldiers, being exempt from paying a fee for using the baths.[219] The presence of imperial freedmen and slaves in varying functions at Vipasca is not a valid argument for or against the mines being public property. Despite being part of the *familia Caesaris,* they did not only conduct the private business of the Roman emperor. They also appear in public

[216] *LMD* l. 34: *liberi bona pr(ocurator) in fiscum cogito (...)*; *LMD* l. 28: *liberum procurator comfiscato (...).*

[217] In classical Roman Private Law the destroyer of private property was only liable for the loss to the owner, cf. Mommsen 1899: 832 f.; Nicholas 1962: 218 f.; Kaser 1971: 619–22; Kaser & Knüttel 2005: 255 f. As Mommsen 1899: 810, 822–5, points out, regulations on the *damnum iniuria* of public property hardly exist.

[218] Brunt 1990*a*: 145 f.; Alpers 1995, *passim.*

[219] *LMV* ll. 23–24:... *Excipiuntur liberti et servi [Caes(aris), qui proc(uratori)] in offi⟨c⟩is erunt vel | commoda percipient, item inpuberes et milites.*

functions, for example, as administrative staff of financial or presidial procurators.[220] The regular presence of military personnel, however, may confirm the view of the *metallum Vipascense* and other mining districts being public property rather than the emperor's own. The Republican military oath, as recorded by Livy and Polybios, included a promise to assemble on the consul's orders, not to leave military service without instructions or desert during a battle, to be obedient and to carry out orders to the best of one's ability. More importantly, Servius added that 'they swore that they would act on behalf of the state'. Towards the end of the fourth century Vegetius reports that the formula 'on behalf of the Roman state' was still included in the military oath. Whether the military oath reflected reality at the time is another issue, but it is significant enough that the reference to the Roman state probably remained in the oath throughout the Principate.[221]

According to Tacitus and Suetonius, shortly after Augustus's death in AD 14 a *libellus* with a *breviarium totius imperii* listing the state's assets, the *opes publicae*, was found amongst four documents deposited with the Vestal virgins. Besides listing the 'vassal'-kingdoms and provinces, the *tributa* and *vectigalia*, as well as necessary expenditures, *necessitates*, and presents, *largitiones*, the document included the number of soldiers serving in the legions, auxiliary units, and fleets.[222] This suggests that Augustus (and his successors) understood the Roman army as a state asset. In his *Res Gestae*, however, Augustus refers to the troops as 'exercitus meus'.[223] Wickert argued that the use of the possessive in describing Augustus' bond with the troops reflected the real and exclusive power of the *princeps* over the legions. Even so, Wickert conceded that this close relationship between emperor and army did not translate into a formal 'ownership'.[224] Augustus and his successors were quite aware of this: in the acts of the *ludi saeculares* (AD 17), the legions are clearly identified as *legiones populi Romani Quiritium*.[225] Cassius Dio provides additional clarification: in his work, Tiberius is quoted as saying to the senate 'the legions are not mine, but the state's'.[226] Moreover, in AD 6 Augustus had the Roman senate create a treasury, the *aerarium militare*, to dispense the pay to soldiers.[227] The *aerarium militare* was fed by a series of newly

[220] For the staff of presidial procurators, cf. Eck 1997*a*: 104 f.; Haensch 1997: 725 f.; for others, cf. Boulvert 1970.

[221] Servius, *Ad Aen.* 8.1: (...), *qui singuli iurabant pro republica se esse facturos* (...); Vegetius, *Mil.* 2.5, cf. Campbell 1984: 19, 23 f.

[222] Tacitus, *Ann.* 1.11.4. Suetonius, *Aug.* 101.4, cf. Speidel 2000*a*: 116.

[223] *RGDA* 30, cf. Alföldy 1987*b*: 23.

[224] Wickert 1954: cols. 2100–3, with discussion of additional sources.

[225] *CIL* VI 32323 = *ILS* 5050 = *AE* 2002: 192.

[226] Dio 57.2.3, cf. Campbell 1984: 25.

[227] *RGDA* 17, cf. Wolters 1999: 180 f.; Speidel 2000*a*: 132.

introduced taxes, such initiatives being a legislative privilege of the Roman senate.[228] There is no doubt that following the consolidation of Augustus's position, the Roman army was regarded as part of the *opes publicae*, financed by a public treasury *de iure* under the control of the Roman senate.[229] Consequently, from a strictly 'constitutional' point of view and in compliance with the figment of a *restitutio rei publicae*, the Roman army belonged to the Roman state.

Hence, Roman jurists understood military personnel by and large to be deployed on state business. This is expressed in the *Digest:* amongst the legal means of a *praetor* the *restitutio in integrum*, restitution of previous status, figures as one of the more prominent instruments. It could be used if one's property or rights had changed whilst away on state business, *rei publicae causa abesse*, for example.[230] As a rule, Roman soldiers and their commanders were regarded as absent *rei publicae causa*, except when on furlough. No other form of deployment is noted.[231] Thus, any deployment of military personnel was theoretically in the interest of the state. This included work for the benefit of local municipalities: Ulpian details the obligation of the provincial governor in supplying military specialists to public building projects.[232] In consequence, the use of soldiers for any private benefit may have been ruled out. The first book of Macer's 'De re militari' quotes the following from the *disciplina Augusti*: 'Although I know it is not unusual for soldiers to perform artisanal/construction work, I nevertheless fear that—should I permit any such thing to be done for my benefit or for yours—no limitations which I could tolerate would be set in this matter.'[233] Augustus, probably addressing a military commander, apparently prohibits the use of military personnel for private works. This is repeated by Tarruntenus Paternus in the later second century: military commanders were not to dispatch soldiers for private works.[234] Epigraphic and papyrological evidence partly corroborates the legal texts. Roman soldiers are found collecting taxes and duties in provinces, policing the countryside, guarding the transport of grain, and improving the public infrastructure through the construction of roads, bridges, canals, and aqueducts—all in the interest of local communities and the Roman state.[235]

[228] Talbert 1984: 375–9; Speidel 2000*a*: 144–6.

[229] Speidel 2000*a*: 126–35.

[230] *Dig.* 4.6.1.1.

[231] *Dig.* 4.6.7, 4.6.32, 4.6.34, 4.6.35.pr., 4.6.35.4, 4.6.35.9, 4.6.40, 4.6.41, 21.2.66.1, 49.16.1.

[232] *Dig.* 1.16.7.1.

[233] *Dig.* 49.16.12.1: *Etsi scio fabrilibus operibus exerceri milites non esse alienum, vereor tamen, si quicquam permisero, quod in usum meum aut tuum fiat, ne modus in ea re non adhibeatur, qui mihi tolerandus.* For translations cf. Brand 1968: 187; O. Robinson, in Watson 1998.

[234] *Dig.* 49.16.12. 1: [...] *ad opus privatum* [...] *militem non mittere.*

[235] Fink 1971: nos. 1, 2, 9, 10, 47, 63; Davies 1974: 33–68; Isaac 1991; Alston 1995: 79–96; Speidel 1996: 78 f. with fn.4. Duties of the *beneficiarii*, cf. Ott 1995: 61–163; Nelis-Clément 2000: 211–68.

In line with the observations made above, the use of soldiers and military specialists for the benefit of any private individual, including the emperor, is prohibited. Thus, when we find Roman centurions seconded to quarries like Karystos, Dokimeion, or Mons Claudianus in order to support or perhaps even implement the organization and management of extractive operations, these districts are most probably to be identified as public property (cf. 5.1.1).

The deployment of Roman army detachments to imperial quarries and mines seems to distinguish these districts from patrimonial estates in North Africa and Asia Minor.[236] Throughout the mid-second to the mid-third century AD, a number of inscriptions from estates in Asia Minor and Thracia document the misconduct of soldiers and imperial personnel passing by: their illegal requisitioning of goods prompted a series of petitions sent to the emperor by the inhabitants of these estates.[237] The preserved texts imply that Roman soldiers were not a regular fixture of the emperor's private estates. Undoubtedly, military help could be requested by the patrimonial procurators in charge to quell unrest, but they did not hold any regular authority over military units.[238] The epigraphic evidence from the patrimonial estates in Roman Africa and Asia suggests that military garrisons were usually absent from private imperial estates: their absence is not necessarily explained solely by the lack of technical and security tasks suitable for a military detachment at these estates. The deployment of the Roman army could probably only occur where there were tasks or duties to be performed in the interest of the Roman state. Given the emperor's unchallenged control of the Roman army, it is possible that this may not always have been the case. Yet, the adherence of most emperors during the first and early second century to the strict division between patrimonial treasury and *aerarium Saturni / fiscus provinciae*, between private and public property, make an 'unconstitutional' deployment of Roman troops for private interests less plausible.

[236] Particularly in the case of Asia Minor, the military garrison under the command of the proconsul only consisted of a few auxiliary units which would hardly allow stationing detachments at the numerous imperial estates alone in Central Phrygia, cf. Strubbe 1975: Eck 1997*b*: 201 f.

[237] Hauken 1998: 34–57, no. 3 (Aga Bey Köyü), 58–73, no. 4 (Kemaliye), 74–139, no. 5 (Skaptopara, Thracia), 140–61, no. 6 (Aragua), 162–8, no. 7 (Kavacik), 203–14, no. 4 (Tabala), 217–43, no. 6 (Takina); and Herrmann 1990.

[238] On procuratorial command over troops, cf. Tacitus, *Ann.* 4.15.2, 15.31, 15.32.3; Dio 57.23.5; Jos. *AJ* 18.158, cf. Brunt 1983: 52 f.; Brunt 1990*b*: 165 f.; Eck 1995*c*: 251–3; Eck 1999; Levick 1999*a*: 135. In addition, *CIL* VIII 10570 = *ILS* 6870, col. II, ll. 9–14 (Suk el-Khemis with Flach 1978: 471 f.), implies the presence of soldiers at the *saltus Burunitanus* to be highly unusual. For the death of a soldier at a *saltus Philomusianus*, cf. 5.1.2.2.

3.3.3. *Metalla* and Convicts

The question of military personnel set aside, the use of convict labour in quarries and mines under imperial control can probably only be explained if we assume that the *metalla* were a publicly owned enterprise rather than a private venture on the part of the emperor. Indications of this are provided mainly by the laws concerning *damnati in metallum* found in the *Digest*. The legal status the *damnati* acquired with their conviction is relevant to our discussion of the legal nature of mines and quarries. Gaius, in book seventeen on the 'Provincial Edict', states that *damnati* condemned to the mines lost not only their freedom, *libertas*, but also forfeited their property, *bona*, to the Roman state.[239] Caracalla and Gordian III stated in rescripts that the *bona* of *damnati in metallum* collected by the *fiscus* were not to be returned, even in the case of the convict being freed by order of the emperor.[240] Contrary to the property of the convict, which fell to the Roman state, the convict was made a *servus poenae*. These 'slaves of the penalty' were neither the property of the *fiscus* nor the property of the emperor. Nor did *servi poenae* enjoy the usual rights of a slave. An edict of Antoninus Pius appears to have ruled that slaves of the *fiscus* could receive an inheritance, while *servi poenae* could not.[241] Moreover, slaves condemned to the *metalla* and saved by the clemency of the emperor—according to Ulpian—were not returned to their owner but could become *servi fisci*.[242] The creation of a special status for the *damnati in metallum* certainly was of a practical nature. The reduction in status of a Roman citizen to a *servus* made the convicts subject to corporal punishment—they could be beaten like slaves.[243] Moreover, *servi poenae*, unlike *servi fisci*, were not sold while condemned to the *metalla*, nor could they buy their freedom or be set free by their former masters like regular slaves. The legal links between the condemned slaves' former owners and the family of a condemned free man were severed, as in theory no *damnatus in metalla* was expected to return from his sentence. This might explain why Roman jurists thought it necessary to introduce a special status apart from those of proper slaves or *servi fisci*.

The fact that former slaves condemned to the mines, but afterwards released, could become *servi fisci*, that is slaves of the Roman state, suggests a connection between *fiscus* (i.e. *aerarium*) and *metalla*. This is particularly clear in one specific case: a female slave condemned to the saltworks for a crime was taken captive by foreign *latrunculi* and deported beyond the *limes*.

239 *Dig.* 28.1.8.4.
240 *CJ* 9.49.4 (Gordian); *CJ* 9.51.2 (Caracalla).
241 *Dig.* 34.8.3.pr., 48.19.17.pr., 29.2.25.2–3, cf. Millar 1992: 183 with fn. 54.
242 *Dig.* 40.5.24.5, cf. also Millar 1992: 183 with fn. 55.
243 Paulus, *Sent.* 3.6.29, *Dig.* 49.14.12.pr., 50.13.5.2.

She then was brought back across the *limes* into the empire and sold. Cocceius Firmus, a *centurio*, bought her on the slave market. He apparently filed a report to the authorities, claiming back the money he paid for the slave. The *fiscus* refunded the payment and the slave was returned to the saltworks.[244] In his study on the *fiscus* and its development, Peter Brunt pointed out that in the *Digest* the term 'fiscus' is used as a mere synonym for *aerarium*, i.e. the public treasury of the Roman state.[245] The story of the female slave thus demonstrates the close relationship between the *fiscus*/*aerarium* and the salt mines which must be regarded as public property.[246] Moreover, this ruling for one specific case seems to have been applied to all *metalla* during the reign of Septimius Severus.[247] Hence, we must conclude that under the Severans, the *metalla* to which convicts were condemned were understood to be the property of the Roman state. Although *metalla Caesariana* are mentioned in the *Digest*, there is no evidence that convicts were condemned to the emperor's mines and quarries or those in private ownership.[248]

3.3.4. Quarries and Imperial Gifts

As detailed in an inscription from Smyrna, emperor Hadrian presented the town with—amongst other benefactions—72 columns of Synnadan marble, 7 porphyry columns, and 20 columns of Numidian marble for its gymnasium.[249] It was accepted that these columns originated from 'imperial quarries', which were part of the emperor's *patrimonium*. This permitted Hadrian to hand out columns as gifts. Moreover, the use of precious marble is believed to have been limited to public buildings and the palaces of the emperor, with only small amounts of marble appearing in private contexts in Italy and throughout the provinces. This emphasizes the exclusiveness of Hadrian's 'donation' of a significant and costly amount of columns from Phrygia, Africa, and, probably, Egypt.[250] The wording of the inscription implies that the senate in Rome had dealt with requests from Smyrna on Hadrian's behest and through the agency of Antonius Polemo, and passed a 'second *senatus consultum*' bestowing the title

[244] Pomponius, *Dig.* 49.15.6.pr. For Marcus Cocceius Firmus, known from Latin inscriptions from Auchendavy, cf. Birley 1953.

[245] Brunt 1990*a*: 139, 143.

[246] Pomponius, *Dig.* 49.15.6.pr., Millar 1992: 183.

[247] Tryphoninus, *Dig.* 49.15.12.17.

[248] *Dig.* 48.13.8.1, 48.19.38.pr. = Paulus, *Sent.* 5.21a.1.

[249] Pausanias 1.18.9 (Athens). *CIG* 3148 = *IGRR* IV 1431 = *IK* 24/1, 697, ll. 33 ff. *SEG* XXXII 1203. Dubois 1908: 34; Fant 1993*a*: 155 f. with fn. 73. For further evidence, cf. Horster 2001: 217 f.

[250] Millar 1992: 184, 420 f.; Fant 1993*a*: 155 f.

of *neokoros* on Smyrna. The inscription lists the *beneficia* by Hadrian as follows: a sacred festival, immunity from taxation, an official panegyrist of the gods, singers of hymns, 1,500,000 denarii(?), and the columns mentioned above. At least in the case of the immunity from taxation, Hadrian is likely to have acted through the Roman senate, *de iure* still the financial 'authority' in Rome, recommending an exemption of taxes for Smyrna to the Roman senate.[251] Hence, it is probable that other benefactions listed were also subject to the sanction of the senate. Provided this line of thought is correct, marble columns may have originated from public properties.

This assumption is partially substantiated by an oddity encountered in the epigraphic evidence of Simitthus. Unlike inscriptions on most other types of marble, the labels on *giallo antico* blocks from Simitthus name Domitian, Hadrian, or Antoninus Pius in the genitive form, indicating the emperor's ownership of the marble, while in inscriptions from the first century and from the reigns of Marcus Aurelius and Commodus such indications of their ownership are missing.[252] Had the quarries at Simitthus been the private property of the emperor, the indication of ownership would hardly have been necessary, certainly not within the quarrying district. Nor was it a requirement to mark the emperor's stones to prevent the exaction of trade taxes (*portoria*) (from which imperial/public goods generally were exempt[253]), as the golden-yellowish Numidian marble of Simitthus was not only a visually distinctive and bulky item, but probably, as Fant suggests, not transported for any other use than public building projects in Rome and the provinces, and for the adornment of the emperor's palaces.[254] In this context, it is of great significance that most of the *giallo antico* blocks from Rome's marble yards are not inscribed with the name of the emperor in the genitive form.[255] Hence, the transport or trade of marble blocks from Simitthus did not necessarily warrant the indication of ownership by the emperor—the legal status of the quarry did. Given the limited use of *giallo antico* outside a mostly public/imperial context, one may tentatively argue that marble not inscribed with the emperor's name in the genitive form belonged to the Roman state. The former was thus marked to differentiate it from the *giallo antico* blocks or columns destined for public use and, hence, public property.[256] The quarries

[251] For precedence, cf. e.g. Dio 54,30,3 (Augustus) and Alpers 1995: 138 (Claudius). On control of state finances by the senate, cf. Talbert 1984: 375–9.

[252] See Appendix (=App.) nos. 778–837.

[253] Engelmann & Knibbe 1989: ll. 58–66 (§§25–6).

[254] Fant 1993*a*: 152–5.

[255] See App. nos. 787, 795–8, 829, 830, 832, 843, 844, 846, 847, 876–80, 883–9, 903–23.

[256] Of course, additional information might have been displayed in painted letters or was provided by lead tags attached to blocks, cf. Antium and Tivoli, App. nos. 925–7.

at Simitthus would consequently have to be seen as public property, at least in legal terms—a view corroborated by the likely presence of condemned convicts at this site.[257]

To return to Hadrian's benefactions for Smyrna, the hypothesis that the 'Numidian' marble quarries at Simitthus and perhaps other quarries were public property would have had only minor legal implications for the emperor's benefactions to Italian or provincial cities. By analogy with the procedure attested for the exemption of taxes, Hadrian either bought the Numidian columns through the *fiscus Caesaris* directly from Simitthus (hence, the indication of his name on marble blocks), or initiated the decision of the senate to provide Smyrna with Numidian columns from its quarries in Africa proconsularis.[258] This decision may have been included in the *senatus consultum* awarding Smyrna the second *neokoros* title and exemption from taxation.[259] For the Smyrneans, however, these complexities of procedure were hardly relevant. It was Hadrian who ultimately provided the precious columns, the tax cuts, and the *neokoros*-title. In their eyes, the role of the Roman senate was marginal.

3.3.5. The Emperor, Ingots, and Public *Metalla*

The emperor could own mines or quarries in a 'private' capacity as he owned vast tracts of land in Italy and the provinces. The prospects of identifying individual mines or quarries owned by the emperor vis-à-vis public extractive ventures are poor, since written evidence on the issue remains scarce. Ulpian records an edict of Antoninus Pius concerning the punishment of thieves of silver or gold from *metalla Caesariana*.[260] The plural *metalla Caesariana* suggests that more than one silver or gold mine was Caesar's property. As the term 'metallum' without an adjective is regularly used throughout the Corpus Iuris Civilis, the employment of the term 'metalla Caesariana' perhaps designated a certain category of extractive operations as the private property of the Roman emperor.[261]

[257] Perhaps originally owned by the Numidian kings, the quarries probably became public property in the aftermath of the Iugurthine War, cf. Lassère 1980: 31, 40 f. with fn. 9; Rakob 1997: 2 f.

[258] Alpers 1995: 138.

[259] *IK* 24/1, 697, ll. 33–42. The *aerarium* could supply financial funding for public building projects in cities, cf. Suetonius, *Claud.* 25.5, and Horster 2001: 210 with fn. 7. Given the involvement of the *aerarium* in public building projects throughout the empire, it is possible that the Roman senate, probably on the initiative of the emperor alone, could occasionally provide or allow access to building material instead of financial funding.

[260] *Dig.* 48.13.8.1; 48.19.38.pr.

[261] Cf. Millar 1992: 182.

Further evidence for mines owned by the princeps is thought to be provided by inscribed lead ingots. Mould marks set in a panel on the face of lead ingots occasionally display the name of the emperor, as a number of ingots found in Sardinia, Italy, Britain, and Gaul show:

Rena Maiore	*Augusti Caesaris Germanicum*	*AE* 2000: 653
Fluminimaggiore	*Caesaris Aug(usti)*	*CIL* X 8073/1, p. 1002
Rome	*Caesaris Aug(usti)*	*CIL* XV 7914
St. Valéry sur Somme	*Neronis Aug(usti) Britan(nicum) l(egio) (secunda)*	*RIB* II 2404.24
Charterhouse	*Imp(eratoris) Vespasiani Aug(usti)*	*RIB* II 2404.11
Richborough	*Imp(eratoris) Nervae Ca[es(aris)]*	*RIB* II 2404.65
Fluminimaggiore	*Imp(eratoris) Caes(aris) Hadr(iani) Aug(usti)*	*CIL* X 8073/2
Pistis	*Imp(eratoris) Caes(aris) Hadr(iani) Aug(usti)*	*AE* 1991: 902a
Cheshunt	*Imp(eratoris) Caes(aris) Hadriani Aug(usti)*	*RIB* II 2404.66
Bath	*Imp(eratoris) Hadriani Aug(usti)*	*RIB* II 2404.14
Aston/Snailbeach/Snead	*Imp(eratoris) Hadriani Aug(usti)*	*RIB* II 2404.29–30
Charterhouse/Bristol	*Imp(eratoris) Caes(aris) Antonini Aug(usti) P[ii] p(atris) p(atriae)*	*RIB* II 2404.16–18
Bruton/Wells/ Charterhouse	*Imp(eratorum) duo(rum) Aug(ustorum) Antonini/ et Veri Armeniacorum*	*RIB* II 2404.19–22
Sassenay	*[Imp(eratoris) Caes(aris) L(ucii) S(eptimii) Severi Pertinacis] Aug(usti) Part(h)ici Adiabenici*	*RIB* II 2404.72a
Lillebonne	*Ị[mp(eratoris) Ca]ẹ[s(aris) L(ucii) S(eptimii)Severi Perti]nacis Ạug(usti) Part(h)ici Adiabenici*	*CIL* XIII 3222

The name of the emperor is given in the genitive case, a *genetivus possesivus.* In some instances the term *Germanicum* or *Britan(nicum)* is added (cf. *AE* 2000: 653; *RIB II* 2404.24), apparently as the adjective attribute to *plumbum,* lead. This is confirmed by the moulded inscriptions on lead ingots from the shipwreck 'Saintes-Maries-de-la-Mer 1' which render *[L.] Flavi(i) Veruclae plumb(um) Germ(anicum).*[262] Consequently, the emperor is primarily identified by the moulded inscription as owner of the lead ingot. These mould marks originate

[262] Rothenhöfer 2003.

from dies set in a trough-shape form in which the lead was cast following roasting and heating, thus separating the elementary lead from the galena (lead ore or lead sulphide). This smelting process of lead ore is assumed to have taken place in the immediate vicinity of the mines. The fact that at least part of the lead produced was allocated to the emperor near the mine gives the impression that the lead mines were directly owned by the emperor. Yet, in the light of observations made above, other alternatives are possible. A number of lead ingots provide a precise indication of origin for the metal in question:

RIB II	Face (moulded in panel)	On sides of ingot
2404.3	*Neronis Aug(usti) ex k(alendis) Ian(uaris)*	Back (stamp): *ex argent(ariis)*
2404.4-10	*Imp(eratoris) Vespasian(i) Aug(usti)*	Front (moulded): *Brit(annicum) ex arg(entariis) Veb(-)*
2404.36	*Imp(eratoris) Domit(iani) Aug(usti) Ger(manici)*	Front (moulded): *Deceang[l(icum)]*
2404.13	*Imp(eratoris) Vesp(asiani) Aug(usti) VIIII Brit(annicum) ex ar(gentariis)*	
2404.39	*Imp(eratoris) Caes(aris) Hadriani Aug(usti) met(alli) Lut(udarensis)*	
AE 1999: 1683	*Imp(eratoris) Domit(iani) Aug(usti) Ger(manici)*	Back (moulded): *met(alli) Dard(anici)*
Piso 2005: fig. 2	*Imp(eratoris) Tr(aiani) me(talli) Ulp(iani)*	Stamps: *C. Ulp. D.S.*

Besides indication of ownership and general origin of the ingot e.g. *(plumbum) Britannicum*, moulded inscriptions can render the source of the lead as *ex argentariis Veb(-)* or *metalli Lutudarensis*, 'from the silver-mines of Veb(-)' or 'from the Lutudarensian mines', simply *Deceanglicum* for Deceanglian lead, or *metalli Dardanici*, for Dardanian mines. In none of these instances is an indication of ownership by the emperor of the mines in question provided.[263] This stands in contrast to one, perhaps even two, moulded inscriptions on

[263] Moulded on the face of ingots from Flintshire and Yorkshire in Roman Britain, the names of the emperors Vespasian, Titus, and Domitian are given in the ablative, as consular dates, i.e. *Imp(eratore) Vesp(asiano) V Aug(usto) T(ito) Imp(eratore) III co(n)s(ulibus)*. The origin of the lead, *Deceangl(icum)* or *Brig(anticum)* is indicated in mould marks on the front of the ingot, *RIB* II 2404, 31–35, 61 f. Perhaps one might see these lead ingots as a yearly tribute owed to the Roman state, cf. Tacitus, *Agr.* 31.2, 32.4 with Neesen 1980: 57. The mould mark *RIB* II 2404.3, *Neronis Aug(usti) ex k(alendis) Ian(uaris)* may be interpreted in the same fashion.

lead ingots from Moesia Superior, on which the emperor's ownership of a mining venture was clearly marked:

AE 1978: 705 = *IMS* IV 136 (Žuč): Face (moulded): *Metallo Imp(eratoris) Aug(usti)*
AE 1978: 706 = *IMS* IV 135 (Jasenovik): Face (moulded): *[Metall]o [Cae]saris Aug(usti)*

Hence, lead ingots originating from a mine owned by the emperor were labelled in accordance with a moulded inscription on the face of the ingot. This adds further weight to the statement that ingots merely displaying the emperor's name must not necessarily originate from mines owned privately by the emperor.

Provided this line of argument is valid, the question arises in whose ownership the mines were from where the emperor derived the ingots. The following inscription may provide a clue:

St Valéry sur Somme NERONIS AVG. BRITAN. L. II *RIB* II 2404.24

The editors of *RIB* II read the abbreviated inscription as follows: *Neronis Aug(usti) Britan(nicum) l(egio) (secundo)* and translate this as '(property) of Nero Augustus, British lead, the second legion (made this)'. Provided the reading and translation are accurate, this would render evidence for the Roman army running extractive operations in the Mendips for a certain period, perhaps with most of the produce intended for military use. As has been argued above, the Roman army was probably not seconded to the private works of the emperor.

The cast moulds for gold ingots discovered on the Magdalensberg with the imperial title *C(ai) Caesaris Aug(usti) Germanici Imp(eratoris) ex Noric(is)* highlight an additional problem.[264] The known sources of gold as recorded by Gernot Piccotini are 30 km or more distant from Magdalensberg and mostly appear to lay outside the *patrimonium regni Norici* outlined by Alföldy (cf. 3.2.2). The gold or gold ore from these primary and secondary deposits must have first been transported over a considerable distance to Magdalensberg and then smelted and cast into ingots. The possibility of metal being cast or even recast away from the mines further complicates any attempt to identify the legal context under which the metal passed into the ownership of the emperor. Consequently, caution is advised in drawing any conclusions regarding the ownership of mines on ingots inscribed with the name of the emperor.

Occasionally, the name of the emperor appears in abbreviated form as a stamp on the ingot. The stamp is applied cold after the ingot has been cast. Most, if not all ingots stamped with the emperor's name are from shipwrecks in the western Mediterranean:

[264] Contra Piccottini 1994*a*: 471–3; Piccottini et al. 2002: 103.

	Mould mark on face	*Stamps*
Saintes-Maries-de-la-Mer 1, a)	[.] FLAVI VERUCLAE PLUMB. GERM.	IMP.CAES.
Saintes-Maries-de-la-Mer 1, b)	no mould mark	IMP.CAES.; L.FL.VE.; EROTIS
Cabrera 6	[P.A]EMILI GALLICI	IMP. CAES.; AVG.
	N.MEVI APRI	IMP. CAES.; AVG.
	Q. CORNVTI	VESP. AVG; AVG.
	L. MANLI	VESP. AVG; AVG.
	C M A	IMP. CAES.; AVG.
	SO VR	VESP. AVG; AVG.
Rena Maiore	AVGVSTI CAESARIS GERMANICUM	L. VAL. RUF; IMP; CHI

The lead ingots from the shipwreck off the coast of Southern France near Saintes-Maries-de-la-Mer are of particular interest. The ingots can be divided into two types: eight of them carrying the mould-mark [.] FLAVI VERUCLAE PLUMB GERM are stamped IMP. CAES. and incised with numerals. Ninety-one ingots do not have a mould-mark, but the stamp of L.FL.VE. for *L(ucii) Fl(avii) Ver(uclae)* appears first on the ingot, followed by the stamp EROTIS and finally IMP.CAES., the latter perhaps simultaneous to the numeral incisions.[265] Recent studies have revealed that these lead ingots probably originated from mines in the Sauerland near Brilon, some 130 km east(!) of the Rhine and were thus designated *plum(bum) Germ(anicum)*.[266] The probable origin and the reading of IMP. CAES. as the titulature of Augustus allows to narrow down the date of production of these ingots to the period 8 BC to AD 9.[267] It was assumed that L. Flavius Verucla was the lessee of mines owned privately by Augustus, as the stamps IMP. CAES. are read as *Imp(eratoris) Caes(aris)*, therefore indicating the ownership of the ingot by the latter. The ingots stamped in this manner have been interpreted as the rent in produce owed to the emperor.[268]

Two observations, however, question this plausible reconstruction. Provided the lead was owned by Caesar, one would expect to find his name in the genitive as a mould mark on the face of the ingot. This is the case with forty-two ingots salvaged from Rena Maiore/Aglientu, a shipwreck off the coast of

[265] Long & Domergue 1996: 818, 832–4.

[266] Peter Rothenhöfer (2003: 178 f., 280 with fn. 15) published a fragment of a lead ingot discovered in Bad Sassendorf-Heppen with the cast mould-mark inscription L FLA[-] on the face and the stamp with the letters L.F.VE. Based on the analysis of lead isotopes, he argued that the ingot from Bad Sassendorf-Heppen and the ingots from the shipwreck of Saintes-Maries-de-la-Mer had the same origin and that the inscription, by analogy with mould-marks from the shipwreck, should be read as 'L. Flavius Verucla'; see also Trincherini et al. 2001.

[267] Eck 2004*a*: 74–6.

[268] Eck 2004*b*: 19–21.

Sardinia which display the name of the emperor in full, *Augusti Caesaris (plumbum) Germanicum*. These may also have come from Brilon in the Sauerland.[269] The product of smelting intended for the emperor thus could be cast as ingots displaying the imperial titles. Hence, the application of the emperor's stamp to ingots mould-marked with the name of a private individual require an alternative explication. Based on the superimposition of stamps, nail holes and incised numerals on ingots, Claude Domergue was able to show that stamps could be applied to the ingots well after they had been cast.[270] In the case of the ingots salvaged from Saintes-Maries-de-la-Mer IMP. CAES. was stamped on after all other imprints and perhaps simultaneously with the incision of numerals. On ingots from other shipwrecks the numerals were regularly incised after transportation on river boats, probably to indicate deviation from the standard weight (100 or 200 Roman pounds) for the calculation of payable duties or *portoria*.[271] The application of a stamp IMP. CAES. thus must have taken place in this context. The stamp most likely indicated that the ingot came into possession or ownership of the emperor and thus its export was free of duty. Whether these ingots had been bought or requisitioned before, during, or after weighing, or whether they were detached from exported cargo as a payment of *vectigal*, *portorium*, or tribute in kind, can not be answered conclusively.

In the light of the previous discussion one may tentatively argue that mould marks or stamps on lead ingots displaying the name of the emperor in the genitive case do not mean that the emperor owned the mines from which the ingots originated.[272] Of course, the strictly legal differentiation between the property of the Roman state and the property of the Roman emperor was irrelevant when it came to actual control. Most important mines and quarries were managed by imperial officials, that is, members of the *familia Caesaris* and equestrian administrators appointed by the emperor. Moreover, the emperor not only directly controlled his own *fiscus Caesaris*, but could indirectly redistribute the public monies in the *aerarium* through the Roman senate, *de facto* controlling public resources as well. In fact, the observance of these legal distinctions was entirely at the discretion of the emperor, for only he and his Palatine bureaux were able to compile the *rationes* from the *provinciae Caesaris* and provide a balance sheet of revenues

[269] Riccardi & Genovesi 2002: 1319–23; Eck 2004*a*: 76; Hanel & Rothenhöfer 2005.

[270] Domergue 1998: 202 f., 206–9.

[271] Domergue 1998: 203, 215 (table 2).

[272] For what purpose the emperor obtained lead ingots can only be guessed at: the use of *fistulae* (lead pipes) in public building projects in Rome, Ostia and throughout Italy or the provinces certainly increased the demand for lead. Given the considerable amount of construction work paid for by the emperor's private treasury (cf. Millar 1992: 189–201; Duncan-Jones 1994: 41 f.; Drexhage, Konen, & Ruffing 2002: 55), the purchase of lead by imperial agents in order to supply the building sites at Rome is plausible.

and expenditures for the whole empire—a *breviarium totius Imperii*.[273] Despite the complete control exercised by the emperor over the financial matters of the Roman state, the legal division between *aerarium* and *fiscus Caesaris* was strictly observed by the majority of Roman emperors during the first century AD and certainly was seen as a trait of good governance, as becomes clear in Suetonius' or Tacitus' histories.[274] These legal differentiations were still respected by Trajan and other emperors of the second century, and we have no reason to suspect that they were not observed on the provincial level.[275]

3.4. SUMMARY

Imperial mining and quarrying districts and territories held a special position amongst the territorial entities within a province. The bronze tablets from Vipasca clearly outline the jurisdiction powers the mining procurator held within the district. Given the use of the term *territorium* and the regulations meted out for the Vipascan district, this suggests the relative autonomy of these districts comparable to colonies or municipalities. Moreover, in Noricum and other Danubian provinces mining zones could be regarded as territorial exclaves in relation to the Illyrian customs district; and the metals extracted there and transported out of the mining districts were subject to a toll or tax. As to whose property the mines and quarries under imperial control were, a number of indications suggest that perhaps most of these extractive operations *strictu sensu* belonged to the Roman state. Quarries like Mons Claudianus, Mons Porphyrites, Tiberiane, and Mons Ophiates in Roman Egypt, Dokimeion in Asia Minor, Simitthus in North Africa, and Karystos on Euboea were probably public property under the emperor's control rather than private domains of the emperor. Similarly, the mining areas under imperial and military control in north-western and southern Spain, Britain, Sardinia, Gaul, Noricum, Pannonia, Dalmatia, and Moesia were mostly state-owned as well. Although the *Digest* mentions *metalla Caesariana*, the survey of imperial extractive operations documented in the provinces indicates that there is insufficient evidence for directly identifying patrimonial mines or quarries. Apart from extractive operations under imperial control, there is occasional written evidence for private and, perhaps, municipal ownership.

[273] Alpers 1995: 259–63; Speidel 2000*a*: 115–18.

[274] Corbier 1974: 682–9; Brunt 1990*a*: 160 f.; Alpers 1995, *passim*; Wolters 1999: 174–202. Only Domitian appears to have 'illegally' diverted a *vectigal* connected with water-rights from the *aerarium* to the *fiscus Caesaris*, cf. Frontinus, *De aquae ductu* 2.118 with Alpers 1995: 243 f.

[275] For Trajan, cf. Bennett 1997: 124 f. The creation of *res privata* as opposed to the *patrimonium* probably had no effect on the general distinction.

4

Imperial Officials and Extractive Operations

After reviewing the available evidence in an attempt to define imperial quarrying and mining 'districts', their configurations and legal status, as well as the organizational challenges which faced the Roman authorities, it is necessary to examine the officials who supervised imperial mines and quarries. These officials were usually members of the *familia Caesaris*, equestrians in the emperor's service, military officers, and possibly even civilian engineers.

4.1. PROCURATORS

4.1.1. Quarrying Procurators in Roman Egypt

The management of the stone quarries at Mons Claudianus, Mons Porphyrites, and Mons Ophiates was the responsibility of one *ἐπίτροπος τῶν μετάλλων* / *procurator metallorum*. Evidence for this assumption is provided by various Greek and Latin inscriptions on stone, as well as letters and notes on ostraca discovered at these quarries: Epaphroditos Sigerianos, imperial slave (δοῦλος) and μισθωτὴς τῶν μετάλλων, had an oblong architrave inscribed, commemorating the building of a temple for Zeus Helios/Sarapis and other gods. The inscription lists Rammius Martialis as prefect of Egypt (AD 117–19), Chresimus (a *libertus Augusti*) as ἐπίτροπος τῶν μετάλλων, and Avitus, a *centurio* of the *cohors I Flavia Cilicum equitata*, as head of the quarries at Mons Claudianus.[1] This list of officials is repeated in a further building inscription from the reign of Hadrian—this time on a temple for Zeus Helios Sarapis at Mons Porphyrites: the same Epaphroditos, μισθωτὴς τῶν μετάλλων, lists Rammius Martialis as *praefectus Aegypti*, a Proculeianus as *centurio* in charge, and Marcus Ulpius Chresimus as ἐπίτροπος.[2] Given that Chresimus is named at Mons Claudianus and Mons Porphyrites in charge of

[1] *I.Pan.* 42 = *CIG* III 4713 f = *IGRR* I/5 1255; for Epaphroditos as imperial slave, cf. *I.Pan.* 21 = *CIG* III 4713 = *IGRR* I/5 1256 (Mons Porphyrites).

[2] *I.Pan.* 21 = *CIG* III 4713 = *IGRR* I/5 1256.

the quarries, his procuratorial function entailed the oversight over both quarries. Additional evidence stems from Mons Claudianus where an Ulpius Himerus is identified as ἐπίτροπος in two unpublished ostraca. Both ostraca are dated by another letter, sent in AD 152/3 to Athenodoros, the *tabularius* of Ulpius Himerus.[3] Perhaps the same Ulpius Himerus, *p[rocurator Augus]ti* is documented in an inscription at Mons Ophiates (AD 150–4).[4] Thus, the procuratorial district of the ἐπίτροπος τῶν μετάλλων must have contained the quarries at Mons Claudianus, Tiberiane, Mons Porphyrites, and Mons Ophiates (cf. 3.2.1).

The epigraphic record of the Egyptian Eastern Desert yields further names of quarrying procurators: perhaps the earliest known ἐπίτροπος recorded at Mons Claudianus is Encolpius. He is named together with the current holders of the offices relevant to Mons Claudianus such as Quintus Accius Optatus, the *centurio,* on an inscribed altar dating to Trajan's reign.[5] The published *ostraca* at Mons Claudianus complete the picture: an Encolpius is mentioned on one of the ostraca while in another fragmentary text an ἐπίτροπος is said to have been notified of the number of camels and the water-supply.[6] Two further *procuratores* appear to be documented in further ostraca from Mons Claudianus: a certain Probus is the addressee in a number of letters sent either by quarry workers and smiths, or by the *praefectus Aegypti* Pomponius Faustianus (AD 185–7); and a Tertullus, ἐπίτροπος (τῶν μετάλλων?) is one of the addressees in a recently published ostracon.[7] A further procurator, Anokanos, emerges from a votive inscription at Mons Porphyrites in AD 137/8.[8] Additional evidence for an ἐπίτροπος was discovered close to Fatireh el-Beida and Abu Zawal on the road from Mons Claudianus to the Nile Valley. A Latin inscription of six lines was engraved between AD 103 and 116, and names Flavius Diadumenus as *procurator Augusti.* The proximity of the inscription to the Mons Claudianus quarries implies that Diadumenus was a *procurator metallorum* as well.[9] The subsequent list of procurators consists mostly of imperial freedmen, as the *nomina gentilia* indicate.[10]

[3] *O.Claud.*inv. 4739, 7737, cf. Cuvigny 1996*c*: 96 f.

[4] *I.Pan* 53 = *AE* 1952: 248 (AD 150–154; Wadi Semna), with Cuvigny 1996*c*: 91–101.

[5] *I.Pan* 38 = *CIG* III 4173[e] = *IGRR* I/5 1254 = *SEG* XV 865 (The altar was commissioned by Apollonios, the ἀρχιτέκτων, and dedicated to Zeus Helios/Sarapis).

[6] Encolpius: *O.Claud.* 19; unknown ἐπίτροπος, cf. *O.Claud.* 380.

[7] *O.Claud.*inv. 7218, 7295, cf. Cuvigny 2002: 242 f., 245–8.

[8] Maxfield & Peacock 1998*a*: 190.

[9] Sidebotham 1996: 190.

[10] Encolpius and Anokanos were probably freedmen, even though we lack substantial evidence to support this view. The *procuratores* attested during Commodus' reign, Probus and Tertullus, do not bear slave-names or *cognomina* derived from Greek slave-names and therefore, unlike their predecessors, might not be imperial freedmen originating from the eastern half of the empire. There is no evidence to suggest that they were members of the *ordo equester*: the vice-curator Rufus does not address these procurators with their full name, as he does in the case of Vibius Alexander, the equestrian *praefectus,* cf. Cuvigny 2002: 245 f.

Fatireh el -Beida	*Mons Claudianus*	*Mons Porphyrites*	*Mons Ophiates*
Flavius Diadumenus, AD 103/116	Encolpius, AD 98–117		
	Chresimos, AD 118	M. Ulpius Chresimus, AD 117/19	
		Anokanos, AD 137/8	
	Ulpius Himerus, AD 152/3		Ulpius Himerus, AD 150–4
	Probus, AD 185/7		
	Tertullus, AD 189		

4.1.2. The *μεταλλάρχης* and the *procurator montis Berenicidis*

Prior to the *ἐπίτροπος τῶν μετάλλων*, the control of the *metalla* in the Egyptian Eastern Desert, was, at least for a short period of time, in the hands of an *ἀρχιμεταλλάρχος* or *μεταλλάρχης*. This office is known from two inscriptions found in Wadi Umm Wikala (Mons Ophiates) and Wadi Hammamat. According to the inscription at Wadi Umm Wikala, a sanctuary was dedicated in AD 10/11 to Pan by Publius Iuventius Agathopus, 'procurator, administrator, benefactor' of all *metalla* in Egypt, and freedman of Publius Iuventius Rufus, himself tribune of the *legio III Cyrenaica*, commander (*ἔπαρχος*) of Berenike, *ἀρχιμεταλλάρχος* of the emerald mines, topaz mines, pearl (fisheries), and all mines of Egypt.[11] The Wadi Hammamat inscription is similar to the Wadi Umm Wikala inscription (with slight differences), but was inscribed some years later on 2 October, AD 18. Again, Publius Iuventius Agathopus, freedman and *προνοητὴς πάντων* (general administrator) for Publius Iuventius Rufus, the *μεταλλάρχης*, of the emerald mines, the topaz mines, the pearl(-fisheries), and of all quarries in Egypt, sets up a shrine for Pan in the name of his master.[12]

We do not appear to have any further attestation of an *ἀρχιμεταλλάρχος* or *μεταλλάρχης* in the papyri and inscriptions of Roman Egypt, nor are we able to delimit precisely the function of the office-holder. Based on the inscriptions mentioned above, Hirschfeld and Dubois thought the *ἀρχιμεταλλάρχος* /

[11] Tholemaios, *curator* of the *cohors Flori*, of the centurionate of Bassus, added his *προσκύνημα* to the inscription, as did the two architects, Mersis and Soter. *I.Pan* 51, cf. Sidebotham et al. 2001: 139.

[12] The inscription lists other leading figures involved in the quarrying procedures at Wadi Hammamat such as the *γραμματεύς* Aryothes, Mersis, the *ἀρχιτέκτων*, and Mammoyais, soldier of the *coh Nigri*, garrisoned at the (quarry) works, cf. *OGIS* II 660 = SB 8580 = *I.Ko.Ko.* 41 (with further commentary). Four years earlier Publius Iuventius Agathopus also had a *προσκύνημα* (adoration) to Pan carved at Wadi Hammamat on 25 October in AD 14, cf. *I.Ko.Ko.* 39 = *SB* 8579.

μεταλλάρχης was in charge of all mines and quarries in Roman Egypt.[13] Whether this actually was the case is difficult to assess. From AD 10/11 to AD 18 the ἀρχιμεταλλάρχος/μεταλλάρχης P. Iuventius Rufus had at least the quarries at Wadi Hammamat and Mons Ophiates under his control and delegated supervisory tasks to his freedman P. Iuventius Agathopus. Whether Agathopus managed both quarries simultaneously or consecutively and whether or not his responsibilities included other quarries remains unknown. The inscriptions of Wadi Hammamat and Wadi Umm Wikala show that Iuventius Rufus' range of tasks as ἀρχιμεταλλάρχος/μεταλλάρχης certainly included further economic ventures: emerald/beryl was mined at Mons Smaragdus which is probably to be located in the area of Wadi Gemal, Wadi Nuqrus, Wadi Sikait, Gebel Zabara, and Gebel Umm Kabu.[14] The archaeological remains at these sites, mainly pottery shards, suggest mining operations from the Roman, if not Ptolemaic era, up until the seventeenth century AD. At Wadi Sikait a mining settlement is attested of which, so far, a precise rendition of its topography has not been provided. An inscription from the year AD 253/4 perhaps documents the presence of a workforce there. However, apart from its proximity to a mine the epigraphic evidence does not directly provide any pointers for the profession of the inscribers.[15] Once more, ostraca provide additional insights: a recently published ostracon discovered at Didymoi, a fortlet on the Koptos–Berenike road, names an emerald worker as a carrier of messages. According to Adam Bülow-Jacobsen, emerald workers occasionally transmitted letters, as they appear to have travelled on a regular basis between Koptos and Senskis/Sikkayt at Mons Smaragdus.[16] Besides the emerald mines, topaz mines were perhaps located on the Zabargad Island in the Red Sea, probably together with the pearl fisheries. The amethyst mines at Gebel Abu Diyeiba and Wadi al-Hudi may have come under the supervision of P. Iuventius Rufus as well.[17] Pliny the Elder claims that many quarries in Egypt had been opened under Augustus and Tiberius, and the archaeological surveys so far seem to support the view that the main extractive operations carried out in the Egyptian desert were nearly exclusively stone quarries.[18] This might be reflected in a slight change in title of P. Iuventius Rufus. In the

[13] Hirschfeld 1905: 172; Dubois 1908: 69. For a different view, cf. Fitzler 1910: 126, 128.

[14] Shaw, Bunbury, & Jameson 1999: 205 f.

[15] *I.Pan* 69.

[16] *Σμαραγταριοί*: *O.Did.*inv. 329, 484, cf. Bülow-Jacobsen 2001: 121, l. 5 with commentary.

[17] Meredith 1957; Shaw & Jameson 1993: 83–7; Sidebotham, et al. 2001: 140 f., with fns. 14, 15.

[18] Pliny, *NH* 36.55. Attention has been drawn to the fact that no special mention of gold mines is made in either of Iuventius' inscriptions. Although gold was extracted in the Eastern Desert, the exploitation was marginal compared to the 'Ptolemaic' and 'Early Arab' periods, and hence might have been of minor importance during the 'Roman/Byzantine' era, cf. Klemm &

Wadi Umm Wikala inscription of AD 10/11 Iuventius Rufus carries the title of *ἀρχιμεταλλάρχος τῆς ζμαράγδου καὶ βαζίου καὶ μαργαρίτου καὶ πάντων τῶν* **μετάλλων** *τῆς Αἰγύπτου*, while in AD 18 his title has been altered to *μεταλλάρχης τῆς ζμαράγδου καὶ βαζίου καὶ μαργαρίτου καὶ* **λατόμων** *πάντων των τῆς Αἰγύπτου*. The replacement of the term *μέταλλα* (a generic term for extractive operations) with *λατόμ(ει)α* (quarries) might describe the nature of extractive operations throughout the Egyptian Eastern Desert more precisely.[19]

Hélène Cuvigny has recently maintained that P. Iuventius Rufus held the post of *μεταλλάρχης* in conjunction with the office of *praefectus montis Berenicidis*.[20] While the Greek inscription of Wadi Umm Wikala names P. Iuventius Rufus as tribune of the *legio III Cyrenaica*, as *praefectus Berenicidis*, and as *ἀρχιμεταλλάρχος* (connected by the conjunction *καί*), the later Wadi Hammamat inscription only identifies him as *μεταλλάρχης*. Given the size of the job at hand, the office of *μεταλλάρχης* might have been held separately, after Iuventius Rufus had obtained the tribunate in *legio III Cyrenaica* and the Berenician prefecture. Ostraca and inscriptions show, however, that the Berenician prefecture was often held jointly with a military command over an auxiliary unit.[21] The exact tasks of the Berenician prefect are debated. Based on the honorific inscription for D. Severius Severus, set up in Solmona/mod. Sulmo in Italy and labelling him as *praefectus praesidiorum et montis Bero-nices*(sic), Kurt Fitzler argued that Severius Severus was the official in charge of a *Mons Berenicides*. Fitzler located the toponym Mons Berenicides at Gebel Zabara, and the *praesidia*, military stations, along the road from Koptos to Berenike. Echoing Fitzler, Lesquier emended the *ἐπίτροπος ὄρους* named in I. Portes 86 (AD 219) to *ἐπίτροπος ὄρους* (*Βερενείκης*); he thought this official was a procurator controlling extraction procedures not only at Gebel Zabara but at all quarrying (and mining) sites in the desert of Berenike. He saw this *ἐπίτροπος ὄρους* as a subaltern official answerable to the

Klemm 1994: 211, with fig. 6 (with wrong caption); Sidebotham et al. 2001. The recent excavations and surveys at Wadi Umm Fawakhir suggest that, despite the low yield of 2–3 gm of gold per metric ton of ore, its exploitation was pursued with vigour during the 5th and 6th cents. AD, supposedly as a result of a high demand for gold during that period (Meyer 1995: 199). This might be symptomatic of most Roman/Byzantine sites in the Egyptian Eastern Desert, which were possibly opened during Late Antiquity rather than the first to third centuries AD, when gold was mostly provided by other more abundant sources (Spain, Dacia).

19 Bernand 1972: 84 f.

20 Cuvigny 2002: 240.

21 *O.Did.*inv. 733; *I.Memnon* 14, cf. Cuvigny 2001: 172, l. 3 with commentary; Bülow-Jacobsen & Cuvigny 2007: 23.

ἐπίτροπος (...) *πάντων τῶν μετάλλων τῆς Αἰγύπτου*, the title P. Iuventius Agathopus bears in the Wadi Umm Wikala inscription.[22]

However, the translation of the Latin *mons* or Greek *ὄρος* as 'mountain' appears to have been misleading. As Hélène Cuvigny has shown, the terms *mons*/*ὄρος* do not necessarily designate a specific mountain or location, but may be better understood as referring to a more extensive area, such as a mountain ridge or desert plateau. Based on the ostraca found in the forts and fortlets along the route from Koptos to Myos Hormos or Berenike among other evidence, Cuvigny has shown that the *praefectus montis Berenicidis* or *ἔπαρχος ὄρους Βερενίκης* was a territorial command held by an equestrian officer. The prefect commanded the auxiliary units scattered between Koptos and Berenike although he himself was probably based at Koptos in the Nile valley (despite what his titulature may indicate).[23] Similarily, the *ἐπίτροπος ὄρους* (*Βερενείκης*) reported by Lesquier—which he thought was a *procurator metallorum*—was not in charge of mining operations, but might have replaced the *praefectus montis Berenicidis* (last attested in AD 216) for a short period.[24] Alternatively, the titles of *praefectus/ἔπαρχος* and *procurator montis (Berenicidis)/ἐπίτροπος ὄρους* (*Βερενείκης*) could be explained as derivates from an official title similar to 'procurator Augusti et praefectus montis Berenicidis'.[25]

It must remain unanswered whether the monitoring of quarrying operations was part of the functions of the procurator and prefect. The *praefectus/procurator montis Berenicidis* is continuously attested from the early first to the early third century AD, whereas the office of *μεταλλάρχης* is known only for the late Augustan/early Tiberian period. Perhaps the latter function was incorporated in the regular tasks of the Berenican prefecture?[26] The Greek ostraca which have been found during surveys and excavations alongside the Koptos–Myos Hormos road have demonstrated the presence of army detachments at the *praesidium* of Persou at Wadi al-Hammamat, known also for its quarrying activity of the so-called '*bekhen*' stone during the Roman period. These *praesidia* and their small garrisons stood under the command of the Berenician prefect, which perhaps suggests a certain responsibility, at least in military terms, for the quarries and mines along the Koptos–Myos Hormos route.[27] However, a direct connection between the prefect and the quarrying

22 *ILS* 2699 = *CIL* IX 3083 (Severius Severus); Fitzler 1910: 100 f., 130 f.; Lesquier 1918: 153 fn. 2, 240 with fn. 4.

23 Bülow-Jacobsen, Cuvigny, & Fournet 1994; Bülow-Jacobsen 1998; Cuvigny 2003*b*: 295–305.

24 Cuvigny 2000*b*: 416 f.

25 Bülow-Jacobsen & Cuvigny 2007: 23 f.

26 Hirschfeld 1905: 471 f.; Fitzler 1910: 128; Bülow-Jacobsen & Cuvigny 2007: 21 with fn. 20.

27 Cuvigny 2003*a*: 280–4.

operations at Wadi Hammamat or Wadi Umm Wikala/Mons Ophiates can not be established.

4.1.3. Quarrying Procurators in Roman Asia and Achaia

An inscription for the *procurator provinciae Frygiae* and *libertus Augusti*, M. Aurelius Marcio, found at Tricomia/mod. Kaimaz (Turkey) lists his previous procuratorial posts, which include the office of *procurator marmorum*. Aurelius Marcio reached the office of a *procurator marmorum* after having served as a *proximus rationum*, and before continuing his career as *procurator provinciae Britanniae, procurator summi choragi*, and, finally, *procurator provinciae Frygiae.*[28] *Proximus rationum* or *proximus a rationibus* was the highest clerical rank an imperial freedman could achieve within the financial administration at Rome, and was usually the starting point for a string of procuratorial posts.[29] Marcio was next appointed to the position of *procurator marmorum*. Hirschfeld and Strubbe argued that this post was the procuratorship of the imperial quarries at Dokimeion.[30] Of course, the omission of any geographical indication following the title *procurator marmorum* may be explained by the relative proximity of the quarries at Dokimeion, assuming the potential readers of the inscription knew what was meant. However, it is problematic that the inscription at Tricomia was commissioned by a *tabularius* named Senecius Collid(us?) in honour of his superior, the *procurator provinciae Frygiae* Aurelius Marcio.[31] Given the function of *proximus rationum* at Rome preceded the post of *procurator marmorum*, it appears likelier that Aurelius Marcio held the latter office at Rome as well. This is corroborated by the epigraphic record attesting further such office holders at Rome (cf. 8.2.2.2.).[32]

Few procurators are known from inscribed labels on quarried blocks and inscriptions found in Phrygia: an otherwise unknown *procurator* Maro appears only in label inscriptions on *pavonazetto* in Bacakale.[33] Furthermore, a *procurator* Irenaeus is documented on *pavonazetto* blocks transported to

[28] *CIL* III 348 = *CIG* 3822 (Tricomia/Kaimaz), cf. Corsten 1997: 89, No. 43.

[29] Weaver 1972: 224, 252–8.

[30] Hirschfeld 1905: 170 fn. 5; Strubbe 1975: 244 fn. 60.

[31] For further inscriptions of an Aurelius Marcio, cf. *MAMA* IV 4 (Afyon). *IGRR* IV 704 (Synnada).

[32] Hirschfeld 1905: 175 with fn. 4.

[33] App. nos. 302, 303, AD 194. A white marble block found in Rome, presumably originating from the white marble quarries at Bacakale, names an Aur(elius) Epity(nchanus?) following a barely readable 'sub cura' mark (App. no. 437). However, the inscription is too fragmentary to ascertain whether or not Aurelius Epitynchanus should be added to the list of procurators at Dokimeion.

Rome.[34] His namesake appears in an inscription at Ginik in Phrygia where he had set up a boundary stone in his capacity as a *procurator.*[35] Provided these Irenaei are one and the same person, this procurator's involvement in boundary disputes suggests that he held authority beyond the mere supervision of the Dokimean quarries.[36] Both Maro and Irenaeus were regarded not as *procuratores marmorum* but *procuratores provinciae Frygiae.*[37] The *provincia Phrygia* was not a 'province' in its own right but a district originating from the unification of several imperial territories, such as *regiones* or *tractus,* into one vast district under one freedman procurator, possibly during Hadrian's reign.[38] The administrative headquarters of this district is to be located at Synnada where a number of *officiales* and procurators are documented.[39] Strabo may confirm a connection between Synnada and the Dokimean quarries: he claims that in Rome the marble from Dokimeion was known as *marmor Synnadicum* or *λίθος Συννάδικος.*[40] One may tentatively suggest that this may reflect some administrative connection in the form of a supervisory function for the imperial quarries in Phrygia. As yet, there is insufficient proof for the inclusion of the Dokimean quarries in a vast imperial private estate.[41]

A supervisory function may also have been exerted by the patrimonial procurators of Achaia: there are *procuratores Augusti* mentioned in the quarry labels found on *cipollino* from Karystos stones at Rome or Ostia, Minic(i)us Sanctus (AD 132) and a C. Cerialis (late Hadrianic period).[42] Minicius Sanctus or his son appear in Egypt as *praefectus Aegypti* in AD 149/50.[43] C. Cerialis

[34] App. nos. 115, 116 (AD 136).
[35] *CIL* III 12237 = *AE* 1897: 73 (Ginik); Burton 1993: 20.
[36] For procurators and territorial disputes, cf. Burton 2000.
[37] Christol & Drew-Bear 1991: 122 f.; Drew-Bear 1994: 814 with fn. 240.
[38] Boulvert 1970: 294 with fn. 209; an alternative term is *dioecesis*, cf. Scheuermann 1957 with further bibliography; Mason 1974: 38; Bleckmann 1997.
[39] *CIL* III 7046, 7048 (*liberti* of *procurator* M. Ulpius Marianus); *MAMA* IV 53 (Hyacinthus, *tabularius*), 54 ([T. Aelius, *Augusti l*]*ibertus, verna procurator*), 62 (Amiantus, *Aug. lib. a commentariis*), 63 (Aur. Aristeinetos, *procurator*); *SEG* XXVIII, 1210 (M. Aurelius Paulos, *adiutor tabulariorum*); *IGRR* IV 704 (Aur. Marcio, *proc.*) cf. Boulvert 1970: 294 with fn. 209 for a list of *procuratores provinciae Phrygiae.*
[40] Strabo 12.8.14.
[41] Strubbe 1975: 230 f., 244, 249, observed that the epigraphic record shows direct involvement of the Roman state in the quarrying business in Phrygia from the late 1st cent. onwards, whereas regular imperial estates seem to come into existence only in the mid-2nd cent. AD. He concluded that the Dokimean and Kurt köy quarries (and presumably their administration) were not incorporated in the imperial estates.
[42] App. nos. 604, 618. In his commentary to *CIL* III 12286 (= no. 618) Th. Mommsen proposed to date the inscription to the years around AD 132, based on the *pr(obator?)* Crescens, who also appears in no. 603, cf. Wheeler 2000: 289 fn. 174.
[43] *PIR*^{2} M 628.

might be the *procurator provinciae Achaiae* under Hadrian.[44] Hans-Georg Pflaum assumed that both Minicius Sanctus and Caius Cerialis were mentioned in the label inscriptions in their capacity as *procuratores provinciae Achaiae*, though there is no direct proof that the former ever held such a position.[45]

4.1.4. Chresimus, *procurator a marmoribus*

The activities of Chresimus, a *procurator a marmoribus*—well attested under the emperors Domitian and Nerva through inscriptions at Ephesos, Miletos, Tralles, and Mylasa—are a riddle.[46] His inscriptions occur in an entirely municipal context, with the exception of one bilingual milestone, found on a road near Mylasa by Louis Robert in the 1930s. As the inscriptions do not reveal for which quarry Chresimus was responsible, it becomes difficult to assess his function. Peter Herrmann, who provided the 'editio princeps' of the Mylasa milestone based on a drawing by Louis Robert, thought that the restoration of roads near Mylasa which took place under the supervision of Chresimus was directly connected with marble production in the area. Herrmann argued that the quarries (mentioned in literary sources and attested archaeologically) in southern Ionia and northern Caria near Alabanda, Miletos, Mylasa, Herakleia on Latmos, Iasos, and Ephesos therefore might have been under imperial control.[47] That Chresimus had access to marble is indicated by an inscription set up in Tralles. He apparently embellished the *cella(?) caldaria* of the gymnasium with coloured(?) marble.[48] Thus, Chresimus' titles *[procurator lapicidin]arum*(?) (Tralles), *procurator a marmoribus* (Mylasa), ἐπίτροπος λατομίων/ε[ιῶν] (Mylasa/Tralles) and his benefaction of coloured marble to Tralles, seem to suggest that his *raison d'être* was closely linked to either the marble trade or the quarrying of marble in this area.

The latter possibility seems less likely. There is no epigraphic evidence to prove Herrmann's assumption that Chresimus was involved in the production and trade of local marble in southern Ionia and northern Caria. Although the Greek translation of his Latin titulature ἐπίτροπος λατομίων (Mylasa)/ἐπίτροπος λατομε[ιῶν] (Tralles) appears to indicate control of

[44] *PIR*² C 670. *AE* 1934: 3 (Korinth).

[45] Pflaum 1960–1: 1001 f., no. 139bis, 1071.

[46] *AE* 1988: 1028 = *SEG* XXXVIII 1073 = Herrmann 1988: 122–5 (Mylasa), Milet, Inv.-Nr. 288. *IK* 13/3, 856 (Ephesos).

[47] Herrmann 1988: 126 f.

[48] Tralles, *CIL* III 7146 = *IK* 36/1, 148, with Liermann 1889: 43.

several quarries, the original Latin title provided by the Mylasa inscription merely names him as *procurator [a] marmoribus* or 'procurator for marble'.[49] There is no further indication that the local quarries were under imperial control. The closest quarries where the presence of imperial officials is attested and with which Chresimus might have been associated are the *pavonazetto* quarries at Dokimeion, or perhaps the quarries at Chios and Teos, which might have been under imperial control. Even if one expects an imperial official of Chresimus' standing to live in the comfortable setting of a town rather than close to a quarry, Ephesos, Miletos, and Tralles are a little too remote from the quarries at Dokimeion, Teos, or Chios. Chresimus' presence in this area, therefore, must be explained otherwise.

Ephesos and Miletos, both harbour towns, were major trading centres, Ephesos being the most important harbour within the province of Asia. Both were positioned near the mouth of the Maeander valley, which allowed access to the inner regions of western Anatolia, particularly Phrygia. The marble from Dokimeion and the quarries in the Upper Tembris valley was presumably exported via Synnada, Apameia, and Laodiceia ad Lycum, down the Maeander to Tralles and then to Miletos or Ephesos.[50] It is tempting to argue that Ephesos and Miletos might have been possible harbours for the reloading of marble products. The edict of Lucius Antoninus Albus, proconsul of the province of Asia, on the stockpiling of wood and stones in the harbour of Ephesos indicates that such activities were going on there. However, linking the imperial marble trade with these activities is not warranted. Nonetheless, Chresimus perhaps was involved in the export of marble.[51]

A further possibility must be considered: Herrmann assumes that Chresimus' headquarters were to be found at Miletos.[52] His activities also allowed him to commission inscriptions at Ephesos and Tralles, probably in connection with projects he sponsored in both towns.[53] Chresimus' willingness to donate money for construction work might reflect a more profound bond than mere fondness for these cities. Various building projects in Ephesos were completed during Domitian's reign: the Artemision underwent a refurbishment under Titus, the Harbour Gymnasium, a temple for the imperial cult, as well as a *hydreion* and other buildings were constructed. These may have been erected in recognition of the title of *neokoros* being bestowed on the city of Ephesos in AD 82 shortly

49 *CIL* III 7146 = *Eph. Ep.* V 61, no. 163.

50 Hirt 2003: 119 fn. 15, with further bibliography.

51 *IK* 11/1, 23, cf. Eck 1997*a*: 123.

52 Herrmann 1988: 127 f. He bases this on the observation that a verb is missing from the inscription there. Hence, the inscription of Chresimus found on a block of grey marble and sculpted as a *tabula ansata* was not a building inscription, but was presumably once part of a funerary monument.

53 Herrmann 1988: 122.

after Domitian's accession.[54] Moreover, construction of the Serapeion at least was begun during Domitian's reign.[55] The 'marble hall' in the Harbour Gymnasium was embellished with thirteen different coloured marble types of which, to my knowledge, none have been identified precisely.[56] Some of the coloured marble most likely came from quarries under imperial control in Asia Minor, Greece, or Africa, and might either have been presented as an imperial 'gift' to the Ephesians, or been bought by the builders.[57] Would Chresimus have been responsible for supplying the capital of the *provincia Asia* and Miletos, Tralles, and other towns, with the required marble from imperial quarries? The answer must await a closer study of marble types used in the public and private town buildings of Asia Minor.[58]

Chresimus' supervision of road construction was most likely not related to his activities as *procurator a marmoribus*. Occasionally imperial officials could be burdened with additional tasks outside their usual responsibilities. In AD 62 the *portorium Asiae*, a public toll on imports and exports, was apparently managed by the patrimonial procurator of Asia, an office usually devoted to the supervision of the emperor's private estates.[59] Moreover, financial and patrimonial procurators and even freedmen are attested as being in charge of road construction within a province, usually the preserve of the governor.[60] Thus, it is possible that besides his usual tasks as *procurator a marmoribus* Chresimus was summoned to supervise the road construction.[61]

4.1.5. Quarrying Procurators at Simitthus

The epigraphic evidence for *procuratores* and other officials in charge of the quarries at Simitthus is abundant. Iulianus, Maximus, and Athenodorus are

[54] Dräger 1993: 150–74; Karwiese 1995: 88 ff.; Halfmann 2001: 39–44.

[55] Halfmann 2001: 42 f.

[56] Benndorf & Heberdey 1898: 63 ff., 71 ff.; Heberdey 1912: 161 ff.; Scherrer 1995: 178.

[57] For marble columns as gift from the emperor, see *IK* 54, 193 (Perge; Antoninus Pius). *IK* 24/1, 697 = *CIG* 3148 = *IGRR* IV 1431 (Smyrna; *c.*AD 124).

[58] The Chresimus from the Mylasa inscription is not to be identified with the Ulpius Chresimus known from Mons Claudianus, cf. Herrmann 1988: 121. For distribution maps of precious marble, cf. Dodge 1988: 73–6.

[59] Eck 1997*a*: 72 with fn. 16, 80 f., 135.

[60] Pékary 1968: 77–86; M. Arruntius Aquila, financial procurator of Galatia, built roads in Lycia-Pamphylia (AD 50): *CIL* III 6737 = *ILS* 215 = *IGRR* III 768; C. Iulius Aquila, financial procurator in Bithynia, a road between Apameia and Nicaea in 58: *CIL* III 346 (p. 976) = *IGRR* III 15; L. Antonius Naso, financial procurator of Bithynia, in 78: *CIL* III 6993 = *ILS* 253 = *IK* 39, 9. *CIL* III 188/3. T. Pri[-] Pacatus, financial procurator, a road in Sinope: *AE* 1986: 646; Iulius Honoratus, patrimonial procurator in Gallia Narbonnensis, during Domitian's reign: *CIL* XII 5432 (p. 857) (Senez).

[61] Hirt 2003.

clearly identified in dated label inscriptions on *giallo antico* blocks as *procuratores Augusti*.[62] Moreover, Agatha, *lib(ertus) Augusti*, Gallus/Iulius Gallus, Nova(tus?) and Hymenaeus are documented in the genitive case following *sub cura* or *caesura* marks.[63] This is a more or less secure indicator of their procuratorial rank, given that Agatha is named *procurator m(armorum?) N(umidicorum)* in a votive inscription at Simitthus. Other procurators such as Primus, Pientius, Alceta, Amyrus, and Tertullus dedicated votive inscriptions at Simitthus and Henchir Frouri, 3 km south-east of Simitthus.[64] Few of these votive inscriptions by procurators can be dated with any precision: Pientius appears to have completed a porticus with cella and cistern which he dedicated to the *populus Simittuensis*. The first line of the inscription commemorates the return of emperor Marcus Aurelius, possibly to Rome, at the end of AD 176 after his journey to the Near East.[65] According to Khanoussi a possible [*procurator* (or *dispensator*?)] *m(armorum) N(umidicorum)* named [-]toris, set up an altar to the Dii Mauri for the welfare of Commodus.[66] Finally, Tertullus set up a slab of Numidian marble, devoted to the recently deceased and consecrated Septimius Severus (AD 211).[67] Thus, the sequence of procurators can partly be reconstructed as follows:

Agatha	AD 137/8, 139(?)–141
(Iulius) Gallus	AD 138
Iulianus	AD 172
Pientius	AD 176(?)
-]toris	AD 180–91)
Maximus	AD 199
Athenodorus	AD 199/200
Tertullus	AD 211

[62] App. nos. 834 (Iulianus), 835 (Maximus), 836–8 (Athenodorus).

[63] App. nos. 804, 806, 811–20 (Agatha); nos. 805, 808, 809 (Iulius Gallus/Gallus); nos. 859, 863, 865(?) (Nova(tus?); no. 866 (Hymenaeus).

[64] *CIL* VIII 14551 = Dubois 1908: no. 97: *[Ag]athae Aug. lib. pro[c.] m.n.* ; *CIL* VIII 10589 = 14552 = Dubois 1908: no. 125: *Primus Aug. lib. proc. m. n.*; *AE* 1994: 1885: *Pientius Aug. lib. proc. m.n.*; *CIL* VIII 25692 = Dubois 1908: no. 77: *Aug. lib. Alceta proc. m.n.*; *AE* 1913: 165 = *ILAfr* 428 (Utica): *Alci[t]a proc. m.n.*; *AE* 1994: 1878 = Kraus 1993*a*: 59: *Aug. lib. [A]lceta proc. m.n.*; *AE* 1994: 1879 = Kraus 1993*a*, p. 60; *Amyrus lib. proc. m. [n.]*; Khanoussi 1996: 1001, no. 5: *Amyrus [*; *AE* 1994: 1880: *[.] Amyrus*; *AE* 1994: 1881: *Tertullus lib. proc. m.n.*; *AE* 1998: 1573: *]toris [—] m.n.*

[65] *AE* 1994: 1885, l.1: *Pro salute et reditu Imp. M. Aureli Antonini*; cf. Kienast 1996: 138.

[66] Khanoussi 1996: 1011, no. 32 = *AE* 1998: 1573. Commodus changes his name to Imp. Caes. L. Aelius Aurelius Commodus Aug. before 29 August 191, cf. Kienast 1996: 148.

[67] *AE* 1994: 1881, cf. Kienast 1996: 157.

The procuratorship at the Simitthus quarries was presumably held by one official at a time. (Iulius) Gallus, however, held the procuratorship in AD 138, the same year Agatha is attested as procurator in label inscriptions. There is currently no epigraphic evidence indicating shared responsibilities by both procurators. It thus appears likely that (Iulius) Gallus replaced Agatha for a short period during 138. Most *procuratores* listed above were *liberti Augusti* and, therefore, members of the *familia Caesaris*. The onomastics and the indications of status provided by the inscriptions confirm this notion.[68] It remains questionable whether or not the procuratorship of 'Numidian marble' was restricted to Simitthus only.[69] The discovery of marble rubble at Simitthus ('Gelber Berg') from different locations suggests that other stones from the region, possibly from Bordj Helal or Ain el Ksir, were dressed here.[70] The dressing of these 'alien' stones within the walls of the quarry district might indicate that the term *procurator m(armorum?) N(umidicorum)* included not only the supervision of the quarries at Simitthus, but also of further quarries in the area.[71]

4.1.6. Mining Procurators in Roman Spain

A series of inscribed stelae from the gold-mining zones of the Duerna Valley set up at Villalís, Luyego, and Corporales give the names of *procuratores Augusti*. The exact title of these *procuratores* is never specified, but in one inscription the abbreviation *met.* can quite safely be read as *met(allorum)*. Besides these imperial freedmen, the inscriptions also mention the commanding *centuriones* and *decuriones* as well as the *immunes* of the legionary and auxiliary detachments based in the mining area.[72] As these stelae were set up on the *dies natalis* of the respective unit (the exact day and consular date of which are indicated), it is possible to construct an absolute chronology for the *procuratores*, military commanders, their units and soldiers seconded to this mining area. The earliest of these inscriptions discovered at Luyego, set up during the reign of Antoninus Pius (AD 138–61), only displays the name of a

[68] Iulius Gallus appears to be the only procurator(?) with proper gentilnomen and cognomen. Moreover, there is no direct indication that he was an imperial freedmen. One could conclude that Iulius Gallus' was an equestrian procurator, cf. *PIR*^{2} I 336. Were this to be the case, Iulius Gallus and Agatha might be viewed as procuratorial partners in a 'collégialité inégale'. However, imperial freedmen could also have *nomina gentilia* not derived from the name of the ruling emperor or dynasty, cf. Weaver 1972: 36 f.

[69] Dubois 1908: 34 f; Kraus 1993*a*: 63.

[70] Röder 1993: 33 f., 52 f.

[71] Dubois 1908: 35 f.

[72] LeRoux 1977: 365–7; LeRoux 1982: 273; Domergue 1990: 293 f.; Ott 1995: 154.

centurio and *decurio* in charge of a vexillation of the *legio VII Gemina*. A mining procurator is not mentioned.[73] The earliest procurator documented at Villalís, an imperial freedman named Hermes, appears on a stela of 10 June, AD 163.[74] The imperial freedman and procurator Zoilus is recorded on two stelae at Villalís, on one set up on 22 April, AD 166, the other more than a year later on 15 October, AD 167.[75] Marcus Aurelius Eutyches is documented twice, at Luyego in a stela set up on 10 June, AD 180, and at Villalís, on a stela erected on 10 June, AD 184, on the same day four years later.[76] The last attested procurator, named on a fragmented inscription at Villalís on 22 April, AD 191, is an Aurelius Firmus. The inscription probably provides the full title of Aurelius Firmus, a *[pr(ocurator)] met(allorum)*.[77] The series of inscriptions ends with this last specimen. The procurators are only referred to by their Greek *cognomen* (Zoilus, Hermes), but from AD 180 on they bear the full *tria nomina* (Marcus Aurelius Eutyches) or their *nomen gentile* and *cognomen* (Aurelius Eutyches, Aurelius Firmus). Further information on their function is not given, but they were apparently all imperial freedmen and procurators. They are not the only imperial mining officials documented in the epigraphic record of northwestern Spain: an M. Ulpius Eutyches, *procurator metalli Albo-c(rarensis)* (a mining district mentioned by Pliny) during Trajan's or Hadrian's reign is named in an inscription of unknown origin.[78]

The presence of *beneficiarii procuratoris Augusti* in the mining areas of the north-west casts light on the close relationship between the financial procuratorship of Asturia and Callaecia and the administrative heads of the various mining zones. A *beneficiarius procuratoris Augusti* is named together with the procurator Hermes on a stelae of Villalís set up 10 June, AD 163.[79] A further stela from Villalís (22 April, AD 175) lists another *beneficiarius procuratoris Augusti*.[80] They were presumably sent there on the orders of the equestrian

[73] *AE* 1967: 229 = *IRLeón* 32 = *EAstorga*, no. 112, with García Bellido 1966: 24. Whether mining procurators were not yet appointed for that area in the mid second century, or whether mining operations at first(?) were run under military supervision only, cannot be ascertained given the fragmentary epigraphic record.

[74] *CIL* II 2552 = *EAstorga* 114; Sánchez-Palencia 1980, fig. 8. Hermes appears once more in an inscription found at Castro de Corporales, which appears to be quite similar to the one from Villalís. Both inscriptions name the same legionary *centurio*, the same procurator and the same auxiliary *decurio*; only the names of the *beneficiarius* and the *signifer* differ. Patrick Le Roux argued that this inscription is a fake; it appears to be a copy of the Villalís inscription, and the sun disc adorning the stone, a typical feature of Iberian funerary inscriptions, does not fit at all with the usual form of votive military inscriptions, cf. *EAstorga* 148, with LeRoux 1984.

[75] *CIL* II 2556 = *EAstorga* 115; *CIL* II 2553 = *EAstorga* 116.

[76] *AE* 1967: 230 = *IRLeón* 34 (Luyego). *AE* 1910: 5 = *EAstorga* 118 (Villalís).

[77] *AE* 1910: 2 = *IRLeón* 41 (Villalís).

[78] *CIL* II 2598; Pliny, *NH* 33.80, cf. Christol & Demougin 1990: 188; Domergue 1990: 294.

[79] *CIL* II 2552 = *EAstorga* 114; Sánchez-Palencia 1980, fig. 8.

[80] *AE* 1910: 1 = *AE* 1928: 176 = *IRLeón* 40.

financial procurators at Astorga. One may assume that their main task was to act as liaison officers to the financial procurators in Astorga.[81] Numerous inscriptions at Asturica Augusta/mod. Astorga document the presence of a *procurator Asturiae et Callaeciae.*[82] The headquarters of this financial procurator appear to have been located at Asturica Augusta in the immediate vicinity of important alluvial gold-mining areas—likely one of the main reasons for the creation of this subprovincial region.[83] Apart from monitoring the imperial mining activities, the financial procurator for Asturia and Callaecia, like his colleagues in other imperial provinces, was to provide the pay for the army units based in his area.[84] Moreover, he was responsible for the collection of tribute and taxes from the provincial population.[85] Perhaps the earliest holder of the financial procuratorship of Asturia and Callaecia, L. Arruntius Maximus, is recorded in office in AD 79.[86] The list of the equestrian incumbents of the procuratorship of *Asturia* and *Callaecia* reaches down to the late second century.[87]

Surprisingly, a Greek honorific inscription, re-edited in 1969 by Christian Habicht, seems to provide information on the last known financial procurator appointed to Asturia and Callaecia—an imperial freedman named Saturninus.[88] The statue base, on which this inscription was engraved, was discovered in the Asklepieion at Pergamon in Roman Asia in 1931. The text lists the procuratorial posts held by Saturninus in inverse order: it begins with

[81] LeRoux 1982: 241; Nelis-Clément 2000: 263.

[82] *CIL* II 2477, 5616 = *ILS* 254, Santos Yanguas 1983: 97; Haensch 1997: 174 f., 488; Alföldy 2000*a*: 63–7.

[83] Haensch 1997: 174 f.; Alföldy 2000*a*: 46. The importance of the Spanish north-west is furthermore highlighted by the constant presence of *legio VII Gemina* at León whose main task most likely was the protection of the mining region. During Hadrian's reign Asturia and Callaecia even received its own *legatus iuridicus* to deal with juridicial matters which may perhaps also have arisen from an increase in economic activity in the area, cf. Alföldy 2000*a*: 42 f., with fn. 70 and further bibliography.

[84] Strabo 3.4.20.

[85] Dio 53.15.3, 54.21, cf. Brunt 1990*b*: 165; Burton 1993: 16; Eck 1997*a*: 132–8. It is not clear whether or not the financial procurator of Asturia and Galicia was subordinate to the *procurator provinciae Hispaniae citerioris*, as LeRoux 1995: 75, suggested, cf. also Christol 1999: 239.

[86] *CIL* II 2477 = *ILS* 254 = Rodríguez Colmenero 1997: 418–20, no. 587 (building inscription, Aquae Flaviae/mod. Chaves), cf. *PIR*² A 1145; Pflaum 1960–1: 1047; Pflaum 1982: 117; Alföldy 2000*a*: 63. Ronald Syme had argued that Pliny the Elder was probably a *procurator Augusti* of *Asturia* and *Callaecia* as well. Pliny's unique knowledge of mining methods used in Spain and of the census numbers of the *conventus* of *Lucus Augusti*, *Bracara Augusta* and *Asturica Augusta* seem to support this hypothesis, cf. Syme 1969: 45 fn. 78. Alföldy (2000*a*: 46 fn. 76), however, remains undecided as to whether Pliny was a procurator for the northwest or 'just' *procurator Augusti Hispaniae citerioris.* Only a few equestrian administrators displayed the full title of *procurator Asturiae et Callaeciae* in their inscriptions

[87] Alföldy 2000*a*: 63–7.

[88] Habicht 1969: 88–92, no. 44, cf. *AE* 1933: 273.

his last procuratorship in Pergamon and ends with his first appointment as procurator of the Quintilian estates (*ἐπίτροπος τάξεως Κυντιλιανῆς*). The Quintilii, Roman senators, were put to death by Commodus in the year 182/3 and their estates, presumably near Rome, were taken over by the imperial administration. Saturninus, therefore, reached his first procuratorship at some point after 183.[89] For the subsequent post of a *procurator a pacticiis* Saturninus remained in Rome for a while,[90] before being moved to Alexandria and appointed *ἐπίτροπος χαρταρεᾶς* or director of the *χαρτηρά*-tax, a tax on papyrus.[91] He then took over the financial procuratorship in Asturia and Galicia in northwestern Spain, perhaps in the years between 193 and 198.[92] The mining district in Vipasca/mod. Aljustrel in Lusitania was Saturninus' next assignment.[93] Following these offices on the Iberian peninsula, Saturninus returned to the eastern half of the empire and took over a procuratorial post in the financial administration of Asia.[94]

The appointment of Saturninus to the procuratorship of Asturia and Galicia, a post normally held by equestrians, appears to be irregular. Hans-Georg Pflaum has argued that Saturninus must have shared this office as a junior partner with an equestrian colleague. He regarded this 'collégialité inégale' as a normal feature of procuratorial offices in the second century.[95] Nevertheless, it probably was not a general rule. Based on an inscribed votive altar set up by Saturninus in Lucus Augusti/mod. Lugo, Patrick Le Roux, and Geza Alföldy have dated his tenure of the financial procuratorship to the years between AD 193 and 198.[96] During this period there is no equestrian procurator documented, nor is there any other epigraphic evidence that the equestrian predecessors of Saturninus held this office in conjunction with an imperial freedman. Given the available evidence, Saturninus might not have been subordinate to an equestrian colleague, but possibly held the financial procuratorship alone. Why this imperial freedman would have been appointed to a post formerly held by members of the equestrian order is not easily explained. This unusual move may have been the result of the

[89] Habicht 1969: 88–92, no. 44. On the Quintilii, cf. Habicht 1969: 89–91; Pflaum 1970: 298; Ricl 1997: 259–65.

[90] On *a pact(ionibus) / a pact(iciis)*, cf. Habicht 1969: 91; Boulvert 1970: 163 with fn. 474; LeRoux 1985: 224; Christol & Demougin 1990: 183.

[91] A tax on papyrus fabrication(?) or perhaps a stamp tax, cf. Lewis 1974: 137–9; Lewis 1989: 42.

[92] Habicht 1969: 92; Pflaum 1970: 298; LeRoux 1985: 223 f.; Christol & Demougin 1990: 169; Alföldy 2000*a*: 46–8 with fn. 78.

[93] Pflaum 1970: 301–4; Christol & Demougin 1990: 190.

[94] Christol & Demougin 1990: 197–207.

[95] Weaver 1972: 279–81; Eck 1997*a*: 90–2.

[96] LeRoux 1985: 223 f.; Alföldy 2000*a*: 46 f. with fn.78.

insecure political situation in Hispania Citerior. During the events of 195–7, the provincial governor of Hispania Citerior, Lucius Novius Rufus, sided with the usurper Clodius Albinus against Septimius Severus. Although the *legio VII Gemina* remained neutral, Novius Rufus and many provincial aristocrats had given financial support to Albinus, and consequently were executed after the Albinians had been crushed.[97] Perhaps during or immediately after this civil war the appointment of an imperial freedman to the post of financial procurator may have appeared to be a sensible measure.[98] However, the probable abolition of the financial procuratorship of Asturia and Galicia soon after Saturninus' time in office, and the disappearance of mining procurators from the epigraphic record of the north-west after 191 may also be a consequence of diminishing returns from the mining operations.[99] Based on the observation that *beneficiarii procuratoris* are no longer attested in the vexillation-inscriptions from Villalís and Luyego after AD 175, one might tentatively argue for a disengagement of the financial procurators from the mining areas in the Duerna Valley, perhaps resulting in growing autonomy for the mining procurators there(?). Provided the chronological sequence of an imperial freedman succeeding equestrian procurators is correct, one perhaps may connect this with the decline of the north-west's economic importance.

After Saturninus' stay in northwestern Spain he was appointed ἐπίτροπος τῶν μετάλλων Οὐλπασκησιων, procurator of the copper and silver mines at Vipasca/mod. Aljustrel in southern Portugal, a post held by imperial freedmen.[100] One of the bronze tablets at Vipasca, the *lex metallis dicta*, appears to be a letter written under Hadrian and addressed to an Ulpius Aelianus, most likely one of Saturninus' predecessors in the mining procuratorship at Vipasca.[101] The name of the sender was not preserved on the tablet: Dieter Flach suggested the financial procurator or *procurator provinciae* of Lusitania as the author of this letter.[102] A further inscription on a statue base set up by the *coloni metalli Vipascensis* honours a Beryllus, *libertus Augusti*. At the time

[97] *HA Sev.* 12.1. Birley 1999: 125 f., fn. 15; Alföldy 2000*a*: 23.

[98] Alföldy 2000*a*: 23, 36. The difficult political circumstances might also help explain why Saturninus commissioned a votive altar at Lucus Augusti, cf. *AE* 1968, 229 = *IRPL* 2, cf. LeRoux 1985: 220–3; Christol & Demougin 1990: 160–9. Caracalla's creation of a shortlived *provincia Hispania Superior* including only the former district of Galicia may perhaps be seen as a delayed outcome of the political events of 195 to 197, and probably was designed to fasten the grip of the ruling imperial dynasty on the essential resources of northwestern Spain, cf. Alföldy 2000*a*: 17–38.

[99] Domergue 1990: 215–24; Alföldy 2000*a*: 48.

[100] On the identity of *μέταλλα Οὐλπασκήνσια* and *metalla Vipascensia*, cf. Christol & Demougin 1990: 170 fn. 60.

[101] Flach 1979: 403.

[102] Flach 1979: 413 f.; Domergue 1983: 131, 170 f.; Domergue 1990: 291, assumes a similar arrangement for the mines around Castulo.

the inscription was set up Beryllus apparently held the post of *procurator*, most likely of the *metalla Vipascensia*. Beryllus was furthermore *ratio[naliu]m vicarius*, a deputy or auxiliary of the *rationales*.[103] Since the reign of Marcus Aurelius the term 'rationales' designated the equestrian offices of *a rationibus* and the subordinate *procurator summarum rationum*, to which Beryllus appears to have been attached as *vicarius*, 'deputy'.[104] He may have been seconded to Vipasca from this post to oversee the restructuring efforts of the Vipasca mines, probably after the Moorish invasions of southern Iberia in AD 172/3.[105] Beryllus' inscription names him also as *restitutor metallorum*, a reference to his success in getting the mines up and running again. He probably redistributed or confirmed property rights related to the mines, ruled in unresolved contractual issues, and perhaps even provided money to those *coloni* or *conductores* suffering financial losses from damage to the mining infrastructure. In this case of rebuilding a mining site it might be possible that the *rationales* took direct interest in the appointment of a special envoy to the troubled mining district. That the tasks of the *procurator a rationibus* included the monitoring of imperial mines is specifically stated towards the end of the first century AD by Statius in a poem on the father of Claudius Etruscus.[106] Given the extraordinary circumstances of Beryllus' appointment, we can neither automatically conclude that mining procurators at Vipasca were normally appointed by the *rationales*, nor that the *rationales* were their immediate superiors.

The epigraphic evidence for the mining administration in Baetica is far from abundant, despite the numerous archaeologically documented mines. The only inscription in Baetica giving the full title of a mining procurator was discovered in Hispalis/mod. Sevilla. The inscribed statue base of white marble was commissioned by the *confectores aeris* at Hispalis(?), probably at the end of the first century AD, in honour of Titus Flavius Polychrysus, imperial freedman and *procurator montis Mariani*.[107] Polychrysus appears to have supervised a private estate which in AD 33 had come under imperial control after the previous owner, Sextus Marius, had been executed.[108] A further procurator to Marius' mines—a Dorotheus, *libertus Augusti* and *procurator massae Marianae*—dedicated a votive monument to the god Silvanus at Ostia on his departure to 'Marianus' estate'.[109] A slave of Sextus Marius had a

[103] *IRPac* 121; *AE* 1908: 233(?). For Beryllus, cf. Domergue 1990: 299–301.

[104] Domergue 1990: 300; Eck 1997*a*: 84 f. On *vicarius*, cf. Weaver 1972: 202.

[105] *HA M. Ant.* 21; Wickert 1931: 835–8; Domergue 1990: 300; Lazzarini 2001: 105.

[106] Statius, *Silv.* 3.3.89–90: *(...) ab auriferis eiectat Hiberia fossis, Dalmatico quod monte nitet (...)*, cf. Dušanić 1989: 154; on the father of Claudius Etruscus, cf. Weaver 1972: 284–94.

[107] *CILA* II 25.

[108] Tacitus, *Ann.* 6.19. Dio 58.22.2 f., cf. Levick 1999: 101.

[109] *CIL* XIV 52 = ILS 1592. On 'massa', cf. *TLL* VIII, p. 430 f., 2.b. with Brunt 1990*c*: 398, fn. 143.

gravestone set up in Corduba, which does suggest that Marius' mines (Cerro Muriano?) and estates were administered from this town.[110] Mining procurators and imperial officials appear only on a few other sites in Roman Baetica. The mining procurator Pudens, *libertus Augusti*, is attested during the reign of Nerva at Rio Tinto where he commissioned an inscribed monument in honour of his emperor.[111] Further procurators might be mentioned on inscribed copper ingots recovered from shipwrecks off southern France. The reading of these strongly abbreviated inscriptions, however, remains unsatisfying and does not yet provide a sound basis for further deliberation.[112] A different official is recorded on the inscribed tin ingots salvaged from the Port-Vendres II shipwreck off the coast of southern France. They were found together with other cargo allegedly originating from Roman Baetica. These are stamped with the letters *L(ucius) Vale(rius), Aug(ustae) l(ibertus) a com(mentariis)*, a freedman of Valeria Messalina, the wife of Claudius and *Augusta* in AD 41/42.[113] Provided the origin of the ship's cargo has been correctly identified, L. Valerius might have been located in Baetica or had the ingots procured while he remained in Rome.[114]

4.1.7. A *procurator metallorum et praediorum* from Sardinia

Apart from the well-documented equestrian and freedmen mining procurators known from Spain, Gaul, and the Danubian provinces, *procuratores metallorum* rarely appear in the epigraphic record of other provinces. An altar dedicated to the nymphs and honouring Q. Baebius Modestus, *procurator Augustorum* and *praefectus provinciae Sardiniae* was set up around AD 211/12 by an imperial freedman Servatus, *procurator metallorum et praediorum*, in a

[110] *CIL* II^2/7, 441 = *CIL* II 2269. Sextus Marius was from Corduba or had close links with the town, cf. Pliny, *NH* 34.4 (*aes Marianum* = *aes Cordubense*). For a *tessera hospitalis* with his name, cf. Eck & Fernández 1991.

[111] *CIL* II 956.

[112] Part of the engraved letters on the copper ingots from the Marseillan shipwreck off southern France were read as *Telesphori* | *Aug(ustorum) n(ostrorum) (duorum) d(ominorum)* | *pr. Felix CCCXXVII*, cf. Laubenheimer-Leenhardt 1973: 53–5, no. 24; Domergue & Rico 2002: 146. A further copper ingot was found in the Museum of Biterrois, the last lines of the inscription reading *L. Caesi Severi* | *p(rocurator) A(ugustorum) (duorum) d(ominorum) n(ostrorum)* | *pr. Fẹl[.]x̣* | *CCXCVIIII* || *XX*, cf. Domergue & Rico 2002: 145. Various readings are available for lines 4–5 of the inscription on the copper ingot from Planier, southern France (*AE* 1963: 109). Besides *pr(ocuratoris) col(oniae) Ono|bensis*, the variations *pr(ocuratoris) O(s)sono|bansis* or *proc(uratoris) S(.)S(.) lono|bensis* have been put forward, cf. Laubenheimer-Leenhardt 1973: 36–7, nos. 16, 78–82; Domergue 1990: 287, 295, with further bibliography; Lazzarini 2001: 104 f.

[113] Colls et al. 1977: 11–18.

[114] Weaver 1972: 241–3, 245–52; Colls et al. 1975: 75–7.

niche of the Nymphaeum in the baths at Aquae Hypsitanae/mod. Caddas on Sardinia.[115] Prior to the appearance of this inscription it was assumed that the responsibility for the imperial mines on Sardinia lay with the governor of the island. Le Bohec argued that the procurator-governor probably did not get personally involved in the mining administration but left that job to a *tabularius provinciae Sardiniae.* The epigraphic record does not provide us with any information concerning when the office of *procurator metallorum et praediorum* was created.[116] Servatus' inscription for the presidial procurator of Sardinia not only reflects his private admiration for the governor, but in all probability was intended to honour him as his former superior in the provincial hierarchy. The double function of *procurator metallorum* and *praediorum* is hitherto unattested: while a number of *procuratores praediorum* are known throughout the empire, Servatus appears to be the only one documented in this double function so far.[117] The combination of both offices was explained by Christer Bruun as a result of the paucity of mines and imperial estates in Sardinia.[118]

4.1.8. *Procuratores aurariarum* in Roman Dacia

The epigraphic record of Ampelum/mod. Zlatna documents a number of procurators in charge of the gold mines in the area of Alburnus Maior/mod. Roșia Montană and perhaps other gold mines throughout Dacia. In his detailed study of the Dacian mining administration, Hans-Christoph Noeske pointed out that ten *procuratores aurarium* are known to have been located in Ampelum, three of whom were imperial freedmen. The inscriptions hardly ever display the full title of the post. Noeske determined that this was the result of the locations in which the inscribed monuments or stone slabs were set up. At Ampelum, the administrative headquarters, the full title of the procurator was known to the inhabitants and therefore did not have to be

[115] *AE* 1998: 671. Another votive stela discovered in the *natatio* of the baths also honours Baebius Modestus, the inscribed monument having been set up by a soldier of the *cohors II*, cf. Benito Serra & Bacco 1998: 1244 f., fn. 101.; Bruun 2001.

[116] Le Bohec 1992*a*: 260 with fn. 32; Bruun 2001: 359 f. For a list of governors, cf. Zucca 2001.

[117] Numerous *praedia* are attested on Sardinia, e.g. *praedia Statoniensia* (*AE* 1981: 475c; Cagliari); *praedia Patulcensium* (*CIL* X 7852; Esterzili); *pr(aedia) Dom(itiae) Luc(illae)* (*CIL* X 8046, 6b; Assemini); *praedia L. Veri Aug(usti)* (*CIL* XV 737; Arbus); *[arcar]ius praedi[orum]* (?) (*AE* 1979: 307; Cagliari); cf. Sotgiu 1988, index, p. 726. Imperial estates are known from the written sources, cf. *AE* 1971: 120 = *AE* 1972: 227: Alexander, *Aug. ser. regionarius*; Meloni 1990: 171–6, 474 f.

[118] Bruun 2001: 357.

displayed at full length in inscriptions.[119] Although the title of these procurators is never given in full, an inscription erected in the second half of the second century or the early third century by a subordinate official provides enough evidence to reconstruct the title given to at least some of the incumbents as *procurator aurariarum Dacicarum.*[120]

The earliest known *procurator aurariarum* is one M. Ulpius Hermias who served during Trajan's reign. His funerary stone, set up by his wife Salonia Palestrice and his freedman Diogenes, commemorates his death at 55 years of age. His corpse was transported back to Rome, from where he might have originated, 'ex indulgentia Aug(usti) n(ostri)'.[121] Romanus, another *libertus Augusti* and *procurator aurariarum*, set up a gravestone for his daughter Aurelia Ianuaria. The name of Romanus' daugther dates his tenure of the procuratorship at Ampelum to Marcus Aurelius' reign or later.[122] A further imperial freedman is Avianus, *subprocurator aurariarum*, whose inscription was discovered at Apulum/mod. Alba Iulis. He dedicated an inscribed votive monument to Iuppiter Tavianus in the late second century.[123] There is no record of the office of *subprocurator* before the AD 180s. Although the introduction of this office prior to this date cannot be excluded, the appearance of the first *subprocurator* coincides with the occurrence of equestrians holding the regular mining procuratorship.[124]

In an inscription on a votive altar, the equestrian L. Macrius Macer gives his function as *procurator Aug(ustorum duorum)*, which in turn allows us to place his tenure of procuratorial office either in the years AD 161–9 of Marcus

[119] Noeske 1977: 299 with fn. 150.

[120] *CIL* III 1297 = *IDR* III/3, 314, cf. Noeske 1977: 352, AMP 12. The *tabul(arius) aur(ariarum) Dacicarum* Neptunalis named in the inscription probably is the same person as the Neptunalis of *CIL* III 1313 = AMP 12a, husband to Ulpia Marci liberta Trophima. The name of his wife provides a rough date for Neptunalis' term as *tabularius.*

[121] *CIL* III 1312 = *ILS* 1593 = *IDR* III/3, 366, cf. Noeske 1977: 347, AMP 1. Noeske 1977: 296. The name of Ulpius Hermias' wife had been understood to refer to the toponym Salona and is taken as an indication that Ulpius Hermias, prior to his procuratorship in Ampelum, had been posted in Dalmatia in a similar function, cf. Patsch 1899: 267 with fn. 5; Patsch 1900: 73; Hirschfeld 1905: 155 fn. 1; Noeske 1977: 347 f. The name Salonia, however, is rather common in Italy and Rome which might suggest that he was perhaps directly transferred from Rome to his new post, cf. Alföldy 1969*a*: 117; Noeske 1977: 348 with fn. 326.

[122] *CIL* III 1622 = *ILS* 1532, cf. Noeske 1977: 348, AMP 2.

[123] *CIL* III 1088, p. 1390 = *IDR* III/3, 228, cf. Noeske 1977: 348, AMP 3. Noeske (1977: 348 f.) points out that the epigraphic formula 'pro salute (et) victori[a] domini n(ostri) sanctissi[mi]' and the fact that Avianus names only one Augustus as his patron appear to indicate a date for his tenure of office after the Marcomanic wars. The use of the formula 'pro salute et victoria domini nostri' is attested from the reign of Commodus onwards, the adjective 'sanctissimus' surfaces mostly during the reign of Septimius Severus. One therefore would tentativley suggest the Severan period as a date for the inscription.

[124] More cautious: Noeske 1977: 349.

Aurelius and Lucius Verus' reign, or in the reign of Septimius Severus and Caracalla (AD 198–211).[125] Around AD 183–5 an inscribed slab from Ampelum, perhaps a building inscription, naming the emperor Commodus, identifies a [M]aximus as procurator. As Noeske points out, the *cognomen* Maximus is not followed by a freedman status indication. Thus it is likely that he too was an equestrian official.[126] The procurator Aelius Sostratus erected a votive inscription to Iuno Regina Minerva 'pro salute et victoria et incolumitate' of Caracalla and his mother Iulia Domna which allows us to date it to the year AD 212 (death of Geta) or AD 214 (Caracalla's visit to Dacia).[127] An inscribed votive altar to Ceres Augusta was set up in January AD 215 by Suriacus, *dispensator aurariarum* and probably an imperial slave, and dedicated by Aulus Senecius [-] Contianus(?), an equestrian procurator (*vir egregius*).[128] C. Aurelius Salvianus, *procurator Augusti nostri*, is named in a building inscription as overseeing some construction work undertaken by the *duumviri* and the *ordo* of Ampelum, possibly on an aqueduct(?). The naming of these municipal institutions points to a date after the elevation of Ampelum to municipal status under Septimius Severus. The erased name of the emperor makes a date between AD 218 and 235 likely.[129] Finally, a Papirius Rufus, *procurator aurariarum*, received an inscribed memorial plaque from his freedmen and heirs. The inscription cannot be dated.[130] Whether or not one M. Iulius Apollinaris can be counted amongst the procurators is uncertain. Noeske argues that the undated dedication of the inscription by Verus and Romanus, two *vernae Augusti* and *vilici*, as well as the location of the inscription, seems to support the view that Apollinaris was an equestrian *procurator*.[131] This provides us with the following list of *procuratores aurariarum*:

procuratores	*subprocurator*
Ulpius Hermias, *lib.Aug.* (Traj.)	
Romanus, *lib. Aug.* (M. Aur.)	
M]aximus (183/185)	
L. Macrius Macer (M. Aur./Sep. Sev.)	
	Avianus, *lib. Aug.* (Sep. Sev.)

[125] *CIL* III 1310 = 12563 = *IDR* III/3, 307, cf. Noeske 1977: 298, 349, AMP 5; *PIR*^{2} M 34.

[126] *IDR* III/3, 281, cf. Noeske 1977: 351, AMP 6.

[127] *CIL* III 7836 = *IDR* III/3, 318, cf. Noeske 1977: 298, 350, AMP 7 with fn. 341; *PIR*^{2} A 265.

[128] *AE* 1959: 308 = *AE* 2003: 1512 = *IDR* III/3, 292, cf. Noeske 1977: 298, 351, AMP 8.

[129] *CIL* III 1293 = *IDR* III/3, 282, cf. Noeske 1977: 298, 351, AMP 9, with fn. 348.

[130] *CIL* III 1311 = *IDR* III/3, 359, cf. Noeske 1977: 298, 351, AMP 10; *PIR*^{2} P 118.

[131] *CIL* III 7837 = *IDR* III/3, 332, cf. Noeske 1977: 298, 352, AMP 11.

Aelius Sostratus (212)
A. Senecius [-] Contianus (215)

Undated
(M. Iulius Apollinaris (?))
Papirius Rufus

Noeske observed that the last freedmen procurators in Ampelum are attested during the reign of Marcus Aurelius or later and the earliest known equestrian procurator is dated to the same period. Based on this, he concluded that the procuratorial post at Ampelum was shared between an equestrian and freedman procurator in an arrangement called 'collégialité inégale' by Hans-Georg Pflaum. Although this sort of unequal arrangement—with the freedman procurator being the subordinate partner to his equestrian colleague—can be detected fairly often in procuratorial offices of the second century, it was probably not a general rule.[132] As there is no written evidence to confirm the view that equestrian and freedman procurators held the same office at Ampelum simultaneously, a piecemeal replacement of freedmen procurators by those of equestrian rank appears more likely.[133] Nevertheless, imperial freedmen of procuratorial rank continued to be involved in the administration of gold mines, albeit as *subprocuratores*. Whether *subprocuratores* had been a regular feature of the gold-mining administration at Ampelum prior to the Marcomannic Wars and whether they remained a regular feature throughout the Severan period cannot be ascertained. Eck argued that the introduction of subprocuratorial offices in general must have been the result of an increasing pressure on the administrative system during the second century.[134]

Beneficiarii procuratoris, attested in four inscriptions, roughly dated to the second half of the second century AD, have been understood by Noeske and Joachim Ott to refer to soldiers under the command of the *procurator aurariarum*.[135] Contrary to this view, Jocelyne Nelis-Clément pointed out the possibility that the *beneficiarii* had been seconded to Ampelum by the financial procurator.[136] Besides these inscriptions at Ampelum, no other

[132] See e.g. Tacitus, *Ann.* 13.1. Pliny, *Ep.* 10.27.85. *AE* 1930: 86. See Pflaum 1970: 305; Noeske 1977: 300 f.; LeRoux 1985: 227–30; Christol & Demougin 1990: 186–90; Domergue 1990: 292 ff.; Christol 1999: 235. For further examples, cf. Weaver 1972: 279–81; Eck 1997*a*: 90–2. For critisism, see Millar 1963*b*.

[133] For the opposite view, cf. Noeske 1977: 300.

[134] Eck 1997*a*: 83 f. A fragmentary inscription (*AE* 1989: 629) might name an *A*]*ụr(elius) Urs*[*us*]|[—*p*]*ṛaepos*[*i*]|[*tus*—*a*]*ụraria*[*r*]|[*um*].

[135] *CIL* III 7833 = 1289 = *IDR* III/3, 300 = *CBFIR* 565. *CIL* III 1295 = *IDR* III/3, 310 = *CBFIR* 566. *AE* 1991: 1344. Noeske 1977: 314; Ott 1995: 152; Nelis-Clément 2000: 370, no. 186.

[136] Nelis-Clément 2000: 263.

procuratorial *beneficiarii* are attested throughout Dacia.[137] This might support the argument that the *beneficiarii* were on the staff of the mining procurator at Ampelum. But in what function? Security issues were dealt with by the *numerus Maurorum Hispanorum* or perhaps the *legio XIII Gemina*, which both appear to have a detachment based at Ampelum.[138] The specific purpose of *beneficiarii* at the administrative headquarters remains obscure. They were not recorded in the mining zones and may not have served as liaisons between the gold mines and the administrative centre. More importantly, given the freedmen status of the procurators at Ampelum prior to the Marcomannic wars, it is difficult to imagine soldiers coming under their command. Thus, the *beneficiarii procuratoris* are more likely to have been seconded to Ampelum as liaisons with the financial procurators—first of Dacia Superior and probably after AD 124 of Dacia Apulensis. The infrequent visit of the latter is indicated by a votive altar at Ampelum commissioned by C. Sempronius Urbanus, whom we know to be procurator of Dacia Apulensis in and around AD 182.[139] Provided this interpretation is correct, one might conclude that the freedmen *procuratores aurariarum* were perhaps subordinate to the financial procurators of Dacia Apulensis prior to the Marcomannic wars.[140] Besides *beneficiarii* sent by the financial procurator, a number of military men, common soldiers (*librarii*) as well as *beneficiarii consulares*, were being seconded by the governor to Ampelum as well as Alburnus Maior during the second century.[141] The epigraphic documentation in the mining districts does not reveal the purpose of these *stationes* of *beneficiarii consulares*.[142]

4.1.9. Mining Procurators in Moesia Superior and Noricum

There is little documentation on mining procurators in Moesia Superior. According to Dušanić, votive altars for the Dea Orcia and maybe Volcanus(?)

[137] Schallmayer et al. 1990: 392–424.

[138] *Numerus Maurorum Hispanorum*, cf. *CIL* III 1294 = *IDR* III/3, 312; *CIL* III 1316 = *IDR* III/3, 335; *IDR* III/3, 302 = *AE* 1971: 383; *legio XIII Gemina*, cf. *CIL* III 1317 = *IDR* III/3, 344; XIII Gemina (tile stamps), cf. *AE* 1911, 37, *AE* 1988: 961.

[139] Pflaum 1960–1: 542 f., no. 200, 1065. The financial procurators for Dacia Apulensis appear to have had their headquarters at Sarmizegetusa which supposedly was located in recent excavations, cf. Piso 1993: 38 with fn. 53, 89 f. There presence at Sarmizegetusa is attested at the earliest from 169 onwards, cf. Haensch 1997: 345 f.

[140] For a list, see Pflaum 1960–1: 1066.

[141] *CBFIR* 567 (Ampelum), 544–7 (Alburnus Maior). After AD 124 the governor of Dacia Superior was based at Apulum, cf. Piso 1993: 37; Haensch 1997: 338–46 with further bibliography.

[142] Ott 1995: 154 f.; Nelis-Clément 2000: 259.

found at Avala recorded a *vir egregius* Simplicius, who must have been a mining procurator. This claim is not further confirmed by the epigraphic record.[143] The epigraphic material of the Kosmaj and Rudnik mining areas provides us with more substantial evidence for mining procurators. Tyrannus, *libertus Augusti* and procurator, is named in a building inscription at Kosmaj, probably during the second half of the second century, whereas Cassius Ligurinus, *procurator Augusti* during Septimius Severus' reign, supervises the repair of a collapsed temple of Terra Mater.[144] The *municipium D(ar)d(anorum)*/mod. Sočanica probably yields three inscriptions referring to mining procurators. A Telesphorus, presumably a mining procurator, oversaw construction work by *coloni arg(entariarum)* during the years AD 136/7.[145] A fragmentary altar, perhaps from the reign of Antoninus Pius, documents a procurator Amandus, a *libertus Augusti.*[146] Moreover, an inscribed base set up for Gordianus III by the *ordo colonorum* names Titienus Verus, *vir egregius* and *procurator mm(etallorum) D(ar)d(anicorum).*[147] The inscriptions at Kosmaj, Rudnik, or Sočanica do not yield any further evidence for other procurators.

Amongst the imperial *procuratores* documented in Noricum none is known to have been placed in charge of the iron and gold mines. Alföldy argued that the presidial procurators of Noricum, attested from Claudius' reign onwards, were, at first, responsible 'for the entire economic administration of the province'.[148] The inscription of an imperial freedman, a Tertius, *libertus Caesaris*, which he dated to the early first century and believed to have originated from the Magdalensberg, was used as evidence that the mines stood under imperial control 'as early as the time of Augustus'.[149] Further indications for the direct control of mining areas by the imperial administration are provided by two votive inscriptions of two imperial slaves at Hohenstein/Pulst to the goddess Noreia: one inscription displays the name of a Chrysanthus, the *vikarius* of Cypaerus, imperial slave under Claudius.[150] A fragmentary votive inscription at Tiffen might give the name of a *Ca[e(saris) n(ostri) (servus)]*—provided Alföldy's reading is correct.[151] Certainly, the mines near Hohenstein

[143] *IMS* I 20 = *CIL* III 1660 = 8151 (p. 1022) (AD 287). *IMS* I 46 = *CIL* III 1661 (AD 272), cf. Dušanić 1977: 77 f.; Dušanić 1990: 588 f.

[144] *IMS* I 103 = *CIL* III 14536 (Kosmaj/Guberevac). *IMS* I 168 = *CIL* III 6313 = 8333 (Rudnik). On Cassius Ligurinus, cf. *PIR*^{2} C 497.

[145] *AE* 1972: 500 = *ILJug* 501 with Dušanić 1971*a*.

[146] *ILJug* 504.

[147] *ILJug* 503.

[148] Alföldy 1974: 79–81, 115.

[149] *CIL* III 4987, cf. Alföldy 1974: 64 with fn. 12, 115 with fn. 82.

[150] *CIL* III 4808, *CIL* III 4807 with Alföldy 1969*b*: 25, no. 34.

[151] *CIL* III 4822 =11505, with Alföldy 1969*b*: 25.

and perhaps near Tiffen were under direct control of the presidial procurators at Virunum from as early as the mid-first century onwards.[152] Based on the epigraphic evidence Alföldy believed that at the beginning of the second century the mines and their administration were leased out to *conductores ferrariarum Noricarum*. An inscription set up for Ti. Claudius Macro by his *vilicus* in Aquileia might be the oldest attestation of a *conductor ferrariarum Noricarum*, as it belongs to the early second century(?).[153] A further building inscription at Hohenstein names the presidial procurator of Noricum, Claudius Paternus Clementianus, attested around AD 120. It was commissioned by a Sabi[nius], perhaps a *conductor ferrariarum Noricarum*(?), according to Alföldy.[154] The inscriptions of other *conductores* found at Tiffen, Hohenstein, and Wieting are difficult to date but are believed to belong to the first half of the second century. Alföldy argued that when the gubernatorial title changed from *procurator Augusti provinciae Noricae* to *procurator Augusti regni Norici* (probably between AD 158 and 160), the Norican mining areas were returned to the 'direct' administration of the presidial procurator.[155] Under Marcus Aurelius these equestrian *procuratores Augusti regni Norici* were soon replaced as governors by senatorial *legati Augusti pro praetore provinciae Noricae*. The tasks of the *procurator Augusti regni Norici* were now limited to the financial administration of the province, of which the mines were one of the main responsibilities.[156] Only three of these financial procurators are attested: one M. Porcius Verus is recorded in a short inscription on a cult relief of Mithras from Celeia, sometime during the reign of Severus Alexander.[157] An inscription of an unknown equestrian, possibly dated to the beginning of the third century, provides evidence for a further ἐπίτροπος Νωρικοῦ.[158] The third procurator is attested in an undated building inscription at Tiffen.[159] A series of inscriptions mentioning imperial slaves and freedmen at Virunum—who according to Alföldy belonged 'to the period of Marcus Aurelius at the earliest'—display their function followed by the term *regnum Noricum* or the abbreviation PRN (*p(atrimonii?) r(egni)*

[152] Alföldy 1974: 115.

[153] *CIL* V 810 (Aquileia). On date, cf. Panciera 1957: 27 f.; Alföldy 1974: 115, 319 fn. 86.

[154] *CIL* III 14362 (p. 2328,197) + 14363 = *AE* 1968: 408 = *ILLPRON* 148 (Hohenstein), cf. Alföldy 1969*b*: 25, no. 35.

[155] Alföldy 1974: 79; for a list of presidial procurators, cf. Winkler 1969: 29–66; Alföldy 1974: 242–7 (Appendix V).

[156] Pflaum 1960–1: 801; Winkler 1969: 139; Alföldy 1974: 164; Ørsted 1985: 232–4.

[157] *CIL* III 5317 (Celeia). He also is known as the tribune of the *cohors I milliaria Hemesenorum* in Lower Pannonia from a votive altar of Szabadegyháza in Hungary, dated by Alföldy to the reign of Severus Alexander. He probably reached the financial procuratorship of Noricum shortly after, cf. *AE* 1965: 10, cf. Devijver 1976–80: 674 f., P 98.

[158] *AE* 1941: 63 (Alexandria).

[159] *AE* 1968: 409 (Tiffen) with Alföldy 1970: 176; Alföldy 1974: 164, 334 fn. 34.

N(orici)) and hence probably were part of an administrative bureau attached to the financial procurator.[160]

4.1.10. Mining Procurators in Dalmatia and Pannonia

4.1.10.1. Procuratores metallorum/argentariarum *of Pannonia and Dalmatia*

A considerable number of partly fragmentary inscriptions from *Domavium*/ mod. Srebrenica—a municipality located amidst the extensive mining area of the Drina Valley (cf. 3.2.5)—document the presence of procurators there. One inscribed monument from Domavium, dated to the early third century AD, contains a very fragmentary version of the equestrian career of C. Iulius Silvanus Melanio which had been reconstructed skilfully by Juan Manuel Abascal and Géza Alföldy. Based on their reconstruction of the text, Melanio was perhaps honoured by the *municipium Domavianum* on receiving the dignity of patron of the province of Dalmatia. The findspot demonstrates Melanio's close links with the municipality and particularly the mining 'industry' in its vicinity. Melanio appears to have held procuratorships concerned with iron and silver mines in numerous provinces probably during the late second century, before becoming financial procurator in Hispania Citerior *c.* AD 198–206. These positions probably included the procuratorship of silver mines in Pannonia and Dalmatia and may explain the origin of the close ties with Domavium.[161] Less enigmatic is a building inscription at Domavium from the reign of Macrinus (AD 218–20) which gives the name of an equestrian *procurator argentariarum*, Valerius Super, as the overseer of the reconstruction of a *macellum*. A further inscription documenting the supply of sufficient water for the bathhouse by Valerius Super can firmly be dated to the

160 *CIL* III 4797 = *ILS* 1506. *CIL* III 4800 = *ILS* 4198. *CIL* III 4798. 4828. Winkler 1969: 141, no. 1, cf. Alföldy 1970: 171; Ørsted 1985: 232 f.

161 *CIL* III 12732, cf. Abascal & Alföldy 1998: 159 f. Nr. 6, 162 f., reconstruct lines 7–8 of the inscription as follows: *proc(uratori) Aug(usti) [fe]ṛṛ(ariarum) ẹt | ạrg(entariarum)(?) per prouin/[cias] XXIII*. . . . J. M. Abascal and G. Alföldy (pp. 162 f.) assume that some of Melanio's mining procuratorships covered more than one province which would explain the 'twenty-three provinces' refered to above. Further inscriptions by Silvanus Melanio have been recovered at Astorga (ae 1968: 229–31 = irpl 2, 13, 14) and Segobriga (*CIL* II 3136) in Spain, where around AD 198–206 he is documented as *procurator Augustorum Hispaniae citerioris*, shortly after the office of *procurator Asturiae et Callaciae* had been abolished. Moreover, an inscription at Lyon (*CIL* XIII 1729), names him in a further unknown procuratorial function, cf. Pflaum 1960–1: 734 f., no. 276; *PIR*^{2} I 581; Fitz 1972: 223 f., no. 6; Devijver 1976–80: 487 f., I 126; Pflaum 1982: 67 f., no. 276; Haensch 1997: 395, 488; Lefebvre 1998: 248 f., 252–4, 262; Alföldy 2000*a*: 65.

year 220.[162] In the later years of Severus Alexander's reign, between AD 229 and 235, the equestrian procurator Iulius Tacitianus appears as the dedicator of an inscribed monument by the *ordo municipii Domavianorum* to the emperor and his mother Iulia Mamaea.[163] According to a further inscription, a collapsed bath house was rebuilt under the supervision of a *procurator argentariarum*, Aurelius Verecundus, in AD 274.[164] A number of rather fragmentary inscriptions document further *procuratores* at Domavium: an M. Arrius [?Iu]nianus is probably to be identified as a [*pro*]*c(urator) Aug(usti)*, while two fragmentary inscriptions record an unknown *pr*]*oc(urator?) ar*[*gentaria*]*rum* and an unknown *v(ir) e(gregius)* [*proc(urator) A*]*ug(usti) n(ostri)*.[165] While these are, as yet, impossible to date, the inscription of an unknown equestrian *procurator*, [–]tus, can be dated to the reign of Gordian III.[166] Even though the full title of this procuratorial office is not provided, it is possible that they all held the same or a similar post. The inscribed monument honouring L. Domitius Eros, *vir egregius*, is the only one to provide the full title of the office, *procurator metallorum Pannonicorum et Delmaticorum*.[167] This allows us to include further equestrian officials who are not directly attested at Domavium, but held the same procuratorial post at some point in their careers. According to an inscribed honorary monument from Lambaesis in Numidia for Ti. Claudius Proculus Cornelianus, this equestrian also held the office of *procurator metallorum Pannonicorum et Dalmaticorum*, perhaps around AD 161/2.[168] Ti. Claudius Xenophon, whose

[162] *CIL* III 12733 = 8363 = *ILS* 5587. *CIL* III 12734, cf. Pflaum 1960–1: 1063; Fitz 1972: 216; Fitz 1993–5: 1090, no. 719.

[163] *CIL* III 12720 = 8359 + 8360 (AD 229–235), cf. *PIR*^{2} I 595; Pflaum 1960–1: 399, 1063; Fitz 1972: 216 f.; Fitz 1993–5: 1093, no. 726.

[164] *CIL* III 12736 with Patsch (AD 274), cf. Pflaum 1960–1: 1063; Mrozek 1968: 47, no. 16; Fitz 1972: 216 f.; Fitz 1990: 189, no. 203; Fitz 1993–5: 405, 1099, no. 738.

[165] *CIL* III 12725 + 14219^1. For M. Arrius [-]nianus, cf. *PIR*^{2} A 1085; Pflaum 1960–1: 399, no. 13, 1063; Mrozek 1968: 47, no. 15; Fitz 1972: 216; Fitz 1990: 189, no. 127; Fitz 1993–5: 1095, no. 231; *CIL* III 12737, cf. Pflaum 1960–1: 1063; Fitz 1972: 216. *CIL* III 12728, cf. Fitz 1993–5: 1097, no.735.

[166] *CIL* III 12724. Pflaum 1960–1: 1063; Fitz 1972: 216; Fitz 1993–5: 1093 f., no. 727.

[167] *CIL* III 8361 = 12721 = *ILS* 1443. On L. Domitius Eros: L. *PIR*^{2} D 145; *RE* V 1427, no. 56; Pflaum 1960–1: 399, 1063; Dobó 1968: 180, no. 282; Fitz 1972: 216; Fitz 1993–5: 1095, no. 729.

[168] *AE* 1956: 123 = 1991: 1691 = 1992: 1866; Pflaum 1960–1: 400; Christol 1990; Le Bohec 1992*b*; Fitz 1993–5: 404. Cornelianus' career can be dated on the basis of the information provided by his career inscription, as he held the office of *procurator(. . .) ad dilectum cum Iulio Vero per Italiam tironum II legionis Italicae*. As a response to the imminent barbarian incursions, Iulius Verus, together with M. Claudius Fronto, was summoned to Rome to recruit two new legions, the II and III Italica, probably around 166/7, cf. *CIL* III 1457 = *ILS* 1097. Prior to that, Cornelianus held the post of *procurator kalendarii Vegetiani*, which, according to Pflaum's hypothesis, is to be dated to before 166. Pflaum dates Cornelianus' post as *procurator provinciae Syriae ad rationes putandas* to the year 158, cf. Pflaum 1955: 126; Pflaum 1960–1: 1049. This possibly leaves the years around 161/2 for his tenure of the Pannonian and Dalmatian mining

career is displayed on an inscribed monument at Ephesos, was appointed *proc(urator) argentariarum Pannoniarum et Dalmatiarum*, probably in the second half of the 180s.[169] Thus, the chronological framework for the known procurators and their tenure of this centenarian procuratorship can be reconstructed as follows:[170]

Ti. Claudius Proculus Cornelianus	*procurator metallorum Pannonicorum et Dalmaticorum*	161/2–164
L. Domitius Eros	*proc. metallorum Pannon(icorum) et Delmat(icorum)* (Domavium)	*c.* late 2nd c.
Ti. Claudius Xenophon	*procurator argentariarum Pannoniarum et Dalmatiarum*	*c.*185–90
C. Iulius Silvanus Melanio	procurator (argentariarum?) (Domavium)	Sep. Sever.
Valerius Super	*procurator argentariarum* (Domavium)	*c.*218–20
Iulius Tacitianus	*procurator Augusti nostri* (Domavium)	*c.*223–35
[–]tus	*procurator eius* (Domavium)	*c.*238–44
Aurelius Verecundus	*proc. argentariarum* (Domavium)	274
Undated:		
N.N.	*[–pr]oc(urator) ar[gentaria]rum* (Domavium)	?
M. Arrius [..]nianus	*procurator Augusti* (Domavium)	?
N.N.	*procurator Augusti nostri* (Domavium)	?

The evidence records the use of the titles *procurator metallorum Pannonicorum et Da/elmaticorum* or *procurator argentariarum Pannoniarum et Dalmatiarum* as early as the reign of Marcus Aurelius. Fitz argued that the term

procuratorship, cf. Pflaum 1960–1: 397–404, no. 164 bis; Dobó 1968: 179, no. 280; Fitz 1972: 221f., no. 4; Corbier 1974: 232, no. 8; Devijver 1976–80: 267 f., C 174; Fitz 1993–5: 723 f., no. 400.

[169] *CIL* III 6575 = 7127 = *ILS* 1421 = *IK* 13, 652. In the years AD 179/80, Xenophon is documented in papyri as ἐπιστράτηγος in Roman Egypt, cf. Thomas 1982: 189, no. 52. According to two inscriptions from Sucidava and Porolissum (*CIL* III 8042 = *IDR* II 188; *AE* 1988: 977), Xenophon subsequently became the financial procurator of Dacia Apulensis, probably not before AD 183. He succeeded C. Sempronius Urbanus, who is attested after 181. It is after this post that he reached the silver mining procuratorship, cf. *PIR*[2] C 1054; Pflaum 1960–1: 590–2, no. 222; Dobó 1968: 179 f., no. 281; Fitz 1972: 222 f., no. 5; Fitz 1993–5: 735 f., no. 419.

[170] Mrozek (1968: 47, no.4) added a Bai[us–] to the list of procurators, but the fragmentary state of the inscription does not allow a secure reconstruction of his career. That Bai[us–] held the Pannonico-Dalmatian mining procuratorship is a pure conjecture, cf. Fitz 1972: 218; Fitz 1993–5: 405.

argentariae and *metalla* were used alternately by the officials in charge.[171] On the contrary, the terminological change might be significant. The inscription of Domitius Eros, which is not securely dated, might belong to the later second century AD rather than the first half of the third.[172] The title used (*proc. metallorum*) would support a date for Domitius Eros holding office during Marcus Aurelius' reign, perhaps after Ti. Claudius Proculus Cornelianus' tenure of the Pannonico-Dalmatian mining procuratorship. If this hypothesis is correct, the titles would have changed during Commodus' reign from *proc. metallorum.* to *proc. argentariarum*—a post first documented in *c.*185–90 (Ti. Claudius Xenophon). Assuming the *procurator metallorum Pannonicorum et Dalmaticorum* supervised not only the silver mines, but also monitored the copper, iron, and gold mines in these provinces (hence, the use of the more general term *metalla*), this change perhaps reflects the deliberate exclusion of mines other than *argentariae* from the responsibilities of this procurator.[173]

4.1.10.2. *Procuratores ferrariarum (Pannonicarum ?)*

The iron mines around Ljubija, for example, might have been set apart and assigned to another administrative department. The epigraphic record of Ljubija and the surrounding area may add weight to this hypothesis: an inscribed altar set up to Terra Mater by the *vilicus* Callimorphus on 21 April, AD 201, names a C. Iulius Agathopus, a *conductor ferrariarum.* Agathopus' exact title is given by an inscription at Mursa as *c(onductor) f(errariarum) Pannonicar(um) itemque provinciar(um) transmarinarum.*[174] On 21 April

[171] Fitz 1993–5: 403 f.

[172] A. Stein (*RE* V 1427: no. 56) believed the inscription to be no earlier than the 2nd cent. Pflaum 1960–1: 399, 1063, suggested the first half of the 3rd cent. without providing a detailed examination. The inscribed monument contains the following text (*CIL* III 12721 = 8361 = *ILS* 1443, with Fitz 1993–5: 1095, no.729/1): *L. Domitio | [.] Eroti, vi|ro ex eques|tribus tur|mis egregio, | proc[u]ratori | metallorum | [P]ann[o]n. | [et] Delm[a]t., mi|[r]ae integritatis | [e]t bonit[a]tis | L. Aur. Ru[s]ticus, | v.e., duce⟨n⟩., amico | praesta[ntissimo].* The monument was commissioned by an L. Aur(elius) Ru[st]icus, *vir egregius*, and *duce⟨n⟩(arius)* for L. Domitius Eros, *vir egregius.* The title *vir egregius* and the pay grade of *ducenarius* for equestrian offices are attested since the reign of Marcus Aurelius, as is the post of *procurator metallorum Pannonicorum et Delmaticorum* (cf. *AE* 1956: 123 = 1991: 1691 = 1992: 1866; Pflaum 1950: 210 ff.; Alföldy 1975: 148; Eck 1997*a*: 101; for criticism, cf. Millar 1963*b*). Attributes such as 'mirae bonitatis et integritatis' do appear mostly throughout the third century and later. Even so, earlier attestations can be cited, cf. *AE* 1972: 70 (AD 176).

[173] For a similar result, see Škegro 2000: 101, 116.

[174] *AE* 1973: 411 = *ILJug* 779; for Mursa cf. Bulat 1989: 36. Based on the inscription from Mursa it has been convincingly argued that the iron mines of Ljubija formed part of Pannonia, cf. Dušanić 1971*b*: 554; Fitz 1993–5: 407 f.

209, an altar for Terra Mater is set up by the same Callimorphus and one T. Flavius Verecundus, *procurator Augustorum*.[175] The series of inscribed altars for Terra Mater, mostly set up on 21 April, at Ljubija, Prijedor, and Briševo yields further procurators: a Iulius [–] during the reign of Elagabalus (AD 218–22); M. Iulius Macer in AD 228; Nicoma[chus] in AD 229; Co[ss]itianus Firmus in AD 248; and an unknown procurator, possibly during the reign of Valerian and Gallienus.[176] In none of the inscriptions is the full procuratorial title given. As the altars to Terra Mater were mostly set up by *vilici officinae ferrariae* in conjunction with the procurators or under their oversight, one may conclude that these procurators carried administrative responsibility for the iron mines around Ljubija. Given the existence of a *conductor ferrariarum* at Ljubija, it is likely that these equestrian officials bore the title *procurator ferrariarum*.

However, it is difficult to delimit the geographical area in which these procurators operated. The votive inscriptions indicate that the procurators either supervised the erection of altars to Terra Mater (*sub cura illius*) or commissioned them, suggesting their frequent presence at Ljubija. Their responsibilities thus might have been restricted to the local mines only. However, bearing in mind the example of Iulius Agathopus, the *conductor ferrariarum Pannoniarum itemque transmarinarum* attested at Ljubija, the geographical scope of the procuratorship may not have been limited to Ljubija, but included other iron mines throughout Pannonia, perhaps even other Danubian provinces (cf. 6.1.3.4).

The inscriptions at Ljubija and surrounds provide us with the following list of *conductores* and *procuratores ferrariarum*:

C. Iulius Agathopus	*conductor ferrariarum (*Mursa: *Pannoniarum itemque transmarinarum)*	AD 201
T. Flavius Verecundus	*procurator Augg.*	AD 209
Iulius[–]	*procurator*	*c.* AD 218–22
M. Iulius Macer	*procurator Aug.*	AD 228
Nic[o]ma[chus]	*procurator Aug.*	AD 229

175 *AE* 1958, 63 = *ILJug* 157 (Ljubija). T. Flavius Verecundus is perhaps identical with his namesake, a centurio legionis XIIII Geminae from Savaria based at Carnuntum (*CIL* III 4416), cf. Fitz 1993–5: 747, no. 433a = 629, with further bibliography.

176 Iulius [-]: *AE* 1973: 412 = *ILJug* 778 (Ljubija), cf. Fitz 1993–5: 1089 f., no.718. M. Iulius Macer: *AE* 1958: 64 = *ILJug* 158 (Ljubija), cf. Pflaum 1982: 122; Fitz 1993–5: 1092, no. 724. Nicoma[chus]: *AE* 1973: 414 = *ILJug* 781 (Ljubija), cf. Pflaum 1982: 122; Fitz 1993–5: 1092 f., no. 725. Co[ss]itianus Firmus: *CIL* III 13240 = *ILJug* 161 (Prijedor), cf. Pflaum 1960–1: 1062; Pflaum 1982: 122; Fitz 1993–5: 1096, no. 733. Anonymus: *CIL* III 13329 = *ILJug* 162 (Briševo). Pflaum 1960–1: 1062; Pflaum 1982: 122; Fitz 1993–5: 1097, no. 736.

Co[ss]itianus Fir[m]us	*procurator Augg.*	AD 248
N. N.	*procurator*	Valerian/Gallienus?

In the light of the numbers of inscriptions recording equestrian procurators at Ljubija/Prijedor and at Domavium it is very likely that the above list and the list of *procuratores metallorum/argentariarum* for Pannonia and Dalmatia are mutually exclusive. Thus, one may tentatively propose that the iron mines in the vicinity of Ljubija—perhaps together with other *ferrariae* in Pannonia and beyond—formed part of an administrative branch different from the one run by the equestrian officials documented at Domavium.

It is not possible to determine on current evidence whether the iron mines had been separately managed early on in the history of the province or whether a separation from the administration of other *metalla* took place at a later time. Even though the earliest altar to Terra Mater at Ljubija originates in AD 201 and procuratorial oversight of the iron mines there is attested throughout the first half of the third century AD, that does not mean the procuratorship did not exist prior to 201. In this context a further inscription is of interest: Flavius Verus Metrobalanus, *procurator Augusti* and *praepositus splendidissimus vectigalis ferrariarum* had an altar set up at Siscia in Pannonia Superior by Asclepiades, the *arcarius stationis Siscianae,* either during the reign of Commodus or Septimius Severus.[177] As Metrobalanus' title clearly indicates he was in charge of the *vectigal* levied on iron mines. Again, no closer geographical specifications are given. Based on the inscription of a *conductor fer(rariarum) N(oricarum) P(annoniarum) D(almatarum)* from Hohenstein in Noricum, András Mócsy concluded that Metrobalanus' procuratorial post also covered the *vectigal* from iron mines in these three provinces.[178] A parallel might be seen in the post of *procurator ad vectigal ferrariarum Gallicarum*, responsible for the *vectigal* on iron mines in the provinces of 'Gaul' (cf. 4.1.12). As yet, there is no evidence to confirm this view and the activities of the *procurator* and *praepositus vectigalis ferrariarum* may have been limited to Pannonia only. Be that as it may, the *ferrariae* of Pannonia, and perhaps Dalmatia and other provinces (cf. 6.1.3.4), were administered separately from the other mines, perhaps as early as the 180s.

Provided the preceding line of argument is correct, this observation coincides with the change in title recorded for the mines in Pannonia and

177 *CIL* III 3953 = *ILS* 3094, cf. *PIR*^{2} F 396; Pflaum 1960–1: 1064; Dobó 1968: 180, no. 283; Fitz 1972: 224 f.; Fitz 1993–5: 738, no. 422.

178 Mócsy 1962: 594.

Dalmatia from *procurator metallorum* to *procurator argentariarum* during the later years of Commodus' reign. This coincidence thus may reflect an exclusion of iron mines from the tasks of the *procurator metallorum* for the three provinces.[179]

4.1.10.3. *Procuratores argentariarum Pannonicarum, and a procurator argentariarum Delmaticarum*

Three *procuratores argentariarum Pannonicarum*, in charge of the silver mines in Pannonia, are known before the reign of Marcus Aurelius. M. Antonius Fabianus is known from an inscription at Viminacium as *procurator argentariarum Pannonicarum*, a post he probably held during the first half of the second century AD.[180] An inscription from Attaleia in Pamphylia furthermore records L. Creperius Paulus' tenure of this procuratorship, perhaps between 146 and 150.[181] L. Septi[–] Petro[nianus], was appointed to the post of *procurator argentariarum Pannonicarum* at some point after 143, probably as successor to Creperius Paulus.[182] Until Marcus Aurelius' reign, when the first *procurator metallorum Pannonicorum et Dalmaticorum* is documented, the Pannonian silver mines were supervised by their own equestrian procurator. Apart from the inscriptions at Domavium for the silver mines of both provinces, a further inscription names a [*procurator*] *arg(entariarum) Delmatica*[*rum*].[183] Based on the titulature, Pflaum is probably right in dating him to the period before the reign of Marcus Aurelius. A further procurator for the mines of Dalmatia(?) may be attested in a fragmentary inscription from Glamoč. However, the title of the former mining procurator, *ex pro* [*cura*]|—[*to*]*ṛẹ metallorum* [—], is not given in full nor can the inscription be dated.[184] Consequently, one may hypothesise that the (silver-) mines in Dalmatia and Pannonia each had two separate procuratorial posts attached to them during the mid-second century.

179 The specifics of this alteration remain obscure, as the exact responsibilities of a *conductor ferrariarum* or a *praepositus vectigalis ferrariarum*—which might well be mutually exclusive—are unknown.

180 *AE* 1905: 152 = *ILS* 9019. On the date of Fabianus' career, cf. Pflaum 1960–1: 352–4, 978, no. 150; Dobó 1968: 178, no. 278; Fitz 1972: 219 f.; Pflaum 1978: 146, no. 2; Fitz 1993–5: 716 f., no. 393; France 2001: 184–91.

181 *AE* 1915: 46 = *RHP* 72. On dating Paulus' career, cf. *PIR*^{2} C 1571; Pflaum 1960–1: 345–7, no. 146; Dobó 1968: 178, no. 277; Fitz 1972: 218, no. 1; Devijver 1976–80: 306 f., no. 255; Fitz 1993–5: 404, 717, no. 394.

182 *AE* 1958: 156 (Caesarea/Mauretania) with Leveau 1973: 161. On the date of Petronianus' procuratorship, cf. Pflaum 1960–1: 975–8, no. 146 bis; Dobó 1968: 469, no. 13; Fitz 1972: 220 f., no. 3; Devijver 1976–80: 89 f., no. 104; Fitz 1993–5: 722 f., no. 399.

183 *CIL* III 12739+12740 = *AE* 1948: 243 = *ILJug* 83 (Tegare, 10 km east of Srebreniča, mid-2nd cent. AD?).

184 *ILJug* 1655 (Glamoč), with D. Sergejevski, *GZM* 39, 1927: 260 fn. 9.

The title of Ti. Claudius Proculus Cornelianus, *procurator metallorum Pannonicorum et Dalmaticorum*, is regarded as demonstrating unification of the various mining districts in Pannonia and Dalmatia under the supervision of one *procurator metallorum* at the beginning of Marcus Aurelius' reign.[185] The epigraphic material does not provide any information on the separate existence of the procuratorships for the *argentariae Pannonicae* or *argentariae Delmaticae* after the mid-second century, which in turn supports the notion of a unified mining administration. The written sources leave us in the dark on the particulars of this measure instituted during Marcus Aurelius' reign. Provided our interpretation of the changes in titles, i.e. from *proc. metallorum* to *proc. argentariarum* of both provinces, is correct, the management of iron mines in Pannonia and perhaps in Dalmatia as well, was excluded from the functions of the procurator for Pannonian and Dalmatian silver mines as early as the 180s.

4.1.11. *Procuratores ferrariarum* in Gaul?

In his landmark study on procuratorial posts in equestrian careers, Hans-Georg Pflaum provided a list of 'procuratores ferrariarum Galliarum'. He believed that at least nine *procuratores* held this office of centenarian rank at Lugdunum/mod. Lyon.[186] His theory seems to be corroborated by two inscribed monuments of *tabularii (rationis) ferrariarum* at Lyon, and by an inscribed statuary base at Lyon honouring Attius Alcimus Felicianus, one of Pflaum's 'procuratores ferrariarum Galliarum', in the early third century. In his inscription, however, Felicianus' procuratorial title is only given as *procurator ferrariarum* without an attributed toponym.[187] Moreover, the inscribed honorific monument of Abbir Cella in Africa for C. Attius Alcimus Felicianus, also fails to provide a geographic area for the iron-mining procuratorship he held.[188] Based on the attestation of Attius Alcimus Felicianus at Lyon, it was assumed that the title *procurator ferrariarum*, without any closer geographical indication, must refer to a procuratorship held in Gaul, irreverent of where the inscription was set up. Hence, a whole series of inscriptions naming

185 Fitz 1993–5: 404 f. For a different view cf. Dušanić 1977: 67.

186 Pflaum 1960–1: 1053; Pflaum 1982: 119, Sablayrolles 1989: 158.

187 *CIL* III 1797 (Lugdunum). Pflaum 1960–1: 843–9, no 327; Pflaum 1978: 169–72, no. 2.

188 *CIL* VIII 23948 (Abbir Cella/Henchir en-Naâm, Africa procos); cf. also *CIL* VIII 822 = 12345 = 23963 = *ILTun* 741 (Bou Cha) cf. also cf. Pflaum 1978: 169–72.

procuratores ferrariarum of centenarian rank were held to be iron-mining procurators in Gaul.[189] Pflaum provides the following list:[190]

C. Caelius Martialis	*procurator ferrariarum*	*c.* AD 112
Publilius Memorialis	*[procurator] Augusti (...) ferrariarum [-*	Trajan
T. Statilius Optatus	*procurator Augusti ferrariarum*	Hadrian
T. Petronius Priscus	*procurator Augusti ferrariarum*	AD 117–93
Q. Domitius Marsianus	*procurator Augusti ad ferrarias*	*c.* AD 175
C. Iulius Silvanus Melanio	*procurator Augusti* (Lugdunum)	AD 198–209
M. Cosconius Fronto	*procurator Augustorum duorum ad vectigal ferrariarum Gallicarum*	AD 198–209
Domitius Antigonus	*[procurator fe]rrariar(um) [-*	AD 212–18
C. Attius Alcimus Felicianus	*procurator ferrariarum* (Lugdunum)	*c.* AD 231

As can be easily seen from the list, the title 'procurator ferrariarum Galliarum' is not directly attested in this form, but is merely a plausible hypothesis by Pflaum. Only M. Cosconius Fronto provides a full title for his post: *procurator Aug(ustorum duorum) ad vectig(al) ferr(ariarum) Gallic(arum).*[191] In comparison to other procuratorships listed in inscriptions which detail the careers of equestrian officials, the absence of a geographical description from the title 'procurator ferrariarum' is suspicious. A reappraisal of the evidence thus seems necessary.

As stated above, the epigraphic record at Lugdunum offers evidence for imperial officials involved in the administration of iron mines during the second century. An Appianus, *libertus Augusti* and *tabularius rationis ferrariarum*, set up a honorific monument for C. Iulius Celsus, financial procurator for Lugdunensis and Aquitania. Celsus may have held this post around AD 135–45.[192] Given Appianus commissioned an inscribed monument for the financial procurator, it appears likely that the latter was his superior. Consequently, the *ratio ferrariarum* was possibly administered within the provincial offices under the supervision of the financial procurator Iulius Celsus rather than being a separate bureau created specificly for the mining

[189] Pflaum 1960–1: 82 f., 291, 558, 707, 734 f.

[190] Pflaum 1960–1: 1053; Pflaum 1982: 119.

[191] *CIL* X 7584 (p. 995) = *ILS* 1359 (Caralis), cf. also *CIL* X 7583 (Caralis). *CIL* X 7860 (Forum Traiani). Pflaum 1978: 138; Pflaum 1982: 119.

[192] *CIL* XIII 1808. The monument is dated by an additional inscription commemorating the admission of Iulius Celsus' son, Maximianus, into the senatorial order under Antoninus Pius, cf. Hirschfeld 1905: 157; Pflaum 1960–1: 253–7, no. 106 bis (135).

administration.[193] The inscribed funerary monument of a further *tabularius ferrariarum* at Lyon, one Aurelius Calocaerus, commissioned during the period of Marcus Aurelius or the Severans, provides no insights into his administrative affiliation.[194] Apart from the doubts arising over the existence of an independent iron-mining bureau at Lugdunum, none of the *tabularii*-inscriptions offers a clue to the geographical location of the *ferrariae* in question. This may be explained by the fact that its readers, the inhabitants of Lugdunum, certainly knew which *ferrariae* were meant. Based on the inscribed monument set up by the *tabularius* Appianus for the financial procurator for Lugdunensis and Aquitania, one may assume that the iron mines in both provinces were probably covered by the *ratio ferrariarum*. There appears to be no further evidence at Lyon for subaltern officials (*librarii, dispensatores, arcarii*, etc.) associated with the administration of mines. In comparison with the gold-mining administration at Ampelum or with other spheres of provincial administration run by equestrian or freedmen procurators, the epigraphic record at Lugdunum provides hardly any information about the iron-mining administration allegedly based there.[195] This calls into question the notion of an independent iron-mining bureau for Gaul located at Lyon.

The honorary monument for the *procurator ferrariarum* Attius Alcimus Felicianus is the only inscribed monument explicitly honouring a procurator of iron mines at Lugdunum. It was commissioned by Cogitatius Iuvenis, *beneficiarius legati legionis I Minerviae* [[*Alexandrinae*?]] for his *patronus* Felicianus.[196] Cogitatius was the *beneficiarius* of the legionary legate of *I Minervia* at Bonn and as such was not directly attached to Felicianus. The nature of Cogitatius' patronal relation with Felicianus—whether on a personal or professionel level—is not clarified by the text. The location of the inscribed monument at Lyon appears to imply that Felicianus was or had been present at Lyon in the stated function as *procurator ferrariarum*. Whatever brought about the erection of this honorary monument—it does not provide unambiguous evidence for a permanent residence of the iron-mining procurator at Lyon.

[193] Compare *EE* VIII, no. 26: *Aurelius Rufus, tabul(arius) provinc(iae) Lusit(aniae) rat(ionis) pat(rimonii)*, according to Boulvert 1970: 126, a *tabularius* in the office of the Lusitanian financial procurator with the task of dealing with the emperor's *patrimonium* single handedly; Sachers 1931: 1966–68; Haensch 1997: 492.

[194] *CIL* XIII 1825.

[195] Cf. Boulvert 1970, *passim*; Haensch 1997, *passim*.

[196] *CIL* XIII 1797 = *CBFIR* 27; On beneficiarii and their patroni, cf. Nelis-Clément 2000: annexe 11, 427–9.

It is striking that none of the other procurators of Pflaum's list have yet surfaced in the epigraphic record at Lyon—with the exception of C. Iulius Silvanus Melanio. As *procurator Augusti* he set up a votive monument in Lyon, however, without providing any indication in what procuratorial function he had been at Lyon.[197] Furthermore, none of the other titles of the listed *procuratores ferrariarum* (with the exception of M. Cosconius Fronto's) allow an association with *ferrariae Gallicae.* One of the earliest documented of these procurators is Publilius Memorialis. Memorialis' inscription from Forum Clodii does not give any indication as to where he held his procuratorship of the iron mines (... *item ferrar*[*iarum*] |—). Pflaum simply assumed that after holding two sexagenarian posts, i.e. *procurator Augusti in Africa* and *procurator M*[*inicia*] *e*, Memorialis must have reached a centenarian iron-mining procuratorship, most likely the one in Gaul.[198] The honorific monument in Corinth for C. Caelius Martialis places the position of *procurator ferrariarum* after his tenure as *procurator provinciae Achaiae.*[199] Martialis reached the procuratorship of the province of Achaia probably shortly after the second Dacian war of Trajan (AD 105–6), before becoming *procurator ferrariarum.*[200] His career does not suggest that the position or he himself was affiliated with the western provinces or with Gaul, and thus can not be used to argue for his tenure of the centenarian iron-mining procuratorship. In an inscription from Bulla Regia in Africa, dating to the reign of Marcus Aurelius, Q. Domitius Marsianus carries the title of *procurator ad ferrarias.*[201] As he held this post after being *procurator Augusti ad census in Gallia accipiendos provinciarum Belgicae* etc. and before being promoted to *procurator patrimonii provinciae Narbonensis*, Sablayrolles assumed that Marsianus remained in Gaul to head the administrative bureau for iron mines at Lugdunum.[202] Sablayrolles argued the same thing in the case of T. Statilius Optatus. Given the geographical proximity of Optatus' posts as *procurator Augusti ad census Britanniae* and *procurator Augusti ad census Gallorum* with that of *procurator ferrariarum*, Optatus must have been in charge of the iron mines in Gaul.[203]

[197] *CIL* XIII 1729, cf. Pflaum 1960–1: 734 f., no. 276; Pflaum 1982: 67 f., no. 276; Abascal & Alföldy 1998: 163.

[198] *CIL* XI 7554 = *ILS* 9195 (Forum Clodii, Etruria). His career has been dated to the reign of Trajan, based on an inscription (*CIL* X 8038) from Corsica from the year AD 77 or 79 naming a homonym—perhaps his father—as procurator of Corsica, cf. Pflaum 1978: 135 f., no. 1, 140.

[199] *AE* 1934: 2 = Corinth VIII /III 63, no. 135.

[200] Pflaum 1960–1: 170, no. 74; Pflaum 1978: 136, no. 2.

[201] *AE* 1960: 167 = *AE* 1962: 183 (Bulla Regia).

[202] Pflaum 1971; Pflaum 1978: 115–24; Pflaum 1982: 119; Sablayrolles 1989: 158.

[203] Sablayrolles 1989: 158, cf. *CIL* VI 31863 = 41272; cf. Pflaum 1960–1: 289–92, no. 119; Pflaum 1978: 136–8, no. 3. Attius Alcimus Felicianus also held the procuratorships for the

Yet, following the office of *procurator ferrariarum*, Optatus did not remain in Gaul, but was appointed *procurator ad patrimonium*, *procurator hereditatium*, *procurator a [rationibus?]* and finally *praefectus annonae*.[204] All these offices were held in Rome, and one thus might reach the opposite conclusion, namely that the post of *procurator ferrariarum* was also held in the capital of the empire. If one follows Sablayrolles, the career of T. Petronius Priscus sheds an additional light on this matter. After having served his *tres militiae*, Priscus was promoted to *procurator Aug(usti) ferrariarum* **et** *annonae Ost-i(ensi)s*, according to a second-century inscription found at Ostia.[205] Based on the conjunction '*et*' Sablayrolles argued that Priscus held both posts at Rome/Ostia.[206] There is therefore only ambivalent evidence for an independent procurator of the iron mines at Lyon and for the presence of a *procurator ferrariarum* there prior to the reign of Septimius Severus. This, together with the absence of a geographical area in the procurator's title, may indicate that the office of *procurator ferrariarum* documented in these inscriptions was held in or near Rome.[207]

For the other entries in Pflaum's list of 'procuratores ferrariarum Galliarum' the epigraphic evidence is inconclusive: the lines following the term *ferrar*[*iarum*] in Publilius Memorialis' inscription from Forum Clodii once may have provided a clear geographical indication, but they are now destroyed.[208] Similarly precarious is our knowledge of the career of Domitius Antigonus. As *legatus legionis XXII Primigeniae* he set up an inscribed monument honouring the emperor Elagabal at Mogontiacum/mod. Mainz around AD 220 and detailing his own career. Apart from other posts, he is listed holding the office of [*procurator fe*]*rrariar*(*um*)[—].[209] Domitius Antigonus' tenure as procurator of the iron mines has been dated to the years between AD

quadragesima Galliarum and for the *annona provinciae Narbonensis* prior to being appointed *procurator ferrariarum*, cf. *CIL* VIII 822 = 12345 = 23963 = *ILTun* 741 and *CIL* VIII 23948, with Pflaum 1978: 169–72.

204 Pflaum 1978: 137 f.

205 *CIL* XIV 4459 = *ILS* 1442 (Ostia), cf. Pflaum 1960–1: 558, no. 212; Pflaum 1978: 138, no. 4.

206 Sablayrolles 1989: 158.

207 Ostia seems a likely candidate as a 'hub' of Roman iron-mining administration: a votive inscription commissioned by Hilarus, *servus socior(um) vect(igalis) ferr(ariarum)* between AD 102–117 at Ostia and the find of an undated 'tessera nummularia' inscribed with *soc(iorum) ferr(ariarum)* in Rome corroborate this assumption (cf. *AE* 1924: 108 = *CIL* XIV 4326 [Ostia]; *AE* 1928: 17a [Rome]). Moreover, a lead *tessera* naming a *statio ferrariarum fori Ostiensis* further highlights the importance of Ostia (*CIL* XIV 4326, p. 773).

208 *CIL* XI 7554 = *ILS* 9195.

209 *AE* 1966: 262 = *AE* 1965: 242, cf. Alföldy 1965*b*: 187–91, with 'Nachträge' in Alföldy 1987*a*: 366 f.

212 and 218, however the lines referring to the geographical region in which he served have been lost.[210]

In consequence, only one *procurator ad vectigal ferrariarum* for Gaul can be identified beyond doubt, to which one may tentatively add the *procuratores* attested at Lyon. These procurators constitute a chronologically and geographically consistent group of procurators which can be linked with the *ferrariae Gallicae* during the reign of the Severans. As for the preceding *procuratores ferrariarum*, a link with Gaul cannot be established (cf. 6.1.3.4). Provided the above observations are correct, this leaves us with the following list:

procuratores ferrariarum in Ostia/Rome (?)		
C. Caelius Martialis	*procurator ferrariarum*	Trajan
T. Statilius Optatus	*procurator Augusti ferrariarum*	Hadrian
T. Petronius Priscus	*procurator Augusti ferrariarum* (Ostia)	mid-2nd c.?
Q. Domitius Marsianus	*procurator Augusti ad ferrarias*	Marcus Aurelius
procurator ferrariarum 'Galliarum'		
M. Cosconius Fronto	*procurator Augustorum duorum ad vectigal ferrariarum Gallicarum*	AD 198–209
(C. Iulius Silvanus Melanio	*procurator Augusti* (Lugdunum)	AD 198–209)
C. Attius Alcimus Felicianus	*procurator ferrariarum* (Lugdunum)	Severus Alexander
procuratores ferrariarum for unknown areas		
[.] Publilius Memorialis	*[procurator] Augusti (...) ferrariarum [*	Trajan
Domitius Antigonus	*[procurator fe]rrariar(um) [*	AD 212–18

4.1.12. Equestrians, Freedmen, and Procuratorial Titles

Following the study of individual mining and quarrying procuratorships, it is necessary to investigate the social status of the office-holders. As for the incumbents of quarrying procuratorships, their social status was

[210] Alföldy 1965*b*: 191; Pflaum 1982: 65 f., no. 249.

homogeneous. The procurators recorded at the quarries were imperial freedmen, either directly responsible for an individual quarry, a quarrying district (e.g. the Egyptian Eastern Desert), or a patrimonial district (e.g. 'provincia' Phrygia) to which the management of quarrying ventures was perhaps added. In the case of Karystos the supervision of the quarries appears to have been the responsibility of the equestrian patrimonial procurators of Achaia, Minicius Sanctus, and C. Cerialis.[211] As yet, the only equestrian official recorded as being directly in charge of quarrying operations is the μεταλλάρχης, who was in control of all quarries in Egypt during the years 14–18 AD. The μεταλλάρχης is not attested further and seems to have eventually been replaced by *procuratores metallorum* of freedmen status (cf. 4.1.2).

As regards the social status of persons holding mining procuratorships, one can discern considerable differences particularly between the two main mining regions of the empire. Procurators of equestrian rank are recorded mainly in the mining districts or territories of the Danube provinces, increasingly so from the mid-second century onwards. In the case of the gold-mining administration at Ampelum in Dacia, imperial freedman are replaced by members of the *ordo equester* after the Marcomannic wars. The former had held this office from Trajan's reign onwards. Imperial freedmen continued to hold the office of *subprocurator* at Ampelum after the Marcomannic wars, though obviously as subordinates to the equestrian procurators (cf. 4.1.9). This change of status of the procurators of the Dacian *aurariae* seems to have been unique, as similar alterations in social status of the office holders are not clearly detectable. The freedmen procurators known from inscriptions at Sočanica may have been replaced by equestrians (such as the *procurator mm(etallorum) D(ar)d(anicorum?)* dating to the mid-third century), after the reign of Antoninus Pius at the earliest. The validity of this observation, however, is called in question by the absence of any precise procuratorial titles of the incumbents. In fact, it can not even be stated with any certainty whether the freedman in question were mining procurators or imperial officials at all (cf. 4.1.10). Alterations in social status are not otherwise attested: the mid-second century procurators for the silver mines of Pannonia and for Dalmatia, respectively, are all of equestrian rank. The posts of *procurator metallorum/argentariarum* for Pannonia and Dalmatia (which were probably established after the merging of the respective offices for Dalmatia and for Pannonia), *procurator ferrariarum* and *procurator Augusti praepositus splendidissimus vectigalis ferrariarum,* were also held by equestrians. Beyond the

[211] App. Nos. 604, 618; PIR² C 670, M 628.

Danubian provinces the evidence for equestrians holding procuratorial mining offices is limited to the *procurator ad vectigal ferrariarum Gallicarum* and the other centenarian *procuratores ferrariarum* (cf. 4.1.12).

The social status of mining procurators of the Iberian Peninsula stands in marked contrast to their equestrian 'counterparts' in the Danubian provinces. The procurators known from the gold-mining areas in north-western Spain are exclusively imperial freedmen, as are the mining procurators recorded in Lusitania, such as the *procuratores metallorum Vipascensium*, or the mining procurators in Roman Baetica (cf. 4.1.7). The Sardinian mining procurator, who was directly subordinate to the presidial procurator of the island, must also be added to the list of freedmen procurators (cf. 4.1.8).

At first glance, the evidence assembled for the Danubian provinces points towards a marked increase in the appointment of equestrian officials to existing or previously unattested mining procuratorships from the mid-second century onwards. Bearing in mind our lacunose source material (not least the result of local epigraphic habits), this phenomenon reflects the general trend in appointment 'policy', observable throughout the imperial administration during the second century AD, towards replacing imperial freedmen with equestrians or relegating freedmen procurators to the position of deputy to the equestrian office holder.[212] Remarkably, this supposed trend is not mirrored in the known evidence from the Iberian peninsula, where freedmen mining procurators are attested for most of the second century AD, particularly in north-western Spain.

This significant difference in appointment policy requires explanation. One approach lies in the comparison of procuratorial titles: the titles of equestrian office holders from the Danube provinces reflect the magnitude of the organizational task at hand. Setting aside the *procuratores aurariarum* at Ampelum, the equestrian *procuratores* documented at Domavium and elsewhere supervised the *metalla* and later the *argentariae* of Pannonia and Dalmatia, while the procurators documented at Ljubija perhaps had charge of the *ferrariae* of one or more Danubian provinces. As regards the western provinces, the *procurator ad vectigal ferrariarum* apparently supervised the iron mines in all of 'Gaul'. The lacunose inscription at Domavium, which charts the career of C. Iulius Silvanus Melanio, provides additional insights. The detailing of Melanio's mining procuratorships in lines 7 f. is of particular interest: . . . *proc(uratori) Aug(usti) [fe]ṛṛ(ariarum) ẹṭ | ạrg(entariarum)(?) per prouin/-[cias] XXIII*. . . . Provided the reading is correct, Melanio, quite obviously, did not hold twenty-three different iron- or silver-mining procuratorships

[212] Eck 1997*a*: 88–94.

during his career. Some of these mining procuratorships likely included two or more provinces, as is the case with the *procurator metallorum Pannonicorum et Dalmaticorum*, the *procurator Augusti praepositus vectigalis ferrariarum* or the *procurator ad vectigal ferrariarum Gallicarum.*[213] In particular, the administration of iron mines was usually divided into vast superprovincial entities, as can be seen in the titles of *conductores ferrariarum*, eg. *conductor fer(arriarum) N(oricarum) P(annoniarum) D(almatarum)*; *manceps ferrariar(um)* [-]I *et provinciae Raetiae itemque Daciarum trium*; *c(onductor) f(errariarum) Pannonicar(um) itemque provinciar(um) transmarinarum.*[214] One thus might venture to conclude that equestrian mining procuratorships in general included one, two or more provinces within which *metalla* or a certain category of mines (e.g. *ferrariae, argentariae*), were under their control (cf. 6.1.3.4).

In contrast to these equestrian procuratorships, the titles of freedman procurators attested on the Iberian peninsula do not indicate the same geographical expanse.[215] The titles of *procurator metallorum Vipascensium*, *procurator metalli Alboc(rarensis)*, or *procurator montis Mariani* identify specific localities of limited geographical scope. We may thus assume that the other freedman procurators recorded on the Iberian peninsula were also put in charge of mining districts or territories of similar size. There is no evidence to date for the existence of a procuratorial post which was responsible for either all or only one certain category of mines within one or all provinces of the Iberian peninsula.[216] The dearth of evidence for a supra-provincial mining authority might be explained by the fact that in the case of the gold-mining zones in northwestern Spain, for example, the financial *procurator Asturiae et Callaeciae* was ultimately responsible for mining operations in his area. This is suggested by the presence of *beneficiarii* of the financial procurator in the mining zones of the Duerna valley. As some scholars have argued, the financial procurator of Lusitania (like his colleague in Asturia and Callaecia) may have exerted similar power over the *metallum Vipascense* and the mining procurator there (cf. 4.1.7). Hence, the freedmen procurators of mining districts on the Iberian peninsula in all probability answered to the equestrian financial or perhaps even the patrimonial procurator of their province. The subordination of the freedman procurator to the equestrian financial or presidial procurator may also be attested in Sardinia:

[213] Abascal & Alföldy 1998: 162 f.

[214] *CIL* III 4809 = *ILS* 1467 = *ILLPRON* 151 (Hohenstein/Noricum); Nuber 1977: 233 f. and Nuber 1985: 131 (Augsburg/Raetia); Bulat 1989: 36 (Mursa/Pannonia)

[215] Mrozek 1968: 48, 54.

[216] Andreau 1989: 98; Domergue 1990: 291; contra: Schönbauer 1929: 103; Täckholm 1937: 102 f.; Mrozek 1968: 55.

the *procurator metallorum et praediorum* commissioned a votive monument honouring his likely superior, the *procurator Augusti* and *praefectus provinciae Sardiniae* (cf. 4.1.8). Prior to the Marcomannic wars, a similar arrangement might have been in place in Dacia: there, *beneficiarii procuratoris* were probably on secondment from the financial procurator of Dacia Apulensis to the freedmen mining procurators at Ampelum. It thus appears likely that freedmen mining procurators were supervised by the equestrian financial procurators of the province. With the replacement of the freedmen procurator at Ampelum by an equestrian official, the subordination of the procurator at Ampelum to the financial procurator of Dacia Apulensis may have also come to an end.[217] It is not possible to determine on the basis of available evidence whether the replacement of imperial freedmen by equestrian officials went hand in hand with a geographical expansion of procuratorial tasks to include all *auraria* within the Dacian provinces.[218]

The picture which emerges from the above analysis is that there were two administrative systems working side by side which controlled imperial mines: (1) freedmen procurators managing regional mining zones under general supervision of the presidial, financial, or patrimonial procurator responsible; (2) equestrian procurators in charge of various *metalla* or specified mines (*ferrariae*, *argentariae*) within one or more provinces.

4.2. SUBALTERN OFFICIALS

4.2.1. Subaltern Officials in Imperial Mines

The epigraphic record from the administrative headquarters at Ampelum in Dacia provides abundant evidence for imperial freedmen and slaves working in the mining administration, offering us a relatively complete picture of the possible composition of this mining *officium*. The personnel employed at the mines were predominantly members of the *familia Caesaris*: an inscribed

[217] Although the subordination of equestrian officials under an equestrian prefect is documented for Roman Egypt (cf. Bowman 1996*a*: 252 f.; Bowman 1996*b*: 66 f.), there is no evidence to suggest that this was also the case with equestrian mining procurators, cf. Eck 1997*a*: 142.

[218] The sequence of known procuratorial titles from Ampelum which would possibly document such an change is of no help as the titles lack any geographical indication. The inscription (*CIL* III 1297 = *IDR* III/3, 314, cf. Noeske 1977: 352, *AMP* 12) displaying *tab(ularius) aur(rariarum) Dacicarum* (from which one might reconstruct the full title) can not be precisely dated. We thus do not know whether the freedman incumbents were called *procuratores aurariarum Dacicarum* or whether this title began to only be used when the post was filled with equestrian officials.

monument honouring the empress Annia Lucilla was commissioned by the *liberti et familia* and the *leguli aurariarum* in AD 165/6.[219] A *dispensator aurariarum*, Suriacus, provided a votive altar for Ceres Augusta and dedicated it to the procurator A. Senecius [-] Contianus(?) in January 215, and another *dispensator*, Callistus, commissioned an altar for Iuppiter Optimus Maximus Aeternus Conservator, probably sometime after the Marcomannic wars.[220] Another votive monument for Isis, which was erected for the benefit of the mining procurator(?) M. Iulius Apollinaris, names a Verus and a Romanus, both *vilici*.[221] Besides *vilici*, a number of *tabularii* surface in the epigraphic record at Ampelum: the imperial freedman Neptunalis, a *tabularius aurariarum Dacicarum*, commissioned a votive inscription and set up a gravestone to his wife Trophima. She had been manumitted by a M. Ulpius, probably at some point during the second half of the second century AD.[222] Furthermore, a certain Zmaragdus, *tabularius (aurariarum)*, erected a votive altar to Deus Aeternus, together with his wife Aurelia Urbica and his daugther Matrona, probably no earlier than the reign of Marcus Aurelius.[223] A further *tabularius*, [Eutyche]s(?), had a sanctuary built at an unknown date for Aesculapius, whereas a Moschus, imperial freedman and *tabularius*, commissioned an altar to Fortuna Domestica, together with his wife Aurelia Asclepiodote; as the *nomen gentile* of Moschus' freeborn wife suggests, this occurred at the end of the second or the beginning of the third century.[224]

The *tabularii aurariarum* at Ampelum were supported by a number of subaltern clerks: Fuscinus, *ab instrumentis tabularii*, had a funerary stone engraved for his wife Sossia Sabina, possibly sometime after the Marcomannic Wars.[225] A further inscription also records *a commentariis* from the late second or early third century AD: Liberalis, a *(libertus) Augustorum duorum a commentariis*, together with his wife Aelia Victoria, commissioned an inscribed funerary monument for his father Syrillio, a former *aedituus*.[226] Iustinus, *subsequens librariorum*, is documented on an inscribed funerary stone set up by his colleague(?), Tertius, in the second half of the second century AD whereas two

219 *CIL* III 1307 = *IDR* III/3, 283 (= AMP 4), cf. Noeske 1977: 306, 349 with commentary.

220 *AE* 1959: 308 = *IDR* III/3, 292. *CIL* III 1301 = *ILS* 3014 = *IDR* III/3, 289 = *AE* 1965: 34 = *AE* 2003: 1512 (= AMP 14), cf. Noeske 1977: 310 f., 353, see commentary for date. Callistus is also known from a further inscription to Iuppiter Optimus Maximus Aeternus Conservator at Apulum (*CIL* III 1085).

221 *CIL* III 7837 (= AMP 11), cf. Noeske 1977: 306–8, 352.

222 *CIL* III 1297 = *ILS* 1594 = *IDR* III/3, 314 (= AMP 12). *CIL* III 1313 (= AMP 12a), cf. Noeske 1977: 309, 352.

223 *CIL* III 1286 = *IDR* III/3, 288 (= AMP 13), cf. Noeske 1977: 309, 352.

224 *AE* 1959: 306 = *IDR* III/3, 280 (= AMP 19), cf. Noeske 1977: 354. *AE* 1992: 1471.

225 *CIL* III 1315 = *IDR* III/3, 364 (= AMP 17), cf. Noeske 1977: 309 f., 354.

226 *AE* 1959: 305 = *IDR* III/3, 365 (= AMP 18), cf. Noeske 1977: 310, 311, 354.

adiutores tabulariorum are named in two undated votive altars at Ampelum.[227] Additional imperial slaves and freedmen also appear in epigraphic documents at Ampelum, but their function is not stated.[228] The Roman army stationed two clerks at Ampelum: M. Aurelius Antoninus, soldier of *legio XIII Gemina* and a *librarius*, is known by his gravestone; P. Helvius Primanus, *miles legionis XIII Geminae* was a *librarius consularis*. Helvius Primanus, and probably M. Aurelius Antoninus as well, were seconded to the mining administration on the orders of the governor. Noeske has dated both inscriptions to the beginning of the third century.[229] Given the difficulties in dating the inscriptions any chronological outline must remain sketchy, as the following table shows:

procuratores	*dispensatores*	*vilici*	*auxiliary personnel*
Ulpius Hermias, *lib. Aug.* (Traj.)			
Romanus, *lib. Aug.* (M. Aur.)			
			Neptunalis, *lib. Aug.* (mid-2nd c.), *tabularius*
			Iustinus, *verna Caes.* (2nd half 2nd c.), *subsequens librariorum*
	Callistus, *dispensator* (after 175)		
			Fuscinus, *verna* (after 175) *ab instrumentis tabularii*
M]aximus (183/5)			
L. Macrius Macer (M. Aur./Sep. Sev.)			
Avianus, *lib. Aug.* (Sep. Sev.) *subprocurator*			

(Continued)

[227] *CIL* III 1314 = *IDR* III/3, 366 (= AMP 16). AE 1944: 27 = *IDR* III/3, 336 (= AMP 20). *CIL* III 1305 = *IDR* III/3, 323 (=AMP 22), cf. Noeske 1977: 309–11, 353–5.

[228] *CIL* III 1303 = *ILS* 3382 = *IDR* III/3, 319 (=AMP 15). *CIL* III 1335 = *IDR* III/3, 372 (= AMP 21). *CIL III* 1300 = *IDR* III/3, 358 (= AMP 23). *CIL* III 1333 = 7842 = *IDR* III/3, 334 (= AMP 24). *AE* 1983: 822. *AE* 1988: 954.

[229] *CIL* III 1317 = *IDR* III/3, 344 (= AMP 32); *CIL* III 1318 = *IDR* III/3, 354 (= AMP 33), cf. Noeske 1977: 310 f., 359.

procuratores	*dispensatores*	*vilici*	*auxiliary personnel*
			Liberalis, *lib. Augg.* (M. Aur./Sep. Sev.), *a commentariis*
Aelius Sostratus (212)			
A. Senecio [-] Contianus (215)—	Suriacus, *disp.* (215)		
			Zmaragdus, *lib. Aug.* (beg. 3rd c.), *tabularius*
			Moschus, *lib. Aug.* (beg. 3rd c.), *tabularius*
			M. Aurelius Antoninus, *miles leg. XIII Gem.* (beg. 3rd c.), *librarius.*
			P. Helvius Primanus, *miles leg. XIII Gem.* (beg. 3rd c.), *librarius consularis*
Undated			
(M. Iulius Apollinaris(?))——		Verus & Romanus *vernae Aug*(?), *vilici*	
			[Eutyche]s, *lib. Aug. tabularius*
			Leonas, *lib. Aug.*(?) *adiutor tabulariorum*
			[?] *lib. Aug.*(?) *adiutor tabulariorum*

The table provides an incomplete picture of the subaltern personnel involved in the administration of the Dacian gold mines. The bulk of evidence appears to date to the period following the Marcomannic wars with a 'peak' at the beginning of the third century. Whether or not this reflects an increase in extractive and therefore administrative activity resulting in the demarcation of the functions of subaltern personnel can not be answered with any certainty on the basis of the available evidence. The increase in inscriptions of subalterns may merely reflect a change in local epigraphic habits. Moreover, the considerable amount of undatable inscriptions prohibits any confident statements regarding an increase in size or functional differentiation of the procuratorial staff. One may note, however, that military personnel were perhaps added to the subalterns at Ampelum sometime during the third century.

Sadly, the discovery of a wide range of inscriptional testimonies at the headquarters of a mining administration, such as at Ampelum, remains an isolated occurrence. The documentation from other mining areas does not provide sufficient material to allow for comparison. In the Danubian provinces, for example, only few imperial slaves and freedmen can be connected with the mining administration and only occasional finds yield unexpected insights: a gravestone discovered at Salona/mod. Split was commissioned by a *Felicissimus dispesator* (sic) for a Thaumastus *Augusti (libertus?) commentariesis* (sic) *aurariarum Delmatarum*. The latter was associated with the administration of Dalmatian gold mines; both perhaps were part of the financial procurator's office based at Salona.[230] A series of inscriptions discovered at Ljubija and dating to the first half of the third century were set up by *vilici officinae ferrariae* either with or on behalf of their *conductores* or *procuratores Augusti*: in AD 201 the *vilicus* Callimorphus honoured the *conductor ferrariarum* C. Iulius Agathopus and set up an inscribed altar to Terra Mater together with a *procurator Augustorum* in AD 209.[231] From this date on, Terra Mater regularly seems to have received an altar from the *vilicus*: the *vilicus officinae ferrariae* Heliodorus provided one in AD 228, an unknown [–]s, *vilicus officinae ferrariae*, in AD 229, as did the *vilicus ferrariarum* Iucundus in 247/8.[232] Although it is not possible to determine the significance of the change in title of *vilici* between 229 and 247/8, one could cautiously use the recorded change to date a further inscription for Terra Mater: the inscribed altar set up by *vilicus officinae ferrariae* Mercurius therefore may have preceded the altar of Iucundus.[233] At Stari Majdan/Sanski Most a number of other *vilici* are documented on altars dedicated to Nemesis Pia and Sedatus Augustus: a Ianuarius and an Aurelius are the recipients of good wishes from a *corpus* and *collegium*.[234] The *metalla Dardanica* at Sočanica in Moesia Superior have not yielded much information on subaltern officials to date. The imperial freedman P. Aelius Menander, a *7 officinarum*, commissioned a funerary stone for his wife. As Menander was a member of the *familia Caesaris* the reading *contrascriptor* for '7' (instead of *centurio*) appears more likely.[235] Perhaps the *primiscrinius officinae* Raesius Pudens, known from an inscription of Montana in Moesia Inferior, must be included in this specific group of imperial clerks.[236]

In the vicinity of Ljubija and Prijedor in the Japra Valley, where iron mines are situated, the inscribed altar of a Calimorphus, *verna Augusti* and

[230] *CIL* III 1997 (Salona, 1st cent. AD ?), cf. Hirschfeld 1905: 154; Wilkes 1969: 272. On the extraction of gold from Dalmatian mines, cf. Statius, *Silv.* 3.3.89 f.; Florus 2.25.12. On Salona as seat of the financial procurator, cf. Haensch 1997: 81, 423–5.

[231] *AE* 1973: 411 = *ILJug* 779 (AD 201). *AE* 1958: 63 = *ILJug* 157 (AD 209).

[232] *AE* 1958: 64 = *ILJug* 158 (Ljubija, AD 228). *AE* 1979: 414 = *ILJug* 781 (Ljubija, AD 229). *CIL* III 13240 = *ILJug* 161 (Briševo, AD 247/8).

[233] *CIL* III 13329 = *ILJug* 162 (Briševo). Further inscriptions, albeit rather fragmented, might possibly refer to *vilici* as well: *AE* 1973: 412 = *ILJug* 778 (Ljubija, AD 211/17).

[234] *ILJug* 775, 776 (Stari Majdan). The nature of the collegium is not known. The corpus might have included miners (*fabri ferrarii*), cf. *ILJug*, p. 107, commentary to no. 775.

[235] *CIL* III 14606 = *ILJug* 1378 (Rudnica/Sočanica). For further bibliography, cf. *ILJug* III, p. 50 with commentary on no. 1378.

[236] *CIL* III 14209, cf. Rankov 1983: 47; Velkov & Aleksandrov 1994: 14, no. 23.

dispensator, was discovered built into the porticus of a Christian church at Crkvine.[237] Although there is no reference made to iron mines, the *dispensator* was probably connected with the mining administration of the *ferrariae*. Apart from the administrative personnel at Virunum, the headquarters of the presidial procurator, a number of imperial freedmen and slaves were seconded to the mining areas of Noricum as well. Chrysanthus, a *servus Augusti* and *vicarius* of emperor Claudius, and another unknown imperial slave (De[cimus?]) set up votive inscriptions to Noreia, found at Hohenstein, where a further votive inscription of a *conductor ferrariarum* was also discovered.[238] The administrative function of these imperial slaves is not revealed.

Few subaltern clerks appear in the inscriptions of Lugdunum/mod. Lyon in Gaul (cf. 4.1.11). An honorific inscription for the provincial procurator of Lugdunensis and Aquitania was set up in the years AD 135–45 by the imperial freedman Appianus, *tabularius rationis ferrariarum*.[239] Aurelius Calocaerus, an imperial freedman, records his position of *tabularius ferrariarum* in a funerary inscription dating to the end of the second century AD or later.[240] Apart from a *ratio ferrariarum* at Lugdunum, members of the *familia Caesaris* are present in few mining areas throughout Gaul, despite the abundancy of iron-mining ventures in this province. At Villefranche-de-Rouerge, in the Département de l'Aveyron, the inscribed slab for a *vilicus* Zmaragdus names the *familia Caesaris* of emperor Tiberius 'quae est in metallis'.[241]

The known mining procurators aside, there is only marginal evidence for administrative staff in the mining zones of the Iberian Peninsula. The *lex metallis dicta* of Vipasca notes that *liberti* and *servi Caesaris* who are *in officis* of the procurator or receive salaries, 'commoda', from the Roman state, are exempted from paying entrance fees for the baths.[242] However, no inscriptions of imperial officials, apart from the known *procurator* Beryllus, have hitherto been discovered at Vipasca, thus severly limiting our knowledge of the administrative personnel who assisted the procurator.[243] Moreover, in Roman Baetica hardly any subaltern officials from mining areas are known.

237 *ILJug* 765.

238 *CIL* III 4807, cf. Alföldy 1969*b*: 25, no. 34. *CIL* III 4808. *CIL* III 4809 = *ILS* 1467. A further imperial slave might be attested in a very fragmentary inscription: *CIL* III 4822 = 11505, cf. Alföldy 1969*b*: 25, no. 34.

239 *CIL* XIII 1808, cf. Hirschfeld 1905: 157; Pflaum 1960–1: 253–7, no. 106 bis (135).

240 *CIL* XIII 1825.

241 *AE* 1892: 23 = *CIL* XIII 1550.

242 *LMD* ll. 23–4: *Excipiuntur liberti et servi [Caes(aris), qui proc(uratori)] in offi⟨c⟩is erunt vel | commoda percipient, (...)*.

243 *AE* 1908: 233, where ll. 4–5 have been read as *vi|[lici]*, might be a copy of the honorary inscription for the procurator Beryllus (*IRPac* 121). Different: Noeske 1977: 308; Lazzarini 2001: 176 f., fn. 357.

The funerary stone of a Privatus, *dispensator et verna Augusti* in Regina/Los Paredones near Casa de Reina (BA), *c.*20 km west of the Azuaga mining area, is the only indication of a state or imperial interest in the extractive operations in this particular area of the Sierra Morena.[244] A connection between our *dispensator* and these local mining activities appears likely, as there is no evidence of other economic reasons for his presence in this remote area. For the mining area of Castulo in Hispania Tarraconensis, the inscription of a M. Ulpius Hermeros, *libertus Augusti* at the El Centenillo mines seem to suggest at least partial imperial control of mining operations there. The inscription, however, does not give any additional information on the rank or function of this imperial freedman.[245] Inscribed tin ingots from the Port-Vendres II shipwreck originating from Roman Baetica name a *L(ucius) Vale(rius), Aug(ustae) l(ibertus) a com(mentariis)* during the reign of Claudius. He may have been either located in Baetica or in Rome and was involved in the production or import of tin(?).[246]

4.2.2. Subaltern Officials in Imperial Quarries

4.2.2.1. *Regular Staff*

Procurators in charge of the quarries in the Egyptian Eastern Desert most likely commanded a staff of accountants and other administrative personnel. The *tabularii* of (Mons) Porphyrites and other *metalla* named in the inscription of their *hospitium* at El-Ashmunein were presumably part of this procuratorial bureau (cf. 3.2.1). *Tabularii* are also documented in the ostraca of Mons Claudianus, together with an οἰκονόμος, *tesserarius, subtesserarius, laccarius*, γραμματεῖς, and κιβαριάται of the φαμελιάριοι, and their assistants.[247] Οἰκονόμοι or *dispensatores* are recorded in further imperial quarries: A *d[[isp(ensator)]]* and *servus Caesaris* is the only imperial official who

[244] *CIL* II^2/7 981.

[245] *AE* 1922: 9.

[246] Weaver 1972: 241–3, 245–52; Colls, et al. 1975: 75–7; Colls, et al. 1977: 11–18.

[247] οἰκονόμος: *O.Claud.* 510, 551, 618; *O.Claud.*inv. 2238, 8175, cf. Cuvigny 2000a: 62 f.; *tabularius, tesserarius*: *O. Claud.* 485, 498, 563, 576, 596; *O.Claud.*inv. 6483, cf. Bülow-Jacobsen 1996: 61; *O.Claud.*inv. 1538, cf. Cuvigny 2000*a*: 64 f.; *subtesserarius: O.Claud.*inv. 1538, cf. Cuvigny 2000*a*: 64; κιβαριάται: *O.Claud.* 3–6, 155, 156, 244, 365, 382, 417–31, 433, 434, 436–48, 450–3, 455, 470, 474, 486, 487, 489, 490–3, 497(?), 518–24, 526, 527, 529–31, 533–5, 539, 558, 562, 570, 572, 577; *O.Claud.*inv. 1158, cf. Cuvigny 2000*a*: 58 f., 62 f., 65; γραμματεῖς: *O. Claud.* 9, 22. Cuvigny 2000*a*: 66; *laccarius*, ἀρτοδότης βοηθὸς κιβαριάτου, βοηθοὶ κιβαριάτου, αἰούτωρ κιβαριάτου: *O.Claud* 3–8, 446–8, 459, 461, 463–5, 469–84, 520–39, 592; *O.Claud.*inv. 3229, 7366, cf. Cuvigny 2000*a*: 60, 66.

appears in the epigraphic record at Dokimeion.[248] A further *dispensator* and *verna Augusti,* Saturninus, is attested at Simitthus, while at Krokeai a votive inscription to Castor and Pollux was commissioned by a *dispensator Augusti,* who most likely was part of the quarrying administration at the *marmor Lacedaemonium* quarries near Krokeai.[249]

The functional grade of *vilicus* is attested at Luna/mod. Carrara, not only for members of the *familia Caesaris* but also for slaves of private individuals. A *vilicus Caesaris* set up a small inscribed votive altar to the Nymphs near Luna at Le Canalie/Bedizzano, perhaps between the end of the first and the mid-third century AD.[250] At Luna a *servus* of the emperor Vespasian dedicated an inscription to Bello.[251] Both inscriptions seem to document the presence of imperial slaves from the Flavian period. This fits well with the evidence from Rome, which attests *tabularii rationis marmorum Lunensium* during the Flavian period. The special account was probably introduced after a significant increase in the demand for Lunensian marble after the fire of Rome in AD 64 (cf. 7.2.3).[252]

The *ταβελλάριοι/tabellarii* are a unique group of people attested quite often in the ostraca of Mons Claudianus.[253] Their basic function was to guarantee the communications between the quarries in the Eastern Desert and the procuratorial headquarters in the Nile valley. They therefore appear to be partly connected with the procuratorial bureau.[254] *Tabellarii* also appear in a military context. They are summoned by the military *curatores* of the fortlets guarding the routes to and from Mons Claudianus.[255] Furthermore, a circular dispatch of the *centurio* Horion during the reign of Hadrian or Antoninus Pius orders the *curatores* of the way-stations to supply any help necessary to the soldiers and the *tabellarii* in pursuit of fugitives, probably local Bedu.[256] The *tabellarii* were also dispatched as guides for travellers and caravans, or

[248] *AE* 1986: 674 = Christol & Drew-Bear 1986: no. 5.

[249] *AE* 1991: 1681 = *AE* 1994: 1883. *CIL* III 493 = *IG* V.1 1569. Le Roy 1961: 212. On Lacedaemonian quarries, cf. Pliny, *NH* 36.55, and Fant 1993*a*: 164 fns. 113, 114.

[250] *AE* 1980: 476. For photo, cf. Angeli Bertinelli 1978: 11, fig. 5.

[251] *CIL* XI 1315.

[252] *AE* 1974: 153. *CIL* VI 8485, 8484. The mark RAM read by Dolci 1998 as *ra(tio) m(armorum),* discovered on a block in the Scalocchiella quarries above the Valle di Colonnata was understood to refer to this new bureau.

[253] Cf. Bingen et al. 1992: 186; Bingen et al. 1997: 296.

[254] The *tabellarii* were mostly used for official messages within the offices responsible for the administration of the *annona* or the *vicesima hereditatium* such as the *statio hereditatium*, the *officium annonae* or the *statio marmorum,* cf. Eck 1995*a*: 69 fns. 56, 57. In Ephesus and at Carthage *tabellarii* seem to be subordinate to the *procurator provinciae Asiae,* cf. Eck 1995*a*: 69 fn. 57; Kolb 2000: 80, 276 f. with 277 fn. 1.

[255] *O.Claud.* 358, 363, 366, 367, 380. For the use of *tabellarii* for private matters: *O. Claud.* 250, 282, 290.

[256] *O.Claud.* 357.

accompanied σκληρουργοί or donkeys to their destination.[257] The socio-legal status of the *tabellarii* at Mons Claudianus is not well known. *Tabellarii* in the imperial service appear to be mostly slaves or freedmen. However, some freeborn *tabellarii Augusti* are also recorded.[258] The titles used by *tabellarii* such as *praepositi*, *decuriones*, *optiones*, or *tesserarii* have been understood to reflect the *tabellarii*'s military organization.[259] Anne Kolb argues that the *tabellarii*, as in other branches of the imperial administration, used titles reminiscent of military functions even though they were not part of the Roman army.[260]

Numerous inscriptions of imperial freedmen and slaves appear at imperial quarries, although they do not always provide an indication of their function. At Simitthus the *Caesaris servi* Abascantus, Corinthius and Ponticus, as well as C. Iulius Crestus Samianus, *Augusti libertus*, were probably involved in the quarrying administration.[261] The names of the *servi(?) Caesaris* are also documented in the label inscriptions on Carrara marble, such as Eros and Ephebus.[262] At Paros an Eros, imperial slave, left an inscription in the quarries, naming himself superintendent of works (ἐργεπιστάτης).[263]

4.2.2.2. *Hymenaeus Thamyrianus 'a lapicidinis Carystiis' and Karystos*

A unique example is provided by the epigraphic evidence of an 'a lapicidinis Carystiis'. Close to the town of Paleochora near Karystos on Euboea a large white marble base was discovered. The base, covered with inscriptions on the front and back, probably dates to the mid-first century.[264] With this monument Hymenaeus Thamyrianus, a *libertus Augusti*, honoured the *dispensator Augusti* and *nutricius* Thamyrus Alexandrianus, as well as members of his own family and acquaintances. A similar inscription by Hymenaeus Thamyrianus in Rome, again honouring Thamyrus and Hymenaeus' family members and friends, identifies Hymenaeus as *Caesaris servus* and *a lapicidinis Carystiis*. The inscription in Rome antedates the marble base in Karystos, as Hymenaeus is not yet manumitted.[265] Hymenaeus' *agnomen*, Thamyrianus, certainly bears testament to a close relationship with Thamyrus. Few imperial

[257] *O.Claud.* 287, 358, 363, 366.
[258] Hirschfeld 1905: 201 f.; Eck 1995*a*: 68; Kolb 2000: 275.
[259] For overview of scholarly opinion, cf. Kolb 2000: 276 fn. 1.
[260] Kolb 2000: 276 with fns. 2 and 3.
[261] App. nos. 867–70, 874, 875, Khanoussi 1996: 998, 1009, 1015.
[262] App. nos. 1157–61, 1172, 1174–6.
[263] *IG* XII.5 253 (cf. add. 312).
[264] *CIL* III 563. 12289 (Paleochora), cf. photo in Pensabene 1994: fig. 286.
[265] *CIL* VI 8486 (Via Ostiense in S. Pauli).

slaves or freedmen carry *agnomina* in reference to their former masters who were imperial slaves themselves. When slaves of imperial slaves became the emperor's own property, the former *vicarii* or 'slaves of slaves' not only reached a higher social status, but their occupational status also improved. This change in status could be expressed by taking a second name ending in *-ianus*, mostly in deference to their former owner. Weaver designates this group as 'vicariani' and provides a rough outline of their career. After reaching the status of 'vicarianus' imperial slaves appear to have reached the rank of a *dispensator*. Following their manumission at about forty years of age these 'vicariani' progressed to the post of *tabularius*. Some of them might even have reached a procuratorial post.[266] In the case of Hymenaeus it is not quite clear what intermediate or higher clerical grade the function of *a lapicidinis Carystiis* was equal to. The Roman inscription identifies Hymenaeus as an imperial slave when holding the post of *a lapicidinis Carystiis*. In accordance with Weaver's reconstructed sequence this post equalled the functional rank of a *dispensator*. In that sense the former *vicarius* to Thamyrus, the *dispensator Augusti*, might have followed in his master's footsteps. In this specific context of the imperial administration *vicarius* not only denotes the 'slave of a slave' but can also signify 'assistant to the *dispensator*'.[267] Perhaps Thamyrus had also been involved in the management of the quarries at Karystos, as Dubois' interpretation of the inscription shows that the names of both Hymenaeus and Thamyrus appear in ligature on a series of *cipollino* products recovered from the Emporium in Rome.[268]

The fact that Hymenaeus commissioned the first inscription in Rome led Rostovtzeff to believe that the *a lapicidinis Carystiis* organized the import of Karystean marble to the city of Rome and thus was based there.[269] The import and stockpiling of *cipollino* marble at different sites throughout Rome would certainly have merited supervision by an imperial official.[270] Hymenaeus' monument in Rome, however, does not give his title as *a(b) marmoribus*, but indicates that he was responsible for *lapicidinae*, quarries, near Karystos and not for the marble (*marmor*) itself.[271] Moreover, the Karystean inscriptions suggest that Hymenaeus was present at the *cipollino* quarries—at least long enough to commission the making of a white marble base. What prompted Hymenaeus to commission two monuments in Rome and in

[266] Weaver 1972: 216–18, 222, 230.
[267] Weaver 1972: 205 f.
[268] Dubois 1908: 117 f.
[269] Hirschfeld 1905: 167 fn. 4, contra Rostovtzeff 1904: 453.
[270] cf. Maischberger 1997.
[271] *TLL* VII/2, 941.

Karystos is not specified in the inscriptions. Given inscribed honorary monuments were set up in presence of those to be honoured, the sequence of events may perhaps be reconstructed as follows. After his promotion to *a lapicidinis Carystiis*, in step with the change in status from 'vicarius' to 'vicarianus', and perhaps on the eve of his departure from Rome (together with his former master Thamyrus), Hymenaeus, now Hymenaeus Thamyrianus, commissioned an inscribed monument in honour of Thamyrus. The second inscribed monument in honour of Thamyrus was set up in Karystos either on the occasion of Hymenaeus' manumission or perhaps Thamyrus' departure from Karystos(?). As the latter inscriptions seemingly indicate, Hymenaeus, as *libertus Augusti*, remained in charge of the quarries at Karystos. However, the epigraphic material does not inform us of his clerical grade.

4.3. 'HEADQUARTERS' OF IMPERIAL EXTRACTIVE OPERATIONS

The accumulation of epigraphic evidence for *officiales* and procurators at one location has been taken to indicate the administrative residence of a mining/quarrying procurator. Not all inscriptions, however, can be employed in definitively pinpointing an administrative centre. In his study of *capita provinciarum* Rudolf Haensch offers a categorization of the epigraphic evidence based on its value in identifying provincial capitals.[272] This categorization (which shall not be reiterated here) can also be applied to distinguish inscribed monuments identifying procuratorial seats of office from inscriptions that only attest the occasional presence of a procurator within a mining or quarrying zone, or those pinpointing merely outposts of his staff. As a preliminary it is of use to provide examples for both ends of the spectrum. The epigraphic record of Ampelum in Roman Dacia shows what an administrative centre of one or more mining or quarrying districts could consist of. Besides ample evidence for subaltern personnel such as *dispensatores*, *tabularii*, *adiutores tabulariorum*, *librarii*, *subsequens librariorum*, and *a commentariis* (whose funerary, votive, and building inscriptions attest their activity on the spot), the funerary and votive inscriptions clearly attest the presence of freedmen and equestrian procurators at Ampelum. This furthermore corroborates the notion of a permanent administrative bureau for the *aurariae Dacicae* located at Ampelum. As of yet, physical evidence for a 'bureau' of this kind is absent.

[272] Haensch 1997: 37–62.

At the other end of the spectrum one may cite the ἐπίτροποι τῶν μετάλλων for the Egyptian Eastern Desert. Despite being named in a number of inscriptions, the ἐπίτροπος presumably did not reside at Mons Claudianus, Mons Porphyrites, or Mons Ophiates. Only a minority of the ostraca published so far were addressed to the ἐπίτροπος directly and no procurator commissioned an inscription in the Eastern Desert. Moreover, he does not appear in the water distribution lists at Mons Claudianus.[273] A *hospitium* of the *tabularii* at El-Ashmunein indicates that the accountants of the 'Porphyrites' quarrying administration were at least temporarily based in the Nile valley, probably close to the *procurator* responsible for the Porphyrites district(?).[274] The necessity of using *tabellarii*, messengers who circulated between the Nile valley and Claudianus and transmitted the procuratorial orders to the foremen in the quarries, strongly suggests that the administrative headquarters for the quarrying operations at Mons Claudianus, Mons Porphyrites, and Mons Ophiates were located somewhere in the Nile valley, or perhaps in Alexandria.[275]

The epigraphic evidence is rarely conclusive for potential sites of administrative residences other than Ampelum. It was assumed that an administrative headquarters for mining operations might have been located at Domavium. Ten procurators are attested in inscriptions from this town, of which two were commissioned by *procuratores argentariarum* to commemorate the restoration of baths.[276] Two further building inscriptions by the *municipium Domavianum* name procurators as supervisors or dedicators of buildings, and a votive inscription of a procurator underlines the presence of equestrian officials at Domavium at least on a short term basis.[277] Moreover, two inscriptions were commissioned to honour procurators, one of which carries the full title of *procurator metallorum Pannonicorum et Delmaticorum.*[278] A further inscribed monument was raised by the *municipium Domavianum* in honour of C. Iulius Silvanus Melanio, likely a former *procurator argentariarum.*[279]

[273] Cuvigny 2005*b*.

[274] Cuvigny 1992*a*: 85f. The procurator might have been located near the headquarters of the ἐπιστράτηγος of the Heptanomia either at Hermoupolis or Antinoupolis, cf. Thomas 1982: 58; Cuvigny 2002: 241.

[275] *O.Claud.*inv. 6483, cf. Thomas 1998: 182.

[276] *CIL* III 12734, 12736. For Domavium as headquarters of the mining administration, cf. Hirschfeld 1905: 153; Dušanić 1977: 66.

[277] *CIL* III 8363 = 12733, 8359 + 8360 = 12720, 12725 = 14219[1].

[278] *CIL* III 8361 = 12721 = *ILS* 1143; *CIL* III 12739 +12740 = *ILJug* 83 = *AE* 1948: 243.

[279] *CIL* III 12737 with Abascal & Alföldy 1998: 159, 162 f. The precise content and context of a further, yet fragmentary inscription (*CIL* III 12728) probably naming a procurator can not be established.

Although there is sufficient evidence for the presence of procurators at Domavium, the epigraphic record reveals no subaltern officials of a mining bureau there.[280] Undoubtedly, Domavium was significant enough in order for the mining administration to warrant the attendance of procurators there on a regular basis. In comparison with Ampelum, however, the absence of *officiales* connected with the mining administration at least challenges the assumption that Domavium was the permanent place of residence for *procurator metallorum/argentariarum* for Pannonia and Dalmatia. The inscription of a *beneficiarius consularis legionis II Adiutricis* from Bölcske provides additional insight on this issue: the inscription lists one of the beneficiarius' postings as *statio argentariarum Pannonicarum e[t Delmaticarum ...]*. The *statio* probably lay within Pannonia Inferior, as the *legio II Adiutrix* was based at Aquincum. Alföldy offered the explanation that the administration of the *argentaria* of both Pannonia and Dalmatia held at least one *statio* per province.[281] A further *statio* was probably located near Domavium at Skelani where numerous *beneficiarii* left their inscriptions.[282] Given the magnitude of the task at hand, one must assume that the procurator perhaps frequented these *stationes* throughout the year. A similar system probably was in use for the *procurator Augusti* and *praepositus vectigalis ferrariarum* for Pannonia: an *arcarius stationis Siscianae* was subordinate to this procurator and further such posts have been assumed elsewhere.[283] In the absence of any subaltern personnel, the main headquarters of these procurators—if any such thing existed—can not be ascertained. Domavium remains the most promising candidate.

There is a great number of inscriptions which record iron-mining procurators at Ljubija, located in the Upper Drina valley (cf. 4.1.10). A number of votive monuments for Terra Mater were erected here by *vilici officinae ferrariae* under the supervision of procurators, or with their participation.[284] Archaeological finds document smelting furnaces and so-called 'administrative' buildings closeby.[285] These inscriptions, however, do not provide enough evidence to regard Ljubija as the headquarters of the procurators of the iron mines. There are no inscriptions set up by other staff of the procuratorial bureau and we lack other examples of the procurators' presence at Ljubija or

[280] *ILJug* 81, 1517, 1522, 1524, 1532, 1543.

[281] *AE* 2003: 1426 with Alföldy 2003: 225 f.

[282] *CBFIR* 471–5.

[283] *CIL* III 3953 = *ILS* 3094; cf. Mócsy 1962: 594; Alföldy 2003: 225 f.

[284] cf. *AE* 1958: 63 = *ILJug* 157; *AE* 1958: 64 = *ILJug* 158; *AE* 1973: 411 = *ILJug* 779; *AE* 1973: 412 = *ILJug* 778; *AE* 1973: 413 = *ILJug* 780; *AE* 1973: 414 = *ILJug* 781.

[285] Škegro 2000: 118–23, with further bibliography.

in the surrounding area. The commissioning of votive monuments naming procurators appears to be connected with the supervisory activity of the *vilici* running the smelting furnaces. We thus may assume that the named procurators were not constantly present at Ljubija but probably had their 'headquarters' elsewhere. A possibility may be Siscia, a *statio* of the *procurator Augusti* and *praepositus vectigalis ferrariarum* and perhaps the seat of the financial procurator for Pannonia Superior.[286]

The funerary inscription set up by a *dispensator* for Thaumastus, *libertus Augusti* and *a commentariis aurariarum Delmatarum*, at Salona and dating perhaps to the first century AD, may place the gold-mining administration there.[287] To date, *procuratores aurariarum Delmatarum* are not recorded. Thus, the *a commentariis aurariarum Delmatarum* was probably attached to the procurator of Dalmatia. This appears to be the case with one inscribed monument set up by a *tabularius rationis ferrariarum* at Lugdunum in honour of the financial procurator for the *provinciae Lugdunensis et Aquitania*. The inscription reflects a professional relationship between the *tabularius* and procurator, with the *tabularius rationis ferrariarum* (for Lugdunensis and Aquitania?) perhaps serving on the staff of the latter. Thus one must assume that a further *tabularius ferrariarum* attested at Lugdunum served on the staff of the financial procurator as well.[288] The epigraphic evidence does not allow us to conclude whether the *procurator ad vectigal ferrariarum Gallicarum* also had his seat at Lugdunum and thus his own staff at his disposal (cf. 4.1.11).

Further sites of procuratorial bureaux are not revealed by the epigraphic record. The statue base set up by the *ordo colonorum* at Sočanica and dedicated by the *procurator mm(etallorum?) DD(ardanorum?)* does not offer sufficient proof for the location of the procuratorial 'headquarters' there.[289] Likewise, the freedmen mining procurators named in the series of inscribed altars commissioned by the army vexillations at Luyego and Villalís in northwestern Spain were perhaps not stationed at the findspot of these inscriptions. We can only assume that the few attested Roman mining procurators in Roman Baetica—a *procurator montis Mariani* known from an inscription at Hispalis, and a procurator as initiator of an honorary monument to emperor Nerva at Rio Tinto—did not reside permanently in the mining zones of the Sierra Morena. At least for the former, the *procurator montis Mariani*,

[286] Haensch 1997: 352 f., remains undecided on the issue whether Poetovio or Siscia functioned as base for the bureau of the financial procurator. For the *praepositus vectigalis ferrariarum*, cf. *CIL* III 3953 = *ILS* 3094.

[287] *CIL* III 1997; Hirschfeld 1905: 154, Haensch 1997: 76–81.

[288] *CIL* XIII 1808; Haensch 1997: 135, 463 f.

[289] *ILJug* 503.

Corduba might be a suitable candidate for the seat of his procuratorship. The funerary stone of a slave of Sextus Marius set up in Corduba does suggest that Marius' mines (Cerro Muriano?) and estates had been run from Corduba prior to their confiscation.[290] Besides being the probable provincial capital of Baetica, Corduba appears to have been an important administrative centre for the mining industry and the metal-trade.[291] The city was not only located at the Guadalquivir, but was also at the head of the main roads north to the mines around Azuaga, Cerro de Malatrigo, Belalcazar, Pozoblanco, and Sisapo/Almadén.[292] Although direct epigraphic evidence for the presence of an imperial mining administration is missing, major mining companies (*societates publicanorum*) left their epigraphic traces in Corduba. Several *liberti*, receiving the gentilicium Argentarius after being set free by the *societas Sisaponensis*, set up an inscription in Corduba, probably during the first century AD.[293] Hence, the *procuratores montis Mariani* perhaps also held their office in Corduba. This does not necessarily conflict with the monument the *confectores aeris* set up at Hispalis. Eck and Haensch have argued that the epigraphic record of Hispalis may attest the seat of the patrimonial procurator there, and the *procurator montis Mariani* may have thus occasionally paid a visit to Hispalis.[294] The evidence for Vipasca as mining headquarters appears to be more promising: the *lex metallis dicta* unmistakeably notes *liberti et servi Caesaris, qui proc(uratori) in officiis erunt*.[295] At least in the case of Vipasca we may assume that the *procurator metallorum Vipascensium* required 'officiales' in accomplishing his various duties and therefore was in need of some form of residence or bureau.[296] The honorary monument for Beryllus at Vipasca and the procuratorial titulature of Saturninus *ἐπίτροπος τῶν μετάλλων Οὐλπασκησίων* point to Vipasca as the likely seat of the freedman mining procurator.[297]

Apart from the administration of the quarries in the Egyptian Eastern Desert, there is limited evidence for the bureaux of procurators charged with the management of quarrying ventures. The inscriptions at Simitthus

[290] *CIL* II^2/7, 441 = *CIL* II 2269. Sextus Marius was from Corduba or had close links with the town, cf. Pliny, *NH* 34.4 (*aes Marianum* = *aes Cordubense*). For a *tessera hospitalis* with his name, cf. Eck & Fernández 1991.

[291] Knapp 1983: 52; Alföldy 1996: 452; Haensch 1997: 178–83, 492.

[292] Sillieres 1990: fig. 18.

[293] *CIL* II^2/7 415a. *CIL* II^2/7 334 = *AE* 1971: 181.

[294] Eck 1994: 564 f.; Haensch 1997: 184 f.

[295] *LMD* ll. 23–4.

[296] Besides Aljustrel, where Domergue (1983: 31f.) believes to have unearthed the remains of a 'House of the Procurator', buildings where the administration for a mining/quarrying district was housed have yet to be identified.

[297] *IRPac* 121; Habicht 1969: no.44.

do name a *disp(ensator) m(armorum?) N(umidicorum?)* and a number of inscribed monuments were comissioned by procurators (cf. 4.1.5). Inscriptions by *officiales* of the quarrying administration have not been found at Simitthus. However, the funerary stone set up by the *procurator* Galata for his *alumna* or foster-daughter Alexandria at Simitthus is testament to a strong affiliation of the procurator with the site, one that goes beyond regular stopovers.[298] A further inscribed monument documents building activity (a porticus with cella and cistern) by a *procurator m(armorum?) N(umidicorum?)* after fulfilment of a vow.[299] Apart from the usual votive inscription set up by *procuratores*, the two inscriptions may confirm their long-term presence at Simitthus. In particular the funerary inscription strongly suggests that the procuratorial headquarters were situated at Simitthus. The scarce epigraphic material from the quarries at Dokimeion is inconclusive. A *d[ispensator]* might be named in a fragmentary inscription, but apart from this subaltern official, no further members of an administrative branch are identified.[300] Based mainly on the epigraphic and literary evidence, Otto Hirschfeld and others have argued for Synnada being the administrative centre of the Dokimean quarries. Even if Strabo tells us that in Italy *pavonazetto* marble from Dokimeion was known as *marmor Synnadicum*, thus suggesting the important role of Synnada in the trade or extraction of this marble, there is no further evidence to substantiate this thesis. Even though the inscriptions found at Synnada record imperial officials, none of these individuals carry a title suggesting a linkage with the quarries at Dokimeion. Synnada was probably the headquarters of the *procurator provinciae Phrygiae* whose involvement in quarrying operations has yet to be proven.[301]

In the light of the preceding survey, one may conclude that the majority of administrative bureaux of mining or quarrying procurators were not located in or near mining or quarrying districts as a matter of course. Thus, inscriptions attesting procurators at such sites may only have reflected occasional visits by these officials. Given the judicial powers procurators possessed, one might imagine an administrative practice similar to the assize system of large provinces like Asia or Hispania Tarraconensis. The provincial governor was required to travel part of the year through the province, visiting the assizes in

[298] *AE* 1991: 1681 = *AE* 1994: 1883.

[299] *AE* 1994: 1885.

[300] *AE* 1986: 674 = Christol & Drew-Bear 1986: no. 5.

[301] Strabo 12.8.14, cf. Hirschfeld 1913: 170; Ward-Perkins 1992*a*: 67 with fn. 25; Hirt 2003: 119 f. For Synnada, cf. *CIL* III 348 = *CIG* 3822. *CIL* III 12237 = *AE* 1897: 73. *AE* 1901: 132a, b, c. *MAMA* IV 52, 55. *MAMA* VI 378.

order to rule on various legal issues of concern to the local communities.[302] In a similar fashion, the equestrian *procuratores metallorum*, *ferrariarum*, *argentariarum*, or *aurariarum* of one or more provinces moved from one mining district to the next in order to deal with the grievances of local occupants of the mines. This is suggested by the evidence for *stationes* of the *vectigal ferrariarum* at Siscia or of the *argentariae Pannonicae et Delmaticae*. Similarly, at least the equestrian *procuratores aurariarum* documented at Ampelum may have travelled throughout Dacia part of the year. They probably spent the rest of the year at the administrative headquarters, directing or supervising the compilation of accounts which they then reported to the Palatine bureaux in Rome.[303]

4.4. SUMMARY

The survey of imperial procurators associated with mining and quarrying ventures—be they imperial freedmen or equestrians in the emperor's service—yielded a variety of functional titles (e.g. *procurator montis Mariani*, *procurator ad vectigab ferrariarum Gallicarum*). We have examined the differences of these titles in relation to the social status of their bearers: whereas equestrian procurators, in general, were placed in charge of one category of mine or all mines throughout one province or more, most freedmen procurators presided over much smaller regions of mining or quarrying areas. In completing their duties, the procurators were heavily dependent on their subaltern staff, the size and composition of which may have varied considerably in accordance to the task at hand. In the light of the disparity of evidence for subaltern staff and our poor knowledge of the location and composition of procuratorial 'headquarters', any observations on correlations of this sort remain hypothetical at best. The inadequacies of our sources set aside, the detailed titles of procurators and subalterns assist us in identifying their main functions within quarrying and mining organizations. This should provide a sufficient basis for determining the role each official played in the administration of these outposts. Before we embark on this task, however, it is necessary to examine the role of another group, who did not form part of the imperial *familia* or the equestrian 'bureaucracy', namely the Roman army.

302 Eck 1997*a*: 119 f.; Haensch 1997: 18–36.

303 To my knowledge there is no firm evidence to assume that Ampelum or sites like Domavium or Ulpiana were incorporated within mining districts. On the so-called municipalization of former mining districts, cf. Noeske 1977: 277–85 with further bibliography.

TABLE 1: Synoptic table of procurators

DACIA	PANNONIA	DALMATIA	MOESIA SUPERIOR	PROC. FERR. (GALL.?)	ASTURIA-CALLAECIA	LUSITANIA/BAETICA
						T. Flavius Polychrysus, *lib. Aug.*, *proc. mont. Marian.* (Flav.)
proc. aurar. (Dac.)	***proc. arg. Pann.***			C. Caelius Martialis, *v.e.*, *proc. ferr.* (Traj.)	M. Ulpius Eutyches, *lib. Aug.*, *proc. met. Alboc.* (Traj./Hadr.)	Pudens, *lib. Aug. proc.*, Rio Tinto, (Nerva)
Ulpius Hermias, *lib. Aug.* (Traj.)	M. Antoninus Fabianus, *v.e.* (1st half 2nd c.)					
		?, *proc. arg. Delm.*	Telesphorus, *proc.*(?) (Sočanica) (136/7)	T. Statilius Optatus, *v.e.*, *proc. ferr.* (Hadr.)		Ulpius Aelianus, *lib. Aug.*, *proc. met. Vipasc.* (Hadr.)
	L. Creperius Paulus, *v.e.* (146/50)		Amandus, *lib. Aug.*, *proc.* (Sočanica) (Ant. Pius)			
Romanus, *lib. Aug.* (M. Aur.)	L. Septi[-] Petro[nianus], *v.e.* (after 143)		Tyrannus, *lib. Aug.*, *proc.* (Kosmaj) (2nd half 2nd c.)	T. Petronius Priscus, *v.e.*, *proc. ferr.* (mid-2nd c.)		
L. Macrius Macer, *v.e.* (M. Aur./S. Sev.)	***proc. met. Pann. et Dalm.***			Q. Domitius Marsianus, *v.e.*, *proc. ferr.* (M. Aur.)	Hermes, *lib. Aug.*, *proc.* (Villalís) (163)	
	T. Claudius Proculus Cornelianus, *v.e.* (161/4)					
	L. Domitius Eros, *v.e.* (late 2nd c.)				Zoilus, *lib. Aug.*, *proc.* (Villalís) (166/7)	Beryllus, *lib. Aug.*, *proc. met. Vipasc.* (*c.*172/3)

	(C. Iulius Silvanus Melanio, *v.e.* (late 2nd c.))					
	proc. ferr. (Pann.)	***proc. arg. Pann. et Dalm***	Cassius Ligurinus, *v.e.*, *proc.* (Rudnik) (S.Sev.)	M. Cosonius Fronto, *v.e.*, *proc. ad vectig. ferr. Gall.* (*c.*198–209)	Aurelius Eutyches, *lib. Aug.*, *proc.* (Villalís) (180/4)	
	Flavius Verus Metrobalanus, *v.e.* (*c.*180/97)	Ti. Claudius Xenophon, *v.e.* (*c.*185/90)				
Ael. Sostratus, *v. e.* (212)	T. Flavius Verecundus, *v.e.* (209)	Valerius Super, *v.e.* (*c.*218/20)			Aurelius Firmus, *[proc.] met.* (Villalís) (191)	Saturninus, *lib. Aug.*, *proc. met. Vipasc.* (*c.*193)
A. Senecius [-] Contianus, *v.e.* (215)	Iulius [-], *v.e.* (218/22)	Iulius Tacitianus, *v.e.* (c. 223/35)		C. Attius Alcimus Felicianus, *v.e.*, *proc. ferr* (Sev. Alex.)		
	M. Iulius Macer, *v.e.* (228)	[-]tus (*c.*238/44)	Titienus Verus, *v.e.*, *proc. m(et.) D(ar)d. (?)* (Sočanica) (238/44)			
	Nic[o]ma[chus], (229) Cossitianus Firmus, *v.e.* (248)					
		Aurelius Verecundus, *v.e.* (274)				

5

The Roman Army and Imperial Extractive Operations

5.1. THE ROMAN ARMY AND IMPERIAL QUARRIES

5.1.1. Roman Officers and the Operative Process

The presence of Roman army officers at imperial quarries is documented in the epigraphic record and, in some rare instances, in label inscriptions on quarried products. This raises the question as to whether or not the direction of quarry operations was divided between civilian administrators/engineers and Roman army officers. Unpublished ostraca from Mons Claudianus are inscribed with reports on the progress of quarrying work. They were sent to the military *praefectus* Antonius Flavianus, probably the commander of the auxiliary unit stationed at Mons Claudianus, by the collective workforce including the ἐργοδότης (foreman).[1] Furthermore, a recently published ostracon documents two copied letters probably written by the vice-curator of the *praesidium* at Mons Claudianus(?) to the *praefectus* and the procurator. Whereas the letter to the *praefectus* Vibius Alexander recounts the outgoing report of the 'vexillation', βήξιλλον, on the 'equipment of the quarries', περὶ ἐπιχρείας τοῦ μετάλλου, and laments the delay of grain supply, the letter to the procurator addresses the lack of personnel (φαμελιάριοι) and animals.[2] Apparently the *praefectus* of the auxiliary unit had to be kept informed about non-military issues concerning the imperial quarries. However, the one official who was able to respond directly to problems relating to the supply of human, animal, and material resources to the quarry was the procurator (ἐπίτροπος τῶν μετάλλων).

The water distribution lists from Mons Claudianus provide information on the hierarchy of military and civilian officers: a *centurio* and a *decurio* are ranked first amongst the recipients of water, the size of rations being left blank. The military personnel in general was supplied with far larger rations

[1] *O.Claud.*inv. 4471, 4852, cf. Cuvigny 2002: 240; Cuvigny 1992*a*: 87.
[2] Cuvigny 2002.

of water than the workforce.[3] From these published lists and other ostraca one might deduce that the quarries at Mons Claudianus stood under the direct command of a *centurio* or, occasionally, of a *decurio*. This notion is supported by inscriptional evidence. At some point after AD 103 Annius Rufus, a *centurio* of *legio XV Apollinaris*, set up an altar in the Sarapis temple. In the accompanying inscription he names himself *praepositus ab Optimo Imperatore Traiano operi marmorum monti Claudiano*.[4] Avitus, *centurio* of the *cohors I Flavia Cilicium equitata*, was put in charge of the quarries '*ὄντος πρὸς τοῖς τοῦ Κλαυδιανοῦ ἔργοις*' during Hadrian's reign. Both are clearly described as being responsible for the quarries, with Annius Rufus being appointed to that position on the orders of the emperor.[5] Further *centuriones* are epigraphically documented at Mons Claudianus and noted in a number of ostraca: Quintus Accius Optatus, *centurio* of an unknown unit, and Valvennius Priscus, *centurio* of *legio XXII (Deiotariana)*, are named.[6] At Mons Porphyrites the *centurio* Proculeianus is mentioned in an inscription commissioned by Epaphroditos, the *μισθωτὴς τῶν μετάλλων* during AD 117–19, and the *centurio* Fanius Severus put up an altar to Isis Myrionyma in AD 137/8.[7] Moreover, a Pankratius, *centurio frumentarius*, is attested at this site. Arrius Iulianus, perhaps a *centurio*(?) in the *cohors III Ituraeorum equitata*, was garrisoned at Mons Ophiates.[8]

Mons Claudianus[9]	*Mons Porphyrites*	*Mons Ophiates*
Crispus(?) (AD 109/10)		
Q. Accius Optatus (AD 110–14)		
Annius Rufus (AD 103–17)		
Valvennius Priscus (AD 113–17)		Proculeianus (AD 117–19)
Iulius Serenus		
Antoninus		
(Maximus?(*c.* AD 100–20))		
Iulius Kelsus		
Licinius Proclus		
Lurius Aquila		
Iulius Aquila		
Avitus (AD 118)	Fanius Severus (AD 137/8)	
Caecilius Vindex (AD 137–45)		Arrius Iulianus (AD 150–4)
Horion		
Alexander		
	Pankratius (AD ?)	

[3] *O.Claud.*inv. 1538+2921, cf. Cuvigny 1992*a*: 86; Cuvigny 2005*b*: 334–6.
[4] *I.Pan* 39. [5] *I.Pan* 42.
[6] *I.Pan* 38, 41. *O.Claud.* 48.
[7] *I.Pan* 21, cf. Maxfield 2001, Table 7.1; Maxfield & Peacock 1998*a*: 190.
[8] *I.Pan* 22 (MP). *I.Pan* 53 = Cuvigny 1996*c*.
[9] Crispus: *O.Claud.* 141. Valvennius Priscus: *I.Pan* 41. *O.Claud.* 49; Iulius Serenus: *O.Claud.* 50, 51, 52; Antoninus: *O.Claud.* 57–71, 148, 149; (Maximus?): *O.Claud.* 72, 73, 146; Iulius

Only few of these officers were legionary *centuriones*, and in the case of Annius Rufus at Mons Claudianus his unit, *legio XV Apollinaris*, was probably never garrisoned in Egypt.[10] Annius Rufus was undoubtedly assigned to the quarries because of his technical expertise.[11] His knowledge of construction and presumably of quarrying procedures made him a suitable candidate for the supervision of the imperial quarries at Mons Claudianus. Moreover, the fact that Annius Rufus was sent to Mons Claudianus on the orders of Trajan strongly supports the view that he was renowned for his technical or organizational expertise. Had an ordinary commander been required for the army detachment at Mons Claudianus, the officer corps of the legions or auxiliary units recorded in Egypt certainly would have sufficed. The transfer, therefore, of a centurio from a different province to Mons Claudianus on the specific orders of the emperor strongly implies that Annius Rufus possessed useful experience managing quarry operations.

Annius Rufus is not an isolated case. Further evidence for the involvement of legionary centurions in quarrying procedures emerges from Karystos on Euboea. A label inscription on a 'cipollino' column found in Rome lists Sergius Longus, *centurio* of the *legio XXII Primigenia*, together with a *probator* Crescens, who shall be of interest further on.[12] One T. Sergius Longus, presumably the same man, only now *centurio* of the *legio XV Apollinaris*, had an inscription for the god Hercules engraved below a niche at a quarry near Karystos.[13] Another inscription on a lead tag, recovered on 'cipollino' marble, names a *Sergius, centurio legionis XV*.[14] Given the rarity of the name 'Sergius Longus' amongst the ranks of the Roman army, the possibility of two officers of different units with the same rank, the same name, and the same

Kelsus: *O.Claud.* 76, 77; Licinius Proclus: *O.Claud.* 78, 363; Lurius Aquila: *O.Claud.* 368; Iulius Aquila: *O.Claud.* 370; Caecilius Vindex: *O.Claud.* 79, 360, 361?, 362?; Horion: *O.Claud.* 286, 357; Alexander: *O.Claud.* 386.

[10] Maxfield 2000: 435. Contrary to the *communis opinio*, Everett Wheeler argued that the *legio XV Apollinaris* was a regular unit posted in Egypt, albeit only posted there for a very short period between AD 106–17 before being transfered to Satala in Cappadocia, cf. Wheeler 2000: 288–93.

[11] Ritterling (1925, 1757) equates him with Annius Rufus, centurion of *legio XIII Gemina*, recorded at the legion's camp at Vindobona/mod. Vienna, cf. *CIL* III 15196[4]. The camp was under construction between AD 98 and AD 107, and a contingent of the *XIII Gemina* was supporting construction efforts of a camp at Brigetio in AD 97 to 101. Apart from acquiring a wide-ranging expertise in construction work at Brigetio and Vindobona, his assignments may also have put Annius Rufus in contact with *legio XV Apollinaris*, which participated in both building projects. Perhaps he was transferred to *XV Apollinaris*, before the *XIII Gemina* was sent to Dacian war in AD 101, cf. Wheeler 2000: 290.

[12] App. no. 618.

[13] *CIL* III 12286 = Dubois 1908: no. 279.

[14] Franke 2000; Wheeler 2000: 288 f.

capabilities being present at Karystos, is less probable than the possibility of both the inscriptions at Rome and Karystos referring to one and the same person.[15] The *probator* Crescens, named in the column label found in Rome, is recorded in another label inscription on Carystean marble from Rome dating to the year AD 132. Consequently, Sergius Longus' presence at Karystos must be dated around 132.[16] His transfer from *legio XXII Primigenia* (in Mogontiacum/mod. Mainz since AD 97) to *legio XV Apollinaris* (in Cappadocia since *c.* AD 117) must have taken place around the same time.[17] From the label inscriptions on *cipollino* and his small monument to Hercules we can conclude that Sergius Longus probably was at Karystos under both legions. None of the named legions, however, was based in Greece, and there is no evidence so far to suggest that Sergius Longus arrived on Euboea in the company of auxiliary troops or a legionary detachment. Sergius Longus was dispatched to the Karystean quarries based on his expertise in quarrying organization. As numerous inscriptions from the Brohltal quarries near Bonn in Germania Inferior demonstrate, an officer or soldier in *legio XXII Primigenia* could easily gain experience in quarrying procedures (cf. 5.1.2). It is therefore possible that the technical experience which Sergius Longus might have gained during his service with *legio XXII Primigenia* prompted his dispatch to the Karystean quarries.

This may also have been the case with two *centuriones* known from numerous label inscriptions on *pavonazetto* blocks discovered in the Bacakale quarries near Dokimeion. Both army officers were probably closely connected with the introduction of the *caesura-officina* system at Bacakale after AD 136. Tullius Saturninus, *centurio* of the *legio XXII Primigenia*, appears in label inscriptions dating to the years 136–8 discovered at Bacakale and *pavonazetto* marble in Rome. As in the case of Sergius Longus, Tullius Saturninus' legion was based at Mogontiacum from AD 97 onwards and appears to have run the Brohltal quarries.[18] The province of Asia where the Dokimean quarries are situated did not have a legionary garrison. Nor did Tullius Saturninus or his successor Aelius Antoninus (AD 147/8) command a contingent of auxiliary or legionary soldiers at Dokimeion. Thus far, no epigraphic evidence has emerged for an army unit being stationed at Bacakale or in its vicinity. The quarry labels on *pavonazetto* marble suggest that the *centuriones* Tullius Saturninus and Aelius Antoninus had *caesurae* to their name—in other

[15] See Mommsen's commentary to *CIL* III 12286.
[16] App. no. 603.
[17] Franke 2000: 98; Wheeler 2000: 288 f., with fn. 174.
[18] Franke 2000: 99 with fn. 43.

words, they supervised one or more extraction sites within the Dokimean quarries (cf. 7.2.1).

Given the fact that these four *centuriones* were separated from their units and sent to quarrying sites, probably for their technical or organizational experience in quarrying procedures, the question remains as to how these officers were picked for these specific tasks. Valerie Maxfield argued that the transfer of legionary *centuriones* from their units to imperial quarries reflected the 'hand of the emperor at work'.[19] This conclusion was based on the inscription of Annius Rufus, centurion of *legio XV Apollinaris* found at Mons Claudianus, where he clearly states that he was appointed *praepositus [—] operi marmorum monti Claudiano* by emperor Trajan (*ab Optimo Imperatore Traiano*).[20] The provincial governor, however, might also have played a major role in demanding the transfer of specialist military personnel to his province. Nonius Datus, *librator* of the Roman army, was requested by the presidial procurator of Mauretania Caesariensis, C. Petronius Celer, on behalf of the town of Saldae. On the orders of the *legatus legionis III Augustae*, Nonius was then transferred from Numidia to Mauretania Caesariensis to sort out a problem with the Saldae aqueduct tunnel.[21] In response to one of Pliny's letters asking for an army engineer (*librator*) for the canal project near Nicomedia, Trajan advised his governor in Pontus-Bithynia to approach the governor of the neighbouring province of Moesia Inferior, P. Calpurnius Macer.[22] Although the evidence is patchy, it appears that the senatorial governors of *provinciae populi Romani* or presidial procurators of *provinciae Caesaris* could ask neighbouring governors with one or more legions at their disposal for the required army specialists. The transfer of Roman centurions from the Rhine or Danube provinces to the quarries in Egypt, Asia, or Achaia may in some instances have been supervised by the emperor, rather than being carried out by the provincial governors themselves. Whether or not the emperor had a central archive of military personnel at his fingertips allowing him to answer requests by provincial governors for military specialists at his own volition is open to debate. The sources only allow for limited insights: according to the *Historia Augusta* (*Sev. Alex.* 21.6–8), Severus Alexander constantly studied full records (which he kept in his bedroom) of the troops' pay and service. Whether further information was as readily available is not stated.[23] Based on Statius (Silv. 5.1.94–100), Eric and Anthony R. Birley

[19] Maxfield 2000: 435. [20] *I.Pan* 39.

[21] *CIL* VIII 2728 = 18122 = *ILS* 5795, cf. Eck 1995*b*: 222 f.; Laporte 1996: 737 f.; Horster 2001: 175.

[22] Pliny, *Ep.* 10.41, 42, 61, 62; cf. Horster 2001: 176.

[23] Millar 1967: 12.

argued that the *ab epistulis* perhaps ran a 'human resources department' or archive concerned especially with the selection of qualified equestrians and military officers for certain tasks.[24] However, no further evidence was provided to substantiate the idea of a central bureau for military personnel. Perhaps emperors like Trajan could provide provincial governors in need of a specialist with a suitable candidate based on their own military experience or on the collective knowledge of their *consilium*. Whatever the decision process may have been, it undoubtedly led Annius Rufus to state that he had been put in charge of Mons Claudianus by the emperor Trajan.

One might conclude from the correspondence between Nonius Datus and the provincial governor concerning the aqueduct of Saldae that the army specialist stood under the command of the provincial governor after his transfer. Werner Eck believes that this would have been the most practical solution. Concerning the issue of pay, the *proconsules* in Asia and Africa were ultimately responsible through their *quaestores* for the payment of the soldiers under their command, as most units in the Roman empire were paid with the revenues from the provinces in which they were stationed. One may assume that the individual *centuriones* transferred to the quarries of Dokimeion or Karystos received their pay from the provincial *quaestor*. However, Eck thinks that issues of promotion, transfer, and retirement remained matters for their unit of origin, as Annius Rufus, Tullius Saturninus, and Sergius Longus continue to see themselves as officers in *legio XV Apollinaris* or *XXII Primigenia*.[25] In fact, a recently published military diploma of AD 115 named not only former members of auxiliary units in Moesia Superior who received Roman citizenship, but also units sent from Moesia Superior on a military expedition to the East. Apparently the records concerning these units remained with their former Moesian commander for the duration of the campaign. In analogy to expeditionary corps soldiers or officers who were seconded to quarries or mines of a different province probably stayed attached to their original unit. Only their pay might have been distributed by the actual commander at their new posting.[26] In this context, Sergius Longus' transfer from the *XXII Primigenia* to the *XV Apollinaris* during his stay at Karystos may have resulted from a vacancy at the latter filled by Longus prior to the completion of his task at the quarries.[27] The assignment of specialist legionary centurions to imperial quarries in the Roman East during the first half of the second century may

[24] Birley 1963–4: 207; Birley 1992: 23 f., 41–54; Eck 1995*e*: 139–41.

[25] Eck 1997*b*: 188 fn. 7. 200.

[26] Eck & Pangerl 2005.

[27] On transfers of centurions, cf. Birley 1963–4: 210.

carry the hallmark of a decision reached at Rome. All four attested centurions are sent from the Roman West to quarries in the Greek-speaking East, in the case of Tullius Saturninus in 136 probably even with the task of reorganizing extraction procedures at Dokimeion. Sergius Longus may have arrived at Karystos in *c.*132, certainly during Hadrian's reign, perhaps with a task similar to Tullius Saturninus'. The use of Latin in noting specifics such as consular dates, numerals, location, contractors, and workshops at Dokimeion and Karystos certainly may have prompted the transfer of officers well aquainted with Latin quarrying terms.

Centuriones frumentarii often appear in the context of mining operations. Pankratius, a *centurio frumentarius* of an unknown unit, was perhaps seconded to Mons Porphyrites for his skills, as was Flavius Mucianus to Luna/Carrara.[28] The presence of the latter is documented by an inscribed monument set up in AD 200 by M. Firmidius Spectatus, *optio* and *frumentarius leg(ionis) II Italicae p(iae) f(idelis)* under the *cura* of Flavius Mucianus, *centurio frumentarius.*[29] Though the main function of the *frumentarii* initially might have been the collection of food for their troops, they were increasingly used for other tasks.[30] Based at the *castra peregrinorum*, the *frumentarii* were not only couriers for the Roman emperor but could be sent out on special missions.[31] It is not known why the *frumentarii* of the *legio II Italica* set up an inscribed monument near Luna, but the reported findspot within the quarries at Carrara suggests a connection with the quarrying business. Another *frumentarius* functioned as an overseer of construction work at Delphi in the name of the emperor Hadrian and in Salona a *centurio frumentarius* of the II Traiana commanded the vexillations of the recently created *legio II (Italica) Pia* and the *legio III (Italica) Concordia* in building an emergency fortification in AD 170.[32] This seems to support the view that two *frumentarii*—temporarily dispatched from the *legio II Italica* based in Lauriacum/Noricum to the *castra peregrinorum* in Rome—were sent to Luna in connection with the Carrara quarries. Pankratius' presence at Mons Porphyrites must be explained along similar lines.[33]

[28] *I.Pan* 24.

[29] *CIL* XI 1322 (Carrara), cf. Banti 1931: 484 f., no. 46.

[30] Austin & Rankov 1995: 136 f., 150.

[31] Clauss 1973: 82–117; Rankov 1990; Birley 1995: 147; Rankov 1999: 29 f.; Kolb 2000: 175, 290–4.

[32] *ILS* 9473 (Delphi); *CIL* III 1980 = *ILS* 2287 (Salona), cf. Zwicky 1944: 76–8; Saxer 1967: 41, no. 71; Clauss 1973: 94; Rankov 1990: 177.

[33] *AE* 1978: 317, 327. *AE* 1990: 353.

5.1.2. Excursus: Quarrying Duty and the Roman Army

Both Tullius Saturninus and Sergius Longus likely gained initial experience in quarrying procedures during their service with the *legio XXII Primigenia* which in AD 97 had its legionary headquarters at Mogontiacum.[34] During the Flavian period, additional Roman fortifications on the Rhine frontier had to be built, and buildings destroyed during the Batavian revolt in AD 69 were restored. The source for most of the stone used in construction was the Brohltal quarries. Commemorating their unpleasant ordeal, most troop detachments sent there erected votive stelae for Hercules Saxanus. From the time of Claudius/Nero to Trajan, a series of troop vexillations were seconded to quarry work, and under Trajan stones were being cut for the colonia Ulpia Traiana/mod. Xanten.[35] The inscriptions are mostly commissioned by *centuriones* of legions (*VI Victrix, X Gemina, XXI Rapax, XXII Primigenia* etc.), auxiliary units (*coh. II Varcianorum, coh. II Asturum, coh. II c.R.* etc.) or the *classis Germanica*, naming their fellow *vexillarii* or *commilitones* or the *vexillatio* consisting of further legions. Despite the location of these quarries in Germania Superior, most army units mentioned were based in Germania Inferior.[36] The appearance of the *classis Germanica* in the Brohltal quarries is explained by the use of this flotilla for transporting quarried blocks, but members of the Roman navy undoubtedly also took part in the quarry work itself. It has been suggested that a detachment under the command of a trierarch worked in the trachyte quarries at Drachenfels in the Siebengebirge to supply the ongoing construction work at colonia Ulpia Traiana.[37] Military vexillations of the *legio VIII Augusta*, the *X Gemina, XIIII Gemina Mars Victrix*, the *XXI Rapax* and a *cohors V* were also working in the tufa quarries of Norroy during the Flavian period, while a vexillation and soldiers of the *legio XXII Primigenia* are documented during Septimius Severus' reign in the tufa quarries at Brunholdisstuhl near Bad Durkheim.[38] At some point in the second century AD a vexillation of the *legio XXX Ulpia* appears to have quarried stone in Idylle near Kruft for construction work at the legionary camp of the *legio I Minervia* in Bononia or the *XXX Ulpia*'s own camp at Vetera.[39] A further inscription by the *coh(ors) XXVI vol(untariorum)*

[34] Franke 2000: 99 with fn. 43.

[35] Saxer 1967: 74, 79 with fn. 440, nos. 194–203, 209, 211–16, 240–5, 248, 250–2.

[36] Bauchhenss 1986: 90.

[37] Saxer 1967: 86, no. 253 = *CIL* XIII 8036 = *ILS* 2907 (AD 160).

[38] Norroy: Saxer 1967: nos. 204–7. *CIL* XIII 4625 = *ILS* 3454, 4624 = *ILS* 3453, 4623 = *ILS* 9120. *AE* 1920: 118. Brunholdisstuhl: Nesselhauf 1937: 73, no. 80, 1–20; Saxer 1967: no. 222.

[39] Nesselhauf 1937: 91, nos. 131–5; Saxer 1967: 82 f., nos. 234–6.

R(omanorum) is known from subterranean tufa quarries near Kruft.[40] Another small quarry near Reinhardsmünster in Alsace displayed an inscription naming the *officina* of the *legio VIII Augusta*.[41] Outside of the German provinces military personnel are also found in other quarries: a *protector consularis* together with a *centurio* of the *coh. III Alpinorum Antoniniana* set up an inscription in the quarries on the island of Brattia/mod. Brac near Salona.[42] Furthermore, vexillations of the *legio II Augusta* and the *legio XX Valeria Victrix* are recorded in the Cumberland quarries in Roman Britain, perhaps supplying (although not exclusively) the construction efforts at Hadrian's Wall.[43] In the immediate vicinity south of Hadrian's Wall, at Coombe Crag, Lodge Crag, and Haltwhistleburn inscriptions record the use of soldiers from *legio VI Victrix* in quarries opened for this specific purpose.[44]

One of the inscriptions of the Cumberland quarries may give the function of a soldier of *legio XX Valeria Victrix* as a *[co]r(nicen)*.[45] The Brohltal quarries also yielded a votive inscription by M. Hellius Secundus, a *tubicen* of the *legio X Gemina pia fidelis* probably from Trajan's reign.[46] The use of the 'signalling corps' in army quarries is moreover documented in the quarry of Arulis/mod. Enesh near Zeugma on the Euphrates. A Rabilius Beliabus(?), *tubicen* of *legio IIII Scythica*, seems to have been an essential part of the legionary workforce there.[47] A number of inscriptions found at the Arulis quarries, engraved on different heights on the rockface, some in the form of *tabulae ansatae*, others inscribed on altars found in niches, seem to indicate that vexillations of the *IIII Scythica* quarried stone there from the beginning of the second century AD to the early third century AD.[48] Oliver Stoll argued that *signiferi* and *tubicines* were used in the quarries on account of their acquaintance with administration and accounting. Michael A. Speidel, however, has proposed that the *tubicines* might have sounded signals in order to indicate the hours or to give warnings about blocks being moved by crane.[49]

The *legio IIII Scythica* took part in a number of building projects: legionary contingents helped to cut a canal between the Orontes and Karasu river in AD 75 and construct a number of bridges over it.[50] At the same time soldiers of the *IIII Scythica* dug canals and tunnels into Mons Coryphaeus at Seleuceia Pieria in order to divert a stream which threatened to silt up the city's

40 *AE* 1922: 62 = Finke 248.
41 *CIL* XIII 5989.
42 *AE* 1979: 448, cf. Wilkes 1969: 228 f.
43 *RIB* 998, 999, 1001, 1005, 1008, 1009, 1014, 1016.
44 *RIB* 1946, 1953, 1680.
45 *RIB* 998.
46 *CIL* III 7694 (Brohltal).
47 *IGLS* I 68.
48 Stoll 1998: 120 f. *IGLS* I 69, 70, 77 = Saxer 1967: 279–81; Speidel 2001: 153 f. with fn. 6.
49 Speidel 1998: 177, no. 4; Stoll 1998: 122.
50 *AE* 1983: 927. On the involvement of the *legio IIII Scythica* in construction work at Antiochia, Dura Europos, and Artaxata, cf. Speidel 2001: 154 f.

harbour.[51] In the spring of AD 73 a detachment of the *legio III Gallica* and the *legio IIII Scythica* constructed a hydraulic installation at Aini. The *legio IIII Scythica* built a bridge over the Karasu river near Habes, possibly during the Flavian period, the building material of which was provided by a quarry in the immediate vicinity. The rockface of the quarry was inscribed by the 'hard-working' *IIII Scythica.*[52]

Although unpopular with most soldiers, quarry work was apparently a regular part of legionary life. A soldier of the *legio III Cyrenaica* based in Petra/Arabia, C. Julius Apollinaris, writes home to his father in Egypt in AD 107 that he is fine, while 'others were breaking stones all day, and doing other things'.[53] After being promoted to the rank of *principalis* and transferred to Bostra, he writes another letter and exclaims: 'I give thanks to Sarapis and Good Fortune, that while all are labouring the whole day through at cutting stones, I as a "principalis" move about doing nothing.'[54] From its stay in Egypt the *legio III Cyrenaica* left a duty roster on papyrus, rendering the assignments of soldiers from October 1–10 in AD 87. The roster gives the posting of C. Iulius Valens for the 3/4 October *ad cunicul(os) cal(carios)/cal(cis)* in a limestone quarry or lime-kilns of unknown location.[55]

Besides legionary units, auxiliary *cohortes* and *alae* also were seconded to quarrying and construction duty in the Nile Valley. Near Eileithyiaspolis/mod. El-Kab, 20 km north of Edfu, on the left bank of the Nile, a sandstone quarry was opened at El Mahamid.[56] We have records of an unknown soldier of a *legio II*, who, in the eleventh year of Hadrian's reign, searched and discovered these sandstone quarries (μέταλλα), and a certain Apollonios of the *ala Commagenorum*, who inscribed his προσκύνημα in the rockface(?) of the quarry.[57] Moreover, in the Gebel el-Toukh quarries on the east side of the Nile, close to Ptolemais Hermiu/mod. El Menshije, where a fine white limestone was quarried, traces of Ptolemaic and Roman extraction were observed in 'Steinbruchgebiet' (quarrying area) 4 and 7.[58] Latin and Greek inscriptions discovered within the confines of 'Steinbruchgebiet' 4 attest to the presence of a military unit there. Q. Caesius Valentius, *decurio* of the *ala Vocontiorum*, received two honorific inscriptions from the *commilitones* under his command.[59] The *commilitones* who were on guard duty, the *cohors (II Hispanorum Cyrenaica) scutata civium Romanorum* and the *cohors III Ituraeorum*, engraved an inscription in the quarries as well.[60] We do find *centuriones* of the *legio XXII*

[51] *IGRR* III 1005 = *IGLS* III 1137.
[52] *IGLS* 66 = Wagner 1977: 522 fn. 38; Speidel 2001: 153.
[53] *P.Mich.* 466, with Speidel 2001: 154.
[54] *P.Mich.* 465.
[55] Davies 1989: 38; Fink 1971: no. 9.
[56] Klemm & Klemm 1992: 228–31.
[57] *I.Th.Sy.* 35.
[58] Klemm & Klemm 1992: 178 f.
[59] *CIL* III 12067, 12068.
[60] *CIL* III 12069. On the *cohors Scutata c.R.*, cf. Spaul 2000: 495.

Deiotariana and the *legio III Cyrenaica*, as well as trierarchs and nauarchs of the *classis Augusta Alexandriana*, setting up a series of altars with votive inscriptions at Akoris. Titus Egnatius Tiberianus, *centurio* of the *legio III Cyrenaica* and *praepositus* of the quarries i.e. ἐπὶ τῆς λατομίας [—], 'from where(?) the pavement of the city of Alexandria derives', dedicated an altar to Zeus Megistos in AD 82/3.[61] The findspot of the altar is unknown, but it has been speculated that the stone originates from the quarries near Akoris in the Gebel el-Teir.[62] The other attested military units perhaps had contingents stationed in the limestone quarries nearby (*c.*10–20 km from Akoris) such as El Saweita, Gebel el-Teir, Tehna, Wadi Sheikh, and in the large quarries of Zawiet Sultan, north of Alabastrine/Alabastronpolis.[63] Based on this assumption, we might have a situation comparable to the Brohltal quarries near Bonn in Germania Superior.[64] Besides inscriptions on stone, Fritz Mitthof recently published a receipt of a soldier who was probably stationed at the Akoris quarries to the πρεσβύτεροι κώμης of Soknopaiou Nesos, in which he acknowledges the requisition of unknown goods for the Akoris quarries.[65] A similar receipt documents the delivery of requisitioned palm fibres from the *komarch* of Philadelphia to a soldier sent by the commander of the Roman army contingent at the Akoris quarries.[66] The unit to which these soldiers belonged is not named. The Roman army certainly supervised the extensive quarrying procedures at Syene where the inscription of an *ala Maur(orum?)*, documents the extraction of columns and pilasters between AD 206 and 210.[67]

Although this short list of quarries run by Roman army and navy units is incomplete, it has become evident that soldiers, as well as officers, were provided with ample opportunity to acquire technical and managerial expertise in quarrying operations. This is unsurprising, as it was part of the governor's task to provide assistance to municipal building projects within his province and deploy soldiers when necessary.[68] In fact, army units were readily deployed for numerous public construction projects such as roads, bridges, aqueducts, etc., which usually included the opening and exploitation of quarries, preferably as close to the building site as possible.[69] The exact procedures for managing the pool of experienced officers, and for dispatching

[61] Bernand 1988, no. 3. For a further soldier of *classis Augusta Alexandriana* at Mons Claudianus (*O.Claud.* 540, 541), cf. Maxfield 2002: 76.

[62] Bernand 1988: 6.

[63] Klemm & Klemm 1992: 78–101.

[64] *I.Akoris* 3, 12, 14, 16, 18–21.

[65] *P.Alex.*inv. 463, cf. Mitthof 1994: 207, 210 f.

[66] *BGU* VII 1612, cf. Mitthof 1994: 207.

[67] *CIL* III 75 = *ILS* 4424.

[68] Ulpian, *Dig.* 1,16,7,1.

[69] For an overview, cf. Horster 2001: 168–87 with Appendices 2 and 3.

them to the imperial quarries, remain unknown. The provincial governors may have requested military officers with special skills from neighbouring provinces, or directly from the emperor.

5.1.3. Deployment and Regular Tasks of Army Units in Imperial Quarries

5.1.3.1. *Egyptian Eastern Desert*

Soldiers of auxiliary units were not only present at the quarries in the Egyptian Eastern Desert, but were posted along the roads as well. The ostraca of Mons Claudianus and inscriptions at Mons Porphyrites or Mons Ophiates document the presence of *cohortes equitatae* or *alae*.[70] The *praefecti* Vibius Alexander and Antonius Flavianus mentioned in the ostraca of Mons Claudianus were probably in charge of one of these units scattered along the desert routes to Mons Porphyrites and Mons Claudianus.[71] Detachments of various auxiliary units garrisoned numerous fortlets along the roads from Koptos to Myos Hormos or Berenike, of which some stood in the immediate vicinity of quarries at Wadi Hammamat and Wadi Umm Fawakhir or Wadi Gemal, the latter near beryl mines.[72]

The majority of the auxiliary soldiers seconded to the different *praesidia* east of the Nile had their main camp in the Thebaid.[73] Thus, the *praefecti* of these units mentioned in the ostraca of Mons Claudianus probably had their

[70] Mons Claudianus: *coh. I. Fl. Cilicum eq.* (AD 118): *I.Pan* 42, 47; *ala Apriana* (Trajan): *O. Claud.*inv. 413; *coh. II Thracum* (Trajan): *O.Claud.*inv. 6038. *O.Claud.* 388 names a Petronius Valens who might be C. Petronius Valens, a *beneficiarius* of the *cohors II Thracum* (*CIL* III 12076 = *SB* 4593); *coh. II Itur. eq.* (AD 144; 222–35): *O.Claud.*inv. 7363. *coh. I. Lus. eq.*: *I.Pan* 56. Maxfield 2000: 429 f.; Maxfield 2001: 152, table 7.2; Maxfield 2002: 80 f., table. The brief of the soldier from *classis Augusta Alexandriana* documented in the two receipts for advance payment (*O. Claud.* 540, 541) is unknown, cf. Maxfield 2002: 76. Mons Porphyrites: *ala Vocontiorum* (AD 113): *AE* 1936: 60; *cohors I Apamenorum equitata* (Antonine period): *I.Pan* 20; Maxfield 2000: 430; Maxfield 2001: 152, table 7.2; Maxfield 2002: 81, table. Mons Ophiates: *coh. Flori* (AD 11): *I. Pan* 51; *coh. III Itur. eq.* (AD 150–152/3): *I.Pan* 53; *ala I Thr. Maur.* (2nd cent. AD): *AE* 1954: 85 = *AE* 1952: 249; *cohors I Lusitanorum equitata* (2nd cent. AD): *I. Pan.* 56; Maxfield 2000: 430; Maxfield 2001: 152, table 7.2; Maxfield 2002: 80, table.

[71] Cuvigny 2002: 240.

[72] Wadi Hammamat/Umm Fawakhir: *Coh. I. Fl. Cilicum eq.* (Domitian): *I.Ko.Ko.* 52; *coh. Nigri*: I.Ko.Ko. 41; *coh. Facundi*: *I.Ko.Ko.* 133; *coh. Flori*: Kayser 1993, no. 1; all AD 11–14. Wadi Gemal:*cohors Nigri Cameresina* (AD 15–18): *O.Tait.* I 245, cf. Maxfield 2000: 426 fn. 105. The ostraca from the military forts at Maximianon/mod. Al-Zarqa and Krokodilo/mod. Al-Muwayh attest the fabrication of stone tools, but do not refer to any extractive operations, cf. Cuvigny 2003*c*: 371–3.

[73] Except for *ala I Thracum Mauretana*, based at Alexandria, cf. Maxfield 2000: 430.

headquarters in the Nile valley. Unfortunately we are unable to ascertain whether they were posted at Syene, the home base of most units documented in the Eastern Desert, or perhaps closer to the supply routes leading out to Mons Claudianus, Mons Porphyrites, and Mons Ophiates.

The auxiliary units were either *cohortes equitatae*, equipped with a significant number of troops on horseback or camel, or were cavalry *alae*. A high level of mobility was an important factor, as the unit detachments executed various tasks along the roads. The ostraca excavated in the *praesidia* (notably Maximianon and Krokodilo) along the Koptos–Myos Hormos road provide some insight into the regular activities of these unit detachments. The rosters for guard duty on the σκόπελος, watchtower, or the *coxae*, the four corner towers of the fort, as well as lists of garrison members, allow one to determine the size of the military garrisons in the *praesidia*. The garrison numbers range between fifteen and twenty soldiers, both foot soldiers and horsemen. The soldiers were probably relieved on a regular basis but the length of duty in the desert is unknown.[74] The garrison stood under the command of a *curator praesidii*. *Curator* was not a regular rank in the Roman army, but merely a role filled by the *principales* (*sesquiplicarii* and *duplicarii*).[75] The main responsibilities of the *curator* and the garrison under his command are explained by documentary evidence, such as circulary notes or letters addressed to the *curator* individually or to the *curatores* along the road by the prefect of Berenike or by *decuriones*, and daily journals. The curator and his garrison were to provide security for travellers and goods on the road section under their surveillance. They also ensured the swift transfer of information by the official courier, and provided water for travellers and animals.[76] The construction of strongholds, as Cuvigny argues, was perhaps necessitated by an upsurge in banditry by the local Bedu. A number of ostraca from Krokodilo and Mons Claudianus attest clashes between the army and the local tribes, who often attacked the forts. For example, in an unpublished ostracon from Mons Claudianus, Petearoeris complains to his father Bekis that he had not eaten for two days out of fear of a barbarian assault. Another text from Krokodilo records that sixty barbarians actually launched an attack on the garrison at Patkoua.[77] In one document, the prefect of Berenike requested the curatores of the *praesidia* to furnish travelling parties with escorts. These military escorts accompanied the travellers for a limited distance from their own *praesidium* to the next and apparently returned to their base after

[74] Cuvigny 2003*b*: 307–9, 311 f.
[75] Breeze & Dobson 1993: 76; Cuvigny 2003*b*: 314 f.
[76] Cuvigny 2003*b*: 315 f., 326–33.
[77] *O.Claud.*inv. 4888. *O.Krok.* 87, cf. Cuvigny 2003*b*: 346–52, esp. 351 f. with further ostraca.

completion of their task. Hence, the escort changed at every *praesidia*. The escorts usually consisted of one or two soldiers, mostly horsemen. Furthermore, patrols, *vestigationes*, could be sent out to monitor the roads and the tribes. Furthermore, circular notes sent by the prefect of Berenike and others were transported by cavalrymen from the *praesidia*, seconded for courier duty. These messengers only carried the document for a certain distance and then handed it over to the next soldier. Water appears not to have been freely available for civilian travellers at the *praesidia*, and certainly not in high quantities. Letters suggest that requests for water went to *curatores* or soldiers and it seems to have been rationed for the members of the Roman army and civilian inhabitants of the forts.[78]

The function of these military installations along the road from Koptos to Myos Hormos was probably not significantly different from the forts along the roads to the quarries at Mons Claudianus and Mons Porphyrites. But since the latter roads were used for quarrying purposes only, there were some variations. As the road to either quarry was a cul-de-sac, the *curatores* of the *praesidia* had to deal with less traffic engaged in private enterprise travelling to the Red Sea and back. The *praesidia* were probably visited by supply caravans going to the quarries, and stone and column transports making the return journey. Other visitors would have included the *tabellarii* of the quarrying administration. Travel was strictly controlled by the soldiers and travellers were issued with a 'pass', πιττάκιον. These passes apparently could be bought at any station or fort, were not issued *ad personam*, and could be used only for a certain period.[79] The issuing of passes/πιττάκια was perhaps necessitated by the scarce water supplies at Mons Claudianus and at the road stations. The amount of water was reported regularly to the *centurio*, who was in charge of rationing the water supply for a population of up to *c.*1,000 civilian and military personnel at Mons Claudianus.[80] Besides this considerable population of long-term residents and a large amount of animals for which the authorities were responsible, the need to supply additional visitors with water surely posed difficulties.[81] The *curatores* of the *praesidia* were required to guarantee the provision of water and the status of water supply was reported regularly to the commanding officer.[82] Perhaps the issuing of πιττάκια provided the authorities with a tool to monitor and restrict arrivals and departures from Mons Claudianus and thus was the basis on which the *praesidia* were informed about the travellers on the road to the Nile Valley. The πιττάκιον

[78] Cuvigny 2003*b*: 326–32 f.
[79] *O.Claud.* 246, 247.
[80] Tomber 1996: 42; Cuvigny 1997*a*: 79.
[81] Maxfield 1996: 11 f.
[82] *O.Claud.* 2.

mentioned in the ostraca should not be confused with the so-called 'laissez-passer'- or 'sauf-conduits'-ostraca found at Claudianus. These ostraca were issued by *centuriones* to the *stationarii* or *curatores* of the *praesidia*, requesting to let a named person or a number of unidentified persons and animals pass.[83] The 'laissez-passer' ostraca were probably not carried by the travellers. Like the circular orders requesting escorts for travellers on the Koptos–Myos Hormos road, these were probably circular orders issued by the *centurio* and copied at the way-stations and *praesidia*, including the fort at Mons Claudianus. The soldiers along the routes were thus informed of the number of individuals and animals travelling on the road to the Nile valley or to Mons Claudianus.[84] Unlike the circular orders from the Koptos–Myos Hormos routes, the *stationarii* or *curatores* were not ordered to dispatch escorts to accompany the travellers. Such assignments do not appear in the published ostraca from Mons Claudianus at any rate.[85] Members of the auxiliary garrisons were, however, sent on a regular patrol, the προβολή, which accompanied travel parties or the regular supply caravan (πορεία).[86] Orders and requests from commanding officers and soldiers, i.e. the *centuriones, decuriones*, and others at Mons Claudianus, to the *curatores* at the *praesidia* and the *stationarii*/ἐπιτηρηταί along the roads—and vice versa—were transmitted by *tabellarii* and the regular cavalry patrol or travelling soldiers.[87] The ostraca also mention a κουράτωρ μετάλ(λ)ου Κλαυδιανοῦ who was in charge of the *praesidium* at the Mons Claudianus quarries. The *curator* stood in regular contact with his colleagues at the *praesidia* or the *stationarii* along the desert roads.[88] The detachment (two officers, thirty footsoldiers, six cavalry, twenty-two recruits) at Mons Claudianus was considerably larger than the garrison of the *praesidia* on the Koptos–Myos Hormos road.[89] The greater numbers of troops attested in the quarries mirrors the amount of additional responsibilities the garrison had. An additional feature of the military presence at Mons Claudianus was the medical infrastructure, which was probably set up and run by the Roman

[83] Rengen 1992; Adams 2001: 173; Maxfield 2001: 160.

[84] Bülow-Jacobsen 1998: 63 f.

[85] Cuvigny 2003*b*: 326.

[86] προβολή: *O.Claud.* 227, 279, 375, 376, 380. πορεία: *O.Claud.* 245, 273, 278, 375, 376. Bingen 1997: 82; Bülow-Jacobsen 1998: 63 f.

[87] Orders and letters to and from *centuriones*: *O.Claud.* 357, 359, 360, 368–70; *decuriones*: *O.Claud.* 363–5; *duplicarii*: *O.Claud.* 124, 125, 144, 366, 367; *sesquiplicarius*: *O.Claud.* 126.

[88] *O.Claud.* 371–6, 378–82; *O.Claud.*inv. 5201, cf. Rengen 1992: 59 f.; Rengen 1997: 219; Maxfield 2001: 159. The rank of the *curator* is not specified, but the function could be held by a *decurio* or normal soldiers, *duplicarii* or *sesquiplicarii*, Bülow-Jacobsen 1997*a*: 208 f. On *duplicarii* and *sesquiplicarii* in general, cf. Le Bohec 1994: 47; Speidel 2000*b*: 69 fn. 27.

[89] Cuvigny 2005*a*: 3.

army.[90] The ill were treated by doctors, ἰατροί, who provided basic treatment, while ἱππιατροί, veterinarians, were brought in for the horses of cavalry units.[91] Sick lists appear on the ostraca, of which some were headed with the title αἶγροι, lat. *aegri.* These lists note the day and month and comprise four to six names.[92] More detailed lists report up to ten, even sixteen, ill or injured. Perhaps a rudimentary 'hospital' existed at Claudianus for which these lists were produced.[93] In addition to treating patients, the doctors also distributed medicine and handed out medicinal appliances to treat injuries.[94] The high risk of (fatal) accidents, injuries, and illnesses connected with the quarrying business is certainly reflected in the written documents. While some ostraca report deaths,[95] others attest a variety of injuries.[96] Not all inhabitants at Mons Claudianus, however, had access to medical treatment. The ostraca appear to name only soldiers or φαμελιάριοι, whereas παγανοί might have been excluded from the medical treatment.[97]

The letters from the ἀντικουράτωρ Rufus to the prefect Vibius Alexander and the procurator Tertullus suggest that supplying the quarries with food, animals, slaves, equipment, etc., may have been the task of the *curator* at the *praesidium* of Mons Claudianus.[98] For certain tasks the army could fall back on the civilian personnel present at the quarries: for the transmission of letters, for example, the *curator* at Mons Claudianus could rely on φαμελιάριοι.[99] Moreover, the curator or *centurio*(?) may have diverted quarry workers from their tasks for the needs of outposts (e.g. to dress millstones).[100] Furthermore the commanding officers on the spot were informed on the whereabouts of quarry workers: in an ostracon Demetras writes to the *decurio* Publius, that Nemonios, a stonemason (σκληρουργός), did not arrive in the quarry of the 'Hydreuma'.[101] Furthermore, one ostracon is headed with the title ἐργάται παγανοὶ ἀτακτήσαντες which translates as 'undisciplined civilian workmen'. The ten people listed did

90 Le Bohec 1994: 52; Wilmanns 1995.

91 *O.Claud.* 120(?). *O.Claud.*inv. 1538+2921, 2055, 2795+3739, 3260, cf. Cuvigny 1992*b*: 106 f.

92 *O.Claud.* 191–210.

93 *O.Claud.* 115, 212–19.

94 *O.Claud.* 120, 174(?), 220.

95 *O.Claud.* 121 (coffins), 211. *O.Claud.*inv. 415, 673, cf. Cuvigny 1992*b*: 108.

96 *O.Claud.* 119, 212, 213, 217, 221–3.

97 Cuvigny 1992*b*: 82; Cuvigny 1997*a*: 20.

98 *O.Claud.*inv. 7295, cf. Cuvigny 2002: 245–8.

99 On use of civilians and φαμελιάριοι cf. *O.Claud.* 229, 266, 371, 372, 374–6. See Bülow-Jacobsen 1998: 64.

100 *O.Claud.* 287, 288.

101 *O.Claud.* 383.

not appear for work. Another list ends with the remark *οὐκ ἀνέβ(ησαν)*: 'did not come'.[102]

In addition to the soldiers the workforce was also seconded for guard-duty at posts. Lists of *vigiles* record the names of the guards, their daily routine and duties. As Adam Bülow-Jacobsen points out, only a few *vigiles* on duty were soldiers; *παγανοί* and *φαμελιάριοι* appear to have been mostly employed in this function.[103] The lists name eight or four men, two guards for each of the four watches during the night.[104] As the lists were found within the fort at Claudianus, the guard posts must have been located close by.[105] The quarries perhaps remained unguarded at night. During the day the working parties could be accompanied by two guards, *τηρηταί*.[106] Distinct from the night-watches were the *σκόπελοι*, manned watchtowers. Their position at Claudianus perhaps ensured the quick exchange of basic information by signalling between the 130 quarries and the central structure at Wadi Umm Hussein. Such towers have also been observed at Mons Porphyrites and Mons Ophiates (cf. 2.1.2, 2.1.4). More complex information was perhaps transmitted by men on *ἀγγαρεία*-duty. The *σκοπελάριοι* and *ἄγγαροι*(?) most likely did not have military backgrounds. The position of watchtowers (*σκόπελοι*) at Claudianus do not imply that they were built in order to monitor the surrounding area, although internal control of convicts might be a possibility.[107]

5.1.3.2. *Simitthus*

The military presence at Simitthus may have consisted partly of soldiers from the *legio III Augusta* and an auxiliary unit, *cohors II Flavia Afrorum equitata*. The evidence for the latter include the tombstone of Cassius Iscoesius Valens, a soldier, and a building(?)-inscription perhaps naming a *praefectus cohortis* of this unit at the time of the Severans.[108] One funerary stone of an L. Flaminius, *miles legionis* of the *III Augusta*, who was killed by the enemy in a skirmish at

[102] *O.Claud.* 384. *O.Claud.*inv. 415, cf. Cuvigny 1992*b*: 108. *O.Claud.* ZPE 1. *O.Claud.*inv. 746, cf. Cuvigny 1992*b*: 102; Cuvigny & Wagner 1986: 63–7.

[103] *Παγανοί*: *O.Claud.* 309–34, 348; *φαμελιάριοι*: *O.Claud.* 337–47, 349–53; soldiers: *O.Claud.* 356. Bülow-Jacobsen 1997*c*.

[104] *O.Claud.* 309–36, cf. Bülow-Jacobsen 1997*c*: 166 f.

[105] *O.Claud.* 337–56.

[106] *O.Claud.*inv. 1252, cf. Bülow-Jacobsen 1992*a*: 57 f.; Peacock 1992: 9; Maxfield 2001: 156.

[107] The excavations yielded a number of ostraca giving the timetable of several posts, cf. *O.Claud.* 175, 304–8 with Cuvigny 1997*b*: 148–50. On watch-towers in the Eastern Desert, Brun, Cuvigny, & Reddé 2003.

[108] *Legio III Augusta* in general, cf. Le Bohec 1989*a*. *Cohors II Flavia eq.*: *AE* 1992: 1821 = Khanoussi 1991: 830 f. *AE* 1994: 1882 = *AE* 1992: 1822 = Khanoussi 1991: 831 f., and photo, p. 832, fig. 5. For the *cohors II Flavia Afrorum equitata*, see Le Bohec 1989*b*: 67–70; Spaul 2000: 462 f.

the *praesidium* in the *saltus Philomusianus*, has been understood to indicate that the quarries at Simitthus were part of this *saltus*.[109] However, there is no evidence to support this view. Flaminius' site of death was not necessarily the same as that where he had been stationed and buried.[110] Besides the tombstone of Flaminius, which dates to the year AD 56–8, the other inscriptions appear to be of a later date.[111] The funerary inscription of L. Venidius Repostus, another legionary soldier of the *III Augusta*, might belong to the AD 160s.[112] A fragmentary inscription seems to attest to the presence of a detachment of *legio III Augusta* at Simitthus under Elagabal (AD 218–22).[113]

The soldiers documented at Simitthus were likely garrisoned there in connection with the quarries. Unlike their colleagues stationed at the remote Mons Claudianus they probably did not have to deal with problems of supply or security of transport. The epigraphic record and archaeological remains at Simitthus, however, only supply fragmentary and circumstantial evidence for the duties of the military personnel. The probable use of convict labour, documented by the large camp north of the quarries, made the presence of military guards necessary, not only in order to guarantee the security of the town of Simitthus close by, but also to prevent convicts from escaping (cf. 2.1.6). However, there is currently no evidence that the military personnel at Simitthus, as at other quarries, was directly involved in the process of quarrying.

5.2. THE ROMAN ARMY AND IMPERIAL MINES

Despite abundant material attesting the presence of military personnel in mining districts throughout the Roman empire, detailed information on the functions of soldiers in mining ventures is not forthcoming. Hence, the subsequent survey of military inscriptions at or near mining sites will merely allow for limited insights into the purpose of their deployment. In fact, the *raison d'être* of some garrisons (Ravna, Montana) may not necessarily be explained by their vicinity to mining operations alone.

[109] *CIL* VIII 14603 = *ILS* 2305.

[110] Khanoussi 1997. See e.g. *AE* 1993: 1572 (Apameia) naming a Vivius(?) Batao who dies whilst debarking at Aigiai in Cilicia. According to his inscription the corpse was taken to Catabolum (Cilicia) and a gravestone set in Apameia-on-the-Orontes in Syria. For critisism of Khanoussi's view, cf. Mackensen 2005: 11 f. with bibliography.

[111] For the date, cf. Lassère 1980: 33 f.; Le Bohec 1989*a*: 157 and fn. 84; Khanoussi 1991: 827.

[112] The name of his *centurio*, a Antonius Clemens, also appears in a list of *centuriones* of *legio III* dedicating collectively at Lambaesis in AD 162, cf. *AE* 1992: 1820 = Khanoussi 1991: 827 f., cf. Le Bohec 1989*a*: 168 f.

[113] *AE* 1992: 1823 = Khanoussi 1991: 833, no. 4, with photo on 832, fig. 6.

5.2.1. Roman Iberia

The series of dated inscriptions by the military vexillations at Villalís and Luyego in northwestern Spain provides us with insight into the internal hierarchy of military detachments seconded to the mining areas and enables us to observe changes in the vexillations and officers dispatched to this area over a limited period of time.

The *vexillatio* of *legio VII Gemina* celebrates the legion's creation on 10 June, whereas the soldiers of *cohors I Gallica* and *cohors I Celtiberorum* honour the formation of their units on 22 April and on 15 October. In the years AD 184 and 191 members of *ala II Flavia* are listed in the inscriptions, suggesting that an additional detachment of a different unit may have been present at Villalís. Perhaps it replaced *cohors I Celtiberorum*, which is not mentioned on the stelae after AD 167. All inscriptions name the highest officer present and the imperial official in charge of the mining district. They list the legionary and auxiliary *centuriones*, and the *decuriones* of the auxiliary units, followed by the *beneficiarii procuratoris Augusti* and finally the *signiferi*, *imaginiferi* and *tesserarii* of the legion or auxiliary cohorts. The *centurio legionis* perhaps held the overall command of the vexillations(?) at first, but after AD 163 legionary centurions disappear from the inscriptions set up for the *vexillatio legionis VII Gemina*. *Centuriones* of *cohors I Gallica* and, after 184, *decuriones* of *ala II Flavia* perhaps took over tasks within the mining district.[114] The *principales* listed in the inscribed stelae of AD 163 to 167 (180?) are a regular feature of most vexillations and therefore did not necessarily apply their administrative capabilities to the mining areas of northwestern Spain. The disappearance of legionary officers and *principales* from the inscriptions at Villalís and Luyego might be of significance, perhaps reflecting a reduction in troop levels from AD 175 onwards.[115] Another noticeable change is the absence of *beneficiarii procuratoris* from the epigraphic record after AD 180. It is not known whether this change reflects the diminishing importance of the gold mines or whether troops were relocated for other reasons, such as the need to fend off the Moorish invaders.[116]

The vexillation seems to have been posted in or near Villalís or Luyego, where the main bulk of inscriptions were found. However, to date no military camp or buildings have been discovered there. In fact, only few military encampments in northwestern Spain are to be found in the immediate

[114] Saxer 1967: 129 f.; LeRoux 1977: 365–7; LeRoux 1982: 273; Domergue 1990: 348–51; Ott 1995: 154.

[115] Domergue 1990: 215–24.

[116] Domergue 1990: 300.

TABLE 2: *Army vexillations at Villalís and Luyego*

Mañanes 1982: 112 =Luyego; *c.*138–161; *leg. VII G. F.*	Iulius Marcus, *centurio*, leg. VII G.F.		Valerius Victoris, *decurio*		
CIL II 2552 =*ILS* 9125 = *CBFIR* 857 =Saxer 1967: 129. Villalís; 10 June 163: *leg. VII G.F.*	Licinius Paternus, *centurio*, leg. (VII G.F.)	Hermes, *lib. Aug.*, *procurator*	Lucretius Paternus, *decurio*, coh. I Celtiberorum	Fabius Marcianus, *b.f.proc. Aug.*	Iulius Iulianus, *signifer*, leg. (VII G.F.)
Mañanes 1982: 148. Corporales, 10 June 163: *leg. VII G.A(ntoniniana?)*	Lice(!)nius Paternus, *centurio*, leg. VII G.A.	Hermes, *(lib.) Aug.* *procurator*	Lucri(!)tius Paternus, *decurio*, coh. I Celtiberorum	Aemulius Aecianus, *b.f.proc. Aug.*	Aulius Aulianus, *signifer*, leg. VII G.A
CIL II 2556 = *CBFIR* 855 = Saxer 1967: 133. Villalís; 22 April 166: *coh.I Gallica*	–	Zoilus, *lib. Aug.*, *procurator*	Valerius Flavius, *centurio*, coh. I Gallica	Valerius Valens, *b.f.proc. Aug.*	Iulius Iulianus, *signifer*, leg. VII G.F.
CIL II 2553 = *ILS* 9127 = *CBFIR* 856 = Saxer 1967: 131. Villalís; 15 October 167: *coh. I Celtiberorum*	–	Zoilus, *lib. Aug.*, *procurator*	Valerius Flavius, *centurio*, coh.I Gallica	Aelius Flavius, *b.f.proc. Aug.*	Iulius Sedulus, *tesserarius*, coh. I Celt-iberorum - Lucretius Maternus, *imaginifer*, leg. VII G.F.
AE 1910 : 1 = *ILS* 9130 = *IRLéon* 40 = *CBFIR* 858 = Saxer 1967, 134. Villalís; 22 April 175: *coh. I Gallica*	–	–	M.Sentius Bucco, *centurio*, coh. I Gallica	Val. Sempro-nianus *b.f.proc. Aug.*	–
AE 1967: 230 = *IRLéon* 40; Luyego; 10 June 180: *leg. VII G.F.*	–	M. Aurelius Eutyches, *lib. Aug.* *procurator*	M. Mes[. . .]ris, *centurio*, coh. I Gallica	–	Avitius Paternus
AE 1910: 5 =*CIL* II 2554 = Saxer 1967:	–	Aurelius, Eutyches *lib. Aug.*,	Valerius Sempronianus *decurio*, ala II	–	–

(Continued)

Table 2: *(Continued)*

130; Villalís; 10 June 184: *leg. VII G.F.*		*procurator*	Flavia		
AE 1910 : 2 = *ILS* 9131; Villalís, 22 April 191: *coh. I Gallica*	–	Aurelius Firmus, *[pr(ocurator)] met(allorum)*	Valerius Marcel[lus], *decurio*, ala II Flavia	–	–

vicinity of the mining zones. An auxiliary camp was discovered at Valdemeda, situated near Roman mining sites in the Rio Eria Valley.[117] Moreover, several training camps were built near Castrocalbon, but never housed an auxiliary unit for a long period.[118] The detachments present in the mining zones of the Duerna valley originated from legionary and auxiliary bases close by.[119] During the Flavian period the former encampment of *legio X Gemina* at Rosinos de Vidriales was turned into a smaller fort for *ala II Flavia Hispanorum c.R.* and remained occupied until the third century AD.[120] The *cohors I Celtiberorum eq. c.R.* had its headquarters in Santa Maria da Cidadela, Sobrado dos Monxes, *c.*40 km west of Lucus Augusti/mod. Lugo, where it is attested from AD 134 onwards.[121] The camp of *cohors I Gallica eq. c.R.* was possibly located at Baños de Bande (Orense), although the *Notitia Dignitatum* attests its camp in Veleia (Iruña, Prov. Alava) at the end of the fourth century AD.[122]

In the Iberian context the distribution of military units provides some interesting insights into the strategic importance of the north-west. Until the mid-first century, the initial legionary and auxiliary units deployed during the Cantabrian wars had been strategically placed to secure the control over the Cantabrians and Asturians. Under the Flavians, the military presence was significantly reduced, leaving only one legion and a few auxiliary units in charge of the Iberian peninsula. Apart from the auxiliary camps of the units mentioned above, *legio VII Gemina* at Legio/mod. León was positioned immediately south of the passes up to Asturia Transmontana and west of the mining areas near Asturica Augusta, being superbly situated to handle any problems arising in the

117 Sánchez-Palencia 1980; Sánchez-Palencia 1986: 228, 234, fig. 2.

118 Jones 1976: 59; LeRoux 1992: 235.

119 LeRoux 1982: 107, 112–14; Balil Illana et al. 1991: 85.

120 LeRoux 1982: 145–7; LeRoux 1992: 233.

121 Caamaño Gesto 1984; Caamaño Gesto 1989; LeRoux 1992: 233. The unit is attested in the region through a *tessera hospitalis* of the *praefectus cohortis* with the *Coelerni* of the *conventus Bracarensis*, cf. *AE* 1972: 282. LeRoux 1982: 149–50; LeRoux 1992: 233.

122 LeRoux 1982: 147; LeRoux 1992: 232–4.

region.[123] Further units, such as *ala Tauriana torquata victrix c.R.*, probably based near Calagurris in the Ebro Valley, and the *ala I Augusta* close to Añavieja (prov. of Clunia), were to be found in the north as well.[124] Although military detachments were placed throughout the three Roman provinces in the Iberian Peninsula, the main military focus clearly lay on the north-west.[125] The Moorish incursions into the southern Iberian peninsula in AD 171/2 revealed the weakness of this military disposition.[126]

The presence of military contingents seems less marked in imperial mines other than those in northwestern Spain. The epigraphic record of Trêsminas contains a few inscriptions of the personnel based there: a soldier of *legio VII Gemina* set up a votive inscription to Iuppiter at the end of the first century, a legionary detachment of the same unit commissioned a further votive monument in AD 130, and *cohors I Gallica eq. c.R.* had an unknown number of soldiers posted there.[127] Apart from Trêsminas, information on army contigents in mining areas is scarce. The Vipasca tablets do refer to *milites* using the baths within the mining district free of charge. However, no inscriptions record a military presence.[128] Funerary stelae of legionary soldiers from other mining sites in Lusitania and Asturia et Callaecia provide evidence for the presence of soldiers in their vicinity.[129] In Roman Baetica army units hardly appear in the mining areas. Only a very fragmented funerary inscription of an *eques legionis* from the *legio IIII Macedonica* dating to the mid-first century records the presence of military personnel at Magacela, a town in the vicinity of the Castuera mining area.[130]

It has been suggested that stamps on ingots provide additional evidence for army units in Spanish mining zones, as lead ingots from the shipwreck off the coast near Comacchio (Ferrara) in northern Italy were thought to be produced by the Roman army in Baetica in the early Augustan period. However, recent metallurgic analysis of the ingots in question locates the source of the lead not in Baetica, but near Carthago Nova, and it was

[123] LeRoux 1982: 144–53, 166; LeRoux 1995: 72.

[124] LeRoux 1992: 232–4.

[125] LeRoux 1995: 72.

[126] Christol & Demougin 1990: 195; Domergue 1990: 299, contra: LeRoux 1982: 173–267, fig. 3; LeRoux 1992: 248–58; LeRoux 1995: 73–5. For *legio VII Gemina* and Septimius Severus, cf. Birley 1999: 126.

[127] *AE* 1980: 582. *CIL* II 2389 = *AE* 1907: 150. *AE* 1907: 151.

[128] *LMV* l. 25.

[129] Near Beccerea/Rio Navia 1 (Lugo), *ILER* 1014 = *IRPL* 75 (San Roman de Cervantes; AD*c*.30–63), cf. LeRoux 1982: 181, no. 37; near Barbantes, Razamonde (Orense) or Lago, Maside (Orense), *ILER* 5997 (29 May AD 79), cf. LeRoux 1982: 196, no. 87; near Ribeira da Aravil, Monforthino (*POR* 17), and Meimoa, Idanha-a-Velha: *AE* 1961: 358 = *AE* 1967: 145; *AE* 1961: 359.

[130] *CIL* II^2/7 964 = *AE* 1994: 885.

convincingly argued that these ingots were probably not produced in a military context.[131]

5.2.2. Germany

Incisions and stamps may reveal a military unit as the owner or perhaps even the 'producer' of an ingot, but they do not *per se* prove their control of mines from which the ingot originated. A lead ingot from the camp at Haltern on the river Lippe, east of the Rhine, weighing 64 kg, had its weight (CCIII or 203 Roman pounds/66.472 kg) and the letters L XIX, the initials of the 19th legion destroyed in the Teutoburg forest in AD 9, carved into its side. Sigmar von Schnurbein concluded that the lead probably originated from a mine in the Eifel run by the legions. His interpretation, however, was based on the assumption that the inscription, LXIX, indicated the producer of the ingot.[132] Even though we cannot rule out that *legio XIX* (re-)cast ingots, the inscription and the find spot do not support Schnurbein's conclusion. Moreover, the ingot may originate elsewhere: recent finds show that, in the Augustan period, mining operations may have taken place near Brilon, some 90 km east-south-east of Haltern, perhaps supplying the Roman army in the 'provincia Germania'.[133] A further lead ingot inscribed with *leg(io/nis) XVI* was discovered in the Eifel region near Bad Münstereifel at Mechernich—an area well known for its Roman mining activity. *Legio XVI Gallica* was based at Novaesium/mod.Neuss from AD 43 to 70/1, *c.*70 km north of the findspot.[134] Again, the inscription identifies a legion either as owner or producer of an ingot. There is, however, no direct proof for a Roman vexillation extracting the lead. The proximity of extractive operations to military installations, such as iron-mining and smelting near frontier forts of Aalen and Buch, may be a weak indication for the army's involvement in mining operations.[135]

5.2.3. Roman Britain

Roman Britain may offer more convincing evidence that army units were in direct control of smelting procedures, perhaps even the exploitation of the ore itself. This is suggested by one lead ingot, discovered at St Valéry sur Somme

[131] Domergue et al. 2006, for older views cf. Berti 1985; Domergue 1987*b*; García-Bellido 1994–5; García-Bellido 1998.

[132] Schnurbein 1971.

[133] Rothenhöfer 2003; Hanel & Rothenhöfer 2005: 53–65.

[134] Horn 1987: 154 f. with p. 156, fig. 91.

[135] Johnson 1987: 13.

and probably originating from the Mendips, with the moulded mark *Neronis Aug(usti) (plumbum) Britan(nicum) l(egio) II (Augusta)*, which could be translated as 'British (lead) of Nero Augustus. Legio II Augusta (made this)'. As already argued above, the inscription may indicate the production of ingots by the *legio II Augusta* for the Roman state (cf. 3.3.5). Further ingots from Britain display the name of army units in moulded marks: a lead ingot found near Mendip Hills renders *Britannic[u]m Aug(usta) ḷẹ(gio) IỊ*. Another found at Caerwent was stamped by the same unit.[136] One of two lead ingots originating from British mines and discovered near Châlon-sur-Saône were stamped with the letters LEG XXII, as well as DOC and BFLIDOC. The latter is read as *b(ene)f(iciarius) l(egionis) I (Minerviae)* DOC. Nelis-Clément tentatively argued that the import of lead ingots into Gaul may have been controlled by a *beneficiarius* based at Gesoriacum/mod.Boulogne.[137]

Inscriptions on lead ingots set aside, archaeological evidence has been interpreted as indicating the involvement of the Roman army in mining operations. Excavations at the 'fortlet' at Charterhouse-on-Mendip appear to show that the Roman army was in control of the silver/lead mines as early as the year 49.[138] The fort at Pumsaint documents military presence at the Dolaucothi gold mines in Wales, certainly during the Flavian period and into the second century.[139] The find of lead and lead ore at Brough-on-Noe and of lead ingots near Brough-on-Humber, which probably came from the lead mines of Derbyshire, suggests that this was the reason for the establishment of a *castellum* at Brough-on-Humber. Moreover, military forts at Brough-upon-Stainmore, Whitley Castle, and Kirkby Thore in Westmoreland may have been constructed to protect the lead mines at Alston Moor. It is a matter of conjecture whether the two lead seals found at Alston Moor with the stamp of *cohors II Nerviorum* and the letters *metal(la?)* on the reverse suggest the exploitation of lead mines by this auxiliary unit based at Epiacum/mod. Whitley Castle. A small wooden fortlet at Nanstallon near Bodmin (Cornwall) has been thought to serve as a guard post for the lead and silver mines in the area.[140]

The excavations at Beauport Park near Bottle have yielded a vast number of tile stamps from the *classis Britannica*.[141] Beauport Park is a site in 'The

[136] *RIB* 2404.2, 25.

[137] Nelis-Clément 2000: 259 f.; *RIB* II 2404.72a and b.

[138] Todd 1996: 4; For the Mendip lead industry, cf. Elkington 1976; Stewart 2002: 179–4. For the forts, cf. Breeze 1987: 16; Johnson 1987: 13, 272.

[139] Jones & Mattingly 1990: 180–4; Jones 1991: 210, with further bibliography; Burnham 1997; Stewart 2002: 134.

[140] LeRoux 1989: 172 with fn. 13.

[141] *RIB* II/5 2481.6 ff.

Weald', an area in East Sussex heavily mined in Roman times for its iron. The mining area covered at least 8 ha and comprised of a bath-house and a slag heap of up to 15 m in height and 0.8 ha in size. The bath-house contained an inscription and 1,641 stamped tiles of the *classis Britannica* were found nearby.[142] The Roman fleet certainly provided the building material, but there is no further epigraphic or archaeological evidence to attest the constant presence of members of the *classis Britannica* there. Thus, it is unknown whether the Roman fleet provided security measures, ran the mines to supply their own needs and/or those of the Roman army in Britain, or guarded the transport of iron ingots off the British Isles.

5.2.4. Danubian Provinces

We are comparatively well informed on the disposition of the army detachment at Montana in Moesia Inferior which might have been charged with protecting mining facilities to its southwest(?) (cf. 3.2.4). The *cohors I Sugambrum veterana equitata*, partly equipped with cavalry, was garrisoned there some time during the first quarter of the second century AD.[143] During Hadrian's reign, a legionary detachment of *legio XI Claudia* replaced the *coh. I Sugambrum*. The detachment consisted of nearly 100 men, a considerable reduction in size from the roughly 500 of the preceding auxiliary unit.[144] Besides the inscriptions of the *vexillatio legionis XI Claudiae* at Montana, a number of inscribed votive monuments were commissioned by members of *legio XI Claudia*, mostly during the second century.[145] A small number of soldiers from *legio I Italica* may have formed their own *vexillatio*: according to an inscription they were party to a 'venatio'.[146] Moreover, officers and *beneficiarii consulares* from this legionary unit were based at Montana as well,

[142] Wilson 1971: 277 f.; Brodribb & Cleere 1988; Jones & Mattingly 1990: 192 f.; Cleere & David 1995: 57–86; Stewart 2002: 138–53. For the inscription, cf. Wright & Hassall 1971: 289, no. 2; Brodribb & Cleere 1988: 261 f. with fig. 11.

[143] Ptolemy, *Geog.* 3.10.1, cf. Rankov 1983: 41 f. with fn. 16. Its presence at Montana is confirmed by two tile stamps of the unit (*CIL* III 12529), cf. Aleksandrov 1977: 59.

[144] *AE* 1987: 867 = *AE* 1999: 1327, cf. Rankov 1983: 53; Velkov & Alexandrov 1988; Velkov & Aleksandrov 1994: 8 f., no. 9. The inscription (found at Eumeneia/mod. Ishekli in Phrygia) attesting the presence of *cohors I Sugambrum* at Montana also reports its temporary departure to Asia: *AE* 1927: 95 = Buckler, Calder, & Cox 1926: 74–8, no. 201.

[145] *AE* 1969/70, 577. *AE* 1985: 746. *AE* 1985: 751 = *AE* 1987: 247. *AE* 1987: 869, 871, 876, 883, 886–8 = Velkov & Aleksandrov 1994, nos. 10, 16, 18, 34, 38, 42, 47, 53, 65, 71; additional inscriptions, cf. Velkov & Aleksandrov 1994, nos. 84, 86, 88, 121, 138, 148.

[146] *AE* 1987: 867, cf. Velkov & Alexandrov 1988; Velkov & Aleksandrov 1994: no. 9.

perhaps from the later second century onwards.[147] Cavalry elements of *legio I Italica* are represented by a *p(rinceps) d(uplicarius)* and a *vexilarius equitum* during Gordian's reign.[148] In the first half of the third century an auxiliary unit, a *numerus civium Romanorum*, is recorded at Montana under the command of *centuriones* of *I Italica* or *XI Claudia*.[149] Rankov assumes that the presence of an additional unit at Montana might be the result of a heightened level of threat to the area and the mining zone.[150] Moreover, according to a fragmentary inscription from a fort near Kamena Riksa, a *cohors Dacorum* may have been based there around AD 204 to provide security for the mines in the Ogosta valley.[151] As already noted above (cf. 3.2.4), military officers of *legio XI Claudia* and *legio I Italica* may have controlled a *regio Montanensium*.

At Timacum Minus(?)/mod. Ravna in Moesia Superior, military personnel are likewise involved in the administration of a *territorium*. A *praefectus territorii* had control of an area which could have incorporated a number of silver and gold mines. The inscription, dated to the late second century AD, suggests that this *praefectus territorii* had his own staff for the administration of this district, consisting of soldiers of *legio VII Claudia*.[152] The camp was alternately occupied by different auxiliary units: a *cohors I Thracum Syriaca* is attested there from AD 70 to the early second century, and after AD 169 a *cohors II Aurelia Dardanorum*.[153]

The Mt Kosmaj region in Moesia Superior was home to several different auxiliary detachments. Excavations in the early twentieth century yielded evidence for a fort at Stojnik.[154] Moreover, an inscribed votive altar discovered there documents the transfer of a tribune from *coh. XVIII Voluntariorum c.R.* to the *cohors I Ulpia Pannoniorum milliaria equitata* (*c.*161–9). The inscription does not clarify which unit was stationed at Mt Kosmaj, but one might expect the newly promoted tribune to have commissioned the altar on his impending departure to the new unit.[155] A *praefectus* of *cohors V (Callaecorum?)*

[147] Rankov 1983: 54 f. *AE* 1985: 743, 744. *AE* 1987: 867, 872, 873, 875, 881, 884. *CIL* III 12371; cf. Velkov & Aleksandrov 1994, nos. 9, 14, 19, 30, 35, 41, 44, 48, 50, 51, 56, 57, 58, 78, 91.

[148] *AE* 1957: 341, cf. Velkov & Aleksandrov 1994, nos. 22, 91; Rankov 1983: 55.

[149] *AE* 1957: 338, and Velkov & Aleksandrov 1994: 3, 20, 96, 132.

[150] Rankov 1983: 58 f.

[151] *CIL* III 12382 (Kamenna Riksa).

[152] *IMS* III/231. *IMS* III/229 = *ILJug* 1297 = *AE* 1934: 185. *IMS* III/230. *IMS* III/233 = *ILJug* 1312 = *AE* 1905: 163. A *centurio* of *legio X Fretensis* (based in Syria-Palaestina), is attested at Timacum Minus, but probably died there during the course of the Marcomannic wars (*IMS* III/2 34 = *ILJug*, 1308 = *AE* 1934: 187).

[153] Coh. I Thrac. Syr.: *IMS* III/2, 29, 37–40; coh. II Aur. Dard.: *IMS* III/2, 7(?)–9, 18–20, 22, 23, 43–52.

[154] Dušanić 1978: 237; Werner 1986: 562.

[155] *IMS* I 97 = *CIL* III 6302 = *ILS* 2606 (Stojnik).

Lucensium set up a further votive altar discovered at Suvodol near Stojnik, perhaps in the years 166 to 169(?).[156] A cippus in grey limestone even documents a *valetudinarium*, a hospital, of *cohors II Aurelia nova milliaria eq. c.R.* for AD 179 in Stojnik.[157] Furthermore, inscriptions record the presence of ordinary auxiliary soldiers of this cohort at Guberevac, Sopot, and Stojnik.[158]

One of the main tasks of the Roman army units in Moesia Superior was to provide security for the mining districts. The army forts at Stojnik and, perhaps, Ravna seem to provide evidence for such a military objective, and it has been argued that the series of forts along the Danube not only guarded the river trade (by which metals were exported) but also the approaches up the tributaries to mining areas like Bor or Mt Kosmaj in the hinterland.[159] Moreover, the creation of six new auxiliary units in the later second century allowed Rome to solve two problems at once. The *Historia Augusta* reports that emperor Marcus Aurelius turned *latrones Dalmatiae atque Dardaniae* into *milites*—the two *cohortes Delmatorum*, two *cohortes Aureliae Dardanorum* and two *cohortes Aureliae Novae* which appear in inscriptions.[160] These native auxiliary units were partially deployed on guard duty in the mining areas of Moesia Superior.[161] This measure not only secured the important mines against barbarian incursion from across the Danube, but probably dealt effectively with the internal threat of brigands as well.

A fragmentary inscription found at Sočanica might attest *cohors I Dardanorum* in the mining area, but to date no further evidence for a military presence there has been forthcoming.[162] The mining area of the Kumanovo plain in southern Moesia Superior has yielded further epigraphic material: a *librarius consularis* and *miles legionis IV Flaviae* received a funerary stone at Lopate; a *strator legati* of *IV Flavia* set up a gravestone for his wife at Lojane in the second half of the second century AD; and at Konjuh the funerary inscription names a *centurio* of a *cohors I*.[163] From the epigraphic evidence one may conclude that some of these soldiers were ordered there mainly on

[156] *CIL* III 14542 = *AE* 1901: 22 = *IMS* I 98, (Suvodol) cf. Dušanić 1978: 106 with fn. 20, 125; Spaul 2000: 87.

[157] *IMS* I 116 = *CIL* III 14437 = *ILS* 9174 (Stojnik).

[158] *IMS* I 112 = *AE* 1910: 98 (Stojnik). *IMS* I 117 = *CIL* III 14541 (Guberevac). *IMS* I 118 = *ILJug* 26 (Guberevac); *IMS* I 119 = *CIL* III 14217[6] = *AE* 1908: 21 (Sopot). *IMS* I 120 (Stojnik). A fragmented inscription might perhaps attest a *(cohors) I Aurelia n[ova]* at Stojnik or a *vet(eranus)* of *(II) Aurelia n(ova)*: *IMS* I 99 = *AE* 1901: 23 = *CIL* III 14545 (Stojnik), with Wilkes 1969: 118 fn. 6, 218 fn.4.

[159] Dušanić 2000: 349; Werner 1986: 561 f.

[160] *HA M.Aur.* 21.7.

[161] Dušanić 1978; Dušanić 2000: 348 f.

[162] Dušanić 1977: 75 with fn. 148; Rankov 1983: 46.

[163] Lopate: *IMS* VI 227. Lojane: *IMS* VI 241 = *AE* 1984: 790 (early 3rd cent. AD). Konjuh: *IMS* VI 236 = *ILJug* 563 = *AE* 1964: 275 = *AE* 1969/70: 459 = *AE* 1979: 562.

administrative duties. The post of *librarius consularis* suggests an official had been commissioned by the governor of the province to undertake administrative work in the region.[164] Whether their presence was related to mining work in the area is not clear from the written evidence.

The strength report of *cohors I Hispanorum veterana*, formerly stationed at Stobi in northern Macedonia, does record an unknown number of its officers or men being sent *in Dardania ad metella* (sic). Given the date of this report, the year 101 or the year 105, these auxiliary soldiers and/or officers probably were not sent to Upper Moesia to provide technical support, but to protect the Dardanian mines during the Dacian War.[165] However, no detailed information on their location and the mines in question is provided by inscriptions.

Despite the considerable size and importance of the Dacian gold-mining areas, the inscribed monuments of military personnel known from Ampelum and Alburnus Maior are few. Apart from inscriptions of *beneficiarii consulares/procuratoris* at Ampelum, two *librarii*, both of *legio XIII Gemina*, appear in the epigraphic record of Ampelum. One of them was seconded there on the orders of the provincial governor—his function is given as *librarius consularis*.[166] The fragmentary inscription of a further active or former member of *XIII Gemina* suggests that there may have been a legionary contingent at Ampelum.[167] In the later second century, additional men and horse were provided by *numerus Maurorum Hispanorum*.[168] Votive inscriptions record two *praefecti*, one soldier of this unit, and a funerary monument for a further soldier.[169] The relative scarcity of evidence for a military contingent based at Ampelum might be explained by the proximity of two large military bases. The camp of *XIII Gemina* at Apulum and the auxiliary camp at Micia were well-positioned to guard the approaches to the widespread *aurariae Dacicae*.[170] There is, however, evidence for construction work by the army at Ampelum, as tiles stamped by *legio XIII Gemina* and *legio IIII Flavia Firma* have been found there.[171] Moreover, stamped tiles of *legio XIII Gemina* have also emerged from the recent surveys and excavations at the Tomuș site at Alburnus Maior.[172] *Centuriones frumentarii* are found in other mining areas,

[164] Austin & Rankov 1995: 152; Rankov 1999: 30.

[165] Fink 1971: no. 63; LeRoux 1989: 172 with fn. 21; Dušanić 2000: 348 f.

[166] *CIL* III 1317 (= AMP 32). *CIL* III 1318 (= AMP 33).

[167] *CIL* III 1319 (= AMP 34).

[168] Noeske 1977: 314. On cavalry in *numeri*, cf. Dixon & Southern 1997: 31.

[169] *CIL* III 1294 = *IDR* III/3 312 (= AMP 36). *CIL* III 1316 = *IDR* III/3 339 (= AMP 38). *IDR* III/3, 302 = *AE* 1971: 383 (= AMP 35). CIL *III* 1149 = *ILS* 3558 (= AMP 37) (Borosbocsárd). On the history of the unit, cf. Petolescu 1983, with older bibliography.

[170] Gudea 1977.

[171] *AE* 1988: 961b, 961c. The latter legion is attested in Dacia during Trajan's wars and appears to have returned to Singidunum in Moesia Superior around 119, cf. Strobel 1984: 88–90; Piso 2000: 208–13.

[172] Damian 2003: 36, 431 f.

most notably Noricum. A *centurio frumentarius* of *legio I Adiutrix* set up a votive altar at Tiffen, while a further *centurio frumentarius* of *legio II Italica* is named in a fragmentary inscription at Feldkirchen.[173] The sources do not provide any detailed information on the function of *centuriones frumentarii* in a mining context. It is noteworthy that these officers were not directly attached to the legion but either to the *castra peregrinorum* in Rome or to the headquarters of the provincial governor.[174]

5.2.5. Other provinces

Military personnel also appear in or near other mining districts. The Misenum fleet may have been connected with the mines of Gonnesa on Sardinia: a soldier of the Misenum fleet is documented in the first half of the second century at the site.[175] Further inscriptions document military personnel in the vicinity of mining ventures of the Sardinian Iglesiente: at the end of the first or beginning of the second century a member of *cohors I Sardorum* is recorded at Grugua, a mining site[176]; and a *centurio* of the same unit is attested at Campingeddus near the mines of San Nicolao.[177]

Further evidence is provided by inscriptions from the eastern half of the empire. The inscription of a soldier of *legio III Cyrenaica* discovered at Wadi Tuweiba, west of Aila was purportedly found near a copper mine.[178]

5.3. THE ROMAN ARMY AND EXTRACTIVE OPERATIONS

Apart from money and various victuals, the Roman army required a constant supply of metals to produce military equipment. In general, army units endeavoured to be self-sufficient, acquiring water, building materials, firewood, leather, ore, and other natural resources in the vicinity of the camp.[179] The manufacture of military equipment was usually carried out by military

[173] *CIL* III 4861 (Feldkirchen), 4787 (Tiffen).

[174] Clauss 1973: 82–113; Rankov 1983: 54.

[175] *CIL* X 7537.

[176] *CIL* X 8321 = Sotgiu 1988: C113. The name of a soldier from an unknown unit is inscribed on a monument at the same site in the second half of the second century, cf. *AE* 1985: 485a = Sotgiu 1988: B115.

[177] *CIL* X 7535.

[178] *AE* 1972: 671 = *AE* 1936: 131; Speidel 1977: 694 f.

[179] Speidel 1996: 72 with further bibliography.

personnel.[180] However, we have less evidence for soldiers running mining ventures to meet the demands of the army. The episode reported by Tacitus, in which Curtius Rufus, governor in Germany, is awarded the triumphal insignia by emperor Claudius for ordering his legionary soldiers to dig up silver ore in the *ager Mattiacus*, is to be understood in the context of the supply of the Roman state with precious metals.[181] Although army units are named in moulded marks and cold stamps on ingots, this should not be regarded as an indication of their direct involvement in the extraction process. The raising of forts and fortlets in the vicinity of lead mining ventures in Britain only demonstrates that the Roman army guarded these facilities. In fact, Roman soldiers actually partaking in iron or lead mining operations, is, with the one exception, not directly attested in literary or epigraphic sources. Although soldiers were not spared the wearisome burden of quarrying for stone, cutting wood, or producing tiles for the construction of various military or public installations, they may not have been required to mine for iron and lead, which were the metals most commonly needed by the army. Tacitus' passage provides further insights on this matter: the soldiers of Curtius Rufus secretly wrote a letter to the emperor Claudius asking for the award of governors with triumphal insignia before their assignment to the provinces. This was obviously intended to curb the enthusiasm of governors in putting their soldiers to work on mining operations or construction tasks, as Corbulo also received triumphal insignia for overseeing the building of a canal between Maas and Rhine.[182] Tacitus describes the difficulties encountered by the soldiers in pursuing the silver lodes and removing rubble in underground mining operations—all for meagre returns.[183] The Tacitean episode conveys the reluctance with which soldiers went about such chores. The fact that the *gravamina* of the soldiers were directed to the emperor may perhaps mark the rarity of using army personnel in mining operations. Unlike the occasional supply of building materials, such as stone and wood, for military and public construction, the constant demand for iron, lead, and copper by the Roman army would require the permanent deployment of a considerable party of soldiers to opencast or underground mining sites. Besides the hazards of working in a subterranean environment, the extent and intensity of labour involved may have excluded soldiers as a constant supply of work force. There is little written evidence to support these assumptions. Mould marks, stamps, or incisions by a legion or an auxiliary

[180] Bishop & Coulston 2006: 233–8.
[181] Tacitus, *Ann.* 11.20.3.
[182] Tacitus, *Ann.* 11.20.2.
[183] Tacitus, *Ann.* 11.20.3.

unit may at most be interpreted as indicating ownership of the lead ingot. However, in the case of the lead ingot found at St Valéry sur Somme, the mould mark might indicate the production of lead ingots by the Roman legion.[184] This military involvement may have been limited to running the smelting ovens and exercising general control over mining procedures. One cannot categorically exclude the possibility of Roman soldiers working in mines (as Tacitus clearly shows); even so, any such assignment probably was only for a short period. Further, there is no written evidence for military engineers providing technical advice on mining infrastructure such as aqueducts, tunnels, galleries, etc. In comparison to the evidence from imperial quarries in Egypt and Asia Minor, one may tentatively argue against the regular employment of military forces to source required metals from mines. The mining activities for the needs of the Roman army were likely delegated or contracted out to civilian personnel.

The main task of the Roman army was undoubtedly to provide protection for the mining and quarrying districts, not only against external enemies, but against brigands within the province. Moreover, based on the judicial powers of the procurators, we can expect soldiers attached to certain mining districts to have assisted in the execution of procuratorial decisions. Like the units known from some of the imperial quarries, the military contingents in imperial mines were often accompanied by a cavalry unit, be that an *ala* or *cohors equitata*. These units probably provided protection for overland transports of gold or silver ingots, as well as the regular traffic to and from the mining districts—a role possibly assigned to the Roman fleet in the case of imperial mines on islands like Sardinia or Britain.[185] The heavy garrisoning of auxiliary troops in Moesia Superior and Inferior, and perhaps Dacia too, may have been an answer to security problems. Security considerations also consolidated the garrison of *legio VII Gemina* at Legio/mod. León in the vicinity of the gold-mining zones of northwestern Spain. In the initial phases after the Roman conquest of Asturia and Galicia, a heavy military presence was required, probably in connection with the forced resettlement of the indigenous population and their subsequent employment in mining operations.

Apart from the protection of mining districts, the Roman army is recorded at Ampelum in Dacia, and possibly at Lopate and Timacum Minus in Moesia Superior as providing *librarii*. Vegetius described the military *librarius* as book-keeper of military *rationes*. Haensch, however, argued that the presence of *librarii* at all levels of military administration suggests they were regular scribes.[186] They either joined the local administration (as at Ampelum) at the

[184] *RIB* II 2404.24.
[185] Rankov 1983: 46.
[186] Veg. 2.7; Haensch 1997: 722.

behest of the provincial governor or, as at Timacum Minus, were party to an *officium* of the *praefectus territorii*. The *principales* of the vexillations present at Villalís and Luyego, such as the *signiferi*, the *imaginifer* of *legio VII Gemina*, and the *tesserarius* of *cohors I Celtiberorum*, were higher ranking soldiers. Their regular functions certainly included administrative tasks: the main responsibility of the *signifer* (and perhaps of the *imaginifer* as well) was running the accounts of the soldiers' savings and the distribution of pay, whereas the *tesserarius* ran the communication within the unit.[187] Perhaps they directed part of their administrative expertise towards the mining operations in northwestern Spain. One might presume similar tasks for the *principales* and officers attested at Montana—provided the *regio Montanensium* included any valuable mining sites of interest. However, *signiferi*, *tesserarii* and others were regular part of vexillations, and their presence at mining sites is more likely explained by the requirements of the military body present.[188]

The *beneficiarii procuratoris* may have served as a link between the financial administration of the province and the mining administration. Probably attached to the office of the financial procurator (rather than the *procuratores metallorum/aurariarum*), these *beneficiarii* most likely were present as liaison officers. The existence of such a connection is evidence enough to suggest that at least some mining procurators, notably those of the gold-mining areas of north-western Spain (Villalís, Luyego) and Dacia (Ampelum/mod. Zlatna), were in direct contact with the financial procurators at Asturica Augusta or Sarmizegetusa. The case of the *beneficiarii consulares* is a bit more obscure. These soldiers are quite often found in the service of provincial governors, carrying out a wide range of tasks within the provincial administration.[189] Hence, their purpose in the context of the imperial mining industry is difficult to assess. Were they stationed in or near mining territories to enforce the collection of rents or taxes on extracted ore exported from mining districts? Were *beneficiarii* based near mines as military police, providing the mining procurator with a body of men to execute his legal decisions? Or were the *beneficiarii consulares* 'liaison' officers between the mining procurators and the provincial governors? *Beneficiarii*, usually senior legionary soldiers, were seconded to the provincial capital to serve at the headquarters of the governor. A number of them, however, were outposted throughout the province in legionary or auxiliary forts, at important road junctions, and

[187] Veg. 2.7; 2,19 f. Breeze 1974: 263 ff.; Davies 1974: 44, 246 fn. 48; Speidel 1996: 60 f., 62 with further bibliography on subject matter.

[188] Saxer 1967: 129 f.

[189] Ott 1995: 82–157; Nelis-Clément & Wiblé 1996: 211–68; Nelis-Clément 2000: 220–59.

most notably in mining areas.[190] Papyri in Roman Egypt document the involvement of *beneficiarii consulares* in the prosecution of minor crimes and acting as liaisons to the *praefectus Aegypti*. It has been argued that *beneficiarii consulares* perhaps fulfilled similar tasks in the mining districts.[191] The thesis of Siegfried De Laet that *beneficiarii consulares* also protected the custom stations of the *portorium* is based on the false observation of an overlap between *stationes of beneficiarii* and customs stations.[192] Hence, the road stations of the *beneficiarii* might not have been exclusively related to the collection of customs, although papyri do reflect an involvement of *beneficiarii* in tax-collection in Roman Egypt.[193] In this context, the absence in the written evidence of any *beneficiarii consulares* or *beneficiarii procuratoris* from imperial quarries in Roman Egypt, Asia, Achaia, Africa proconsularis has to be noted.[194] Consequently, we do not know whether the staff of every provincial governor included *beneficiarii* and how many a governor could deploy. The presence of two legions in Egypt and numerous auxiliary units would supply an ample basis from which to promote *beneficiarii* to the governor's service at Mons Ophiates, Mons Claudianus, or Mons Porphyrites.[195]

Administrative tasks were also performed by army officers and soldiers seconded to the imperial quarries. Apart from the regular military duties, such as protecting the quarries and the supply routes, securing the water supply and communication, army personnel were also involved in procuring basic needs such as food, water, tools, and animals for the workforce, particularly in the quarries of the Egyptian Eastern Desert. Moreover, progress reports from the quarry workers were sent to the *praefecti* of the auxiliary units garrisoned in the quarries. It remains doubtful whether army officials in other imperial quarries fulfilled a range of functions comparable to those of their colleagues in Roman Egypt. Certainly, in provinces like Asia or Africa, quarries and their supply routes did not have to be protected to the extent the extractive operations in the Eastern Desert had to be, but other problems arose from the proximity of these quarries to inhabited areas. At Simitthus the main task of the army might have been the segregation of the convicts condemned to the quarries from the town. The Roman army probably also

[190] Austin & Rankov 1995: 195–204; Haensch 1997: 721; Rankov 1999: 28 f.; Nelis-Clément 2000: 259–64.

[191] Austin & Rankov 1995: 195 f., 201.

[192] De Laet 1949: 449; Austin & Rankov 1995: 196.

[193] Dise 1995: 76; Ott 1995: 129–42; Eck 1997*a*: 131 f.

[194] It remains uncertain whether the *stationarii* mentioned in the ostraca at Mons Claudianus were *beneficiarii*. *O.Claud.* 50–73. 80–2; Haensch 1997: 522.

[195] Austin & Rankov 1995: 154; Haensch 1997: 721.

provided the medical infrastructure for soldiers and imperial personnel, as the 'hospital' at Mons Claudianus and the *valetudinarium* at Mt Kosmaj suggest. Apart from guarding mining and quarrying districts and providing personnel for administrative purposes, the Roman army—on the orders of the emperor—appears to have dispatched skilled quarrying engineers to the imperial quarries. The Roman army had a large supply of well trained and experienced engineers at its disposal because soldiers were regularly employed in various construction and quarrying activities (mostly connected with specific building projects). The secondment of individual *centuriones frumentarii*, as well as legionary *centuriones* halfway across the empire suggests that they were deployed to the imperial quarries based on their administrative and/or technical expertise and experience in the quarrying industry.

Army detachments deployed on guard duty, and possibly for the technical support of imperial mining and quarrying establishments, usually came from the same provinces where the mines and quarries were located, with the exception of the administrative or technical specialists brought in for specific tasks. *Procuratores metallorum* could rely on military commanders and officials to assume additional administrative functions if required to do so. As the legions and/or auxiliary units stood under the command of the provincial governor, he probably ordered their deployment to the imperial quarries or mines. Moreover, the provincial governors also remained responsible for the pay and supply of the troops with basic provisions.[196] This underlines a strong interdependence of provincial administration and imperial mines and quarries, an interdependence which emerges in other areas as well.

[196] *P.Dura* 64 = RMR 91 = *ChLA* VI 319; *P.Dura* 129; *O.Bu Njem* 75–9, cf. Fink 1971: 383 f.; Mitthof 2001: 47–50.

6

Imperial Officials and the Allocation of Responsibilities

The responsibilities of the heads of mines or quarries would have included an assessment of the geological, topographical, technological, and geographical constraints and challenges. Military personnel provided security and undertook engineering tasks in imperial mines and quarries, and may have even had managerial roles. The majority of remaining organizational measures were probably undertaken by imperial procurators and their staff. In view of the fragmentary evidence, it will not be possible to allocate organizational tasks to specific personnel (*procurator, tabularius, dispensator,* etc.). Therefore, comparisons will be necessary with other branches of the Roman provincial administration in order to assess the range of activities undertaken by these personnel. Additionally, it is questionable whether each official focused on a specific set of tasks without transgressing into other administrative areas, or whether everyday business was conducted less rigidly. The titles of administrative personnel seem to indicate the former, but appearances may be deceptive.

6.1. PROCURATORS

6.1.1. General Observations

The term 'procurator' appears early on in the legal and literary sources of the Roman Republic and initially describes 'liberti' or 'liberi' who acted as legal representatives of a principal (e.g. proprietor of an estate, owner of a trade business) in various matters of economic concern, such as single transactions or the management of a business or estate.[1] Within certain legal boundaries, the procurator could act on behalf of the principal in various enterprises of legal consequence. The basis of his power was either a *iussum* (in the case of a

[1] Eck 2001*d*: 366.

procurator omnium rerum as general representative in the absence of the principal from Italy) or a *mandatum* (in the case of a *procurator cui mandatum est* as representative in a specified undertaking in the provinces). This included the representation of the principal in court, the purchase of ownership and property rights for the principal, taking up a loan, the rent and lease of property, and the conclusion of obligations.[2] Liability issues arising from the conduct of business set aside, the procurators were also required to keep track of financial costs and income and to provide the principal with a detailed account (*rationem reddere*). In return for their services, the procurators were disbursed by the principal through a *salarium* or *honorarium*.[3]

The legal framework established for procurators of private individuals may at first not have differed significantly from that of procurators dealing with the private estates of the emperor during the Principate. Imperial procurators were, however, also concerned with the management of properties and financial transactions of the Roman state, irrespective of their social status as imperial freedmen or equestrians. On being sent to their provincial posts, imperial freedmen as well as equestrian procurators received *mandata*, a set of directives.[4] Furthermore, as a consequence of their employment in the administration, procurators of some provinces (perhaps of *provinciae Caesaris*) were probably vested with judicial powers from the reign of Augustus onwards. These rights may have been consolidated by Claudius in AD 53.[5] The judicial responsibilities appear to have covered issues of fiscal concern, though with a mandate of the governor the procurator could hear other cases as well. He predominantly served as a judge in cases where the *fiscus* appeared as plaintiff or defendant, or in cases between third parties which resulted from fiscal claims.[6] According to the legal texts, procurators could also seize property for the fiscus, sell it, request the payment of monies due to the fiscus or collect fines imposed by other authorities.[7] In general, they were understood to be *rei publicae causae absentes*.[8] These judicial powers and the administrative activities in the name of the Roman state distinguished imperial procurators clearly from their private counterparts.

[2] Schäfer 1998: 28–36 with sources and bibliography.

[3] Schäfer 1998: 37–87, esp. 78–87.

[4] Dio 53.15.3–4, 57.23; Tacitus, *Ann.* 4.15; *SEG* XVII 755 = *IGLS* V 1998; Pflaum 1982: 21 f., no. 49 bis; Eich 2005: 106–18; *mandata* in general, cf. Burton 1976: Millar 1992: 642 f.; Eck 1997*a*: 117.

[5] Brunt 1990*b*: 165–7 with sources and bibliography.

[6] Brunt 1990*b*: 171.

[7] Brunt 1990*b*: 173.

[8] Paulus, *Dig.* 4.6.35.2, cf. Eich 2005: 288.

The responsibilities and functions of procurators in the different branches of mining and quarrying administration are occasionally highlighted by epigraphic and papyrological material, but they hardly provide a complete picture of activities. By analysing the organizational challenges which faced the management of imperial mines and quarries, one can glimpse the range of responsibilities a procurator might have shouldered.

6.1.2. Responsibilities of Quarrying Procurators

Since the organization of imperial quarries was constrained by the geology and topography of the area, the challenges which faced the head of a quarry were numerous and complex (cf. 2.1.6, 2.1.7). The written documentation for Mons Claudianus and the Nile valley allows the reconstruction of some of the strategies employed by such officials in dealing with the most prevalent problems.

6.1.2.1. *Division of Organizational Tasks*

A list of the specific organizational tasks of the *ἐπίτροπος τῶν μετάλλων* at Mons Claudianus cannot be compiled with ease. Apart from the fact that the procurators hardly appear in the ostraca which have been published to date, a further problem arises from the fact that it is not clear how activities were divided between the procurator and the commander of the military garrison on the site. Two letters from Mons Claudianus exemplify this problem.[9] A published ostracon contains two copies of letters sent on the same day by the *ἀντικουράτωρ* Rufus (a replacement for the regular curator of the fort at Mons Claudianus, a Iulius Silvanus), to the prefect Vibius Alexander, and to Tertullus, the *ἐπίτροπος* (*τῶν μετάλλων*). In the letter to the prefect Vibius Alexander, Rufus reminds him of a previous report the 'vexillation'/*βήξιλλον* (i.e. the military detachment at Mons Claudianus) had filed on the *ἐπιχρεία τοῦ μετάλλου*, 'the equipment of the quarry', and on the delivery of grain (for the soldiers) and of 'liquid nutriments', i.e oil and wine, being late.[10] The prefect Vibius Alexander was presumably a commander of an auxiliary unit based in the Thebaid. The prefect, however, not only took care of issues relating to the auxiliary unit under his command but, as in

[9] *O.Claud.*inv. 7295, cf. Cuvigny 2002: 245–8.

[10] Cuvigny 2002: 245 f. Whether 'the equipment of the quarry' refers to military equipment of the vexillation at Mons Claudianus or to quarrying tools remains unclear.

the case of Vibius Alexander's predecessor Antonius Flavianus, seems also to have received reports on the progress of work by the quarrymen.[11] In the second letter, addressed to the procurator Tertullus, Rufus does not mention the problems in grain supply, but compiles a detailed list of items left behind by his predecessor, the curator Iulius Silvanus. Moreover, the procurator is told of significant shortages in slaves and donkeys, and of empty cisterns resulting from the lack of personnel.[12] The fact that a military ἀντικουράτωρ of the fort at Mons Claudianus sent a letter to the quarrying procurator may show that the division between 'military' and 'civilian' spheres was not clear-cut. Whereas the military commanders were kept informed about the progress of work in the quarries, the procurator apparently received specific requests for provisions, material and animals for the quarry workers through military officials.

Apart from the logistical support for the personnel at Mons Claudianus, the ostraca also attest the direct interest of the procurator in the amount of quarried products available. One document names a procurator, probably Ulpius Himerus (AD 152/3), who demanded information on the stock of quarried blocks in his own *caesura* at Mons Claudianus through his *tabularius* Athenodoros.[13] The quarry labels found on blocks and columns there provide further evidence, with the ligatured mark CIP being read as *c(aesura) ⟨H⟩i-(meri) p(rocuratoris)*.[14] Other ostraca confirm that *procuratores* had *caesurae* in their name.[15] A *caesura* was an administrative term describing one or more quarrying sites within a district where marble was extracted on behalf of a procurator or a contractor (cf. 7.2.3). The quarry labels on *giallo antico* from Simitthus document personal names of procurators in a similar context. These observations imply that the procuratorial responsibilities also included an output-related involvement in the quarrying process. In other words, the registration of incoming orders, their translation into extracted and refined blocks or columns, and the keeping of records on outgoing produce to an agreed destination is likely to have been the responsibility of the procurator. On occasion, similar tasks may have fallen to the military specialists seconded

[11] *O.Claud.*inv. 4471, 4852, cf. Cuvigny 2002: 240.

[12] Cuvigny 2002: 246.

[13] *O.Claud.*inv. 6483, cf. Bülow-Jacobsen 1996: 61 (translated from Danish): 'Athenodoros to the most honourable Sokrates, greetings! I have been ordered by my master, the *procurator*, to give you notice immediately that you should inform him on the stones still lying in the quarry of his caesura (until?) the 31st of December. Inform me thus quickly, so that I can write a letter to him in addition(?) to your letter. I hope you are well.'

[14] Quarries 19: App. nos. 971–5; 23: no. 994; 36: no. 1005; 93: nos. 1113–15; 113: no. 1130. Valerie Maxfield and Wilfried van Rengen believe the lower case b to be a ligatured LD, cf. Bülow-Jacobsen 1996: 61; Peacock & Maxfield 1997: 223.

[15] Bülow-Jacobsen 1996: 60.

to the quarries on the orders of the emperor or the praefecti who received quarrying reports from the work force at Mons Claudianus (cf. 5.1.1).

6.1.2.2. *Recruitment and Pay of the Workforce*

As the letter to the procurator Tertullus suggests, the recruitment and supply of quarry workers and other personnel were the tasks of the *ἐπίτροποι τῶν μετάλλων* and their staff. The workforce was divided into two distinct groups, the *παγανοί* and *φαμελιάριοι*, suggesting different modes of recruitment and supply of labourers for the quarries. The term *παγανός* or *paganus* usually describes non-members of the Roman army, in other words, civilians.[16] Within the context of quarry operations, this term identified free and skilled labourers, hired for their expertise from renowned quarrying areas such as Syene, Alexandria, the Thebaid, and the Fayoum.[17] A water distribution list renders the exact numbers for the different groups of *σκληρουργοί* (stonemasons): 210 Alexandrians, 130 Syenites, six Arsinoites, and three Memphites.[18]

Παγανοί and *φαμελιάριοι* probably both received a money-payment (*ὀψώνιον*), and additional food rations: *παγανοί* earned 1 artaba (40 l?) of wheat plus a wine-ration, members of the *familia* 1 artaba of *σῖτος* (bread), lentils, and oil, as well as an annual subsidy for clothing.[19] The *παγανοί* had to buy oil, lentils and other goods from the Nile valley through their *κιβαριάται*. The prices for these items owed by the workers and the amounts for reimbursement of advances on pay and payments to *collegia* were simply deducted from their wages. According to Hélène Cuvigny, the payment of wages to *παγανοί* was administered differently to that of the *φαμελιάριοι*, creating two sets of distinct documents: the *ἐντολαί* for the *παγανοί*, and receipts for advanced payments for the *φαμελιάριοι*. Both groups of documents are attested during the second century AD, but the majority date to the period of AD 136–53.[20] Any reconstruction of the pay scale of the *φαμελιάριοι* is difficult to ascertain, as the receipts for advanced payment do not mention the amount of the wage. However, the *ἐντολαί* do include the wage paid per month to the *παγανοί*. The pay-scales for *παγανοί* can be reconstructed:

[16] Cuvigny 2000*a*: 11–13.
[17] Cuvigny 1996*a*: 139; Cuvigny 1998: 87.
[18] *O.Claud.*inv. 1538 + 2921. Cuvigny 2005*b*: 328 f.
[19] Cuvigny 2000*a*: 41, 43.
[20] Cuvigny 2000*a*: 6–10, 22, notes a change in administrative practise, as from AD 136 onwards the receipts for advanced payment are being dated. This might coincide with the changes in epigraphic formula observed on *pavonazetto* and *giallo antico* from Dokimeion and Simitthus (cf. 7.2.8).

47 drachmas appears to be the highest amount a common worker could expect. Others received 37 drachmas and 4 obols, or even less, 28 drachmas per month. Promotion within the pay-scale was a matter of seniority rather than qualification. It seems that around AD 151 wages were increased, reaching 48 drachmas and 38 drachmas, 38 drachmas 2 obols, 38 drachmas 4 obols.[21] Compared with the usual wages paid in the Nile valley (25 drachmas per month), the wages paid at Mons Claudianus were certainly above average.[22]

The money and the grain for the *παγανοί* were not distributed at Mons Claudianus, but handed out at a central location in the Nile valley. Every month, a group of delegates 'elected' from amongst the *παγανοί*, the *κιβαριάται*, travelled to the Nile valley and dealt with the payment of wages and the supply of food rations. Their co-workers handed them *ἐντολαί*, or instructions as to what items the *κιβαριάται* had to purchase with part of their wages. The *κιβαριάται* most likely noted these instructions on a central papyrus roll and took it with them on their trip to Egypt. Some *παγανοί* asked for their grain rations to be handed to their relatives at Kaine or Apollonopolis mikra;[23] the wives and mothers of the quarry workers then produced bread loaves which were sent on to the Desert quarries with the *κιβαριάται*.[24]

Hélène Cuvigny argued that the *φαμελιάριοι* were not exclusively imperial slaves or freedmen i.e. members of the *familia Caesaris*. The names of some *φαμελιάριοι* displayed in the documents also included patronyms. This has been taken to suggest that freeborn individuals could be part of the *familia Caesaris*.[25] In general, Roman jurists understood the term *familia* not to include slaves and freedmen alone. Ulpian states that free men (*liberi*) as well as slaves of a third party (*servi alieni*) are considered part of the *familia* of the tax-farmer (*publicanus*), provided they are involved in collecting the tax.[26] Paulus further explains that free individuals working in a fiscal function become part of the fiscal *familia*, without losing their free status.[27] The

21 Cuvigny 1996*a*: 140 f.

22 Drew-Bear & Naour 1990: 425–9.

23 Cuvigny 1998: 88 f., 91.

24 *O.Claud.* 3–8. Cuvigny 1996*b*: 25; Cuvigny 1998: 87. For further *κιβαριάται* cf. *O.Claud.* 156, 365. For the archive of Petenouphis, *κιβαριάτης* at Tiberiane and his position within the quarrying community: *O.Claud.* 243–54. *O.Claud.*inv. 4457, 5134, 5281, 5307, 6153, cf. Bülow-Jacobsen 1997*b*: 69–80.

25 Cuvigny 2000*a*: 25 f.

26 Ulpian, *Dig.* 39.4.1.5: 'Familiae nomen hic non tantum ad seruos publicanorum referemus, verum et qui in numero familiarum sunt publicani, sive igitur liberi sint sive servi alieni, qui publicanis in eo vectigali ministrant, hoc edicto continebuntur.'

27 Paul, *Sent.* 5.1.3: 'Descriptio ingenuorum ex officio fisci inter fiscalem familiam facta, ingenuitati non praeiudicat'; cf. also Cicero, *Brut.* 85; Cicero, *Pro Caec.* 58. For further evidence, cf. Cuvigny 2000*a*: 26–8.

onomastic evidence for freeborn φαμελιάριοι is partly corroborated by the fact that they received a 'salary'.[28] The main genre of documents related to the φαμελιάριοι are the receipts for advance payment.[29] Salaries and rations were distributed to them at the end of the month, as a reward for work done during that period. At the time of entering the imperial service, the φαμελιάριοι had to go into debt, if they did not command sufficient savings on which to live until the end of the month. Most of them therefore received part of their monthly payment in advance and signed a receipt for this transaction.[30] Unlike the παγανοί, their wages and food rations were administered by their own κιβαριάται who were appointed by the authorities. The analysis of the onomastic material suggests that φαμελιάριοι not only were recruited in Egypt, but originated from Asia Minor, Syria, or Palestine. In contrast to the παγανοί, a number of φαμελιάριοι were brought in from overseas, perhaps even from other imperial quarries, to Mons Claudianus.[31] The recently published ostracon detailing the distribution of water to the work force and military personnel at Mons Claudianus lists a number of highly specific personnel who provided for the φαμελιάριοι. The text attests a guardian of the *cellae*/dwellings of the *familia* (τηρητὴς κελλῶν φαμιλίας), and a guardian of bread of the *familia* (τηρητὴς ἄρτου φαμιλίας). Provided the text is complete, one may observe that a similar function, i.e. guards for the *cellae* of the *pagani*, does exist, whereas a guard for their supplies is not named.[32]

6.1.2.3. *Φαμελιάριοι, παγανοί, and Operative Processes at Mons Claudianus*

The reason for distinguishing between two groups of workers is not quite clear, as both παγανοί and φαμελιάριοι were incorporated in the same working parties sent to the quarries. A closer look at the division of labour, however, shows that φαμελιάριοι and παγανοί mostly fulfilled different functions. The necessity of supplying water to the inhabitants of Mons Claudianus (*c.*1,000 military personnel and quarry workers), produced a set of documents which provide a unique insight into this specific organization.[33] The texts list water rations according to function, and hence, provide a blueprint of its hierarchical structure. The largest rations of water were probably apportioned to the *centurio*, the *decurio*, and the architect(s).[34] *O.Claud.*inv. 1538 provides a clear

28 Cuvigny 2000*a*: 25.
29 Cuvigny 1992*b*: 103; Cuvigny 1996*b*: 21; Cuvigny 2000*a*.
30 Cuvigny 2000*a*: 46–50.
31 Cuvigny 1996*b*: 20; Cuvigny 2000*a*: 32 f.
32 Cuvigny 2005*b*: 316.
33 Cuvigny 1992*a*: 85 f.; Cuvigny 2005*b*.
34 *O.Claud.*inv. 1538 + 2921, cf. Cuvigny 2005*b*: 348–50; Cuvigny 1992*a*: 86.

categorization of the work force, auxiliary personnel, and soldiers stationed at Mons Claudianus based on the amounts of water distributed. A Fannius, perhaps a *centurio* (?), and Sansnos, ἐργοδότης, were each given 1 *keramion* (6.5 l) of water. Soldiers and cavalrymen, veterinaries, and the ἀναμετρητὴς μαρμάρων received five-sixths of a *keramion* (5.4 l). The largest group of workers and other personnel, e.g. foremen, *tirones*, and the παγανοί in general, received half a *keramion* of water (3.25 l), while φαμελιάριοι apparently received only a third (2.16 l).[35]

Apart from the Roman army personnel, the highest position within the work force was held by the ἀρχιτέκτονες (engineers), followed by the ἀναμετρητὴς μαρμάρων, the ἐργοδόται, the παγανοί, and finally the φαμελιάριοι. The ἀρχιτέκτονες appear to have been of importance in organizing the operative processes in the quarries.[36] At Mons Claudianus, Apollonios of Alexandria commissioned an altar during Trajan's reign, and a Herakleides had an extracted column and a rockface inscribed during the same period.[37] The latter's name is also found in an inscription on a granodiorite(?) column in Rome which gives the name of the *praefectus Aegypti* M. Rutilius Lupus (AD 113–17). Moreover, Herakleides' name appears in the water distribution list.[38] The architects did not necessarily remain at the same quarrying site: Apollonios was apparently moved from Mons Claudianus to Mons Porphyrites.[39] Whether such relocations happened on a regular basis remains unknown.[40]

The ἀρχιτέκτονες headed the working parties in the quarries. A fragmentary ostracon lists the team of the ἀρχιτέκτων Apollonios: twenty ἐργάται (workers), five ἀκισκλάριοι (adze specialists), four φαρμαξάριοι (hardeners), ten σφυροκόποι (hammermen), and two further men.[41] From other ostraca we hear of far larger teams, consisting of thirty-six stonemasons, three smiths,

[35] Cuvigny 2005*b*: 336 f., 348.

[36] Two architects, Mersis and Soter, appear in the earliest Roman inscription (AD 10/11) found at Mons Ophiates/mod. Wadi Umm Wikala, and the same Mersis is again named in inscriptions at the Paneion in Wadi Hammamat (*I.Pan* 51; *I.Ko.Ko.* 41).

[37] *I.Pan* 38 = Bernand 1992: 40, 41. Perhaps both are mentioned in one of the water distribution lists, cf. Cuvigny 1992*a*: 85.

[38] *IG* XIV 2421 = *IGRR* I 550. Bingen 1992: 47 f.; Cuvigny 2005*b*: 315. Inscribed sherds of wine amphora imply that architects were the recipients of wine deliveries, *O.Claud.* CdE 1, cf. Bingen 1992: 42 f.; *O.Claud.* 15–41; *O.Claud.*inv. 2755, 4338, 4391, cf. Bingen 1992: 43. App. no. 1071 (*Ιερω(νύμου ?)*).

[39] *O.Claud.* 17. The fluctuation of personnel between Claudianus and Porphyrites seems to have been quite common, cf. *O.Claud.* 143.

[40] For ἀρχιτέκτονες at Gebel Gulab/Syene, cf. Nachtergael 1996: 335. I.*Th.Sy.* 107 (Silsilis). *SB* 4534 (Gebel Toukh). *SB* 5682, 5683, 5684 (Ombos), cf. Bernand 1969: 265 f.; Nachtergael 1996: 337 with fn. 11. Hermoupolis Magna, cf. Bernand 1999: no. 81, with further bibliography.

[41] *O.Claud.* 15.

six bellowsmen and hardeners, two guards, one foreman, and one hammerman. In total forty-nine men are sent to a quarry on one day.[42] Adam Bülow-Jacobsen summarizes the contents of a further unpublished ostracon naming the work force sent to the Myrismos quarry at Mons Claudianus. The text makes a distinction between παγανοί and φαμελιάριοι: one ἐργοδότης (foreman), thirty-five σκληρουργοί (stonemasons), one σφυροκόπος (hammerman), three χαλκεῖς (smiths), two φυσηταί (bellowsmen), all in all forty-two παγανοί. And of the *familia*: one σφυροκόπος (hammerman), three φαρμαξάριοι (hardeners), one κασιότης (crane-specialist), one overseer(?) of side-wedges, one overseer(?) of chisels(?), one δεκάνος, fifty-nine ἐργάται (workers).[43] However, the published ostraca do not shed any light on the particular responsibilities of ἀρχιτέκτονες as regards the quarries at Mons Claudianus.

In the case of the ἀναμετρητὴς μαρμάρων named in the water distribution list *O.Claud.*inv. 1538, Cuvigny assumes that they basically controlled the marble extracted by the work force.[44] In this context the observation made by Valerie Maxfield in the quarries of Mons Claudianus has to be considered: a group of inscriptions beginning with the letters *Π°* for πό(δες)(?), indicate measurements in cubic feet.[45] These measurements apparently do not refer to the size of the unhewn blocks, but to the quarried block in its dressed state. The probable size of the blocks was noted on the stone itself. As an unpublished ostracon proves, a note was made for the archive; it gives the dimensions, location, and the N-number of a stone as a reference.[46] Moreover, there is evidence from Wadi Hammamat to suggest a similar administrative procedure being followed in the Basanites quarries: an ostracon gives a list of block-measurements in Roman feet (πόδ(ες)).[47] Perhaps the technical term ἀναμετρητὴς μαρμάρων refers to the specific activity, also observable in Gaul, of measuring and estimating the size and future volume of marble blocks.[48]

The ἐργοδόται are mentioned frequently in the ostraca at Mons Claudianus.[49] The work lists suggest that the ἐργοδόται oversaw working parties between twenty-six and eighty-six men strong. According to one water distribution list, nine quarries were being exploited simultaneously by working parties under the command of seven ἐργοδόται. Hence, ἐργοδόται could

42 *O.Claud.*inv. 1252, cf. Bülow-Jacobsen 1992*a*: 58; Peacock 1992: 9; Cuvigny 2005*b*: 330.
43 *O.Claud.*inv. 3385, cf. Bülow-Jacobsen 1996: 40; Cuvigny 2005*b*: 330.
44 Cuvigny 2005*b*: 341.
45 App. nos. 936, 937, 960, 990, 996, 1079, 1096, 1097, 1099, 1120.
46 Bülow-Jacobsen 1996: 51 f.; Peacock & Maxfield 1997: 224.
47 Kayser 1993: 126 f. and no. 20.
48 For Gaul, cf. Monthel & Lambert 2002: 111 f.
49 Cuvigny 1992*a*: 84 f., 87 f.

supervise work in more than one quarry.[50] The ἐργοδόται do not seem to be ranked more highly than their fellow workers: they do not appear first in the list of workers nor in the letters written to the *procurator metallorum* or the *praefectus*. Together with the σκληρουργοί and the smiths, the ἐργοδόται report to both officials on the progress in quarry work at Mons Claudianus.[51] Whether ἐργοδόται were primarily recruited from amongst the παγανοί or φαμελιάριοι is not discernable from the written evidence.

The stonemasons, σκληρουργοί, probably account for the largest group amongst the workforce at Mons Claudianus, and are widely attested in other quarries throughout Roman Egypt.[52] The water distribution records provide exact numbers: 212 σκληρουργοί are listed as coming from Alexandria, 130 from Syene/Assuan, and others from Fayoum and Memphis, all of them παγανοί.[53] In the lists of workmen under the command of an ἀρχιτέκτων or an ἐργοδότης, the σκληρουργοί usually made up the largest group of workmen with up to 100 individuals working on one day in a quarry. The stone-masons most likely were 'multi-task' workers. They set wedge holes and dressed the extracted blocks. Adam Bülow-Jacobsen has argued that most of them were used for other tasks as well, such as giving stones their rough form or helping with transport preparations.[54]

Certain functions, such as those of the πλατεάριοι or the ἀκουάριοι, may have been reserved for the φαμελιάριοι:[55] the water distribution lists usually name one, rarely two πλατεάριοι, who may have been responsible for the construction(?) and maintenance of the slipways;[56] water carriers, ἀκουάριοι, carried drinking water or water for metallurgical purposes.[57] However, technical specialists such as ἀκισκλάριοι, φαρμαξάριοι and σφυροκόποι were

[50] Cuvigny 1992*a*: 74, 78, 85 f., and App., no. 1067. For ἐργοδόται at Wadi Hammamat, cf. *I.Ko.Ko.* 48, 112.

[51] Cuvigny 1992*a*. For a similar letter, cf. *SB* 9230 (3rd cent. AD, Syene ?).

[52] Further σκληρουργοί, cf. Wadi Hammamat: *I.Ko.Ko* 49, 75, 105, and Kayser 1993: nos. 11. 14; *SB* 9230, cf. also Cuvigny 2003*a*: 280–4. Krokodilo: K 527 with Cuvigny 2003*c*: 371 f. El-Boueib: *I.Ko.Ko.* 158. Syene: *O.Eleph.* DAIK 66, cf. Locher 1999: 71.

[53] *O.Claud.*inv. 1538 + 2921, cf. Bülow-Jacobsen 1997*d*: 140 with fn. 3; Cuvigny 1992*a*: 85; Cuvigny 2005*b*: 328 f.

[54] For the size of a quarry, cf. Peacock & Maxfield 1997: 178–89. For other uses of the workforce of σκληρουργοί: Bülow-Jacobsen 1996: 43.

[55] Other functions not directly related to quarrying such as baker, wagon steward, and *tabellarius* were probably exclusively held by φαμελιάριοι (*O.Claud.*inv. 1538, 4524, 7334. 8686, cf. Cuvigny 2000*a*: 30).

[56] For πλατεάριοι: *O.Claud.*inv. 6467, cf. Cuvigny 2000*a*: 29, 307; Bülow-Jacobsen 1996: 46 f.; Cuvigny 2005*b*: 342.

[57] For ἀκουάριοι: *O.Claud.* 212; *O.Claud.*inv. 6438, cf. Cuvigny 2000*a*: 29. For water-tanks in quarry huts, cf. Peacock & Maxfield 1997: 234. The carrying of waterskins (ἀσκοφορία) may

φαμελιάριοι as well.[58] Ἀκισκλάριοι, or *aciscularii*, were well trained in the use of the *acisculus*, a kind of adze with a long shaft. Given the fact that fewer ἀκισκλάριοι than σκληρουργοί appear in the work lists, the ἀκισκλάριοι were perhaps responsible for delicate quarrying tasks.[59] Σφυροκόποι, hammermen, who were sometimes παγανοί or φαμελιάριοι, were probably specialists in the use of sledgehammers.[60] Παρασφενάριοι, translated by Bülow-Jacobsen as 'side-wedgemen', are only found in small numbers compared with the σκληρουργοί.[61] Side-wedges, παρασφήνια, were used to enhance the effect of the iron-wedges used at Mons Claudianus. The side-wedges were probably iron plates, holding the wedge firmly in its position. Bülow-Jacobsen argues that the παρασφενάριοι were in charge of breaking away the stone from the rockface.[62] The supply and repair of iron-headed tools for these workmen required the presence of metalworkers in the quarries. While iron bars were brought on a regular basis by camel, the quarry workers needed the tips of their adzes and picks resharpened every hour.[63] With thirty-six stonemasons working in the quarries, and having their tools refitted with a new tip every hour, three smiths (χαλκεῖς, σφυροκόποι), six bellowsmen (φυσηταί) and hardeners (φαρμαξάριοι) had to accompany the workforce.[64] The work lists reveal that these metal-working specialists were provided by both παγανοί and φαμελιάριοι.[65]

have been the task of φαμελιάριοι, as was the supply of water for animals (ποτισμός), the collection of wood as well as providing services such as running the bath at Mons Claudianus, cf. *O.Claud.* 126–9; *O.Claud.*inv. 8512, 8696, 8815, cf. Cuvigny 2000*a*: 29. On the ἀσκοφορία, see also Cuvigny 2005*b*: 332.

58 Cuvigny 2000*a*: 29.

59 Ἀκισκλάριοι: *O.Claud.* 15, 23, 130, 132, 162; *O.Claud.*inv. 746, 1064(?), 1125, 1168, 1190, 1550, 1814, 1870, 2072, 2575, cf. Bülow-Jacobsen 1992*b*: 117. The sick-lists mention ἀκισκλάριοι as well: *O.Claud.* 212. Their specific task required leg protectors (*ocreae*), made of old water-skins, cf. Bülow-Jacobsen 1996: 45. *O.Claud.*inv. 539, cf. Bülow-Jacobsen 1992*a*: 55.

60 *O.Claud.* 15, 213, cf. Cuvigny 2000*a*: 29.

61 Bülow-Jacobsen 1992*a*: 57.

62 Bülow-Jacobsen 1996: 34.

63 Camel-transport of iron bars: *O.Claud.* 27–34. Bülow-Jacobsen 1996: 48 f; Bülow-Jacobsen 1997*d*: 140 f.; Peacock & Maxfield 1997: 250.

64 *O.Claud.*inv. 1252, cf. Bülow-Jacobsen 1992*a*: 58. The repair of broken tools was apparently done by the main smithy at the fort, where military needs in weaponry were served as well. Bülow-Jacobsen 1997*d*: 140 f., believes the smithy in quarry 92 to be a στομωτήριον, a place where 'steel', στόμωμα, was welded. According to an unpublished ostracon στόμωμα was not produced at Mons Claudianus itself, but imported (*O.Claud.*inv. 4852 + 5398), cf. also Bülow-Jacobsen 1996: 49; Bülow-Jacobsen & Cuvigny 2007: 17. For further χαλκεῖς at Wadi Hammamat, cf. *I.Ko.Ko.* 91, 113, 127, 131.

65 Further specialized functions are, for example, κασιότης, a 'crane-constructor', cf. Bülow-Jacobsen 1996: 46. For ropes and fibre, cf. *O.Claud.* 129, 133, 376. Peacock & Maxfield 1997: 259 fig. 7.1. Φαλαγγάριοι were perhaps responsible for rollers (φαλάγγες) and for the transport of blocks or columns down slipways (*O.Claud.* 20, cf. Bülow-Jacobsen 1996: 46). For an ἀρχιμηχανικός, a 'chief-engineer', at El Hosh, cf. *I.Portes* 117, 118. Klemm & Klemm 1992: 236 f.; μηκαν(άρ)ιος at Gebel Silsilis, cf. *I.Th.Sy.* 159.

The detailed terminology for specific functions within the operative process certainly attests to a high degree of labour division and specialisation at Mons Claudianus. While the *φαμελιάριοι* provided specialists for specific tasks and low-skilled labour for auxiliary duties, few of the *παγανοί* were technical specialists; most of them were *σκληρουργοί*. No *παγανοί* have been documented in auxiliary duties, nor are they recruited for other tasks at Mons Claudianus, such as collecting wood, guarding the quarries as *vigiles*, manning the *σκόπελοι*, carrying letters and other goods to the way-stations and from the Nile valley to Mons Claudianus. Together with the *φαμελιάριοι*, they did however participate in the collection, storage, and distribution of water.[66] The division of the quarry workers into *παγανοί* and *φαμελιάριοι* might hint at distinct recruiting mechanisms or terms of employment utilized by the authorities. Given the fact that the *παγανοί* are left alone in organizing the transport of grain and wages from the Nile valley to Mons Claudianus, they were probably hired on the basis of a work contract different from that of the *φαμελιάριοι*. The members of the *familia Caesaris* aside, *φαμελιάριοι* not only received wages and better food rations, but also medical attention.[67] The fact that *φαμελιάριοι* are often found in activities not directly related to the extraction of marble, but connected with the protection and provisioning of the quarrying community, might suggest that they were posted at Mons Claudianus longer than *παγανοί*. One therefore might assume that the latter were only recruited for a limited period. Hélène Cuvigny explained the division of the workforce by arguing that the *παγανοί* had originally been provided by contractors of quarrying work.[68] There is only one recorded *μισθωτὴς τῶν μετάλλων* at Mons Claudianus and Mons Porphyrites, an imperial slave(!) Epaphroditos Sigerianos, who must have contracted out quarry work during Hadrian's reign.[69] The division of the workforce into *παγανοί* and *φαμελιάριοι*, however, seems to have been a feature at Mons Claudianus throughout the second century AD, and therefore was not necessarily connected with the employment of contractors. Perhaps the contracting out of quarry work was also handled by the procurator of the eastern Egyptian quarries, though no direct evidence for this exists.

Whatever the precise arrangements might have been, given the range of procuratorial powers it is likely that the quarrying procurator (and his staff)

[66] For *φαμελιάριοι* and their tasks, c.f. *O.Claud.* 115, 226 (with 229), 257, 264, 270, 271, 273, 274, 372, 374–6, 382, 387; *O.Claud.*inv. 7029, 7032, 7036, 7042, 7045, 7054, cf. Bingen 1997: 101–3. 105.

[67] For medical supply, cf. Cuvigny 1992*b*: 82; Cuvigny 1997*a*: 20.

[68] Cuvigny 2000*a*: 22 f.

[69] *I.Pan* 21, 42.

hired and paid the workforce required for the operative process, while the responsibility for organizing the operative process probably lay with the resident ἀρχιτέκτων. Consequently, we must assume that the calculation of food rations and fodder, and of the animals required for certain transport tasks, were part of the procuratorial duties as well. Moreover, further functions attested in the water distribution list recall the LMV tablet from Vipasca: two shoemakers (σκυτεῖς) and a barber (κουρεύς) are named as recipients of water. The LMV demonstrates that the provision of services to the work forces was contracted out to shoemakers and barbers at Vipasca.[70] It is not known whether these or other functions were contracted out by the procurator at Mons Claudianus.

6.1.2.4. *The Supply of Provisions*

The workforce in the Egyptian Eastern Desert had to be supplied with food and water. Apart from the monthly supply of grain rations as part of the wages for the παγανοί and φαμελιάριοι, the analysis of botanical and faunal remains from the middens gives us an idea of the range of imported goods.[71] The roughly 22,000 recovered bones provide evidence that various domesticated animals, such as donkeys, pigs, camels, goats, sheep, chickens, and horses, were imported for food and work. The bones of horses were found in far fewer numbers than donkey or pig bones, which suggests that these animals did not figure as main attractions on the daily menu at Mons Claudianus, though butchery marks were found on camel and horse bones as well.[72] Dorcas, gazelle, ibex, goose, and sandgrouse were the most favoured hunting game. Fish and molluscs were imported mainly from the Red Sea, but some freshwater fish from the Nile valley, such as catfish, are represented as well.[73] The botanical remains include cereals such as six-row barley and hard wheat. Both were consumed, but wheat grain is attested in far smaller quantities. Pulses and fruits (some originating from India or Persia) found their way to the quarries as well. Some items only appear in small numbers, while dates, olives, grapes, figs, and watermelons were staple goods. Different nuts, oil plants, and olive oil were imported in large quantities. Herbs and spices enriched the food and some of them were probably pot plants kept at Claudianus itself. Pepper, an import from India, was a luxury item even in Egypt and thus is only rarely attested at Mons Claudianus. The seeds for cabbage, beet, cress, lettuce, chicory, as well as mint, basil, and rue were found

[70] *LMV* ll. 32–7.
[71] Van der Veen 1998; Hamilton-Dyer 2001; Van der Veen 2001.
[72] *O.Claud.*inv. 7276, cf. Hamilton-Dyer 2001: 298 f.
[73] Van der Veen 1998: 103 f. and table 1; Hamilton-Dyer 2001: 298–301.

in the garbage heaps, confirming the assumption that vegetables were grown in the middle of the desert.[74] According to M. Van der Veen, this range of foodstuffs would have guaranteed a balanced diet of carbohydrates, protein, sugars, fats, minerals, and vitamins.[75] Despite this abundance, the workforce probably did not have the benefit of a food supply covering all their nutritional needs.[76]

Private and official letters help us to determine what goods circulated amongst which segments of the quarry population, as well as providing additional evidence for imports and exports. Grain and grain products, meat, fish, and other nutriments were imported, while vegetables could be grown locally. Certain imports were for hygienic or medical use only.[77] As Hélène Cuvigny has pointed out, recipients of vegetables and meat or fish probably were members of the Roman army. Since they received higher pay than the skilled workers, they were able to afford additional foodstuffs for their diet. Vegetables were apparently grown at some of the desert outposts by soldiers or *φαμελιάριοι* for their own use.[78] The texts certainly suggest that there were significant differences in the provisioning of the population at Claudianus, the military and *φαμελιάριοι* being far better supplied than the *παγανοί*.

A small number of papyri do refer to the supply of grain and fodder to the imperial quarries or to soldiers stationed there. An order of barley for the animals transporting a fifty-foot column was addressed by an unknown official to the *στρατηγός* of the Heptakomia, Apollonius, in the year AD 96. The official asks Apollonius to assist his representative, Chaeremon, in collecting all the barley available in his nome. Moreover, the barley had to be supplied to Kaine rather quickly.[79] In AD 96 a *Καίσαρος οἰκονόμου οὐικάριος* (*vicarius dispensatoris Caesaris*) sent an acknowledgment to the *σιτολόγος* of Magdola and other towns of the Polemon division that he received 2,089 artaba of barley at the harbour of Kaine. The barley was for the military garrisons of the Thebais and the workers at the quarries between the Nile and the Red Sea.[80] The barley may not have been intended for the workers or the

[74] Van der Veen 2001: 220 f. Mons Claudianus yielded amphorae from different Mediterranean oil production sites, cf. Tomber 1996: 39–49.

[75] Van der Veen 1998: 108.

[76] Cuvigny 1997*a*: 76 f.

[77] Van der Veen 2001: 218 f., and indices in Bingen et al. 1997; Bingen et al. 1992; Cuvigny 2000*a*.

[78] Van der Veen 1998: 110. The ostraca seem to imply that these products were circulating in a military context. *O.Claud.* 255–78, cf. Bingen 1997: 81 f.

[79] *P. Giss.* 69 (AD 118), cf. Peña 1989: 126–32; Adams 2001: 178 f.

[80] *SB* 12169 = Youtie 1978*b*, cf. Adams 2001: 177 f., 179.; Mitthof 2001: 299 f., no. 3. Contrary to Peña 1989: 128 fn. 9, Fritz Mitthof argues that the Kaine mentioned in the text is not the

soldiers to eat, but to be used as animal feed instead.[81] Calpurnius Isidorus, *στρατηγός* of the Themistes and Polemon districts of the Arsinoite nome wrote a reply to a letter from the *praefectus Aegypti* Aurelius Septimius Heraclitus, in which he asseses the state of the grain supply. His letter states how much grain collected during the harvest of the 22nd year of Caracalla (AD 214) was stored in the granaries and how much had been forwarded to the troops in the Thebaid, the men in the Porphyrites and Claudianus quarries, or distributed for other uses.[82] The grain is likely to have been barley intended as fodder for the animals at Mons Claudianus and Mons Porphyrites.[83] The wheat for the *παγανοί* and the *φαμελιάριοι* at Mons Claudianus, which we know of from the *ἐντολαί* and the receipts for advanced payments from the years AD 136–46, was probably acquired similarly. Mitthof might be correct in assuming that the distribution of a grain ration to soldiers (and hence to quarry workers), separate from the regular payment of a *stipendium*, was in place as early as Trajan's or Hadrian's reign.[84]

The collection of barley and wheat for the quarry workers in the Egyptian Eastern Desert was likely organized on similar lines to the grain supply of the Roman army in Egypt. We can construct an outline of the supply system for the Roman army in the last quarter of the second century AD based on a group of texts known as the archive of Demarion, dating to the years AD 184–6. According to Mitthof's reconstruction, the *praefectus Aegypti* ordered the requisition of a certain amount of barley to supply the needs of one mounted unit for a year. This requisition was not limited to the immediate vicinity of the unit's camp, but included different nomes, perhaps of the whole Egyptian province. The *στρατηγοί* were informed how much barley their nome should supply; the authorities of the nomes (*πραγματικοὶ τοῦ νομοῦ*) divided this amongst the villages. Based on this division, the village authorities (*πρεσβύτηροι*), provided the barley. Furthermore, the *στρατηγοί* were ordered by the provincial administration to appoint a board consisting of *honoratiores* from the capital of the nome. They were to receive an amount of money from the public treasury and to distribute it to the villages which contributed barley. The barley was then collected from the villages by a soldier of the unit. In return, the villagers were handed a receipt which they immediately passed on to the *στρατηγός* of their nome. In response to this receipt, the board of *honoratiores* paid out a sum of public money for the villages' barley

Kainepolis at the head of the roads leading up to the imperial quarries in the Eastern Desert, but a town of the same name in the Arsinoite nome.

[81] Mitthof 2001: 299.
[82] *P.Oxy.* XLV 3243.
[83] Mitthof 2001: 334 f.
[84] Mitthof 2001: 310 f.

contribution.[85] A receipt for requistion of unknown goods handed to the πρεσβύτεροι of Soknopaiou Nesos from a soldier stationed in the Akoris quarries in the years AD 195–8 might refer to such a requisition of barley.[86] A similar document attests the requisition of palm fibres by a member of a military unit at the quarries of Akoris in the third century AD.[87] The requisition of barley saw a significant change under the Severans. On the orders of the *praefectus Aegypti*, the στρατηγός and the βασιλικὸς γραμματεύς of a nome advised officials in some villages (i.e. the σιτολόγοι, the keepers of the local granaries) to hand over a predetermined amount of barley to soldiers of a specific unit, who had been sent to collect the grain needed. Unlike the previous scheme, there was no payment for the requisitioned barley.[88] The above-mentioned response of the στρατηγός Calpurnius Isidorus to the prefect of Egypt may have been written after the conclusion of such a barley requisition.[89]

The evidence for the supply of wheat to soldiers is rather scarce. It seems that military units received their wheat (collected through regular taxation), from *dispensatores* of the provincial administration.[90] Regular taxation provided most wheat for the needs of the Roman army, and perhaps of the quarry workers in the Eastern Desert as well. It has been suggested, however, that the forced sales of wheat to the provincial authorities throughout the second century AD document the incapacity of the grain taxation to guarantee a continuous supply.[91] In the first half of the second century, the workers at Mons Claudianus, the παγανοί as well as the φαμελιάριοι, received wheat rations from their employers, the imperial quarrying authorities, and they are mentioned together with troop units as the recipients of barley shipments. Both factors strongly suggest that the imperial quarries were supplied with grain through the regular mechanisms of grain taxation and requisition, as were the army units stationed at the quarries in the Eastern Desert. Whether or not this model applies to other provinces cannot be gleaned from the known literary or epigraphic evidence.

Unlike other goods, water was not imported, but was collected from springs and wells throughout the Eastern Desert.[92] The scarcity of water demanded strict measures. Lists of water-rations confirm that members of

[85] *SB* 14155–62; *P.Amh.* II 107, 108; *P.Bodl.* I 14; *BGU* III 807, cf. Mitthof 2001: 43, 314–317.
[86] *P.Alex.*inv. 463, cf. Mitthof 1994: 207.
[87] *BGU* VII 1612, cf. Mitthof 1994: 207.
[88] Mitthof 2001: 44.
[89] *P.Oxy.* XLV 3243, cf. Mitthof 2001: 44, 334 f.
[90] *P.Oxy.* IV 735.
[91] Mitthof 2001: 41 f. with fn. 20. For regular grain-taxation, cf. Wallace 1938: 31–46.
[92] Maxfield 2001: 143.

the workforce did not receive more than 2–3 litres per day, an insufficient supply for a working adult under these extreme climatic conditions.[93] The water supply was the main concern of military officials based along the way-stations, and the military correspondence provides evidence of severe water shortages and problems of storage.[94] About 750 people at Mons Claudianus had to be supplied with 2,250 l of water per day.[95] Unpublished ostraca mention a *ὑδροφορία*, including *παγανοί* as well as *φαμελιάριοι*. An *ἐργαζομένη δεκανία*, a 'work gang on duty', was in charge of collecting water from five to six wells, using up to eighty camels. Camels owned by the imperial authorites and by civilians were required to carry the water.[96]

6.1.2.5. *The Requisition of Animals*

The large distance between the imperial quarries and the main transport waterways presented the quarrying authorities with a further problem, namely the supply of transport animals. A letter sent by the *anticurator* Rufus to the *procurator metallorum* Tertullus reports on the numbers of donkeys still at Mons Claudianus, implying that the procurator was responsible for the supply of such animals.[97] But there are few references to transport issues in the published ostraca of Mons Claudianus. Nothing is said about the hauling of quarried blocks and columns or who would have carried out this task. However, heavy carts which commuted between Claudianus and Kaine, transporting stones one way and provisions the other, are mentioned in the unpublished ostraca.[98] Animals were required to haul columns and other quarried items down to the Nile, and a caravan of camels, the *πορεία*, came up to Mons Claudianus every month, supplying the required goods to the quarries. Provisions and letters were also brought in by donkeys.[99]

The ostraca record the use of camels for the provision of water. In one text *κάμηλοι κυριακοί*, imperial camels, are distinguished from *κάμηλοι παγανικοί*,

[93] *O.Claud.*inv. 1538 + 2921, 3666, cf. Cuvigny 1997*a*: 77; Cuvigny 2005*b*.

[94] On water shortage, see *O.Claud.* 362, 380. Antistius Flaccus, based at the way-station Raima, reports abundant amounts of water (*O.Claud.* 2). The archive of Successus mentions water-skins used to supply the quarries with water (*O.Claud.* 126–9).

[95] *O.Claud.*inv 1538 + 2921, cf. Cuvigny 2005*b*.

[96] *O.Claud.* 134, 362, cf. Bülow-Jacobsen 1992*b*: 119; Rengen 1997: 203 f.

[97] *O.Claud.*inv. 7295, cf. Cuvigny 2002: 245 f.

[98] The texts mention wagons with two and four wheels. A special vehicle with twelve wheels is recorded as well. One ostracon reports the arrival of a *ἅμαξα*, a chariot usually carrying granite or other materials, cf. *O.Claud.* 362. Rengen 1997: 203; Bülow-Jacobsen 1998: 64; Hamilton-Dyer 2001: 297.

[99] On official and unofficial transports of goods and letters by camel-drivers, cf. *O.Claud.* 140, 142, 162, 224, 243, 248, 267–9, 273, 274, 300; Adams 2001: 177.

'civilian' camels.[100] While the imperial camels had probably been purchased by the quarrying administration, the κάμηλοι παγανικοί were undoubtedly requisitioned from the provincial population. Two declarations dating to the year AD 163 document the requisition of camels for tasks in the imperial quarries. They were sent by the owners of camels from Soknopaiou Nesos to the στρατηγός Stephanos and the βασιλικὸς γραμματεύς Zoilos, of the division/(*meris*) of Herakleides in the Arsinoite nome. The declarations stated how many camels they owned, how many were available for requisition, and how many had been requisitioned. Both texts noted that camels had been requisitioned to haul porphyry columns on the orders of the *praefectus Aegypti* Annius Syriacus.[101] Although the *praefectus Aegypti* authorized the requisition of animals, it was the task of the local administration to carry out these orders.[102] While the main role of the ἐπιστράτηγοι was to appoint officials to requisition animals, the main control of this requisition system fell to the στρατηγοί and the βασιλικὸς γραμματεύς. Apart from the annual census of livestock, designed to enhance the effectiveness of collecting taxes on animals, animal owners were obliged to submit declarations, ἀπογραφαί, of their animals to the state. Moreover, village scribes compiled lists of eligible liturgists for allotted tasks.[103] Thanks to the census, the στρατηγοί had a complete list of animals available within the nome. In one case, the *praefectus Aegypti*, probably following a request (e.g. from the imperial quarries), gave permission for a requisition on submission of a *diploma*. The village scribes requested the animals from the liturgists and the animals then were gathered and sent from the nome to the Eastern Desert.[104]

The liturgy system of Roman Egypt, which ensured the constant availability of animals for various transports, surely has parallels in other provinces. A Greek and Latin bilingual inscription found at Sülümenli on the road from Dokimeion to Synnada might shed some light on the transport burdens which the provincials faced. The inscription displays the record of hearings held by the imperial procurators Aurelius Threptus, Philocurius and Novellius in the years AD 200(?), 213, and 237 on the subject of a dispute between the villagers of Anossa and Antimacheia. The issue at stake was the just distribution of the *angareia*, the burden of providing oxen for the *cursus publicus*, amongst both villages.[105] Based on an initial hearing held at Anossa

[100] *O.Claud.* 362, cf. Rengen 1997: 203 f.

[101] *BGU* III 762 (AD 163). *P.Lond* II 328 (AD 163), cf. Adams 2001: 175. For further examples, cf. Leone 1988: 155–77; Sijpesteijn 1993: 29, no. 2; Leone 1998: 63–7.

[102] Adams 2001: 180–3.

[103] Adams 2001: 181. For a census, cf. *P.Oxy.* XVII 2118 (AD 156).

[104] *SB* 12168 (AD 130?), cf. Youtie 1978*a*; Adams 2001: 182.

[105] Frend 1956: 53–6.

around AD 200(?), the procurator Aurelius Threptus decided to distribute the *angareia* according to the ratio of the tax sum collected both from Anossa and Antimacheia. This ruling was enforced by Aurelius Symphoros, an *optio* of Threptus.[106] The case was heard again in AD 213 by the procurator Philocurius and in AD 237 by the procurator Novellius, and Threptus' initial regulation was upheld. A *stationarius* was supplied to ensure that the ruling was implemented fairly. It has been argued that the *angareia*-duty was not meant for the *cursus velox* but rather for the *cursus clabularius*, the transport of heavy items.[107] Some scholars believe that marble blocks from the imperial quarries of Dokimeion, amongst other things, were transported with the *cursus clabularius*.[108] Given the archaeological evidence, a similar system might have been in place for the imperial quarries at Simitthus as well.[109]

6.1.2.6. *Imperial Quarries and the Provincial Administration*

The procurator's office could rely on the administrative structure already in place for the acquisition of animals, grain, or any other goods required. Hence, the amounts of grain or numbers of animals were calculated on the basis of monthly rations and additional requests, and the information was compiled and transmitted to the *praefectus Aegypti*. He then set the administrative apparatus in motion in order to supply the required staple goods or beasts of burden. As we have seen, the papyri suggest that Egyptian prefects played a pivotal role in the logistical support of the quarries; grain was collected, apparently on the orders of the *praefectus Aegypti*, for the soldiers and quarry workers in the Egyptian Eastern Desert.[110] Moreover, animals were requisitioned for various transport tasks, again on the orders of *praefectus Aegypti*.[111] The *praefectus Aegypti* appears on a regular basis in inscriptions found in the Eastern Desert. The inscriptions of the *μισθωτὴς τῶν μετάλλων*, Epaphroditos, from Mons Porphyrites and Mons Claudianus both name Rammius Martialis as the prefect of Egypt (AD 117–19). An inscription from the Hydreuma at Mons Claudianus was set up in the year AD 109 for Sulpicius

[106] *SEG* XVI 754 = Frend 1956: 46 f., ll. 1–29; Mitchell 1999: 38 f.; Levick 2000: 57–60, translation.

[107] Frend 1956: 54; Zawadzki 1960: 90.

[108] Pékary 1968: 136; Kolb 2000: 96 f., 154 f. On the export route of *pavonazetto* from Dokimeion, cf. Hirt 2004: 119–21 with further bibliography.

[109] Winckler 1895; Salama 1951: 61; cf. 2.1.6.

[110] *SB* 12169 = Youtie 1978*b* (AD 96). *P.Giss.* 69 = Peña 1989. *P.Oxy.* XLV 3243 (AD 214/215). *P.Oxy.* IV 735 (AD 205). Adams 2001: 177 f.; Mitthof 2001: 330, 334 f.

[111] *BGU* III 762 (AD 163). *P.Lond* II 328 (AD 163). On Annius Syriacus, cf. Thomasson 1984: 351, no. 63. On requisition of animals, cf. Adams 2001: 180–3.

Similis, *praefectus Aegypti* from AD 107–12.[112] A number of ostraca confirm a visit in person of the latter prefect (*παρουσία τοῦ ἡγεμόνος*) to the Mons Claudianus quarries.[113] Moreover, a hitherto unpublished ostracon found at Mons Claudianus is presumably a copy (the original probably was a papyrus) of an undated letter by the *praefectus Aegypti* Pomponius Faustianus (AD 185–7) to Probus, a procurator known from other unpublished ostraca. The letter seems to refer to (construction?) activities taking place in the Mons Claudianus quarries.[114]

The involvement of the *praefectus Aegypti* in extractive operations and the visits to the remote imperial quarries were presumably not just out of personal interest, but perhaps part of his duties. A small number of quarries in the Nile Valley were opened on the orders of the *praefectus Aegypti*, perhaps for specific purposes, such as building projects. In AD 89–91 we find the prefect Mettius Rufus ordering stone for a basement, for which a quarry at Gebel Toukh was opened.[115] A Latin inscription from Syene confirms our notion that the *praefectus Aegypti* could be responsible for the opening of new quarries on some occasions. Aurelius Heraclides, *decurio* in the *ala I Thracum Mauretana*, inscribed a column between AD 206 and 209, commemorating the 'discovery' of *novae lapidicinae* near Philae. Under the auspices of Ti. Claudius Subatianus Aquila, *praefectus Aegypti* (AD 206–9) numerous pilasters and columns were extracted.[116] Unfortunately, there is no further evidence to identify the prefects who opened other quarries in the Nile valley. Near Eileithyiaspolis/mod. El-Kab, 20 km north of Edfu, on the left bank of the Nile, a sandstone quarry was opened at El Mahamid.[117] According to the inscription of an unknown soldier of a *legio II* (*Traiana*), these sandstone quarries had been opened in the eleventh year of Hadrian's reign (AD 127).[118] Other quarries might also have been worked at the behest of the *praefectus Aegypti*: The porphyrite deposits at Gebel Dokhan were discovered by a Caius Cominius Leugas, who in the year AD 18 set up an inscribed stelae commemorating his find.[119] According to his inscription he encountered not only porphyry stone, but a stone he calls *knekites*, as well as black porphyry and

[112] *I.Pan* 37, 42.

[113] *O.Claud.* 130. According to Bülow-Jacobsen 1992*b*: 117, and Cuvigny 2000*a*: 9 fn. 28, other texts (*O.Claud.*inv. 369, 2212) may refer to a visit of the prefect Sulpicius Similis who sets up an altar in the temple, cf. *I.Pan* 37, AD 108/9.

[114] *O.Claud.*inv. 7295, cf. Cuvigny 2002: 243. For activities under Commodus, cf. Maxfield & Peacock 2001*b*: 433.

[115] *I.Pan* 16 = *I.Metr.* 116.

[116] *CIL* III 75 = *ILS* 4424.

[117] Klemm & Klemm 1992: 228–31.

[118] *I.Th.Sy.* 35.

[119] *SEG* XLV 2097 = *AE* 1995: 1615 = Rengen 1995: 242 f.

polychrome stones.[120] However, it is not stated who ordered Cominius Leugas to venture into the middle of the Egyptian Eastern Desert.[121] Pliny the Elder possibly sheds some light on the issue: he offers an interesting passage on the 'procurator' Vitrasius Pollio. Pollio imported statues made of red porphyry from Egypt and presented them to the emperor Claudius, who did not take any liking to the purple stone. Pliny claims that after this episode no one tried to imitate Vitrasius Pollio.[122] A Caius Vitrasius Pollio appears as *praefectus Aegypti* in AD 39–41 and is probably the same man. Presumably during his tenure of the prefecture in Egypt Pollio was shown the red porphyry from Mons Porphyrites, which he then mistakenly thought worthy of presenting to the emperor.[123] If Pliny's account is correct, it would imply that Pollio's predecessors as prefect opened up quarries and sent samples of the stone found there to Rome.

Like all provincial governors, the *praefectus Aegypti* had the right to condemn convicted provincials to the quarries for a number of years. However, it seems as if only the alabaster quarries in the Nile valley and perhaps Mons Porphyrites were regularly supplied with convict labour by the *praefectus Aegypti*.[124] Subatianus Aquila, the prefect in AD 206–9, appears at least in two routine Greek documents which inform the relevant *στρατηγοί* of the release of convicts who had been condemned to work in the alabaster quarries. This presumably happened in order to guarantee that the former convict was re-enrolled in the taxation lists of their *ἰδία* by the *ἀμφοδογραμματεῖς*.[125] A fragmentary letter in Latin from one of Subatianus Aquila's predecessors, C. Avidius Heliodorus (AD 137–42), to the official in charge of the quarries near Alabastrine, orders the release of the convict Petesuchos, who had been condemned by the *praefectus Aegypti* M. Petronius Mamertinus (AD 133–7) five years before.[126] According to Flavius Josephus, Titus had prisoners from the Jewish war older than 17 years of age sent to the quarries in Egypt; the *praefectus Aegypti* was likely required to distribute these prisoners of war to the various quarries.[127] Besides this, he also had to supply the prisoners (perhaps not only within Egypt?) with clothing. Towards the end of the year

[120] On the difficulties of identifying the *knekites* stone, cf. Rengen 1995: 243–5.

[121] C. Cominius Leugas appears to be a freedman, although he does not give his social status. The cognomen 'Leugas' does not appear often as a slave name: Lörincz 2000: 24, 'Leuga': *CIL* III 2073 (Dalmatia). 'Leuganus': *CIL* III 1158 (Dalmatia), from Clunia/Hispania citerior.

[122] Pliny, *NH* 36.57.

[123] Thomasson 1984: 344, no. 20; *PIR* V 524.

[124] Mons Porphyrites(?): Aelius Aristides, *Or.* 36.67.

[125] Hagedorn & Shelton 1975: 228, and Haensch 2000: 259 f.

[126] Bastianini 1986.

[127] Josephus, *BJ* 6.418.

AD 153, the *praefectus Aegypti* ordered 6,700 στολαί for prisoners on behalf of the Roman state (δημόσιος ἱματισμός). The guilds of weavers in the Arsinoite nome had to manufacture these garments and were paid by the public bankers (δημοσίων τραπεζῖται) in advance.[128]

Governors of other provinces also sent convicts to work in *metalla*; condemnation *in metallum* was one of the commoner punishments meted out by Roman magistrates for numerous crimes. Roman law envisaged a wide range of punishments for criminals, the majority of which were designed for *servi* or *humiliores* and included hard labour and corporal punishment, namely flogging and beating.[129] Condemnation *in metallum* was considered to be a capital punishment combined with a loss of citizenship or reputation/status.[130] In comparison to extreme capital penalties—such as being sent to the gallows, burnt alive, beheaded, or crucified—condemnation *in metallum* was thought a less severe form of punishment, even though a man so condemned might view it as comparable to a death sentence.[131] In general, condemnation to a mine or quarry was literally a lifetime sentence.[132] Yet, there is evidence for the reduction of the *poena metalli* to a certain length of time: if the *damnatio in metallum* was meted out without defining the length of the sentence, the predetermined period was to be ten years; a rescript of Hadrian states that men were not to be condemned in *opus metalli* (a lighter form of punishment) in perpetuity.[133] Antoninus Pius even extended clemency to convicts who were found to be too unfit for work owing to sickness or advanced age.[134] Despite the occasional leniency, condemnations *in metallum* for perpetuity were still meted out in late antiquity.[135]

The nature of this punishment called for the *damnatus* to be restrained in his movements. A man who had been condemned *in metallum* was either fitted with heavy chains, or if condemned *in opus metalli* was burdened with lighter fetters.[136] However, there is no record of any other differences between the two punishments. The condemnation *in opus metalli* was understood to be a lighter form of punishment than condemnation to a *metallum* and

[128] *P.Graux* 30, Kol. VII, cf. Kambitsis 1997.

[129] In general, cf. Mommsen 1899: 47 f., 960–3; Millar 1984; Gustafson 1994; Gustafson 1997. See Pliny, *Ep.* 2.11.8, 10.58.1–3. Sueton, *Cal.* 29, for legal mistreatment of senators or equestrians.

[130] *Dig.* 48.19.2.pr., 48.19.28.1, 50.13.5.3. Paulus, *Sent.* 5.17.2.

[131] *Dig.* 48.19.28.1.

[132] See e.g. *CTh* 2.14.1, 16.5.40.

[133] *Dig.* 48.19.23.pr., cf. Paulus, *Sent. Int.* 3.6.2; *Dig.* 48.19.28.6. Ulpian, *Coll.* 11.8.3.

[134] *Dig.* 48.19.22.pr.

[135] *CTh* 2.14.1, 16.5.40.

[136] *Dig.* 48.19.8.6.

apparently limited in time. This view is supported by Ulpian who states that a fugitive from *opus metalli* was to be condemned *in metallum*, increasing the severity of his punishment.[137] Apart from the *damnatio in metallum* and the *damnatio in opus metalli*, Ulpian refers to a less drastic form of punishment, the *ministerium metallicorum*, which was a punitive measure for women. This may have been a kind of auxiliary mining or quarrying duty, and was not as physically demanding as other hard labour punishments in the *metalla*.[138] Convicts sent *in metallum* were usually branded, the mark being painfully placed on the middle of their forehead.[139]

Literary sources also document the use of *damnati* in mines or quarries. Aelius Aristides describes in gloomy terms what awaited the convicts condemned to an Egyptian porphyry quarry; the heat and the lack of water prevented convicts from escaping as they feared being 'burned' alive.[140] In his speech to his Caledoni, Calgacus cried that he would rather be killed in battle against the Roman invaders than face deportation to *metalla*—the fate of many prisoners of war.[141] Flavius Josephus reports that Titus sent prisoners from the Jewish War to the quarries in Egypt.[142] Other sources, mostly patristic, refer to Christians condemned to unspecified *metalla* in Numidia, Sardinia, Cilicia, Pontus, Egypt (Thebaid), or Palestine (Phaeno) in the third and early fourth centuries.[143]

Despite the abundant legal and literary sources, the use of *damnati* in quarries or mines is not documented in the epigraphic record. Provided Aelius Aristides' account of convicts sent to the quarries of Mons Porphyrites in the Egyptian Eastern Desert is correct, one would expect there to be at least some archaeological traces.[144] To date, nothing seems to directly attest the presence of convicts, although the σκόπελοι at Mons Claudianus and Mons

[137] *Dig.* 48.19.8.6. A fate deemed even worse by Ulpian than the ordinary *metalla* (*poena metalli*) awaited the convicts in the *calcaria*, lime quarries, or in the *sulpuraria*, sulphur mines (*Dig.* 48.19.8.10).

[138] *Dig.* 48.19.8.8.

[139] Jones 1987.

[140] Aelius Aristides, *Or.* 36.67, cf. Fitzler 1910: 119–21; Klein 1988: 42 f.

[141] Tacitus, *Agr.* 32.4, cf. Lassandro 1995.

[142] Josephus, *BJ* 6.418.

[143] Cyprian, *Ep.* 76, 77, 79. Eusebius, *De mart. Pal.* 7.3–4, 8.1, 8.13, 13.1–4. Eusebius, *HE* 9.1, 4.23.10, 8.12.10. *Ap. Const.* 8.10.15; cf. Gustafson 1994: 422. For the *Passio Sanctorum Quattuor Coronatorum*, cf. Wattenbach 1870; Täckholm 1937: 136 f.; Geerlings 1983; Peacock 1995, with further references. For mining in Late Antiquity, cf. Edmondson 1989. For the recent excavations at Faynan/anc. Phaeno, cf. Barker et al. 1997; Barker et al. 1998; Barker et al. 1999; Barker et al. 2000.

[144] For Christian convicts sent to the porphyry quarries in Egypt, cf. Eusebius, *De mart. Pal.* 8.1, and perhaps the *Passio Sanctorum Quattuor Coronatorum*.

Porphyrites could also be interpreted as watchtowers for guarding convicts. The compound of the quarries at Simitthus/mod. Chemtou is the only architectural structure which has been interpreted as a 'prison camp' (cf. 2.1.6).

6.1.2.7. *Summary*

The administration of the imperial quarries in Egypt and their supply with human and material resources probably required the *praefectus*' attention on a regular basis. The imperial quarries relied heavily on the provincial administration for the supply of provisions and the requisition of animals. The procurators of these quarries therefore were not left on their own in acquiring basic materials for extractive operations. The heavy reliance on the support of the provincial administration in attaining access to these public resources underscores the special status of these quarries in the Eastern Desert. The recurring involvement of the *praefectus Aegypti* may suggest that he exercised a supervisory control over the quarries.

However, we have little evidence which shows how the management of the quarries in other provinces was integrated in the provincial administration as a whole. As prisoners and convicts were sent to the *metalla* by provincial governors, and provincials were required to supply beasts of burden under the framework of *liturgia*, it is possible to suggest that the administration had at least a limited involvement in extractive operations. Naturally, this requires us to assume that the planning of work procedures, the supply of a workforce, and means of transport were the responsibility of imperial officials. The quarrying procurator in the Nile valley was probably required to assess the needs of quarries under his jurisdiction in the Eastern Desert, and inform the provincial authorities of their requirements. However, it is difficult to document these administrative activities precisely, since the majority of them involve the Nile valley and are thus beyond the scope of the Claudianus ostraca. There are a few references to the activities of the procurator in the ostraca to provide us with some idea of his duties. The procurator and his staff at Mons Claudianus apparently kept books on the produce of the quarries, and possibly managed incoming orders, and ensured that the stone required was quarried and exported. It cannot be ascertained, however, whether they were directly involved in the arrangement of work procedures as such. This was probably the task of the military officers or civil engineers present at the quarries. Moreover, the recruitement, allocation and payment of the workforce, be that *παγανοί* or *φαμελιάριοι*, the supply of food rations as well as raw materials, tools, ropes, animals, etc., were in all likelihood supervised by the quarrying procurators in charge, undoubtedly with the help of his own staff and the provincial administration. Special tasks, such as the recruitement of the work force from other quarrying areas in the Nile Valley, may have been delegated to civilian contractors.

Given the specific circumstances of the quarries in the Egyptian Eastern Desert, in particular their location far from the populated and fertile Nile valley, procurators in charge of imperial quarries in other provinces may not have faced the same range of responsibilities as their Egyptian colleagues. The use of convicts may have resulted in similar procuratorial responsibilities, namely the task of supplying food, garments and shelter to the work force. The proximity to populated areas, however, required different measures in organizing extractive procedures, thus producing a different 'portfolio' of tasks for the procurator (cf. 2.1.8).

6.1.3. Function and Responsibilities of Mining Procurators

6.1.3.1. *The Mining Procurators and the Vipasca Tablets*

The main sources of information on the duties of mining procurators are two fragmentary bronze tablets unearthed by the Companhia Mineira Transtagana in 1876 and 1906 at Vipasca/mod. Aljustrel in Lusitania.[145] The first tablet, usually refered to as the *lex metalli Vipascensis*, lists a series of regulations for the contractors of fee collections and other services within the *metallum Vipascense*. The initial regulations were probably written on at least four bronze tablets, three of which have been found. The reverse side of the bronze tablet displays a copy of the text on the front, albeit not in the same layout. The text on the reverse has been moved down a number of lines, probably as a result of emendations made to the 'original' text on the front.[146] The procurator contracted out monopolies for the provision of different services *intra fines metalli Vipascensis*, i.e. within the mining district of Vipasca under his supervision.[147] The first paragraph of the *lex metalli Vipascensis* regulates the *centesima argentariae stipulationis*. As Dieter Flach argues, one per cent of the profit made on an item sold in an auction was collected by the broker (*argentarius*) from the seller as a fee for his services, the terms of which the seller (*venditor*) had agreed to by oral contract

[145] For most recent studies, cf. Domergue 1983; Flach 1979, with extensive discussion of the older studies; cf. now Lazzarini 2001.

[146] Flach 1979: 402; Domergue 1983: 43–6.

[147] Flach 1979: 428 f.

(*stipulatio*). If this interpretation is correct, the term *conductor* used in this passage refers not to a contractor of tax on auctions but to the contractor of broker services at Vipasca.[148] Sales executed by the procurator in the name of the emperor were exempt from this fee and, in the case of mining rights being sold, were collected from the buyer.[149] The sale or lease of items or mining pits by the procurator in the name of the *fiscus* was also exempt from the *scriptura praeconii* (a fee on items put up for sale, advertised by the *praeco*, a herald or crier, and collected from the seller).[150] Both passages show the *procurator metallorum* as a vendor not only of mining rights and mining pits to interested buyers but also of other miscellaneous items.[151] Besides the taxation of sales and the collection of fees for the advertisement of sales, the procurator contracted out the baths to a *conductor balinei*. Moreover, shoemakers, barbers, fullers, and the collectors of slag and stone splinters were, like all *conductores*, guaranteed a monopoly in providing their specific services within the mining district.[152] Apart from contracting out services, most inhabitants of the mining district could be summoned by the *procurator metallorum* for *munera*; teachers were exempt from such duties.[153]

Further insights are provided by the second tablet found at Vipasca in 1906, the *lex metallis dicta* (*LMD*).[154] It probably once covered at least three tablets, of which only the second has been discovered.[155] The *LMD* is a letter addressed to an Ulpius Aelianus, and details the regulations on the 'sale' of mining pits. The remaining text allows a glimpse at the procurator's main responsibilities: (i) superintendence of the seizure and sale of mines and mining rights to *occupatores* or *coloni* and their *socii*, respectively, and (ii) the implementation of mining regulations (cf. 7.1.1). Furthermore, the procurator had jurisdiction within his *territorium*. The theft of ores, the sabotage of the mining infrastructure and the breach of security regulations

[148] Flach 1979: 429–34, esp. 430 with fn. 104; for a different view, cf. Domergue 1983: 68–73.

[149] *LMV*, l. 2: 'exceptis iis, quas proc(urator) metallorum iu[ssu imp(eratoris) faciet', l. 3: 'Conductor ex pretio puteorum, quos proc(urator) metallorum vendet, cent[esimam ab emptore exigito]'; cf. Flach 1979: 407. The text clearly states that auctions conducted by the *procurator metallorum* on the orders of the emperor were exempt from the collection of taxes. Hence, the procurator regularly held auctions.

[150] *LMV* ll. 13–14: 'Si quas [res proc(urator) metallorum nomine] fisci ven|det locabitve'. ll. 15: 'Puteorum, quos proc(urator) metallorum vendiderit'; cf. Flach 1979: 408, 433 f.; Domergue 1983: 73–9.

[151] This probably included possessions of extradited individuals who had been convicted of deliberately ignoring some of the very strict security regulations or had stolen ore, cf. *LMD*.

[152] Flach 1979: 408 f., 434–8; Domergue 1983: 79–99.

[153] *LMV* l. 57: 'Ludi magistros a proc(uratore) metallorum immunes es[se oportet]'. Hirschfeld 1905: 160; Flach 1979: 409; Domergue 1983: 98 f.; Mangas 1999: 248–50.

[154] For procuratorial jurisdiction and its development, cf. Millar 1964: esp. 186; Millar 1965; Brunt 1990*b*: esp. 183 f.; Eck 1997*c*: 161–4.

[155] Flach 1979: 400.

were punished severely. Whereas slaves were flogged, kept in chains, and had to be sold outside of the mining district, perpetrators of a higher status lost their property to the *fiscus* and were evicted from the mining district.[156] The judicial powers vested in the mining procurator also provided him with the necessary means to enforce the swift and continuous exploitation of metal ores. As the *lex metallis dicta* seems to imply, its strict regulations were crafted to force the *occupatores* or *coloni* to keep their mining operations more or less up and running without intermission.

The *procurator metallorum* at Vipasca appears to have been the head of a quasi-municipal community with a clearly defined territory.[157] He sold the mining rights to interested individuals or *societates*, farmed out monopolies for a range of services which were provided within the mining district, and could probably also summon some of the inhabitants for *munera*. Within the mining district, procuratorial jurisdiction was not only a vital tool for enforcing the mining regulations, but was also instrumental in achieving the procurator's main objective—the unhindered and continuous exploitation of the copper and silver mines.

6.1.3.2. *Iberian Mining Districts*

This combination of powers and responsibilities which appears in the tablets, was probably not unique to the *procurator metallorum* at Vipasca. Claude Domergue stressed that the mining regulations of the second tablet (*LMD*) appear to have applied to other hardrock mines as well: the specific geological constraints at Vipasca are found at other mines in southeastern Iberia such as Concepción, El Lagunazo, Tharsis, Herrerías, Rio Tinto, Serra de Caveira, La Zarza, and Sotiel Coronada. These specific conditions favour the sale or lease of mining pits to private individuals.[158]

Unlike the hardrock silver and copper mines at Vipasca, the geology and topography of the alluvial gold-mining zones south-west of Astorga in northwestern Spain required the application of water power on a grand scale. The visible remnants of this technology reflect the efforts of a workforce directed by a single authority.[159] As yet, there is no evidence for *societates publicanorum* or *conductores* running mining ventures on a grand scale in

[156] *LMD* ll. 27–8, 32–5, 40–2, cf. Flach 1979: 404; Domergue 1983: 146 f.

[157] Brunt 1990*b*: 183 f.

[158] Domergue 1983: 151–66, esp. 158 f. At La Zarza and Sotiel Coronada up to 100 and 150 double-shafted mining pits have been found. Such mining pits are assumed to have been in use at Vipasca as well, although the evidence has been removed by the extensive mining at the beginning of the 20th cent. On these mines, cf. Domergue 1987*a*: H6, 7, 9, 12, 19, 30, 36, 40, 43.

[159] Domergue 1990: 299, 303–6.

this region. It is likely that the Roman state chose to exploit the alluvial gold resources directly, rather than compartmentalizing the mining zones and contracting out plots to numerous small-scale contractors, or even sell them to Vipasca-style 'coloni'.[160] Moreover, the resettlement of indigenous communities in order to supply labour for the mines supports the view that the state was directly involved. Based on the survey results of the mining landscape west of Asturica Augusta, it has been argued that with the establishment of Roman rule in northwestern Spain the socio-economic structure of the area was fundamentally reshaped to serve the needs of the Roman mining industry.[161] This process may be observed in the change of settlement patterns.

The pre-Roman landscape was dominated by 'castros', fortified settlements usually positioned on hilltops.[162] The distribution of these pre-Roman hill forts shows that the distances between them never were less than 8 km. Most settlements developed not far from favourable farmland, pastures, woodlands or iron mines, allowing for self-sufficiency.[163] A significant change is detectable from the early first century onwards, when the old pre-Roman *castros* were abandoned. This indigenous form of settlement was reused, albeit in a different topographical and socio-economic context. These Roman 'castros' increased in number, resulting in a far higher density of settlements than in pre-Roman times.[164] Most settlements now stood no more than *c*.2 km apart, clustered around mining sites or places showing signs of metallurgical activity. A further group of settlements seems not to be related to the mining areas, but was positioned close to the agriculturally fertile soils of the area. In addition to the indigenous settlement forms, distinctively Roman building structures were also found in the mining zones. In the light of the increase in settlements (and population?), it is questionable whether the Roman mining settlements or 'castros' were able to maintain self-sufficiency.[165] Although the Roman 'castros' kept their main feature, the wooden palisade, the interior spatial organization was altered. Compared to pre-Roman settlements, the Roman 'castros' were now dominated by open spaces, whereas the 'orthogonal' grid of its streets was

[160] Orejas & Sánchez-Palencia 2002: 589–95 with further bibliography. Contra: Brunt 1990*c*: 397 f.

[161] Domergue & Martin 1977; Domergue & Sillieres 1977; Domergue & Hérail 1978; Sánchez-Palencia & Perez Garcia 1983; Domergue 1986; Fernández-Posse & Sánchez-Palencia 1988; Hérail & Pérez García 1989; Sánchez-Palencia et al. 1990: 262–3; Orejas 1994; Orejas 1996: 103–16, 170–84, 186–9; Sánchez-Palencia Ramos et al. 1996: 107–13; Sastre Prats 1998: 18–24; Orejas & Sastre Prats 1999: 171–83; Orejas & Sánchez-Palencia 2002.

[162] Orejas 1994: 252–4.

[163] Orejas 1994: 250–5; Orejas 1996: 55–101, esp. 100–1.

[164] Orejas 1996: lam. 10.

[165] Orejas 1994: 265–6; Orejas 1996: 176–81.

oriented NE–SW.[166] The 'castros' in the area remained settled until the mid-second century AD.

This vivid archaeological picture is partly corroborated by the literary and epigraphic evidence. Florus (2.32.59–60) tells us that, in Augustus' reign, the Asturians were forced to leave the hillforts, resettle in the plain, and were put to work in the gold mines. A number of bronze tablets found throughout northwestern Spain may reflect the Roman reorganization of indigenous settlement patterns: The *tabula Paemeiobrigensis*, discovered only recently at Bembibre, mirrors Augustan policy in the immediate aftermath of Roman occupation. The tablet refers to two edicts from 14 and 15 February 15 BC, in which Augustus regulates the status of the *castellani Paemeiobrigenses ex gente Susarrorum* and the *castellani Aiiobrigiaecini ex gente Gigurrorum*. In the first edict Augustus declared the perpetual immunity of the *Paemeiobrigenses* from tribute and services (perhaps including corvée labour in mines) as a reward for their loyalty to Rome during a recent upheaval. In the second edict the *castellani Aiiobrigiaecini*—in observance of their wishes—were separated from the *gens Gigurrorum*, rejoined with the *castellani Paemeiobrigenses* and required to provide tribute and services together with the *gens Susarrorum*. The term used to describe the communities, *castellani*, most likely refers to the inhabitants of 'castros'.[167] The text affirms the territorial realignments taking place under the auspices of Roman legates during and after the pacification of the *provincia Transduriana*.[168] Further bronze tablets, known as *tesserae hopistales*, contracts regarding hospitality (*hospitium privatum*) amongst private individuals or communities, may perhaps be a result of a forced relocation.[169] It has been argued that the significant increase in movements of families or members of a *castellum* or tribe required a legal instrument to secure the settlement of these individuals in a new environment. Given the fact that these movements probably had been forced or encouraged by the Roman state, and that these documents appear nearly exclusively in the mining areas of the north-west, Ines Sastre Prats believed that this new legal device was introduced by the Roman authorities to bolster the effects of their resettlement policy.[170]

[166] Fernández-Posse & Sánchez-Palencia 1985; Fernández-Posse & Sánchez-Palencia 1988. For similar developments in the Cuenca del Boeza area east of Ponferrada/León, cf. Sánchez-Palencia & Perez Garcia 1983: 240.

[167] For list of attested *castella*, cf. Sastre Prats 1998: 80–1, figs.

[168] Alföldy 2000*b*: 197. On the *tabula Paemeiobrigensis*, cf. Balboa de Paz 1999; Costabile & Licandro 2000.

[169] Kierdorf 1967: 1234; LeRoux 1995: 90–2.

[170] See mainly Sastre Prats 1998, with further bibliography. On the evidence: *tabula Lougeiorum*, cf. Dopico Caínzos 1988; *tabula Zoelarum*, cf. *CIL* II 2633; El Caurel tablet, cf. *IRPL* 55; Castromao tablet, cf. *AE* 1973: 295 = *AE* 1972: 282.

Provided our interpretation of the archaeological and written evidence is correct, the forced reorganization of the entire social landscape for the purpose of accomodating a new mining industry on a vast scale certainly underscores the tenacity of the Roman authorities in exploiting the metallic resources of northwestern Spain. Undoubtedly the officials in charge of this mining area were reliant on the provincial administration in several ways. Forced labour and, at a later stage, free labour had to be put to work in an effective manner, paid and/or supplied with food rations and other goods; tools, wood, charcoal, and other items had to be provided; aqueducts and furnaces planned and built for the extraction and smelting procedures; the import of material resources and the export of gold ingots had to be organized; and the security of these transports guaranteed.[171] These would have been major logistical undertakings for the mining management in the north-west.

Despite these insights, we are not well informed on organizational strategy and the allocation of administrative responsibilities in the early phase of gold exploitation in northwestern Spain.[172] Nor do we have information concerning the division of tasks between the freedmen mining procurators and the equestrian financial procurators for Asturia and Callaecia, both of which are attested in the Flavian period at the earliest. It remains merely an assumption that the mining procurators documented at Vilalís and Luyego held judicial powers comparable to those at Vipasca. Were this to be the case, the focus perhaps lay more on the enforcement of order and security within the mining district rather than the supervision of miners and their obligations. In addition, a procurator in the north-west could have dealt with the recruitement, provision, and payment of a workforce, as well as a whole range of technical problems, particularly the construction and maintenance of a vast water supply system for the mining operations, as documented at Las Médulas (2.2.1.1).

In general, the responsibilities of freedmen *procuratores metallorum* could therefore have differed significantly, depending on geological constraints and the organizational strategy chosen. One thus could argue that freedmen mining procurators either organized the extractive process by farming out plots to small-scale contractors, by sale of plots to private individuals, or ran

[171] There is no textual evidence on how the local population was forced to participate in mining operatios. Almudena Orejas has suggested that the resettled villagers would provide work as tribute (*operae*), cf. Orejas & Sánchez-Palencia 2002: 593.

[172] As Florus (2.25.12) reports, the governor C. Vibius Postumus put the local population to use in gold-mining operations in the vrbas Valley after the Pannonian mutiny, cf. Wilkes 1969: 272; Zaninović 1977: 796; Kienast 1999: 403. In analogy, one might see the *legati Augusti* reported by the *tabula Paemeiobrigensis* for a *provincia Transduriana* and by Strabo (3.4.20) for Asturia and Callaecia in charge of restructuring the newly gained territories for economic ends.

mines directly by the means of a labour force. We cannot be sure whether the Roman army officers and soldiers stationed in these areas were required to undertake technical tasks or oversee the arrangement and command of work procedures. These differences set aside, the activities of freedmen procurators recorded for the Iberian peninsula were limited to well-defined local districts or regions within the provinces (4.1.13).

6.1.3.3. *Procurators and Mining Districts in Danubian Provinces*

Naturally, it is impossible to be sure whether these observations apply to other known freedmen mining procurators in the Roman empire. The freedmen *procuratores aurariarum* recorded at Ampelum prior to the Marcomannic wars may also have held responsibilities limited to a local district, probably including Alburnus Maior/mod. Roșia Montană. The designation 'aurariae Dacicae', which appears on an inscription by a *tabularius* at Ampelum, does not provide any indication as to the precise geographical scope of the procurator's responsibilities.[173] The inscriptions of *beneficiarii procuratoris* at Ampelum may however indicate the subordination of the freedmen procurators there to equestrian financial procurators of Dacia Apulensis, just like the procurators at Villalís and Luyego (cf. 4.1.9).[174] Provided this assumption is correct, the freedman procurator may likely have supervised the gold mines in a local district, or in Dacia Apulensis at the most.

The wooden tablets from Alburnus Maior in Dacia shed some light on the local regime of Roman gold-mining there. The tablets found in Roman shafts and tunnels in the late eighteenth and early nineteenth century date to the mid-second century and not later than AD 167.[175] The tablets thus mirror the situation at Alburnus Maior prior to the Marcomannic wars and may have been discarded or hidden in the mines during the upheaval.[176] The range of topics covered by the texts is limited, with most containing contracts. Given the different findspots, they were not all part of one archive.[177] Four of the twenty-five tablets found seem to refer to mining ventures at Alburnus Maior. On 19 May AD 164 an Aurelius Adiutor, most likely a Roman citizen, hired the services of a Memmius Asclepi of peregrine status for *opus aurarium*, 'work in the gold mines'.[178] Another work contract names a *Titus Beusantis qui et Bradua* employing the mining worker *[–]s Restitutus agnomine Senioris* for

[173] *CIL* III 1297 = *ILS* 1594 and *CIL* III 1313, cf. Noeske 1977: 352.
[174] Nelis-Clément 2000: 263 f.
[175] Mrozek 1977: 97; *CIL* III p. 921–60.
[176] On Alburnus Maior, cf. Piso 1993: 97 f.; Piso 2004: 301 f.; on Marcomannic wars, cf. Birley 1987: 155–79.
[177] Noeske 1977: 336 f.
[178] *CIL* III p. 948, cf. Noeske 1977: 398–403.

opus aurarium.[179] A contract of 23 October AD 163 was concluded between a Socratio Socrati⟨oni⟩s and the worker L. Ulpius Valerius at Alburnus Maior, but does not explicitly name 'work in gold mines' as the purpose of hire.[180] This evidence lends support to the argument that at least some sections of the mines at Alburnus Maior were worked by a labour force employed by private entrepreneurs who ran extractive operations in agreement with the mining authorities. Furthermore, a Iulius Alexander is named in four different wooden tablets as creditor and debitor.[181] Noeske concluded that Iulius Alexander was a banker/broker of some sort, reminiscent perhaps of the *argentarius* documented in the Vipasca tablets.[182] Although Iulius Alexander might have provided the financial capital for some of the mining contractors at Alburnus Maior, there is no evidence to suggest that the mining administration had awarded him the banking monopoly, or that Vipasca-style regulations for the collection of a broker's fee had been implemented.

The available epigraphic evidence prior to the Marcomannic wars implies that the freedman *procurator aurariarum* at Ampelum delegated part of the underground mining activities at Alburnus Maior to private individuals or companies, whatever the contractual basis. The other part of the mines were perhaps worked 'directly' by the Roman state. For instance, the relatively high numbers of Dalmatians at Alburnus Maior and the mention in one of the wooden tablets of a *vicus Pirustarum* as part of Alburnus Maior, named after the Dardanian tribe of the *Pirustae*, has been understood to refer to a forced resettlement of these skilled miners by the Roman state in order to exploit the gold veins in Dacia.[183] There is, however, no firm evidence for a deliberate deportation of skilled mining workers from Dardania. The possibility of a voluntary migration by Pirustae (perhaps in response to incentives, i.e. high wages, offered by the Roman state) may not altogether be excluded.[184] Like the situation in northwestern Spain, the movement of supposedly qualified men to Alburnus Maior might hint at the direct exploitation of part of the gold mines by the Roman state, perhaps in the early years of Roman rule. As of yet, the epigraphic record of Ampelum and Alburnus Maior does not provide any direct confirmation of this view.[185] One thus can merely presume

[179] *IDR* I 43 = *CIL* III p. 949, XII; cf. Noeske 1977: 403 f.

[180] *CIL* III p. 948, cf. Noeske 1977: 396–8 with 400 fn. 700.

[181] *IDR* I 33 = *CIL* III pp. 930–2, III; *IDR* I 35 = *CIL* III pp. 934 f.,V; *IDR* I 43 = *CIL* III p. 949, XII; *IDR* I 44 = *CIL* III pp. 950 f., XIII; cf. Noeske 1977: 389–91, 404–6.

[182] Noeske 1977: 340.

[183] *CIL* III pp. 944 ff.,VIII, with Mrozek 1969: 141 f.; Mrozek 1977: 99; Noeske 1977: 275 f.

[184] Cuvigny 1996*a*.

[185] The term *leguli aurariarum* recorded at Ampelum (*CIL* III 1307, AD 165/6) and Alburnus Maior (*CIL* III 1260, late 2nd cent. AD) seems to refer to the free labour present in the gold mines also attested in the *tabulae ceratae*, cf. Hirschfeld 1905: 155, 175 fn. 3; Mrozek 1969: 30; Noeske 1977: 349 with fn. 333.

that up to the Marcomannic wars, freedmen procurators at Ampelum had sections of the gold mines at Alburnus Maior directly exploited by an imported workforce. Exploitation with the help of private entrepreneurs on a contractual basis, perhaps even by the sale of mining plots to interested parties as at Vipasca, may also have been a common sight in other mining zones of the Danube provinces. The use of the term *colonus* evokes the establishment of an exploitation system similar to Vipasca; however, there is no evidence to date to corroborate this assumption (cf. 7.1.2).

Unlike the Iberian peninsula or Sardinia, where freedmen mining procurators are recorded holding office until the late second century, their colleagues in the Danubian provinces were apparently mostly phased out or 'demoted' to *subprocuratores* during the reign of Marcus Aurelius.[186] As argued above, freedmen procurators at Ampelum seem to have been replaced by equestrian *procuratores aurariarum.* Along with this change in status the scope of procuratorial duties might have widened as well: whereas freedmen procurators may have been limited to a local district, equestrian procurators were probably responsible for the gold mines on a provincial scale, perhaps even for all three Dacian provinces. A change in social status and thus in scale of responsibilities(?) similar to Ampelum can not be traced as clearly elsewhere: the epigraphic record at Sočanica names a Titienus Verus, a *procurator mm(etallorum) D(ar)d(anicorum?).* Unlike his predecessors, Verus is an equestrian who performed a *dedicatio* of a statue base commissioned by the *ordo colonorum* during the reign of Gordianus III.[187] Unfortunately, we do not know when this change in status occured or whether the equestrian official possessed similar responsibilities as the freedmen procurators documented at Sočanica. The dedication of an inscribed monument would at least imply the procurator's temporary presence there. A further equestrian procurator, Cassius Ligurinus, is recorded on an inscribed slab at Rudnik overseeing the construction of a temple to Terra Mater during the reign of Septimius Severus, providing evidence of his occasional visit to the silver, copper and lead mines at Rudnik.[188] The *dedicatio* of buildings or *cura* over construction work required the temporary presence of these equestrian procurators on site. This, however, does not preclude their official responsibilities covering other mining areas or all mines within the province of Moesia Superior.

As for the responsibilities of the *procuratores* for silver mines in Pannonia and Dalmatia, we are left in the dark as well. We can only assume that their line of work included the superintendence over the collection of *vectigalia*

[186] *CIL* III 1088, p. 1390 = *IDR* III/3, 228, cf. Noeske 1977: 348, AMP 3. Sardinia: *AE* 1998: 671; Bruun 2001.

[187] *ILJug* 503. Dušanić 2004*a* cites a fragmentary inscription, the traces of which he reads as naming a procurator of Commodus at Sočanica.

[188] *IMS* I 168. Whether *IMS* I 103 was set up by two *coloni* is not stated in the inscription, cf. Dušanić 1976: 101.

and/or the direct management of mining districts. A recently published inscription of a *beneficiarius consularis* from Bölcske (*AE* 2003: 1426) identifies amongst his *stationes* a *statio arg(entariarum) Pann(onicarum) ẹ[t Del(maticarum)]*. Like the *statio* at Siscia for the *procurator Augusti* and *praepositus vectigalis ferrariarum*, Alföldy assumed that these equestrian mining procurators held several *stationes* in each province, which they might have frequented on a regular basis.[189] This is supported by the nature of epigraphic evidence from Domavium (cf. 4.1.11). If this argument is correct, a direct involvement in the running of mining ventures is less likely. The *procuratores metallorum/argentariarum* for Pannonia and Dalmatia thus probably functioned as supervisors: they likely possessed judicial powers to deal with conflicts between *conductores* and miners and border issues between mining districts and local communities.

6.3.1.4. *Procuratores ferrariarum*

The erection of an altar to Terra Mater at Ljubija by the local *vilicus officinae ferrariae* and the *procurator* meant that the latter occasionally visited the iron mines. Even though the precise title of these equestrian officials can not be gleaned from the inscriptions, it could be reconstructed as *procurator ferrariarum* of the Pannonian and perhaps other Danubian provinces, similar to the *conductor ferrariarum Pannonicarum itemque provinciarum transmarinarum* documented at Ljubija in AD 201. The task of a *conductor ferrariarum* at Ljubija is not explained by our sources. Given the geographic scope of responsibilities expressed in his title, the collection of a *vectigal* from local mining occupants in various iron-mining areas throughout the Pannonian and 'overseas' provinces was probably his main responsibility.[190] Like the

[189] Alföldy 2003.

[190] Hirschfeld 1905: 152. The term 'vectigal' is primarily understood to mean the revenue from public land or rents from 'state property', but is also used to describe 'indirect' taxes on goods, cf. Mommsen 1887: 434 fn. 3; Pékary 1979; Burton 1999. On the sale/lease of public land, cf. Lintott 1993: 80–2; Rathbone 2003. Ulpian, *Dig.* 50.16.17.1, defines public *vectigalia* as income for the *fiscus*, i.e. the *aerarium Saturni*, amongst which he counts the *(vectigal) metallorum*. This is corroborated by the literary evidence for *vectigalia* on mining ventures, cf. Livy 34.21.7 (M. Porcius Cato imposed *vectigalia* on silver and iron mines during his governorship of Spain in 195 BC); Livy 39.24.2, 42.12.9, 44.18.3, 44.29.11 (Roman senate exacts *vectigal* on Macedonian iron and copper mines payable to the Roman State); Strabo 3.2.10 = Polybios 34.9.8–11 (lessees of silver mines near Carthago Nova provide silver worth 27,000 drachmas per day to the Roman state); Pliny, *NH* 33.118, 34.165 (the cinnabar mines of Sisapo in Baetica are one of the most important sources for the *vectigalia populi Romani*; two lead mines, *metallum Sama/lutariense* and *metallum Antonianum*, each contracted out by the Roman state for 200,000 denarii in *vectigalia*, achieve 255,000 denarii per year, and 400,000 sesterces (or denarii?) after being contracted out a second time).

system of farming out land and pastures, the *vectigal* on mines is believed to be a form of rent payable to the Roman state, collected either by officials or *conductores* to whom collection was farmed out.[191] However, it cannot be completely ruled out that the *conductor* and his company ran the extractive operations directly.[192] On the basis of the evidence from Ljubija, Hirschfeld had argued that imperial procurators replaced *conductores* by AD 209 and dealt with occupants of mines directly.[193] Moderating Hirschfeld's position, Brunt argued the *procuratores* and their administrative apparatus did not function exclusively as 'superintending local managers of settlements like Vipasca', but also acknowledged that they might have supervised 'the activities of contractors, perhaps corporations, interested in the entire business of mining throughout the areas concerned'.[194]

Both concepts might apply to the administration of the iron mines near Ljubija: the exaction of a *vectigal* by the *conductor ferrariarum* from private individuals running the mines (most likely through the agency of the local *vilicus*) may have been monitored by a procurator appointed specifically to supervise the collection and receipt of *vectigal ferrariarum*. This is suggested by the title of Flavius Verus Metrobalanus as *procurator Augusti* and *praepositus splendidissimus vectigalis ferrariarum*. His subaltern of the *statio Sisciana* was the 'treasurer' in charge of the *arca* into which the revenues from the *vectigal ferrariarum* collected by the *conductores* were paid.[195] The principal functions of a procuratorial *praepositus vectigalis ferrariarum* involved 'outsourcing' and supervising the collection of revenue from public mines. As Brunt assumed, the procurators received the 'instalments due under the contracts', accounted for the sums received and disbursed, and may have audited the accounts of the *conductores*.[196] After AD 209, the *procuratores ferrariarum* seemingly took a more direct interest in the running of the smelting ovens (*officinae ferrariarum*) under supervision by the local *vilici* around Ljubija. In contrast to the *praepositus vectigalis ferrariarum*, the titles of the equestrian iron-mining procurators at Ljubija do not reveal any information regarding their function. Thus one can merely speculate whether they continued to monitor the collection of *vectigal* from the occupiers of

[191] Rostovtzeff 1904: 445 f.; Ørsted 1985: 221 f.; Brunt 1990*c*: 394 f.

[192] Brunt 1990*c*: 401.

[193] Hirschfeld 1905: 152 f.

[194] Hirschfeld 1905: 153–9; Brunt 1983: 46; Brunt 1990*c*: 401 f.; Eck 1997*a*: 136.

[195] Based on the inscribed altar set up at Siscia in Pannonia Superior by his *arcarius stationis Siscianae*, Metrobalanus perhaps was contemporary with the *conductor* attested at Ljubija, cf. *CIL* III 3953 = *ILS* 3094 (Siscia), cf. *PIR*² F 396; Pflaum 1960–1: 1064; Dobó 1968: 180, no. 283; Fitz 1972: 224 f.; Fitz 1993–5: 738, no. 422.

[196] Brunt 1990*c*: 385.

mines or whether their brief titles meant they in fact possessed a wider range of administrative responsibilities.

The *procurator ad vectigal ferrariarum Gallicarum*, M. Cosconius Fronto, was probably burdened with tasks similar to those of the *praepositus vectigalis ferrariarum* documented at Siscia. Unlike the Danube provinces, however, there is no evidence for *conductores* contracting the collection of *vectigal ferrariarum* in one or more provinces of Gaul. The scarce evidence appears to pertain mostly to *conductores* who collected *vectigal* on a local level.[197] The documented title of a *promagister f[errariarum] provincia[rum] Narbonens(is) Lu[gdun(ensis)] Aquitanic(ae) Belgi[cae]* in an inscription at Arelate/mod. Arles is perhaps the only indication of a *societas ferrariarum* operating on a super-provincial level within Gaul.[198] As regards the procuratorial administration of iron mines throughout Gaul, we can only speculate whether Attius Alcimus Felicianus, the *procurator ferrariarum* documented at Lyon during the reign of Severus Alexander, controlled the iron mines within a geographical area similar to the *procurator ad vectigal ferrariarum Gallicarum*.

In this context, two inscriptions of a further equestrian procurator, C. Iulius Silvanus Melanio, may warrant consideration. Firstly, as *procurator Augusti* he commissioned an inscribed altar at Lyon without any indication of his precise function.[199] Secondly, his fragmentary inscription at Domavium, in particular lines 7 to 8 have tentatively been read *proc(uratori) Aug(usti) [fe]ṛṛ(ariarum) / ẹṭ ạrg(entariarum)(?) per prouin/[cias] XXIII*. Under the provision that these lacunose lines have been pieced together correctly we wish to offer the following hypothetical line of argument. The text implies that during his procuratorial career he had held sway over iron and silver(?) mines in twenty-three provinces. In accordance with a conventional equestrian career, he must have held at least two, perhaps more mining procuratorships, each post covering the iron or silver mines of several provinces.[200] Melanio was quite certainly a centenarian *procurator argentariarum* of both Pannonian provinces and Dalmatia.[201] As there is only one procuratorship for

[197] Cf. *cond(uctor) ferrar(iarum) ripae dextrae* (*CIL* XII 4398, Narbo) or the *manceps* of the *vectigal massae ferrariarum Memmiae Sosandridis c(larissimae) f(eminae)* (*CIL* XIII 1811 = *ILS* 8641 [Lugudunum]). For the former, cf. Desjardins 1867–85: 414 f.; Hirschfeld 1905: 158 with fn. 5; Sablayrolles 1989: 159 fn. 23 with further bibliography.

[198] Pflaum 1960–1: 118–23, no. 52, 1053, based on the reading of *CIL* XII 671 by Benoit 1932: 138–41.

[199] *CIL* XIII 1729; Pflaum 1960–1: 734 f., no. 276; Pflaum 1982: 67 f., no. 276; Abascal & Alföldy 1998: 163.

[200] *CIL* III 12732, cf. Fitz 1972: 223 f., no. 6; Devijver 1976–80: 487 f., I 126; Abascal & Alföldy 1998: 162 f.

[201] The inscription *CIL* III 12732 from Domavium, according to the reconstruction of Abascal & Alföldy 1998: 162 f., distinguishes between *[fe]rrar(ia)* and *arg(entaria)*. The singling out of

silver mines hitherto accounted for in the late second century AD, the remaining two or three mining procuratorships held by Melanio likely covered iron mines throughout the remaining twenty provinces. Inscriptions from the late second/early third century have provided little evidence to date for equestrian posts with a mandate to supervise iron-mining. The few examples we do have include a procuratorship for iron mines of unknown rank at Ljubija, a *procurator Augusti* and *praepositus splendidissimus vectigalis ferrariarum* at Siscia, and the centenarian *procurator (ad vectigal) ferrariarium (Gallicarum)* at Lyon. As regards the latter procuratorial post, the title *procurator... ad vectigal ferrariarum Gallicarum* does not provide any clue as to the precise geographical extent of the procurator's jurisdiction. Given the inscription of a *promagister f[errariarum] provincia[rum] Narbonens[is] Lu[gdun(ensis)] Aquitanic(ae) Belgic(ae)* from Arles, however, it is plausible that procuratorial jurisdiction may have covered more than 'just' the Tres Galliae. A suitable template may be provided by the customs-district of the *quadragesima Galliarum* which included the Tres Galliae as well as Narbonensis, both Germanies, the Alpes Maritimae, Alpes Cottiae, and Alpes Graiae et Poeninae. One might argue hypothetically that the jurisdiction of the *procurator ferrariarum* probably covered the same nine provinces.[202] C. Iulius Silvanus Melanio perhaps commissioned his votive monument at Lyon as the incumbent *procurator (ad vectigal) ferrariarum (Gallicarum)* under Septimius Severus, a post later held by Attius Alcimus Felicianus.[203]

Let us return to C. Iulius Silvanus Melanio's career: of the iron and silver mines in twenty-three provinces, his post as procurator for silver-mines covered three provinces (Dalmatia, Pannonia Inferior and Superior), whereas the post held as *procurator (ad vectigal) ferrariarum (Gallicarum)*, in all probability, included nine provinces. Thus, we are left with the remainder of eleven provinces. Melanio possibly held an additional procuratorial post dealing with administration of *ferrariae.* The evidence for further equestrian

silver-mines may imply that C. Iulius Silvanus Melanio must have held the post of *procurator argentariarum Pannonicarum et Delmaticarum* at Domavium, thus after the reign of Commodus (cf. 4.1.11).

[202] Abascal & Alföldy 1998: 162 f. The proposed size of these 'dominions' are not unheard of: a *procurator familiarum gladiatoriarum* held responsibility *per Asiam Bithyniam Galatiam Cappadociam Lyciam Pamphyliam Ciliciam Cyprum Pontum Paphlagoniam,* cf. Pflaum 1960–1: no. 1973; Abascal & Alföldy 1998: 162.

[203] Whether or not M. Cosconius Fronto was Melanio's predecessor or successor is not of interest here; however, one might point out that Cosconius Fronto also held office between 198 and 209 AD. Pflaum (1960–1: 1053, and 1982: 119) lets Cosconius Fronto follow Iulius Silvanus Melanio. Abascal & Alföldy 1998: 163, rightly point out that Melanio may have commissioned the altar at Lyon later on in his career, perhaps when he held the ducenarian post of *procurator provinciarum Galliarum Lugdunensis et Aquitanicae.*

iron-mining procuratorships is limited. Apart from the *procuratores (ferrariarum)* attested at Ljubija, only one further post, that of *procurator Augusti praepositus vectigalis ferrariarum* at Siscia, is hitherto attested (cf. 4.1.11). Both the procurators at Ljubija and Siscia possibly held the same post. Based on the premise that our equestrian *procuratores ferrariarum* contracted out the collection of 'vectigal' to *conductores* in different provinces, the list of provinces after the designation of *conductor ferrariarum* may allow one to tentatively outline this procuratorial 'dominion'. The inscribed monument of Agathopus at Mursa records his position as *c(onductor) f(errariarum) Pannoniar(um) itemque provinciarum transmarinarum.*[204] Furthermore, an inscription from Hohenstein in Noricum names a *con(ductor) fer(rariarum) N(oricarum) P(annoniarum) D(almatarum).*[205] Hence, the *procurator Augusti praepositus vectigalis ferrariarum* at Siscia and, later, the *procuratores ferrariarum* at Ljubija may have controlled the iron mines in at least four provinces, namely Noricum, Upper and Lower Pannonia, and Dalmatia. Further provinces may have been included: an inscribed sarcophagus from Augusta Vindelicum/mod.Augsburg in Raetia dating to the early third century supposedly names a *manceps ferrariar(um) [—] I et provinciae Raetiae itemque Daciarum trium.*[206] Thus, the procurator's 'dominion' would amount to at least eight or possibly even eleven provinces, if one wishes to include both Upper and Lower Moesia, and Thracia.[207] A possible template for this superprovincial administration is the customs-district of the *publicum portorii Illyrici* and *publicum portorii ripae Thraciae* which included Dalmatia, Raetia, Noricum, Upper and Lower Pannonia, Upper and Lower Moesia, and the three provinces of Dacia.[208] One could argue that the procuratorship of iron mines of the late second/early third century documented at Ljubija and Siscia may have covered the same provinces, and thus was responsible for all of Illyricum. Such a position could be regarded a precursor to the late fourth century *comes metallorum per Illyricum.*[209]

There is no written evidence to date to corroborate this hypothetical construct, and one cannot exclude the possibility that iron-mining

[204] *AE* 1973: 411 = *ILJug* 779 (Ljubija); Bulat 1989: 36 (Mursa).

[205] *CIL* III 4809 = *ILS* 1467 = *ILLPRON* 51.

[206] Nuber 1985. Whether the *conductores ferrariarum* known from an inscribed early 3rd-cent. altar from Teliucu inferior/Romania also held sway over all Dacian provinces remains unknown, cf. *IDR* III/3, 37.

[207] Abascal & Alföldy 1998: 163, have argued for two further separate iron-mining procuratorships, one for the three Dacian provinces, the other for both Moesia Inferior and Superior, Thracia, Macedonia, and Epirus.

[208] On the expanse of 'Illyricum', cf. App. *Ill.* 6. 29 f.; Zon. 8.19.8; Vittinghoff 1953: 359–61; Alföldy 2004: 208, 212 f.

[209] *Not. Dig. Or.* XIII 11.

procurators at Ljubija and Siscia covered a provincial or super-provincial entity of far smaller proportions. A possible blueprint might again be provided by the late second/early third century procuratorships for the *publicum portorii Illyrici*: inscriptions record a *procurator publici portorii vectigalis Illyrici per Moesiam inferiorem et Dacias Tres* and a *procurator publici portorii vectigalis Illyrici per Raetiam et Noricum et Dalmatiam et utramque Panno- niam et Moesiam Superiorem*.[210] In consequence, the series of mining procuratorships of C. Iulius Silvanus Melanio's career may have included one additional iron-mining procuratorship.

To sum up, C. Iulius Silvanus' hypothetical procuratorial career in mining administration might have covered three or four posts, the order of which remains elusive; he most likely was *procurator argentariarum Pannoniarum et Dalmatiarum*, perhaps *procurator (ad vectigal) ferrariarum (Gallicarum)*, and may have held either an unattested post of *procurator ferrariarum 'Illyrici'* or two further procuratorships for iron mines in super-provincial zones of smaller size within Illyricum.

In the light of the Domavian inscription of C. Iulius Silvanus Melanio, the administration of iron mines in the late second and early third century AD may have consisted of two possible blocks, 'Gaul' and 'Illyricum'—provided the analogy with the customs-districts of the *quadragesima Galliarum* and the *publicum portorii Illyrici* is a valid one. This hypothetical assumption raises the question of the purpose of such super-provincial iron-mining districts: part of the answer may be provided by the inscribed monument for the *procurator ferrariarum* C. Attius Alcimus Felicianus at Lugdunum, which was commissioned by a *beneficiarius* of the *legatus legionis I Minerviae* during the reign of Severus Alexander. Unfortunately, the inscription does not tell us why the soldier of a legion garrisoned on the Rhine at Bononia/mod.Bonn honoured a *procurator ferrariarum* by setting up a monument at Lyon. However, the Roman army was undoubtedly in constant need of iron, the supply of which could have been burdened on the *procurator ferrariarum*. As argued above, the army units were in most cases unable, perhaps even unwilling, to supply their needs by tapping local iron sources (cf. 5.2). One thus might argue for a special authority which coordinated the supply of resources to military units along the Rhine frontier. The *procuratores ferrariarum* at Ljubija and the *procurator Augusti praepositus vectigalis ferrariarum* at Siscia possibly filled a similar role in the supply of military units along the Danube

[210] Mócsy 1962: 594; Alföldy 2003: 225 f.; Alföldy 2004: 212 f. For the *procuratores publici portorii vectigalis Illyrici*, cf. Vittinghoff 1953: 361; Pflaum 1960–1: 1059, 1065. A further super-provincial administrative entity in the area is the *vehiculatio Pannoniae utriusque et Moesiae Superioris et Norici*, cf. Pflaum 1960–1: 647–9.

frontier—the *exercitus Illyriciani*—within their respective section(s).[211] Thus, the responsibilities of the *procuratores ferrariarum* for Gaul and 'Illyricum' may perhaps have gone beyond overseeing the collection of rents or proceeds from iron mines.

The tentative conclusions on the administration of iron mines in the late second/early third century AD rely heavily on conjectures based on our knowledge of the expanse and organizational subdivisions of customs-districts. Further parallels might be drawn regarding the development of iron-mining administration in the 'Illyrian' provinces, chronicled in particular in the series of votive altars to Terra Mater of Ljubija and surrounds. By AD 209 *conductores* disappear from the epigraphic record, their position taken over by *procuratores.* Based on this observation, one might argue for the intensified involvement of the equestrian procurators in iron-mining operations at Ljubija and elsewhere throughout their super-provincial 'domain'. At about the same time administrative modifications can be detected in the *portorium Illyrici* and the *quadragesima Galliarum. Conductores* and their personnel are replaced by members of the *familia Caesaris* between the reigns of Commodus and Severus Alexander.[212] Our sources provide no information on why this change was carried out, and its effects remain somewhat obscure. It has been argued, however, that the direct exaction of these duties by imperial officials would have resulted in a far higher revenue than farming out their collection. The desire for a significant increase in revenues was understood to be a result of a financial crisis in the later second century, mainly a result of a considerable growth in military expenditure in particular.[213] The replacement of *conductores* by imperial personnel at Ljubija could have been prompted by a similar rationale.

The discussion of the administration of iron mines has so far centred on the period of the late second and early third century AD. Prior to this period, the known *procuratores ferrariarum* can not be sufficiently linked to a particular super-provincial iron-mining district, as the procuratorial titles do not provide a geographical area, nor do the findspots of the inscriptions offer any further clues. Based on the inscribed monuments for Attius Alcimus Felicianus at Lyon and Abbir Cella/ Henchir en-Naâm in Africa Proconsularis, where his functional title is rendered as 'procurator ferrariarum' only, Pflaum conjectured that the other known equestrians holding the title of 'procurator ferrariarum' must have been responsible for iron mines

[211] Alföldy 2004: 214.

[212] De Laet 1949: 393, 403–8; Vittinghoff 1953: 361; Brunt 1990*c*: 416; Eck 1997*a*: 136 f.; France 2001: 430–8; Lo Cascio 2005: 154.

[213] France 2001: 436–8; for a different view, Brunt 1990*c*: 415 f.; Lo Cascio 2005: 153.

throughout Gaul.[214] This attribution, however, is not necessarily warranted: the absence of any firm indication for an iron-mining procuratorship for 'Gaul' prior to the reign of Septimius Severus (when this procuratorial title is first documented) and the lack of evidence for an iron-mining 'bureau' in Lyon independent of the regular provincial administration questions whether this argument is valid.

Quite to the contrary, the functional title of *procurator ferrariarum* might designate an office held in Rome. The brevity of the procuratorial title in numerous inscriptions featuring equestrian careers is striking, especially in inscriptions where geographical descriptions of other procuratorial functions are recorded in detail (cf. 4.1.12). The absence of any geographical region in career inscriptions can refer to procuratorial offices held in Rome or in the proximity to the emperor. This might be the case with the *procurator ferrariarum*. If this conjecture is warranted, what was the 'job description' of this procuratorial office? There is a large amount of evidence at Ostia and Rome for activities related to iron-mining. This includes an honorary monument for T. Petronius Priscus, *procurator ferrariarum et annonae Ostiensis*; an undated lead *tessera* naming a *statio ferrariarum fori Ostiensis*; an inscribed monument of AD 102–17 commissioned by a *servus socior(um) vect(igalis) ferr(ariarum)*; an undated 'tessera nummularia' discovered in Rome inscribed with *soc(iorum) ferr(ariarum)*; this implies a possible contemporaneity of *socii ferrariarum* and *procuratores ferrariarum*.[215] One might argue further that the procurators oversaw the *socii ferrariarum* who were involved in the collection of *vectigal* from iron mines or in running mining districts. If Brunt is correct, the procurators 'may have leased the contracts to publicans' which included a regular audit of the latter's accounts; one thus may see the *procurator ferrariarum* at Rome in a similar function.[216] Concerning the expanse of their jurisdiction, the *procuratores XX hereditatium* may provide a suitable model: besides the *procuratores* for different provinces, e.g. *procurator XX hereditatium provinciae Syriae Palaestinae, per provincias Narbonensem et Aquitanicam*, or *per Asiam*, a *procurator XX hereditatium*—his function is regularly given without a geographical indication—was responsible for the collection of the 5 per cent inheritance tax in Rome and Italy.[217] By analogy one might

[214] *CIL* VIII 822 = 12345 = 23963 = *ILTun* 741, cf. also *CIL* VIII 23948, cf. Pflaum 1978: 169–72.

[215] *statio ferr.*: *CIL* XIV 4326, p. 773 (Ostia); *socii vectigalis ferr.*: *AE* 1924: 108 = *CIL* XIV 4326 (Ostia); *socii ferr.*: *AE* 1928: 17a (Rome); T. Petronius Priscus: *CIL* XIV 4459 = *ILS* 1442 (Ostia), cf. Pflaum 1960–1: 558, no. 212; Meiggs 1973: 302; Pflaum 1978: 138, no. 4; Sablayrolles 1989: 158.

[216] Brunt 1990*c*: 385.

[217] Eck 1979: 132; Eck 1997*a*: 80.

argue that the centenarian *procurator ferrariarum* at Rome only oversaw the iron mines of a limited area, possibly the Italian peninsula. It is, however, rather doubtful that the authority of these procurators was limited to Italy: Even though Pliny the Elder identifies iron mines on Elba and near Sulmona/ mod. Sulmo, and iron processing sites such as Comum, there was probably not sufficient iron-mining activity to warrant the creation of an equestrian procuratorship specifically for the Italian peninsula.[218] The mandate of a *procurator ferrariarum* at Rome may therefore have included a vaster 'domain'.

The social status of the *conductores ferrariarum* who appear in the epigraphic record of Noricum may provide a clue. The municipal elite of Noricum was not able to acquire a lease on iron mines, and these leases fell nearly exclusively to contractors from Aquileia, i.e. to entrepreneurs from outside of Noricum.[219] Furthermore, the inscribed monument to Isis Noreia, set up by the procurators of Q. Septueius Clemens for their superior at the sanctuary at Hohenstein during the second century, names Clemens as *conductor ferrariarum N(oricarum) P(annonicarum) D(almaticarum)*.[220] Septueius Clemens is unlikely to have collected leases from each and every financial or presidial procurator of the listed provinces. There is no evidence that the presidial procurator of Noricum awarded these contracts; this explains, in part, the exclusion of the Norican provincial elite.[221] It is plausible to suggest that Clemens, like his Aquilean colleagues, may have acquired these contracts centrally—perhaps from the *procurator ferrariarum* in Rome. During the Principate important revenues still were generally farmed out at Rome.[222] The bureaux for various customs duties, such as offices for the *quadragesima Galliarum*, the *quattuor publica Africae*, and perhaps *publicum portorii Illyrici* may have been based at Rome during the first century and apparently answered to the *a rationibus*.[223] Hence, it is not implausible to assume that by the second century companies interested in the collection of *vectigal* and/or the running of iron mines in one or more provinces throughout the empire were to address the *procurator ferrariarum* in Rome. A rather

[218] Pliny, *NH* 34.142, 144. Davies 1935: 63–74.

[219] Scherrer 2002: 29. It has to be pointed out that the elite of Aquileia traditionally maintained close economic ties with the *regnum Noricum* since the Middle Republic.

[220] *CIL* III 4809 = *ILS* 1467 = *ILLPRON* 151.

[221] The fragmentary building inscription from Hohenstein (*CIL* III 14362, p. 2328,197 + 14363 = *AE* 1968: 408 = *ILLPRON* 148) naming the presidial procurator Claudius Paternus Clementianus was commissioned by a Sabi[nius—], who is understood to be a *conductor ferrariarum*, cf. Alföldy 1969*b*: 25 f.; the inscription, however, provides no secure indication to argue a direct administrative connection of *conductores* with the presidial procurator.

[222] Brunt 1990*c*: 377.

[223] De Laet 1949: 373 f.; Vittinghoff 1953: 390 f.; France 2001: 400–10.

fragmentary inscription from Arelate, listing the career of an unknown equestrian, names a *pro mag(ister) f[err(ariarum)] provincia[rum] Narbonens(is), Lu[gdun(ensis)], Aquitanic(ae) Belgi[c(ae)]*, a representative of the company contracting out the running of iron mines within parts of Gaul in *c.* 75 AD.[224] It is plausible to assume that iron mines within these and other provinces must have been contracted out centrally at Rome, even though there is no evidence for a procuratorship of iron mines dating already to this period.

Our sources do not tell us why iron mines in particular were thought to be important enough to be administered by an equestrian procurator. There are, however, infrequent indications for a unique valuation of iron and iron mines within the Roman empire. The early second century *lex metalli Vipascensis*, for example, refers to a *lex ferrariarum* in accordance with which the shoemaker is to sell nails within the mining district.[225] A tax on nails and the use of iron are the subject of a rescript by Hadrian to the Aphrodisians.[226] Paulus (*Dig.* 39.4.11.pr.) relates that the sale of iron to barbarians is strictly forbidden and subject to capital punishment. Moreover, a rescript of Caracalla (*Dig.* 39.4.16.11) exempts any *dominus praedii*, landowner, from penalty whose *colonus* or slave produced iron (*ferrum facere*) on his estate without his knowledge and not in compliance with the law (*illicite*); the term 'ferrum facere' may imply the mining for and/or smelting of iron ore.[227] It is tempting to conjecture from this evidence that the 'creation' of a procuratorship for iron mines at Rome by the early second century and the lease of iron mines in one or more provinces to *promagistri, socii* or *conductores ferrariarum* must be seen in the context of an otherwise unknown general reform of iron-mining administration which occurred as early as the Flavian period. However, this would go beyond the available evidence. The underlying rationale of these legal restrictions and regulations on iron production and trade, besides the tendency to safeguard iron from unsolicited procurement, remains mostly obscure.

To sum up this tentative model of organizational arrangements, the responsibilities of a hypothetical *procurator ferrariarum* at Rome may have covered all relevant iron-mining regions throughout the empire during the

[224] *CIL* XII 671 with Benoit 1932: 138–41. Pflaum 1960–1: 118–23, no. 52, 1053. On *promagistri*, cf. Brunt 1990*c*: 366; Aubert 1994: 327, 344–6.

[225] *LMV* ll. 34. Whether the *lex ferrariarum* corresponded with the mining law mentioned in the *lex portorii Asiae* is unknown: Based on a νόμος γεωρυχικός (§34) ore or metal exported from the province of Asia to Rome were taxed. §27 may indicate that the export of mining samples may have been tax free (cf. Engelmann & Knibbe 1989: 167 and §§ 25 f; for a possible parallel cf. Mateo 2001: 53 f. with Pliny, *NH* 33.118).

[226] Reynolds 1982: 115–18, no. 15, ll. 8–9; Oliver 1989: 166, no. 69.

[227] Herz 2005: 28.

second century AD. By the end of the second century, one might postulate a division of responsibilities. This division may have occured by the reign of Septimius Severus at the latest when there is evidence for the occasional presence of a *procurator (ad vectigal) ferrariarum (Gallicarum)* at Lyon and for *procuratores ferrariarum* and a *procurator Augusti praepositus splendidissimus vectigalis ferrariarum* perhaps for the Danubian provinces at Ljubija and Siscia. Additional evidence for the date of this administrative division may be provided by the change of procuratorial titles of the imperial officials who administered the (silver) mines of Pannonia and Dalmatia: provided the dating of the relevant inscriptions is not off the mark, the change from 'procurator *metallorum* Pannonicorum et Dalmaticorum' to 'procurator *argentariarum* Pannoniarum et Dalmatiarum' could indicate the exclusion of *ferrariae* from the brief of the procurators at Domavium towards the end of Commodus' reign. The earliest 'procurator ferrariarum' hitherto attested for the region, the *procurator Augusti praepositus splendidissimus vectigalis ferrariarum* Flavius Verus Metrobalanus, may also be dated to Commodus' or Septimius Severus' reign (cf. 4.1.11). Furthermore, the practice of partitioning and regionalizing administrative functions would be in line with the general trend towards diversification and growth of the non-senatorial administration throughout the second century AD.

As of yet, straightforward data in support of this hypothetical explanatory model remains elusive. Most evidence is circumstantial and patchy at best (the Domavian inscription of Silvanus Melanio being a prime example), providing a barely reliable basis for interpretation. The precise administrative arrangements currently remain unknown.[228]

6.3.1.5. *Summary*

The functions and responsibilities of mining procurators could vary dramatically from district to district, owing to the range of different mining operations with varying geological and topographical constraints and organizational implications (opencast and underground mining), and the varying technological and organizational strategies available for the exploitation of ore deposits (hydraulic technology, lease of mines). While at Vipasca the procurator seems mainly to have enforced the laws which concerned the occupants/owners of copper and silver mines, the procurators of the alluvial gold-mining districts in the Spanish north-west were probably directly involved in various organizational tasks such as the provision of a workforce, construction of an infrastructure for the hydraulic mining technology, and so

[228] Eck 1997*a*: 79–88, 135 f.

forth. Mining procurators for the *aurariae Dacicae* or various *metalla* in the Danube provinces faced a combination of responsibilities resulting predominantly from the collection of *vectigalia* from mines by proxy, to a lesser degree perhaps from the direct exploitation of mines.

6.1.4. Procuratorial Careers and Qualifications

The variety and range of responsibilities entrusted to the officials appointed to these posts raises the question as to whether the holders of these procuratorial offices came to the job well prepared and whether this resulted in specialization in their careers. The career structure of quarrying and mining procurators can only be reconstructed for a few equestrian office holders of mining procuratorships. In the case of the *procuratores argentariarum/metallorum* for Pannonia and/or Dalmatia, it is possible to glean sufficient information from the epigraphic sources.[229]

M. Antonius Fabianus[230]	• *proc. XL Galliarum et portus*	• *proc. argentariarum Pannonicarum* • *conductor portorii Illyrici*
L. Creperius Paulus[231]		• *proc. argentariarum Pannonicarum* • *idiologus* in Egypt
L. Septi[—] Petro[nianus][232]	• *archistator praef(ecti) Aegypti* • *praef(ectus) classis [—]* • *a commentariis praefecti praetorio* • *proc. Moesiae inferioris*	• *proc. argentariarum Pannonicarum* • *proc. Mauretaniae Caesariensis*
Ti. Claudius Proculus Cornelianus[233]	• *proc. ad rationibus putandas Syriae*	• *proc. metallorum Pannonicorum et Dalmaticorum* • *proc. kalendarii Vegetiani in Hispania* • *proc. regionis Thevestinae* • *proc. quadragesimae publicorum Africae*

229 Fitz 1993–5: 403–6, 695 f., 700–2, 716–48, nos. 393, 394, 399, 400, 419, 420, 423, 426.
230 *AE* 1905: 152 = *ILS* 9019, Viminacium.
231 *AE* 1915: 46, Attaleia.
232 *AE* 1958: 156, Caesarea/Mauretania.
233 *AE* 1956: 123, Lambaesis.

Ti. Claudius Xenophon[234]	• *proc. viarum* • *proc. ad epistrategiam septem nomorum et Arsinoitum* • *proc. Daciae Apulensis*	• *proc. argentariarum Pannoniarum et Dalmatiarum* • *proc. Illyricus per Moesiam Inferiorem et Dacias tres* • *subpraefectus annonae urbis* • *proc. provinciae Asiae* • *proc. Augusti ad bona cogenda in Africa*

Our information on the careers of *procuratores ferrariarum* is similarly abundant:[235]

Publilius Memorialis[236]	• *praefectus coh. III Cyrenaicae sag.* • *tribunus milit. leg. X. Fretensis* • *praefectus gentis Numidarum dilectator tironum ex Numidia lectorum* • *proc. Augusti in Africae, item M[iniciae]*	• *item (proc.) ferrariarum*
C. Caelius Martialis[237]	• *praefectus coh. I Raetorum* • *tribunus leg. XIII Geminae* • *proc. provinciae Achaiae*	• *proc. ferrariarum*
T. Statilius Optatus[238]	• *praefectus coh. I Lucensium* • *tribunus leg. VI Ferratae* • *tribunus leg. VI Victricis* • *praefectus alae Afrorum* • *proc. ad census Britanniae* • *proc. Augusti ad census Gallorum*	• *proc. Augusti ferrariarum* • *proc. Augusti ad patrimonium* • *proc. Augusti hereditatium* • *proc. a rationibus* • *praefectus annonae*
Q. Domitius Marsianus[239]	• *praefectus militum* • *proc. Augusti ad census in Gallia accipiendos provinciae Belgicae per regiones Tungrorum et Frisauonum et Germaniae inferioris et Batavorum*	• *proc. Augusti ad ferrarias* • *proc. Augusti patrimonii provinciae Narbonensis*

234 *CIL* III 7127 = *ILS* 1421 (Ephesos). *CIL* III 8042 = *IDR* II 188 (Sucidava/Dacia). *AE* 1988: 977 (Porolissum/Dacia).

235 Pflaum 1978: 135–43; Sablayrolles 1989: 157–9.

236 *CIL* XI 7554 = *ILS* 9195.

237 Corinth VIII/3, 135.

238 *CIL* VI 31863 = *ILS* 9011.

239 *AE* 1962: 183a.

T. Petronius Priscus[240]	• *praefectus coh III [–]* • *tribunus leg. VII Geminae felicis* • *praefectus alae II Pannoniorum* • *proc. Augusti annonae Ostis*	• *proc. Augusti ferrariarum*
M. Cosconius Fronto[241]	• *praef coh. I [–]* • *tribunus mil. leg. I Italicae* • *proc. Augustorum duorum ad vectigal XX hereditatium per Pontum et Bithyniam et Pontum mediterraneum et Paphlagoniam* • *subpraefectus annonae urbicae*	• *proc. Augustorum ad vectigal ferrariarum Gallicarum* • *proc. Augustorum duorum et praefectus provinciae Sardiniae*
C. Attius Alcimus Felicianus[242]	• *advocatus fisci provinciarum XI* • *proc. alimentorum per Transpadum Histriam Liburniam* • *proc. per Flaminiam Umbriam Picenum* • *vice proc. XX Galliarum* • *proc. privatae per Salariam Tiburtinam Valeriam Tusciam* • *proc. annonae provinciae Narbonensis*	• *proc. ferrariarum* • *proc. hereditatium Romae* • *curator operis amphitheatri* • *magister summarum rationum* • *magister summae privatae* • *praefectus annonae* • *vice praefectus vigilum* • *vice praefectus*

The list of offices held before and after appointment to the mining procuratorship in Gaul/Rome or Dalmatia/Pannonia does not suggest that these equestrians had acquired a specific knowledge of mining operations prior to their appointment. During their tenure, however, procurators of mining districts certainly had the opportunity to deepen their knowledge in mining affairs—provided they wished to do so. Galen refers to his visit to a copper mine at Soli on Cyprus run by a procurator and friend of his. This ἐπίτροπος τῶν μετάλλων apparently gave Galen a tour of the mines and let him collect slag and samples from the mining shafts, displaying detailed knowledge of the mining geology.[243] Whether this was a requirement of the job or whether Galen's friend had a specific interest in geology is not clear.

240 *CIL* XIV 4459. 241 *CIL* X 7583, 7584 = *ILS* 1358, 1359.

242 *CIL* VIII 23948. *CIL* VIII 822, cf. 12345 = 23963 = *ILS* 1347 add., p. CLXXIV.

243 Galen XII 214–41 Kühn, XIV 7 Kühn, cf. Millar 1992: 185.

Even so, experience gained during a tenure as mining procurator was not necessarily required for subsequent postings. Equestrian procurators were transferred to quite different branches of the administration.[244] No recognizable efforts were made by the imperial authorities to have them specialize in the management of imperial mines or mining provinces—with the exception perhaps of one distinct case: C. Iulius Silvanus Melanio. Provided Abascal and Alföldy's reconstruction of the rather fragmentary inscription at Domavium is correct, Melanio's career might have included a number of posts in the mining administration. Lines 7 to 8 of this inscription read *proc(uratori) Aug(usti) [fe]ṛṛ(ariarum) | ẹṭ ạrg(entariarum)(?) per prouin|[cias] XXIII.* He thus must have held at least two, perhaps more procuratorships of the iron and silver mines over several provinces.[245] An inscription documented at Lugdunum/mod. Lyon giving his name provides circumstantial evidence in support of this notion, even though it remains unclear what procuratorial post he held there.[246] One thus might tentatively argue a certain 'specialization' of Melanio in the field of mining administration. The majority of evidence concerning the careers of equestrian mining procurators, however, suggests such promotions were a rare exception. This can be noted as regards other procuratorships as well: even though mining undoubtedly played a dominant part in the administrative district of northwestern Spain, the known equestrian procurators for Asturia and Callaecia were not equipped with any specific knowledge of mining when promoted to this financial procuratorship (mostly after their *tres militiae*), nor would they have been able to put their knowledge to use in their following appointment.[247] There might have been a rare exception during the reign of Antoninus Pius. After his tenure as procurator of Asturia and Callaecia, M. Bassaeus Rufus advanced to the post of *procurator regni Norici*—a post undoubtedly including the

[244] See in general Brunt 1983: 48–52; Eck 2001*c*.

[245] CIL III 12732, cf. Fitz 1972: 223 f., no. 6; Devijver 1976–80: 487 f., I 126; Abascal & Alföldy 1998: 162 f.

[246] Pflaum 1960–1: 734 f., no. 276; Pflaum 1982: 67 f., no. 276: Abascal & Alföldy 1998: 163.

[247] Cf. Alföldy 2000*a*: 39–51, 63–7. L. Arruntius Maximus (AD 79): *CIL* II 2477 = *ILS* 254 = Rodríguez Colmenero 1997: 418–20, no. 587, cf. *PIR*² A 1145; Alföldy 2000*a*: 63; Pflaum 1960–1: 1047; Pflaum 1982: 117. Q. Petronius Modestus (AD 96): *CIL* V 534 = *ILS* 1379; *Inscr.It.* X 4, 34, cf. Pflaum 1960–1: 147, no. 63; *PIR*² P no. 292. D. Iulius Capito (AD 114/5): *CIL* XII 1855 = *ILS* 1380; *CIL* XII 1869 = *ILS* 6997; *CIL* II 1870, cf. Pflaum 1960–1: 175 f., no. 79; *PIR*² J 244. Sex. Truttedius Clemens (first half 2nd cent. AD): *CIL* II 2643 = *IRPL* 121; *AE* 1985: 374; *CIL* VI 2968; *CIL* IX 5931, cf. Alföldy 2000*a*: 64; Dobson 1978: 274, no. 161; Pflaum 1960–1: 567, no. 216. Calpurnius Quadratus (2nd cent. AD): *CIL* II 2642 = *IRPL* 115, pl. XCII, cf. Alföldy 2000*a*: 63 f.; Haensch 1997: 488; Pflaum 1982: 117; Tranoy 1981: 185. C. Iunius Flavianus (Hadrian/Ant. Pius): Pflaum 1960–1: 320 ff., no. 134; *CIL* VI 1620 = *ILS* 1342 in *CIL* VI 8,3, add. et corr. [C. Iulius(?)] Flaccus Aelianus (AD 161/169 or 177/180): *CIL* VI 1599 + 31828 = *ILS* 1326 = *CIL* VI 41141, cf. Pflaum 1960–1: 389–91, no. 162. P. Aelius Hilarianus (AD 185–192): *CIL* II 5678, cf. Alföldy 2000*a*: 64 f.; Domergue 1990: 289; Haensch 1997: 489; Pflaum 1981: 353 f.; Tranoy 1981: 183.

superintendence of iron mines there.[248] There is no firm evidence, however, that experience gained in mining regions such as Asturia-Callaecia or Noricum was taken into account. The evidence remains scanty at best. In general, skills or knowledge acquired in a certain branch of administration were not as important for promotion as seniority, grade, or, indeed, letters of recommendation.[249] The processing of any information on the equestrian office holders beyond rank, seniority, and current posting would prequire a centralized bureau for personnel. Whether or not the *ab epistulis*—as is implied by a poem of Statius (5.1.94–100)—was, apart from his regular duties, also concerned with managing the personnel of the imperial administration (in particular for army officers and equestrians), can not be proven satisfactorily.[250]

This same 'inefficiency' in the use of human resources can also be seen in the careers of imperial freedmen. Here the evidence is rather slim, and, apart from the career of Saturninus, we have no knowledge of the careers of freedmen procurators attested in the mining districts of the Danube provinces and the Iberian peninsula. On the procuratorial level, careers of freedmen were perhaps comparable to those of equestrians inasmuch as there is evidence for a *cursus*.[251] Like the careers of most equestrians and freedmen in the 'civil service' of the Roman empire, Saturninus' appointment to the financial procuratorship in Asturia and Galicia was not based on any previous expertise in administering mines.[252] Saturninus certainly knew how to run an imperial estate and had gained insights into special fiscal bureaux in Rome, Egypt, and Asia. He possessed experience in conducting judicial procedures in fiscal cases, a good grasp of accountancy and managing a complex organized system.[253] Saturninus, however, did not have knowledge of Roman mining technology or management.[254] In his position as financial procurator of Asturia and Galicia he undoubtedly came into contact with the mining industry and acquired some valuable knowledge along the way. According to the late second-century vexillation inscriptions from Luyego and Villalís, the freedmen mining procurators could stay in office for up to four years or more, which implies that they probably had enough time to become acquainted with the specific requirements of a mining district (cf. 5.2). It is believed that one of Saturninus' equestrian

[248] *AE* 1968: 227 = *IRPL* 7. AE 1968: 228 = *IRPL* 3. *CIL* VI 41278, cf. Nony 1970: 196; Tranoy 1981: 183; Alföldy 2000*a*: 65.

[249] Fronto, *Ad M. Caes.* 5.37, p. 87 Nab, cf. Pflaum 1971: 349–51; Weaver 1972: 268 f.; Birley 1992: 25 f.; Eck 1995*e*: 138 f.; Eck 1997*a*: 98 f.

[250] Birley 1992: 23 f. 41–54; Eck 1995*e*: 139 f.

[251] Fronto, *Ad M. Caes.* 5.37, p. 87 Nab; Weaver 1972: 267–81; Boulvert 1974: 119–80; Eck 1997*a*: 105.

[252] Eck 1997*a*: 97; Eck 2001*c*.

[253] On jurisdiction of procurators, cf. Brunt 1990*b*.

[254] Christol 1999: 243, contra: Hirschfeld 1905: 162.

predecessors in this office, Pliny the Elder, had certainly gained a deep enough insight to provide a reasonably accurate description of Roman mining procedures and of the financial profits made from these mines.[255] This knowledge might have been decisive in choosing Saturninus for the post in Vipasca.[256]

The picture gained from this evidence implies that equestrian procurators, and probably those of freedman status as well, possessed good general knowledge of fiscal jurisdiction and experience in managing a complex organization when appointed to mining procuratorships. Freedmen procurators in particular had sufficient time to gain experience in dealing with the special organizational problems of mines or quarries. Moreover, their longer tenure of office guaranteed a certain continuity of management.[257] Given the numerous posts equestrian procurators held during their careers, we might suspect that the average tenure of a mining procuratorship was around two or three years.[258] Even so, the freedmen and equestrian procurators were unlikely to profit from their expertise by taking up similar posts after their mining procuratorships.

6.2. SUBALTERN STAFF

In light of the rather short tenures of the equestrian procurators, subaltern officials likely guaranteed continuous administrative support for extractive operations.[259] The available epigraphic evidence rarely helps us to define the responsibilities and precise functions of these subaltern officials within the organization of mining and quarrying operations. Only in the case of Mons Claudianus does the internal correspondence between imperial officials yield sufficient information to allow us to assess their role within the administration of the quarries. Athenodoros, a *tabularius* for the procurator Ulpius Himerus (AD 152/3), appears in a number of ostraca from Mons Claudianus, and helped to manage the stock of quarried blocks.[260] According to Cuvigny, Athenodoros remained in regular contact with the smaller quarries at Tiberiane through the *curator* Nepheros, a member of the Roman army, and the *tesserarius* Kallistratos, the latter presumably a civilian (despite his occupational grade). While Nepheros writes to Athenodoros about chronic problems

255 Pliny, *NH* 33.66–78, cf. Syme 1969; Haensch 1997: 403 fn. 23, 489; Alföldy 2000*a*: 45 fn. 76.
256 For a further procuratorial career of a freedman, cf. Aurelius Marcio (4.1.3).
257 Brunt 1983: 51; Eck 1997*a*: 93, 104 f.
258 Eck 1997*a*: 97.
259 Eck 1997*a*: 104.
260 *O.Claud.*inv. 6483, cf. Bülow-Jacobsen 1996: 61.

with water supplies and provisions, Kallistratos reports on the arrival and departure of men and animals, and on the distribution of food. The ostraca reveal the position of the *tesserarii* within the organizational hierarchy: Cuvigny argues that the *curator* and the *tabularius* communicated on the same level, while the *tesserarius* Kallistratos appears as a subaltern to Athenodoros.[261] Moreover, the water distribution lists attest the post of a *subtesserarius*/*ὑποτεσσαράριος*, an assistant to the *tesserarius*.[262] *Tesserarii* were perhaps recruited from amongst the κιβαριάται of the φαμιλία, as the appointment of a former κιβαριάτης to the function of *tesserarius* suggests.[263] Whether or not the γραμματεῖς mentioned in the Claudianus ostraca are to be equated with the *tesserarii*, as Cuvigny suggests, is uncertain.[264] Although *tesserarii* do appear as addressees in a number of receipts for pay advance and debt of φαμελιάριοι, their exact function remains unknown.[265]

The provision of food for the φαμελιάριοι was apparently the main responsibility of the κιβαριάται of the *familia Caesaris* (cf. 6.1.2.1). There only was one κιβαριάτης at Mons Claudianus for the φαμελιάριοι, who presumably had been appointed by the quarrying administration. He managed the salary and rations which the φαμελιάριοι received. His stocks were replenished on a monthly basis, and used for advance payments to the φαμελιάριοι or handed to their creditors. The stocks of the κιβαριάτης, however, were not always sufficient to meet the demands for advance payments. In these situations 'services parallèles' could supply additional resources: the *caesarianus* Gaion and the *laccarius* Serapion seem to have provided supplementary advance payments. In particular, the κιβαριάτης Menelaos encountered difficulties in supplying the necessary grain for indebted φαμελιάριοι in the second half of AD 137 and received an unknown quantity of *matia* and half-*artaba* from the *laccarius* Serapion. Perhaps in response to this supply crisis, the function of an ἀρτοδότης βοηθὸς κιβαριάτου, a 'distributor of bread' and assistant to the κιβαριάτης, was created in AD 139.[266] Βοηθοὶ κιβαριάτου and even an αἰούτωρ κιβαριάτου, assistants to the κιβαριάτης, are attested at Mons Claudianus and Tiberiane.[267]

Another official named in the receipts for advanced payments and in the ἐντολαί of the παγανοί at Mons Claudianus is the οἰκονόμος.[268] The οἰκονόμοι

[261] Cuvigny 2000*a*: 65.

[262] They receive the same amount of water as the παγανοί, cf. *O.Claud.*inv. 1538, cf. Cuvigny 2000*a*: 64.

[263] *O.Claud.*inv. 1158, cf. Cuvigny 2000*a*: 65.

[264] Cuvigny 2000*a*: 66. For γραμματεῖς at Wadi Hammamat, cf. Kayser 1993: nos. 12, 18.

[265] *O.Claud.* 9, 22, 485, 498, 563, 576, 596.

[266] Cuvigny 2000*a*: 59–61, 66.

[267] Alkimos (AD 110–111): *O.Claud.* 3–8. Sarapodoros (AD 145): *O.Claud.* 520–39; *O.Claud.* inv. 7366. αἰούτωρ: *O.Claud.*inv. 3229, cf. Cuvigny 2000*a*: 60.

[268] *O.Claud.* 510, 551, 618. Cuvigny 2000*a*: 62.

dealt mainly with equipment issues. They also were involved in supplying bread and grain at a time when the office of the *κιβαριάτης* had been abolished (after AD 147).[269] The function of an *οἰκονόμος* is not easily described. Its Latin equivalents are *vilicus* or *dispensator*.[270] In the case of the *οἰκονόμος* Isidorus, who also is named as *caesarianus*, the translation as *dispensator* or expansion to (*Καίσαρος*) *οἰκονόμος* appears more likely, particularly since this function is quite often attested in imperial quarries.

Inscriptional evidence tells us little about the subalterns in most other imperial extractive operations except their title. Any study of these functional grades therefore must be dependent on information collected from other branches of the imperial administration. The staff of the *procurator aurariarum* at Ampelum includes a wide range of subaltern personnel. Apart from the *procuratores* (freedmen and equestrians), we find *dispensatores*, *tabularii*, an *ab instrumentis tabularii*, *adiutores tabulariorum*, a *subsequens librariorum*, an *a commentariis*, and *librarii* at Ampelum, the administrative headquarters of the *auraria Dacicarum*. The *librarii* are military personnel, whereas the former are slaves and freedmen of the *familia Caesaris*. The diversity of the procuratorial staff is reminiscent of the personnel attached to presidial procurators, or financial and patrimonial procurators. The functions of the *officiales*, i.e. the members of the procuratorial *officium* at Ampelum, certainly indicate complex accounting procedures, as well as the filing and copying of various documents. These administrative operations undoubtedly called for an archive of some sort, in which documents relating to mining operations could be stored. The composition of the archive might be deduced from the administrative and judicial topics a *procurator aurariarum Daciarum* at Ampelum might have had to cover. The gubernatorial archive, as reconstructed by Haensch, may function as a template for a procuratorial archive.[271] As noted above (cf. 6.1.3.3), the *tabulae ceratae* found in one of the mining shafts at Alburnus Maior suggest that at least part of the mines or individual mining shafts were contracted out, whereas other *aurariae* in Dacia were perhaps operated directly by the state. Furthermore, most of these documents are work or sales contracts, which in case of a legal action by one of the contract parties formed the basis of the legal deliberations by the magistrate or official hearing the case. One may assume that court cases involving the gold mines were heard by the *procurator aurariarum* at Ampelum (cf. 6.1.1).

[269] *O.Claud.*inv. 2238, 8175, cf. Cuvigny 2000*a*: 63.

[270] *O.Claud.* 551. Cuvigny 2000*a*: 63 f.

[271] Haensch 1992.

The Vipasca tablets provide a limited account of the procurator's duties at the *metallum Vipascense*, where mines were sold to interested individuals or *societates*, and monopolies for a range of services were farmed out. These activities must have resulted in a set of written documents: a record of mining pits listed the current owners, as the acquisition of mines had to be reported to the procurator.[272] We may also assume that copies of contracts were stored and the date of their conclusion was noted. This procedure was undoubtedly necessitated by the procurator's judicial responsibilities within the mining territory. Consequently, the minutes of procuratorial decisions on various matters (not only judicial) concerning the mines and other activities must have been filed. We might furthermore expect that petitions (?) and letters to the procurator were stored. The presence of an *a commentariis* at Ampelum suggests that these procuratorial records were known as *commentarii*. Haensch has shown that gubernatorial *commentarii* not only included legal decisions but covered the whole range of the governor's daily activities, forming a log of his term in office. The governor's activities may not have been directly copied to the *commentarii*, but may have been organized according to matter and importance in different volumes. This procedure certainly facilitated the retrieval of procuratorial decisions for specific fiscal cases.[273]

In addition to these central documents, registers of incoming and outgoing letters, as well as records of correspondence with the emperor, the towns and other officials, are attested in the provincial administration, and probably existed for mining procurators as well.[274] Moreover, the various payments made to the *fiscus* by the contractors or occupants of mining pits (cf. 7.1) were probably noted in accounts and receipts for these payments written by the clerks. Further financial gains and expenses from the mining ventures may have been entered in a balance sheet. In mining ventures or quarries directly operated by the state, payments to workers and contractors of quarry work had to be accounted for. The quantity and quality of quarried blocks appears to have been recorded, as the quarry labels on numerous granite and marble blocks of different origin attest (cf. 7.2). Working contracts comparable to those known from Alburnus Maior must have been archived, and, in the case of the quarries in the Eastern Desert, the expenditures for the supply with water, tools and provisions certainly had to be accounted for as well. The quarrying administration there probably also kept a record of animals requisitioned for transport duty (cf. 6.1.2.5). Moreover, the condemnation of criminals to mines

[272] *LMD* ll.21 f.
[273] Haensch 1992: 237–45 with sources and bibliography.
[274] Haensch 1992: 245–63.

and quarries perhaps was partly recorded at procuratorial offices as well. Haensch cautiously assumed that the soldiers guarding the alabaster quarries in Egypt kept records of the date and length of a convict's sentence, and reported to the *praefectus Aegypti* any discharge from the quarries.[275]

Any attempt at reconstructing an archive must remain speculative, as we have no documents which originate from the archive of specific mining or quarrying sites. The mere existence of an archive, however, is implied by inscriptions of subaltern officials named *tabularii*, who handled *tabulae*, writing tablets. The known *tabularii* are predominantly *liberti*. The *tabularii*, a 'basic clerical post in a departmental office or *tabularium*', certainly functioned as keeper of documents and controlled part of the financial transactions.[276] They wrote the receipts for payments and compiled the accounts for creditors and debitors of an imperial domain. Furthermore, they appear to have reported the financial situation of the imperial domain to their superiors.[277] At Mons Claudianus this is partly exemplified by the *tabularius* Athenodoros; he appears to be responsible for the management of the marble stock, is informed about water and food supply, as well as the number of personnel present at Mons Claudianus and Tiberiane.[278] Athenodoros apparently functioned as the proxy of the procurator in the Eastern Desert. Other *tabularii*, however, connected with an administrative bureau at Ampelum or the provincial administration at Lugdunum, might have kept the books on expenditure and revenue from *conductores* of *vectigalia* or extractive operations, basically acting as the accountants of the mining procurator. By analogy with the function of the *dispensator*, the actual tasks of a *tabularius* might have varied depending on which mine or quarry he was posted to.

The *a commentariis*, mostly imperial freedmen, dealt with the *commentarii*, a set of documents containing legal decisions and letters by the provincial governor, basically protocols, and other administrative records (*acta*). Their responsibilities also covered the safe custody of original documents. The documents had to be available to make copies which the *a commentariis* was required to authenticate.[279] The two documented *a commentariis* for the *aurariae Delmatae* and for the *aurariae Dacicae* probably both filled such a purpose. The latter, however, may have been attached to the bureau of the financial procurator for Dalmatia at Salona/mod. Split. The administrative affiliation of *L(ucius) Vale(rius), Aug(ustae) l(ibertus) a com(mentariis)*,

275 Haensch 1992: 281. 276 Weaver 1972: 241.

277 Hirschfeld 1905: 58–64; Sachers 1932; Boulvert 1970: 420–5; Weaver 1972: 241–3; Noeske 1977: 309; Haensch 1997: 725 f.

278 Cuvigny 2000*a*: 64 f.

279 Premerstein 1901; Hirschfeld 1905: 63; Boulvert 1970: 425–7; Weaver 1972: 241–3; Noeske 1977: 310. On gubernatorial archives, cf. Haensch 1992.

known from stamps on tin ingots from the Port Vendres II shipwreck, is uncertain; he may have supervised the export of goods.

The following subaltern officials are documented exclusively at Ampelum: the *adiutores tabulariorum* were subordinate to the *tabularii* and, as their title suggests, aided the *tabularii* in their tasks. The junior clerical grade of *adiutor*, mostly imperial slaves, provided an opportunity for promotion to the office of *tabularius* later on in their career and after their manumission.[280] The *ab instrumentis tabularii* was in charge of the *instrumenta*, the accounts on rents from the leases and all documents related to the imperial domain or the mining district.[281] The task of the *librarii* was probably to keep files and participate in managing the financial accounts. More likely they were used as regular scribes.[282] In the case of Ampelum, the *librarii* were seconded from the Roman army on the orders of the provincial governor. One inscription also names a *subsequens librariorum*, an attendant to the *librarii*(?), a term hardly known in administrative contexts.

Apart from the limited use of army personnel, the subalterns were members of the *familia Caesaris*, of which the *tabularius* and *a commentariis* belonged to the highest intermediate clerical grades within the imperial administration, held by freedmen as well as slaves. Belonging to the same grade, but perhaps less concerned with archival duties, was the *dispensator*. The Roman jurist Gaius provides a definition for the term 'dispensator': 'servi, quibus permittitur administratio pecuniae, dispensatores appellati sunt'. Thus, they were basically slaves placed in charge of the distribution of monies.[283] Their main task was the collection and distribution of payments, which necessitated the formal approval of the imperial procurator.[284] Gérard Boulvert observed that imperial *dispensatores* could make contracts on behalf of the *fiscus*, although only in a limited capacity and at the request of the procurator.[285] Given our lack of knowledge about this office, the actual function of *dispensatores* within the mining administration is not clear. One might imagine, for example, a *dispensator aurariarum Dacicarum* making out or receiving payments from *conductores* of services and fee collections, as well as from the occupants of Dacian gold mines, whatever the contractual basis of the latter's involvement in mining operations (cf. 7.1.2).[286] The documents from Mons Claudianus provide us with some idea of their purpose in the

[280] Sachers 1932: 1972; Weaver 1972: 231–240; Noeske 1977: 309.
[281] Hirschfeld 1905: 64 with fn. 1; Steinwenter 1916; Boulvert 1970: 426; Noeske 1977: 309 f.
[282] Bilabel 1927; Noeske 1977: 310 f.; Haensch 1997: 722.
[283] Gaius, *Inst.* 1.123.
[284] Aubert 1994: 197 f. with further bibliography; Haensch 1997: 726.
[285] Boulvert 1970: 429–33; Weaver 1972: 202; Aubert 1994: 197 with fn. 296.
[286] Noeske 1977: 310.

context of the quarrying administration. Provided the term *οἰκονόμος* can be equated with *dispensator*, these officials were involved in making payments to the *παγανοί*, helped supply bread and grain to the *φαμελιάριοι* and were partially responsible for quarrying equipment.[287] The *dispensatores* in other quarries may have performed similar tasks. Even so, differences in geological and topographical constraints and the choice of exploitation regimes certainly influenced the range of tasks required of a *dispensator*. In the case of *dispensatores* recorded at Dokimeion and Krokeai, one may speculate whether they were responsible for making payments after receiving marble blocks by the *caesura*-holders (cf. 7.2.6). Perhaps the *dispensatores* at Ampelum and elsewhere made out payments in the context of such work contracts.

A further official attested in a number of imperial mining districts is the *vilicus*. On the basis of the information which follows the title *vilicus* in the votive inscriptions to Terra Mater from Ljubija, these imperial slaves appear to have run the *officina ferrariae* (cf. 4.2.1). The *officina ferrariae* is understood to be the smelting furnace for metal ore. According to the *lex metallis dicta*, the *coloni* were obliged to submit the extracted ore to such *officinae*, most likely under the supervision of *vilici*. The title of the seven (*contrascriptor*? *vilicus*?) *officinarum* from Sočanica indicates that more than one furnace might have been run by imperial officials.[288] Private and imperial *vilici* also appear at Carrara in Italy where they probably held other functions (cf. 4.2.2). In general, *vilici* not only played an important part in the financial administration of provinces, such as in the collection of taxes and other revenues, but were also used in more general managerial functions.[289] The managerial tasks *vilici* are known to have carried out at private landed estates might provide an appropriate basis for assessing the role of *vilici* in mining and quarrying environments. Writers on Roman agriculture laid out the managerial functions of a *vilicus*: he was to keep track of the yields of all produce, the work done by members of the household in and out of the estate, register all the transactions in produce and money and the sale of produce. He was also required to buy equipment and slaves, hire additional free labour where required, and to let contracts for specific work to be done.[290] One may assume that a *vilicus* at Luna/Carrara or a *vilicus officinae* fulfilled similar functions, although a number of tasks might have been carried out by their superiors. Like his private counterparts, the imperial *vilicus* attested at marble quarries

[287] Cuvigny 2000*a*: 62–4.

[288] Noeske 1977: 307–8. For an alternative view on the role of the *vilicus ferrariarum*, cf. Dušanić 1977: 84 with fn. 85, 88 with fn. 223.

[289] Schneider 1958; Boulvert 1970: 433 f.; Weaver 1972: 202; Noeske 1977: 306–8; Aubert 1994: 136–8, 169–75.

[290] Aubert 1994: 170 f. with sources.

of Carrara may have kept track of the workforce, supplied workers and equipment, and monitored extractive operations. The *vilicus officinae* at Ljubija, it may be assumed, was to control the ore brought to the furnace and the refined produce leaving it, perhaps setting aside part of the produce for the Roman state.

Most of the subaltern officials probably remained at their posts or at the same location for a number of years. In comparison to the rather short periods spent by imperial freedmen and equestrians in the procuratorial office (cf. 6.1.4), a certain Callimorphus, for example, is attested at Ljubija as *vilicus* in AD 201 and in AD 209, which indicates that he remained in the same office for eight years.[291] Moreover, as the evidence from Mons Claudianus suggests, administrative officials were promoted through the 'ranks' while remaining at the same site.[292] The longer tenure of office by these subaltern officials certainly allowed them to obtain a certain amount of expertise in mining and quarrying administration. Hence, the procurators could rely on an experienced staff, which was well equipped to process the exchange of information between the procuratorial office and the extractive operations or the provincial administration. The ostraca at Mons Claudianus provide evidence for the transfer of experienced officials, e.g. ἀρχιτέκτονες, from one quarry to another within the quarrying territory of the Egyptian Eastern Desert.[293] One might expect that on this organizational level, the φαμελιάριοι could specialize in their functions, with their extensive knowledge being put to use effectively by keeping skilled individuals within the sub-organization. Consequently, a large group of members of the *familia Caesaris* probably remained in the Egyptian Eastern Desert, with most of them not being promoted to higher posts outside of the quarrying administration.

6.3. THE FUNCTION OF PROCURATORS AND IMPERIAL PERSONNEL

The inscriptions of the mining administration at Ampelum perhaps provide the most complete picture of the size and diversity of the administrative staff at the disposal of the procurator. At the other end of the spectrum, the ostraca from Mons Claudianus document the activities of imperial subaltern officials,

[291] *AE* 1973: 411 = *ILJug* 779 (AD 201). *AE* 1958: 63 = *ILJug* 157 (AD 209).
[292] The appointment of the κιβαριάτης Magios to the rank of *tesserarius* is a good example of this phenomenon, cf. Cuvigny 2000*a*: 65.
[293] *O.Claud.* 17, 143.

not at the administrative headquarters, but at an individual quarry. Owing to the sparsity of written evidence for adminstrative bureaux at other imperial mines and quarries, it is prudent not to assume that these sites boasted an administrative staff of comparable size and diversity as Ampelum or Mons Claudianus. Geological and topographical constraints, technological demands, as well as the type of exploitation, i.e. the decision to contract out or sell operations or have them exploited by a workforce, might have significantly influenced the size and make-up of an administrative bureau.

The organizational measures could leave a considerable 'paper trail' which was dealt with by the equestrian and freedman procurators and their staff. The numerous functional titles of subalterns attested at administrative headquarters like Ampelum or at the quarries and mines themselves appear to cover most of the accounting, registration, recording, filing, or authentication tasks connected with mining or quarrying activities. The subaltern clerks and accountants guaranteed administrative continuity as procurators came and went. The short tenure of mining procuratorships held by equestrians and imperial freedmen, which possibly lasted between one and four years, does suggest that these office-holders would have acquired some experience in the administration of mines and quarries. However, these procuratorships did not require an in-depth knowledge of mining or quarrying procedures. In the published ostraca from Mons Claudianus, the procurator is rarely noted and is mostly involved in supplying the workforce with the required provisions, materials, and animals, as well as keeping track of the marble-stock. His main function was probably to compile and pass on orders and requests for provisions and marble. The planning and execution of operative processes (i.e. the quarrying operations as such), were probably the task of the (civilian or military) engineers and foremen on the spot.

The mining procurator of the Vipasca mining district appears to have been even less involved in the technicalities of the everyday extraction process. By supervising the *occupatio* of mining pits, the burden of organizing and supplying the process of extraction, was 'outsourced' to civilian entrepreneurs. The procurator's responsibility was hence focused on the implementation of the regulations and the legal enforcement of a continuous and uninterrupted exploitation of the mines. No in-depth knowledge of geology or extraction technology was necessary, as the implementation of the mining regulations was likely controlled not by the procurator himself but by his subordinates. Specific activities like the smelting of ores, for example, were carried out under the supervision of *vilici*. In the case of extractive operations being run directly by the Roman state, mining or quarrying procurators could be entrusted with the supply of human and material resources to these operations, as well as hauling and storing the extracted material. For part of these organizational tasks the procurators could rely on the provincial administration. In Roman Egypt the requisition of transport animals, as well as the exaction of grain from the provincial population for the workforce, soldiers and animals, profited from the well-established administrative mechanisms under the control of the *praefectus Aegypti.* This perhaps applied to other provinces as well: Simitthus apparently received convicts condemned by provincial governors *in metallum.* In the initial stages after the conquest of northwestern Spain or Pannonia, the officials in charge of the mining zones were even provided with an ideal source of labour by the provincial authorities, as the indigenous population was forcibly resettled to allow the quick start-up of the alluvial gold-mining operations.

In the light of the evidence for iron-mining in Gaul and the Danubian provinces, the Roman state may have contracted iron mines to private individuals and companies in return for a *vectigal*, the collection of which was contracted out to *conductores.* The epigraphic documentation for *procuratores (vectigalis) ferrariarum* implies that the *conductores ferrariarum* were supervised by equestrian officials; the latter possibly covered vast provincial 'conglomerates', such as the provinces of 'Gaul' and perhaps part of the Danubian provinces.

7

Private Partners to Imperial Operations: *Occupatores/Coloni* and *Conductores*

The organization of extractive operations in remote areas required careful planning by the authorities. Where circumstances permitted, all possible steps were taken to reduce the burden on the imperial officials. In line with the organizational principle central to Roman administration, mining or quarrying operations or particular tasks were handed over to private 'entrepreneurs' or corporations. This principle certainly had its benefits: organizational issues, such as the supply and pay of a workforce or the collection of rents or fees from the occupants, could be contracted out to private individuals who thus shouldered a considerable amount of the organizational burden. The specific arrangements, mostly related to work procedures, are in part reflected in numerous quarry labels on blocks and columns excavated at Rome's marble yards and discovered at the quarries. Mould marks or stamps on lead ingots are less revealing for mining arrangements involving contractors. Significant insights into a particular disposition of private 'partners' in a public context are provided by the second Vipasca tablet (*LMD*).

7.1. MINES, *OCCUPATORES/COLONI* AND *CONDUCTORES*

Unlike imperial quarries, the contractual involvement of private 'entrepreneurs' or companies in imperial mining operations is rarely attested. Labels, stamps, and marks on ingots are believed to provide information on this issue; yet, as we shall see below, there are significant limitations to their value as a source. One thus has to rely on the fragmentary epigraphic record for further insights. A unique glimpse of possible 'outsourcing' dispositions is provided by the Vipasca tablets which demonstrate how the organizational arrangements pertaining to the mining infrastructure were handed over to private *occupatores/coloni* or *conductores*.

7.1.1. The Vipasca Tablets: *Occupatores* and *Coloni*

The *lex metallis dicta* (*LMD*), the second tablet discovered at Vipasca in 1906, provides the most detailed insight into mining arrangements within an imperial *metallum*. The *LMD* probably once covered three tablets, of which only the second has survived. The loss of the adjoining columns of the remaining bronze tablet complicates the reconstruction of its original legal content.[1] The first lines of the recovered text (ll. 1–4) introduce the reader of the *LMD* to the consequences of not paying the *pretium* to the mining procurator in time.[2] The following lines deal with the 'sale' of silver mines and the restrictions laid down for the *occupatores* and *coloni* as regards the formation of '*societates*' and the exchange of mines amongst *coloni*. The section also lays out security measures in order to ensure the unhindered exploitation of the mines. The legal procedure of gaining access to mining pits is not mentioned in this part of the regulations, but some elements of this transaction are reflected in the surviving passages.

Two opposing views on the legal nature of the mining regulations have emerged since the tablets were discovered. Some scholars, most recently Dieter Flach and Claude Domergue, assumed that the ore extracted was partitioned between the *fiscus* and the occupant of the mine. Others, such as Alvaro d'Ors or Antonio Mateo, understood the text to mean that the occupant gained full ownership rights over the mine and its contents after they had paid the money for half of the mine to the *fiscus*.[3] These opposing views can be traced to distinct interpretations of the term *pars dimidia ad fiscum pertinens* (*pdfp*) and varying interpretations of the acquisition of 'proprietas' over *pdfp* by the occupant of the mine. The passages of *LMD* (= Vip. II) at the centre of controversy are §§1, 2, and 5 of the second column, which are paraphrased as follows:

- The first four lines of the remaining text (§1) centre on the payment of a price (*pretium*) for the half belonging to the *fiscus*: the occupant of the mine is required to pay the *pretium* for the half of the *fiscus* to the procurator before smelting ore. If the occupant does not pay the *pretium* before smelting ore he loses his part (*pars occupatoris*) and the whole mining pit (*puteus universus*) is sold by the procurator. Anyone who reports an occupant to the authorities for smelting ore prior to paying the *pretium* is to receive one fourth of the sum owed.[4]

[1] Flach 1979: 400; Domergue 1983: 111 f.

[2] For the edition of the text, cf. Flach 1979.

[3] A lucid summary of the debate and criticism of the position of Flach and Domergue is provided by Mateo 2001: 87–131, whose arguments are supported by Domergue 2004.

[4] *LMD* ll. 1–4: 'Aug(usti) praesens numerato. Qui ita non fecerit et convictus erit prius coxisse venam quam pretium, sicut su | {su}pra scriptum est, solvisse, pars occupatoris commissa

- The next three lines (§2) refer to silver mines (*putei argentarii*): these are to be run according to an unknown regulation set out in the preceding column(s) of the law. Ownership (*proprietas*) of the part belonging to the fiscus is acquired by offering a price first and paying 4,000 sesterces to the *fiscus*. These prices are in accordance with the 'liberalitas' of emperor Hadrian.[5]
- A further passage (§5) regulates procedure in the event that a mining pit (*puteus*) sold by the *fiscus* has not been worked continuously for six months. The right of occupation (*ius occupandi*) passes on to a third party. As usual (*ex more*), once ore is extracted from the *puteus* by the new occupant, half of it belongs to the *fiscus*.[6]

The first few lines of the *LMD* provide insight into the constituent parts of a *puteus* or mining pit: it can be inferred from §1 that a mine (*puteus universus*) was partitioned into a *pars occupatoris*, the part of the occupant, and a *pars dimidia ad fiscum pertinens* (*pdfp*), a 'half' owned by the *fiscus*, i.e. the Roman state.[7] The *pdfp* is not to be understood as a spatial or territorial entity within a mining shaft or tunnel. The use of 'dimidia' in the spatial context would not make much sense and terms like *pars fisci* or *pars occupatoris* would suffice to distinguish two separate areas of one mine. Therefore, *pdfp* must refer to a measurable entity, a dividable mass, most likely the contents extracted or to be extracted from the mine. This interpretation is corroborated by §5: as soon as ore is discovered in the *puteus* half of it becomes the property of the *fiscus*.[8] One therefore may argue that the *pars dimidia ad fiscum pertinens* must designate half of the ore. In consequence, the other half of ore became the possession of the occupant.[9]

esto et puteum universum proc(urator) metallorum | vendito. Is qui probaverit ante colonum venam coxisse quam pretium partis dimidiae ad fiscum pertinen|tis numerasse, partem quartam accipito.' For text and translation cf. Flach 1979: 403, 405.

[5] *LMD* ll. 4–7: 'Putei argentari ex forma exerceri debent, quae | hac lege continetur; quorum pretia secundum liberalitatem sacratissimi Imp(eratoris) Hadriani Aug(usti) obser|vabuntur ita ut ad eum pertineat proprietas partis, quae ad fiscum pertinebit, qui primus pretium puteo fecerit | et sestertia quattuor milia nummun fisco intulerit.' For text and translation, cf. Flach 1979: 403, 405.

[6] *LMD* ll. 11–12: 'Puteum a fisco venditum continuis sex mensibus intermissum alii occupandi ius | [es]to ita ut, cum venae ex eo proferentur, ex more pars dimidia fisco salva sit.' For text and translation, cf. Flach 1979: 403, 405.

[7] Flach 1979: 417 with older bibliography; Domergue 1983: 124–6; Andreau 1990: 102.

[8] *LMD* ll. 11–12: 'Puteum a fisco venditum continuis sex mensibus intermissum alii occupandi ius | [es]to ita ut, cum venae ex eo proferentur, ex more pars dimidia fisco salva sit.' Flach 1979: 416; Domergue 1983: 139. For a different interpretaion, cf. Mateo 2001: 131–9.

[9] Flach 1979: 415 f., argued that the *proprietas* was a right of possession rather than a right of ownership, as part of the extracted ore still was handed to the *fiscus* after the *pretium* had been paid. In Flach 1989: 135 f., and Flach 1990: 120, he then changed and refined his view claiming that *proprietas* must have meant ownership of the *pars dimidia ad fiscum pertinens*, while the *pars occupatoris* was only the *possessio* of the occupant.

Hence, the interpretation of these paragraphs allows one to identify the constituent elements of a mine at Vipasca as follows. The mine (*puteus*) consisted of a *pars occupatoris*—presumably including any mining structures, shafts, and tunnels made or constructed by the current occupant, as well as half of the ore—and a *pars ad fiscum pertinens*, comprising half of the ore. Provided this reconstruction is correct, both entities are integral parts of the *puteus*. However, the *pdfp* only becomes measurable upon extraction of the ore from the *puteus*. This legal construct—upon which the *fiscus* based its claim of ownership, 'proprietas', of the *pars dimidia*—only applied when the occupant hit an ore lode (§5). In return for a *pretium* the fiscus transfered 'proprietas' of its half to the occupant (§2). This transfer affected the ores in possession of the occupant as well: with payment of the *pretium* the occupant apparently was free to smelt all the ore extracted from his *puteus* (§1).

Given its importance in *LMD*, the term 'pretium' merits further deliberation. In §8 the *coloni* are permitted to sell *partes puteorum*, which have been sold by the *fiscus* and for which the *pretium* has been paid, for the highest price attainable. The *pars putei* must be the former *pdfp*, which upon payment of the *pretium* becomes the property of the *colonus*. This suggests that with one single payment of the *pretium* the *fiscus* surrenders any rights to the ore, yet strictly limits any future transfers of ownership rights to local *coloni*. In fact, with the sale of the *pars dimidia* now owned by the *colonus*, all of the mine, including the *pars occupatoris*, can be transfered into the ownership of another *colonus*.[10]

The process by which an agreement on the *pretium* between *fiscus* and a potential buyer was reached is only attested for certain types of mine. *LMD* §2 allows limited insights regarding the *pdfp* of silvermines and their sale by the fiscus: the 'proprietas' of the *pdfp* is aquired by the person who first offers a price (*pretium*) for the *puteus* and pays 4,000 sesterces.[11] In my view, the relevant passage on the silver mines might refer to an auction rather than a simple sale.[12] Possible indications are provided by the *lex metalli Vipascensis*, which mentions auctions held within the *fines metalli Vipascensis*, some of them by the *procurator metallorum* on behalf of the emperor.[13] Was the *proprietas* of silver mines auctioned off as well? The wording of the passage might suggest just that: 'ita

[10] *LMD* ll. 20–1: 'Colonis inter se eas quoque partes puteorum, quas | a fisco emerint et pretium solverint, vendere, quanti quis potuerit, liceto.' For the interpretation of this passage, cf. Mateo 2001: 161–6.

[11] *LMD* ll. 5–7.

[12] For a different view, cf. Flach 1979: 415, 420–2, with discussion of older opinions bibliography on this issue; Domergue 1983: 127 f.; Mateo 2001: 156–60; Domergue 2004: 228.

[13] *LMV* ll. 1–3: 'Conductor ea[rum stipulationum, quae ob auctio]|nem intra fines metalli Vipascensis fient, exceptis iis, quas proc(urator) metallorum iu[ssu imp(eratoris) faciet, centesimam a vendito]|re accipito.'

ut ad eum pertineat proprietas partis, quae ad fiscum pertinebit, qui ***primus*** pretium puteo fecerit ***et*** sestertia quattuor milia nummum fisco intulerit.' It stresses that the *pretium* has to be offered first in order to acquire *proprietas* over the *pars dimidia ad fiscum pertinens*—a provision which does not make much sense in a regular contract of sale. An auction better fits the context; Domergue and Flach pointed out that the formula *pretium facere* is used in literary sources in the context of auctions.[14] Both scholars dismissed this possibility as the term 'qui primus pretium puteo fecerit' does not seem to suggest a regular auctioneering procedure. However, the auction might have been held similar to a 'Dutch auction', where the auctioneer starts with the highest price, slowly reducing it step by step, until someone makes the first bid.[15]

Furthermore, the special regulation as regards silver mines does not necessarily specify whether the *pretium* for silver mines is the 4,000 sesterces payable to the *fiscus*, as has been assumed.[16] The conjunction '*et*' separates the act of offering a *pretium* for the silver mine as the first bidder from the actual payment of 4,000 sesterces to the *fiscus*. An explanation for this distinction might be that the *fiscus* wanted 4,000 sesterces to be paid immediately, while the *pretium* (minus the 4,000 HS?), like the *pretium* for copper mines, would be paid as soon as the first ore was to be smelted.[17] Moreover, the 4,000 sesterces had to be paid whatever the price achieved at the auction was, and hence might be understood as the minimum price or 'knockdown' price for which the *fiscus* was willing to give the *pdfp* away. Hadrian's 'liberalitas' might therefore have been to reduce the sum to be paid immediately to the *fiscus* to 4,000 sesterces, lowering the benchmark for possible 'buyers'. As soon as silver ore was extracted, it was easier to take up a loan with the local banker (cf. *LMV* ll. 1–9) for the rest of the *pretium*.[18]

This auctioning procedure was perhaps limited to *putei argentarii* owned by the *fiscus*, in which silver ore had already been discovered or even extracted—hence the adjective 'argentarius'. Any potential buyer of a silver mine would have been highly reluctant to pay the considerable sum of 4,000 sesterces in advance without having some form of guarantee that silver ore would be found

[14] Martial 1.85.7, 7.17.8, 9.59.20; Varro, *Rust.* 2.9.7; Plautus, *Pers.* 5.86; Propertius 4.5.29. Domergue 1983: 126 f.; Flach 1990: 120 f.

[15] Whether or not such 'Dutch auctions' were known in the Roman empire is of course debatable; yet given our limited knowledge of Roman auction procedures, it cannot be completely discounted. The auctioning procedures of the period are not well known, but appear to resemble the more common model with the highest bidder being awarded the auctioned property, cf. Leist 1896; Kaser 1971: 547 with fn. 16. For a different view, cf. Flach 1979: 420 f.; Domergue 1983: 127.

[16] Flach 1979: 415, 420.

[17] Fitzler 1910: 90 with fn. 4, with Flach 1979: 421, 423 on older bibliography.

[18] For a different view Flach 1979: 414, 423; Domergue 1983: 134–7.

in sufficient quantities. Thus, the existence of silver ore was a crucial prerequisite for the sale of the *puteus*. Provided this conclusion is correct, one could acquire 'proprietas' of *pdfp* and therefore of the whole mine (*puteus universus*) directly by buying the *pdfp* of a silver mine at an auction, paying 4,000 HS in advance and, after commencing mining operations and extracting silver ore, the rest of the *pretium*. Other existing mines could also be bought from the *fiscus*, as *LMD* §1 indicates: the procurator sold *putei* from which ore had been extracted, removed and smelted prior to payment of the *pretium*.[19]

A different procedure for aquiring ownership of *pdfp* and consequently, a mine, can be reconstructed from various passages in *LMD* and *LMV*. The last paragraph in the *LMV* sets out the regulations for the collection of a *vectigal pittaciarum*, a fee in return for a *pittacium*, a certificate or written confirmation. This fee had to be paid two days after a *puteus* or *locus putei* had been occupied, at the latest, in order to maintain legal right of possession ($\neq$ownership) over the mine.[20] According to Flach, this act of *occupatio* was followed by an *adsignatio*, as becomes clear from the last section of the *LMD* (§18): 'eos puteos, quos occupaverit adsignatosve acceperit'. Flach argued that the *occupatio* preceded the *adsignatio* and the occupied mining pit was officially assigned to the occupant with the payment for the *pittacium*, a 'registration fee'.[21] According to *LMD* §18, these assigned plots had clearly defined limits (*fines putei*) beyond which the occupant was not to expand his mining operations. As soon as ore had been found and extracted the occupant was to pay a *pretium* for the *pdfp*. Following payment of the *pretium*, the occupant aquired full ownership of the mine and its contents and was allowed to process the metal.[22]

[19] *LMD* ll. 2–3: 'puteum universum proc(urator) metallorum vendito'.

[20] *LMV* ll. 58–60: 'Usurpationes puteorum sive pittaciarum. Qui intra fi[nes metalli Vipascensis puteum locum]|que putei iuris retinendi causa usurpabit occupabitve e lege metallis dicta, b[iduo proximo quam usurpaverit occupa]|verit apud conductorem socium actoremve huius vectigalis profiteatu[r] (...).' The text differentiates between 'usurpare' and 'occupare' which might refer to different objects held by the individuals. Furthermore, the text distinguishes *puteus* from *locus putei*. Whereas the *locus putei* might be a pristine plot of land within the mining district hitherto unscathed by mining works, the *puteus* must indicate an already existing mine. The term 'usurpare', may refer to a person 'taking into use' an already existing mine (*puteus*), whereas 'occupare' could refer to the seizure of a plot of land to be mined (*locus putei*). Without the complete *lex* any interpretation of this difference remains hypothetical, cf. Domergue 1983: 100–2.

[21] *LMD* ll. 42–5, cf. Flach 1979: 439 f. with discussion of older scholarly opinions. Domergue 1983: 145, 168 f., sees the *adsignatio* as an assignment to *coloni* of silver mines which have no occupant. He argues that such a regulation was probably not limited to silver mines alone. Flach 1985: 202; Flach 1989: 135; Flach 1990: 119 later altered his view, adding that mines which had been unworked for six months were assigned to new occupants by the *procurator metallorum*.

[22] *LMD* §§ 1, 5.

The terms *occupator* and *colonus* could refer to one and the same person, probably in different phases of their legal relationship with the *fiscus*.[23] The distinct meanings of the terms *occupare* ('to seize') and *colere* ('to till') allows for greater precision: the person who had acquired the right to occupy a mine or mining plot by paying a fee for the *pittacium*, set up a mining infrastructure, or had begun mining, were probably known as *occupatores*. As soon as ore was extracted, they were termed *coloni*.[24]

As becomes clear from the *LMD*, the right of ownership by the *colonus* and the right of possession by the *occupator* were curtailed by the authorities.[25] The rights of property or possession could easily be lost by *occupatores* as well as *coloni*. If the *pretium* was not paid in time, i.e. before the *colonus* smelted ores, the whole mine returned to the fiscal administration. Anyone who reported the *colonus* to the authorities was rewarded with one quarter of the price for which the mine had been sold.[26] The *ius occupandi* of a mine could revert back to the *fiscus* even before ores were smelted if certain conditions were not met. The *occupator* had to run the mine (for which the *pretium* had not been paid?) without interruption for longer than nine days, after an initial period of twenty-five days in which the *occupator* was allowed to acquire the means necessary for running a mine.[27] If ores were found in one mine, the *colonus* was obliged to continue his efforts *sine intermissione* in 'the other four mines' (for which the *pretium* had not yet been paid). The *lex metallis dicta* refers specifically to one of five mines, which Flach interprets as a fifth of all mines run by one *occupator/colonus*. If ore was extracted in less than a fifth of the mines owned by one *occupator/colonus*, he was allowed to interrupt work in each of his mines for up to nine days. In the case of ore being extracted in a fifth or more of the mines, the *occupator/colonus* was deemed financially fit to increase his investments for uninterrupted mining in the other *putei*.[28] If these conditions could not be met, the *fiscus* awarded the *ius occupandi* to another party. Moreover, if a mine was sold by the *fiscus* (i.e. the *pretium* had been paid), and not been worked for six months, the *ius occupandi* also fell to a third party. If the new occupier hit an ore lode, the *fiscus* could claim *pdfp*.[29]

In order that one entrepreneur was not burdened with all the costs of running a mine, the *occupator* of a mine was allowed to have *socii*, who shared

[23] Flach 1979: 417 does not see *occupatores* and *coloni* as different groups, but acknowledges that *occupatores* and *coloni* could be the same person, cf. also Domergue 1983: 128–31.

[24] Domergue 1983: 128 ff.

[25] Flach 1979: 415 f.; Flach 1989: 135 f.; Flach 1990: 120.

[26] *LMD* ll. 1–4, cf. Flach 1979: 424.

[27] *LMD* ll. 9–11.

[28] Flach 1979: 426 f.

[29] *LMD* ll. 11–12, cf. Domergue 1983: 138 f.; Flach 1979: 426.

the costs and the expenses (as well as the profits as soon as ores were found).[30] If all the costs were not shared in accordance to the size of the share, a detailed procedure could be set in motion to enforce a fair distribution of costs.[31] A *colonus* was also able to force his *socii* to cover some part of the expenses made *bona fide.*[32] The *coloni* could sell a mine they had bought from the *fiscus* at any price they saw fit. However, they could only sell it to other *coloni* and under the proviso of having paid the *pretium* for the *pars dimidia* to the *fiscus.* The vendor and buyer had to declare their intentions to the *procurator metallorum.* Through this procedure the *fiscus* was able to verify whether the *colonus* selling the mine had paid the *pretium.* Moreover, if the *colonus* was a *debitor fisci* he was strictly prohibited from transferring the ownership of a mine by a gift.[33]

Besides minimizing interruptions in the extraction work, a number of regulations in the *lex metallis dicta* were aimed at securing mining pits against deliberate or unintentional destruction. The pits had to be secured with wooden pillars or left-over stone pillars, and they were not to intersect with the *cuniculus* (a drainage tunnel for ground-water).[34] The procurator could permit galleries for the extraction of ore being cut into the rock from the *cuniculus* as long as the restrictions concerning size and length were adhered to.[35] Sabotage of mining installations or any other breach of these security regulations were severely punished.[36] As regards the ore extracted from the *putei,* additional regulations applied. *LMD* §1 sanctioned the smelting of ore prior to payment of the *pretium* for the *pdfp.* In the prevention of this illicit activity the mining authorities, however, appear to have relied, at least in part, on the local community at Vipasca. Anyone who notified the authorities of illegalities taking place was awarded a fourth of the sum owed to the *fiscus.* §9 sheds light on further restrictions on the transfer of ore to the smelting oven. Extracted ore stockpiled near the mines was to be transported to the furnaces (*officinae*) by its owners during daylight hours. If someone was caught removing ore stockpiled near the mine after sunset, he would be fined 1,000 sesterces, payable to the *fiscus.*[37] Moreover, thieves caught stealing

30 Flach 1979: 427 f.; Domergue 1983: 141 f.

31 *LMD* ll. 13–18.

32 *LMD* ll. 19–20.

33 *LMD* ll. 20–3, cf. Domergue 1983: 142–6; Mateo 2001: 161–6.

34 *LMD* ll. 29–30, 35–6, 42–5. For a discussion of the technological implications of these regulations and the mining situation at Vipasca, cf. Domergue 1983: 148–66.

35 *LMD* ll. 37–9.

36 *LMD* ll. 32–5, 40–2.

37 *LMD* ll. 23–5: 'Venas, quae ad puteos prolatae | [i]acebunt, ab ortu solis in occasum ii, quorum erunt, in officinas vehere debebunt; qui post occa|[s]um solis vel noctu venas a puteis sustulisse convictus erit, (sestertium) (mille) nummos fisco inferre debeto. For text and translation,

ore were punished according to their status.[38] These sections of the *LMD* give the perception that the mining authorities were trying to ensure that there was transparency about the amount of ore being extracted. The *LMD* does not tell us why this was the case: perhaps the authorities required exact data on the mineral output of the district as a basis for setting the *pretium* to be paid to the *fiscus*. Moreover, the date of payment was the same as the discovery of ore. If a *colonus* had moved ore to the smelting furnace in secret (apparently these furnaces were not controlled by the mining authorities), he would be able to avoid or at least delay the payment of *pretium*. Apart from the silver mines, there is no indication of how the *pretium* for the other mines (most likely copper) was established.

The security regulations and the strict regimentation of interruptions of the mining activities were undoubtedly aimed at enforcing a continuous and secure extraction of ores by the occupants.[39] In regulating nearly every aspect of the extractive procedure, the administrative body in charge revealed its immense mining experience. However, a certain reluctance in transferring full ownership of mines to private individuals seems to underlie these regulations. Domergue correctly identified the exceptional nature of a *metallum* in which both copper and silver was mined, a feature unique for the Iberian peninsula and detectable only in the pyriteous lodes mined in the south-west.[40] Hence, the regulations recovered at Vipasca might have been drafted specifically for the copper/silver mines of the southwestern Iberian peninsula.[41] Provided this interpretation is correct, the regulations of the *LMD* were not just limited to the copper/silver mines at Aljustrel itself, but probably extended to contemporaneous mining establishments throughout the southwestern Iberian peninsula.

7.1.2. Other Imperial Mining Districts: Owners, Contractors, and the Workforce

Apart from the Vipasca tablets, inscriptions from the mining districts hardly ever include the terms *colonus* or *occupator*. At Vipasca itself we find *coloni*

cf. Flach 1979: 404, 406. These regulations have hitherto been understood to underpin the hypothesis that half of the extracted ore was collected by the *fiscus* (cf. Flach 1979: 422 f.) Most recently Mateo (2001: 139–45) again pointed out the improbability of a *colonus* buying 'proprietas' of the *pars dimidia* and handing in half of the extracted ore to the *fiscus*.

38 *LMD* ll. 26–8, cf. Domergue 1983: 146 f.

39 Domergue 1983: 140.

40 Domergue 1983: 158 f.

41 Domergue 1983: 161.

(...) metalli Vipascensis setting up an inscribed base and statue for their hero Beryllus, the *restitutor metallorum.*[42] *Coloni* also are named outside the Iberian peninsula: P. Fundianus Eutyches and P. Aelius Mucianus commissioned an inscription at Rudnik in Moesia Superior.[43] Furthermore, *coloni* of the *arg-[entariae Dardanicae]* near mod. Sočanica commemorated the construction of a sanctuary during the years AD 136/7.[44] A gravestone from Grugua near one of Sardinia's argentiferous lead mines provides the name of a *col(onus)*, Silvanus.[45] Provided the named *coloni* are to be associated with a mining district, the question arises as to whether legal regulations similar to those at Vipasca also applied, i.e. whether the mines had been sold to these *coloni* as well. Given the lack of epigraphic material from other mining districts, a satisfying answer cannot yet be provided.

The find of twenty-five wooden tablets between 1786 and 1855 in the vicinity of Alburnus Maior/mod. Roșia Montană in Dacia throws some light on the internal organization of mining operations there. Although the tablets appear not to belong to a single archive, the same person can appear on different tablets in different functions, i.e. as witness of or party to a contract. Three tablets are work contracts:[46] Socratio Socratis, an 'Oriental', Aurelius Adiutor, and a 'Titus Beusantis qui et Bradua', an Illyrian of peregrine status, are explicitly identified as employers ('conductores' in a *locatio–conductio* agreement), whereas L. Ulpius Valerius, Memmius Asclepi and [-]Restitutus agnomine Senioris appear to have offered their services as employees ('locatores'). Two of these three work contracts clearly state the purpose of hire—'opus aurarium'.[47] What the term *opus aurarium* entailed is not quite clear. It appears, though, to have required working in a shaft or mining tunnel, as the *conductor* (according to the contract) was not liable to pay the wage in the case of flooding.[48] This suggests that *opus aurarium* included underground mining, a work environment prone to floodings and problems with the water table. These and the other wooden tablets only refer to contracts amongst private citizens. By nature they keep us in the dark about the legal standing of those hiring free labour for *opus aurarium*: perhaps a system similar to

[42] *IRPac.* 121. *AE* 1908: 233.

[43] *IMS* I 168 = *CIL* III 6313 = 8333 (Rudnik). Hirschfeld 1905: 153, questioned whether the named *coloni* actually were part of the mining district at Rudnik.

[44] *AE* 1972: 500 = *ILJug* 501 with Dušanić 1971*a*, cf. *ILJug* 503 with Čerškov 1970.

[45] Sotgiu 1988: B 106.

[46] *IDR* I 40–2 = *CIL* III, p. 948, IX. 948, X. 949, XI; cf. Mrozek 1969: 141; Noeske 1977: 396–404.

[47] *IDR* I 41 = *CIL* III, p. 948, X, l. 3: *opere aurario*; *IDR* I, 42 = *CIL* III, p. 949, XI, l. 3: *opere au⟨ra⟩rio.* Täckholm (1937: 113 f.) mistook the term 'conductor' as identifying the contractor or mines, not the employer in the workcontract.

[48] *IDR* I 41 = *CIL* III, p. 948, X, ll.7–8: *[si laborem] | [fl]uor impedierit....*

Vipasca applied, with *coloni* acquiring property of a mine and its contents. Alternatively, given the sensitivity of the mined metal for the Roman economy, the gold mines at Alburnus Maior may have been farmed out to individuals or small companies in return for a rent in kind or money. It is even worth considering an arrangement in which private entrepreneurs may have been contracted to execute mining tasks in return for a fixed payment. Either way, the contractor/lessee would need to supply or hire his own workforce for carrying out any menial tasks.

Noeske assumes that these miners at Alburnus Maior were probably part of a *collegium funeraticium* mentioned in one of the tablets.[49] Apart from this *collegium*, the tablets also provide an insight into other activities within the mining district. Iulius Alexander, a man of Eastern origin, appears in four tablets and is involved in offering financial support in the period between AD 162 and 167. The recipients of his money, for which Iulius Alexander received interest in return, were mostly of peregrine status: an Anduena Batonis received 140 denarii, an Alexander Caricci 60 denarii, and a Lupus Carentis got 50 denarii as a 'depositum irregulare'.[50] Other financially solvent inhabitants of the mining district purchased or traded slaves or bought and sold houses. The people involved were mostly of peregrine status such as Dasius Verzonis, Maximus Batonis, Dasius Breucus, Bellicus Alexandri, Ingenuus Callisti, and Plator Acceptianus.[51]

Occasionally the inscriptions of Ampelum and Alburnus Maior allow a closer identification of individual roles within the mining community.[52] The only private individual at Alburnus Maior who indicates his profession in a votive inscription is M. Aurelius Maximus, a *legulus auri*.[53] *Leguli aurariarum* are also named in an inscription at Ampelum, honouring Annia Lucilla and dating to the year AD 165/6.[54] The term 'legulus' is only rarely attested in Latin literature: in an agricultural context the term is used to describe 'pickers'.[55] Hence, the term 'leguli aurariarum' might be understood to identify menial labourers, which would explain the rarity of this term in the epigraphic evidence. In fact, the gravestone of M. Aurelius Maximus appears to be the notable exception. Although one should refrain from placing too much emphasis on a single inscription, one could argue that the *leguli aurariarum*

49 Noeske 1977: 343 f. For the *collegium funeraticum*, cf. *IDR* I 31 = *CIL* III, p. 924–7.

50 Mrozek 1969: 144; Noeske 1977: 340, 389–91, 404–6. *IDR* I 33 = *CIL* III, p. 930–2, III. *IDR* I 35 = *CIL* III, p. 934 f., V. *IDR* I 43 = *CIL* III, p. 949, XII. *IDR* I 44 = *CIL* III p. 950 f., XIII.

51 Noeske 1977: 344 f., 392 f., 409. *IDR* I 48 = *CIL* III, p. 954, XVII. *IDR* I 36 = *CIL* III, p. 937, VI.

52 On this problem, cf. Mrozek 1969: 141.

53 *CIL* III 1260 = *IDR* III/3, 390.

54 *CIL* III 1307 = *IDR* III/3, 283.

55 Cato, *Agr.* 64.1, 144.3, 164.3; Varro, *Ling.* 5.94, 6.66; *TLL* VII/2, p. 1134.

may have been better paid than their fruit-picking namesakes: according to Hélène Cuvigny's comparison of wages at Mons Claudianus and Alburnus Maior, the work contracts of ordinary miners show they received above-average pay. This would have allowed individuals to participate in a funerary 'collegium' which might have paid for an inscribed stelae of its late member.[56] The labour force employed in Dacian gold mines thus might have been called *leguli* or *aurileguli*.[57] Some scholars have also equated the term '*legulus*' with '*colonus*'.[58] The fact that *leguli aurariarum* share in commissioning an inscribed monument to a member of the imperial family at Ampelum lends support to such a conclusion. However, this rests on the assumption that the mining organisation at Alburnus Minor was similar to Vipasca, an assumption for which there is no direct support in the epigraphic record.

The composition of the population at Ampelum and Alburnus Maior provides additional insights. Noeske's study of inscriptions revealed that in the period prior to the Marcomannic Wars, its population originated mainly from Italy, as well as Dalmatia, Pannonia, or other Danubian provinces. Strong social ties existed with the legionary camp at Apulum, and the municipal elite of the colonia Dacica Sarmizegetusa.[59] While the old entrepreneurs continued to pursue their interests in the mining district after the Marcomannic wars, new arrivals from the upper echelons of Dalmatia's society (the case of T. Aurelius Aper, *princeps* of Splonum is instructive) and, most notably, of wealthy 'Orientals' from Syria and Bithynia (some of them commissioning Greek inscriptions) added to the entrepreneurial make-up of Ampelum's society.[60] Unlike the inscriptions found at Ampelum, those of Alburnus Maior are smaller in size and yield only sparse information.[61] We therefore might be looking at lower strata of Roman society working at Alburnus Maior. Apart from this difference, the composition of the population mostly resembles that of Ampelum, although recently discovered inscriptions establish Illyrians from Dalmatia of peregrine status as the largest and most coherent immigrant community at Alburnus Maior.[62]

The economic discrepancies detectable between Ampelum and Alburnus Maior may further reflect a difference of involvement in the mining business. Noeske argues that elite citizens resident at Ampelum perhaps owned or

[56] Cuvigny 1996*a*: 141.

[57] *TLL* II, p. 1501.

[58] Rostovtzeff 1904: 449 f.; Mrozek 1969: 151; Noeske 1977: 349 fn. 333; Andreau 1990: 106.

[59] Noeske 1977: 321 f.

[60] Noeske 1977: 327–9; for T. Aurelius Aper, cf. *CIL* III 1322 = *ILS* 7153 = *AE* 1968: 443. On freedmen and slaves, cf. Noeske 1977: 321, 325–7; Wollmann 1989: 110.

[61] Noeske 1977: 329.

[62] Mrozek 1969: 141–3 (on the Pirustae); Noeske 1977: 334 f.; Wollmann 1989: 110–13; Damian 2003:147–89.

leased gold mines from the state or provided financial services for those contracting out mines. Hence, the members of a *collegium aurariarum* recorded at Ampelum were probably contractors or owners of mines.[63] The inhabitants at Alburnus Maior, on the other hand, were perhaps less wealthy miners or workers employed in mining operations.

Migrational patterns similar to those at Ampelum and Alburnus Maior have been observed at other mining districts in the Danubian provinces. Domavium records a considerable influx of 'Orientals' and people from southern Dalmatia, whereas 'Italians' barely surface in the epigraphic record. The growth in population allowed Domavium to first reach municipal, and later colonial status.[64] In the case of the *municipium Dardanorum*/mod. Sočanica, Mócsy argued that the social elite of this mining area partly consisted of freedmen standing in for private entrepreneurs involved in the local mining industry.[65]

The effect of mining districts on migrational patterns can also be observed in the Iberian peninsula. Immigration both from within and outside the region has been detected in the mining areas in southern Spain.[66] At Rio Tinto, we find immigrants originating from Emerita(?), Talabriga and Olisipo in Lusitania. The mines at La Zarza were the focal point of arrivals from even further away: two *Limici* from the *castellum Berense* near Bracara Augusta in northwestern Spain had travelled to these mines. Moreover, workers or entrepreneurs from Clunia in the north-east of the Iberian peninsula, were able to make money at mines in the Sierra Morena, like Nava de Ricomalillo or El Centenillo.[67] In the mining areas of the Spanish north-west, similar patterns can be observed.[68] The votive or funerary inscriptions found at or in the vicinity of Roman mining establishments in Asturia demonstrate the presence of members of different Galician tribes in this area. These individuals probably migrated freely from the western coast of Spain over a distance of up to 200 km. The fact that members of the same tribe appear at completely different places throughout northwestern Spain (up to 100 km apart) might partly corroborate our view of free migration rather than a forced resettlement.[69] Apart from the Galician tribes, the *conventus Cluniensis* proved

[63] *CIL* III 941. Noeske 1977: 331.

[64] Alföldy 1965*a*: 154 f., 186 (Italians), 188 f. (Orientals), 198.

[65] Mócsy 1970: 90.

[66] Domergue 1990: 337–48; Haley 1991: 89–94.

[67] Domergue 1990: 339, tables XVI & XVII, nos. 2, 9–13, 15, 25, 45.

[68] Domergue 1990: 339, tables XVI & XVII; Haley 1991: 94–9.

[69] The latter probably took place immediately after the conquest between 25 and 16 BC, cf. Florus 2.32.59–60. Individuals from Galician and other northwestern tribes such as the *Celtici Supertamarci, Interamici, Zoelae, Bibali, Argaeli(?), Seurri,* and *Limici,* migrated to the mining zones, cf. Haley 1991: 96; Balil Illana et al. 1991.

to be a significant area of recruitment. *Clunienses* appear in the epigraphic record at Trêsminas and Santa Colomba in the Turienzo Valley, bear Romanized names, and were certainly attracted by paid work or other economic incentives provided by the mining districts.[70] Miners with the *origo Cluniensis* or *Uxamensis* are quite prominent in other mining areas of the Iberian peninsula. As already noted above, they also appear fairly often in the mining areas of Baetica and of Lusitania.[71] Although Domergue compiled a list of fifty-seven personal names recorded in the epigraphic documents of mining zones of the Iberian Peninsula, it is still very difficult to determine what these people actually did.[72] Only a few inscriptions give us an insight into their activities. For example, slaves and freedmen of mining companies like the *societas Sisaponensis* documented at Corduba certainly attest that these people were involved in directing or operating various mines.[73] A young man named Vegetus may have worked in the gold mines at Mons Marianus or provided auxiliary services, but his funerary inscription merely informs us that he died there and his remains were transferred to Conimbriga/mod. Condixa-a-Velha in Lusitania where his grieving parents laid him to rest.[74]

7.1.3. Ingots and Mining Arrangements

Apart from imperial officials and soldiers, the epigraphic record rarely allows us to identify precisely the functions of civilians present at mining operations. It has been argued, however, that the mould marks on metal ingots identify contractors of mines, and that the onomastic material does provide some insights into the organizational policies the Roman state implemented during the Republic and Principate.

Most Republican lead ingots found in southern Spain or on shipwrecks have cast mould marks which reveal different groups of ingot-owners: individuals, small companies of two or more people (often members of the same family), and large-scale 'societates'.[75] Domergue argued that the majority of

[70] Haley 1991: 94–6; Wahl 1993: 238–40. It is not easy to prove a direct connection between inscriptions of migrants located in the vicinity of mines and the mining business. However, there seems to be hardly any other economic incentive to move from the Spanish Meseta or the Galician coastland to *Asturia* and *Asturia Transmontana* except for the opportunities provided by the gold mines, cf. Tranoy 1981: 234–41. 251 f.; Haley 1991: 71–3.

[71] On the migration of Uxamenses and Clunienses, cf. García Merino 1973: 11; Haley 1991: 110–11.

[72] Domergue 1990: 338.

[73] *CIL* II^2/7 441 = *CIL* II 2269. CIL II^2/7 334 = *AE* 1971: 181. *CIL* II^2/7 415a.

[74] Alarcão & Etienne 1976: no. 32.

[75] Domergue 1990: 253, 254–7 (Table X). 258 f.; Domergue 2004: 225. On *societates*, cf. Kaser 1971: 572–6.

inscribed lead ingots from the late second and early first century BC originated from the mines of the Sierra de Cartagena.[76] This is corroborated by the observation that names of individuals and family companies which appear on the ingots also emerge in the epigraphic record of Carthago Nova/mod. Cartagena.[77] The study of personal names attested on these early ingots revealed mostly *nomina gentilia* of Italian origin, of which some were borne by known personalities, mostly *homines novi* of the Roman nobility, possibly represented by their freedmen in Carthago Nova.[78] In comparison, large-scale *societates* seem nearly absent from the Sierra de Cartagena:[79] a *soc(ietas) argent(ariarum) fod(inarum) mont(is) Ilucro(nensis?)* or *soc(ietas) montis argentarii Ilucro(nensis?)* may have mined for silver on *mons Ilucro(nensis?)*, a region situated near the mine of Coto Fortuna, close to the mines at Mazarrón, c. 20 km east of Carthago Nova. At Mazarrón itself the mines were possibly exploited by a *s(ocietas) m(ontis) F(icarensis)*.[80] During the first century BC large-scale *societates* must have been active in the Sierra Morena as lead ingots from shipwrecks and lead seals have been found in the mining area.[81] The lead objects were perhaps used to seal bags of ore which were transported from the mines to the smelting furnaces where most of them were found. Whereas one side of the seal was marked with moulded letters or initials, the other side displays numbers or an image. Similar inscriptions can be found on other objects, such as coins or bronze seals.[82] Lead or bronze seals showing the initials S.B.A., S.BA., S.C., S.C.C. were discovered in mines throughout Southern Spain. These abbreviations are expanded to *s(ocietas) Ba(edronensis)* or *Ba(eculonensis)*, *s(ocietas) B(. . .) A(rgentifodinarium)*, and

[76] Domergue 1990: 266 f. However, only one marked ingot (carrying the mark M.P. ROSCIEIS) was found in the mines on the Sierra de Cartagena, cf. Domergue 1990: 256 (table X, no. 1038), 266.

[77] Vives y Escudero 1924: 34, no. 5 pl. XXX, 5; Beltrán Martínez 1939: 27–9; Domergue 1966: 45; Koch 1987: 128 f.; and list by Domergue 1990: 265 f.

[78] Domergue 1990: 321–30. The harbour of Cartagena yielded further ingots inscribed with the names of individuals connected with the local metals trade, cf. Domergue 1990: 266.

[79] Strabo 3.2.10 = Polybios 34.9.8–11. Cicero, *Leg. agr.* 1.5, 2.51. Domergue 1990: 267, 269 f. On an inscribed ingot found in a mining area near Castulo, cf. Domergue 1990: 265 fn. 30.

[80] At Mazarrón a lead object carrying the inscription MONT. ARGENT. was found: Domergue 1990: 260 with fn. 12. For the mines, cf. Domergue 1987*a*, MU10 (Mazarrón), MU12 (Coto Fortuna). According to Domergue, the archaeological remains at Coto Fortuna, dated to the end of the Roman Republic/beginning of the Principate, suggest a vast mining operation worthy of a large-scale publican company, cf. Domergue 1987*a*: 397–405. On *s(ocietas) m(ontis) F(icarensis)*, cf. Domergue 2004: 226 fn. 27 with further bibliography.

[81] The shipwreck Cabrera D (*c.* AD 1–15), for example, yielded 21 lead ingots, carrying mould marks with the names ANTEROS EROS, L. IUNII DVO, T.L. OSCA and a *soc(ietas)vesc(. . . ?)*, and 700 amphora. The pottery in particular suggest that the ingots originate from a ship sailing from Roman Baetica, loaded with olive oil and fish sauce. Thus, the lead ingots probably came from the Sierra Morena, cf. Domergue 1990: 260 f., fn. 15; Parker 1992: no. 126.

[82] Domergue 1971: 358–3, fig. 35; Domergue 1990: 261.

s(ocietas) C(astulonensis).[83] Based on recent finds of ingots at the entrance of the bay of Cartagena, of which one carried the abbreviation SOC. BALIAR., the mark S.BA. has been read as *s(ocietas) Ba(learica).*[84] The *s(ocietas) C(astulonensis)* might have operated several mines in the Sierra Morena, one of which was located at El Centenillo. These mines yielded tools and coins marked with the letters of the *s(ocietas) C(astulonensis).*[85]

These mould-marks on ingots and seals from southern Spain have been interpreted on the basis of the few literary sources. The silver mines near Carthago Nova and Baetica generated *vectigalia* for the Roman state as early as the beginning of the second century BC. In 195 BC one of the first governors, M. Porcius Cato, introduced these rents on iron and silver mines.[86] Strabo, quoting Polybios, recounts that 40,000 miners extracted silver worth 27,000 drachmas per day from the mines near Carthago Nova.[87] Diodorus, based on Posidonius' account, describes a genuine 'goldrush' during the second century BC with many Italians flocking to the mines in southern Spain and making significant profits.[88] As these sources on the Republican exploitation of mines on public grounds are more enigmatic than enlightening, there is room for numerous interpretations:

- J. S. Richardson argued that the mines on public land in southern Spain were initially contracted out to small-scale lease holders, but not along the lines of a *locatio censoria.*[89] He believed that a profit of 27,000 drachmas per day could not have been made if *societates publicanorum* had run the mining operations in the name of the Roman state. These companies, according to Richardson, paid a lump sum to the censors at Rome in return for the right to extract metals from the public mines during a *lustrum*, a five year period.[90] Based on Diodorus' account of Italian immigrants flocking to the mines, Richardson envisaged small-scale contractors exploiting plots of the mining area which were allocated to them, probably for a fixed rent payable to the Roman state.
- Claude Domergue argued that Diodorus' account did not refer to lessees of public mines but to private miners, as the term ἰδιῶται used by Diodorus must refer to private mining ventures rather than state-owned mines. Furthermore, Domergue believed that the *locatio censoria* of public lands

[83] Domergue 1987*a*: CO 56, 81, 82, 102; J 12, 20; Domergue 1990: 261 f.
[84] Poveda Navarro 2000: 297.
[85] Domergue 1990: 262 f.
[86] Livy 34.21.7. On Catos taxes, cf. Domergue 1990: 242 fn. 12.
[87] Strabo 3.2.10 = Polybios 34.9.8–11.
[88] Diod. 5.36.
[89] Richardson 1976: 137; Domergue 1990: 246 f.
[90] Lintott 1993: 88 f.

was not reserved for *societates publicanorum* alone, but that small-scale lessees could be subjected to the conditions of a *locatio censoria* as well.[91] He points out that authors like Livy, Strabo, and Pliny the Elder record that the state generally contracted out mines to *societates publicanorum*.[92] Domergue did not entirely exclude the possibility of *publicani* being present in the mining zones of southern Spain at an early stage. However, instead of large *societates publicanorum*, he imagined small-scale *publicani* as lessees mining near Carthago Nova under a *locatio censoria*.[93]

- Antonio Mateo recently offered a compelling third interpretation of the literary evidence. He argued that different exploitation 'regimes' applied, depending on the geological conditions and abundance of the metal. The opencast or underground mining of rare metals on public grounds was farmed out to *societates publicanorum*.[94] On the other hand, the abundant copper, iron, tin, silver, and lead mines of southern Spain allowed for individuals to mine in small pits. Hence, Mateo believes that in the latter case, the right to mine on public land may have been granted to any occupant (*occupator*) of the land in return for a rent, a *vectigal*, to the Roman state. The collection of these *vectigalia*, instituted by Cato in 195 BC, may have been farmed out to publican companies.[95] The influx of Italians to the Iberian peninsula reported by Diodorus thus is likely to have occurred in this context.[96] Perhaps under dire circumstances (under Sulla?) the silver mines near Carthago Nova were sold by the Roman state to the occupiers, a change in status duly noted by Strabo.[97]

Mateo's model for the Republican era should probably apply, at least in part, to the Principate as well: the epigraphic and literary sources demonstrate the continuation of *societates* during the first century AD for Roman Baetica. Pliny the Elder notes the mining activities of a *societas Sisaponensis* in his day,

[91] Diod. 5.36.2. Strabo 3.2.10. Domergue 1990: 247 f.

[92] On Macedonian gold and silver mines, cf. Livy 39.24.2, 45.18.3–5. Domergue 1990: 243–4, 248.

[93] Richardson 1976: 144; Domergue 1990: 248.

[94] Strabo 4.6.7 (gold mines of the Salassi); Pliny, *NH* 33.78; Strabo 5.1.12 (gold mines of Victumulae in the territory of Vercellae); Strabo 12.3.40 (realgar mine at Mt Sandaracurgium). Pliny, *NH* 34.118, 34.165 (Sisapo); cf. Mateo 2001: 43–55. For criticism of Mateo's position, cf. Domergue 2004: 223–5.

[95] According to Mateo's reading of the passage by Livy on Macedonian mines, the Roman senate may have decided against *publicani* as collectors of *vectigal* from mines and landed estates, not against publican companies running mines directly, cf. Mateo 2001: 43–65, esp. 58 f.; Domergue 2004: 223.

[96] Diod. 5.36.

[97] Strabo 3.2.10; Frank 1959: 157, 256–8; Mateo 2001: 66–71.

a company already known to Cicero.[98] At Sisapo/mod. La Bienvenida and in Almadén the company extracted silver and *minium* (cinnabar).[99] Furthermore, inscriptions within Roman Baetica document the presence of the *societas Sisaponensis*, particularly in Corduba: *liberti*, receiving the *nomen gentile* Argentarius subsequent to manumission by the *societas Sisaponensis*, appear on an inscribed monument.[100] Apart from former slaves of this company, an inscribed slab found near Corduba refers to a *servitus viae*, a servitude imposed by the *societas Sisaponensis* on a road which led from Corduba to the *montes s(ocietatis?) S(isaponensis?)*.[101] Moreover, a funerary inscription from Ostia refers to a *vilicus* of the *socii Sisaponenses* and a C. Miniarius(!) Atimetus, *procurator sociorum miniariarum* at Rome, might be associated with the same company.[102] The latter inscription may provide an idea of the scope of tasks such companies could cover: apparently the *societas Sisaponensis* not only mined cinnabar, but may even have controlled the export of its produce to Rome as well. The practice of naming freedmen after the mining company which had owned them is documented for another company as well: a *M(arcus) Aerarius Telemachus, medicus*, freedman and former slave of *socii aerar(ii)* or *aerar(iarum) (fodinarum?)* received a similar type of *nomen gentile*.[103]

Large *societates* involved in mining on public lands are also attested elsewhere. Lead *tesserae* dating from the second century BC to the first third of the first century AD have been found at the mining village of Lascours in Gallia Narbonensis, close to the copper and lead-silver mines of the Orb valley. Some of these *tesserae* were inscribed on one side with the letters SOC. ARG., on the other side with ROT., to be read as *soc(ietas) arg(entifodinarum) Rot(enensium/s)*.[104] The realgar mines at Mt Sandaracurgium in Pontus were also contracted to *publicani*, whereas a Republican *lex censoria* forbade

[98] Pliny, *NH* 33.118; Cicero, *Phil.* 2.19.48, cf. Domergue 1990: 270 fn. 40; Vitruvius 7.9.4 reports that *minium* had also been mined near Ephesos, cf. Brunt 1990*c*: 398 with fn. 142. Mateo 2001: 53 f., has pointed out a parallel between Pliny's report on *minium* exploitation and the *lex portorii Asiae* (Engelmann & Knibbe 1989: §34, 78–81): Pliny claims that 2,000 pounds per annum of unsmelted and unrefined cinnabar ore were delivered to Rome under seal (*vena signata*) and sold at a fixed price of 70 sesterces per pound. It was forbidden by law (*statuta lege*) to alter the minium ore and the price of sale. The *lex portorii* records the taxation of ore exported to Rome under the provisions of a νόμος γεωρυχικός, a 'mining law'. The ore in question is believed to be *minium*.

[99] Strabo 3.2.3; Pliny, *NH* 33.118, 121; cf. Fernández Ochoa et al. 1994.

[100] *CIL* II²/7 415a.

[101] *AE* 1995: 846 = *CIL* II²/7 699a (Corduba).

[102] *CIL* X 3964 = *ILS* 1875 (Ostia). *CIL* VI 9634 (pp. 3470, 3895) = *ILS* 1876, cf. Domergue 1990: 268–71, 277 fn. 60; Mateo 2001: 51 f. fn. 68.

[103] Koch 1987; contra: Dardaine 1983.

[104] Domergue 1990: 261 fn. 16.

publicani of the *aurifodina Victumularum* (mod. Bessa) in the territory of Vercellae in northern Italy to employ more than 5,000 workmen.[105] The gold-mines of the Salassi in the Aosta Valley had been farmed out to *publicani* prior to their final conquest in 25 BC.[106] And Pliny reports on 'black lead' mines (*plumbum nigrum*), i.e. the *metallum Samariense/Salutariense* and the *metallum Antonianum*, being rented for 200,000 or 255,000 *denarii per annum*, which suggests that large *societates* probably ran mining operations in each *metallum*.[107] Apart from the literary sources and inscriptions on stone, large *societates* are barely named in mould marks or stamps on ingots from Roman Baetica during the Principate.[108]

Inscriptions on lead 'pigs' and 'bun-shaped' copper ingots from Roman Britain and Gaul provide additional evidence for *societates* possibly involved in mining operations: the copper ingots were nearly exclusively found in North Wales stamped either with private names or, in one case, with the letters *socio(rum) Romae*. Cast mould marks on lead ingots read *socior(um) Lut(udarensium) Brit(annicum) ex arg(entariis)*, i.e. 'British lead from the silver mines of the *socii* at Lutudarum (Derbyshire lead mining area)'.[109] Further, the cold stamp of *soc(ii) Novaec(-)* is recorded on ingots from the Mendip Hills.[110] A lead ingot from Anse Saint Gervais carries the mould mark *sociorum plumb(um) Ger(manicum)*.[111] Setting aside those naming companies, most mould marks on lead ingots dated to the Principate provide the personal names of individuals: C. Nipius Ascanus is named on an ingot from Carmel in Clwyd/North Wales, L. Aruconius Verecundus, P. Rubrius Abascantus, C. Iulius Protus, or Ti. Claudius Tr(iferna?), are named on lead ingots from Lutudarum in Derbyshire.[112] The latter name, Ti. Cl(audius) Trifer(na), was also stamped on the left or right end of ingots from the Mendip Hills.[113]

[105] Strabo, 12.3.40; Pliny, *NH* 33.78. For the localization of Vercellae, cf. Cavalieri Manasse et al. 1982: 76–7, also Strabo 5.1.12. *CIL* V, p. 715. Strabo 6.2.10 and Diod. 5.10, name an 'alum' mine on Lipara generating revenues for Rome.

[106] Strabo 4.6.7; Dio 53.25.

[107] Pliny *NH* 34.165. Whether or not the gold mines of Asturia had been contracted out, as Brunt (1990*c*: 397 f.) believes, is difficult to prove. Although Pliny (*NH* 33.78) reports that 20,000 pounds (22,500,000 denarii) of gold were produced a year, there is no direct indication that *societates* necessarily were present in the north-west.

[108] An exception might be the cast mould mark *soc(ietas) Vesc(-)* on lead ingots found on shipwreck Cabrera D (1–15 AD), cf. Domergue 1990: 260 f., fn. 15; Parker 1992: no. 126.

[109] *RIB* II/1, pp. 34–7 (copper); *RIB* II/1, 2404, 53–5, 57–60 (lead).

[110] *RIB* II/1, 2404, 5, 6, 11, 13.

[111] *AE* 1959: 124. Rothenhöfer 2003: 280.

[112] *RIB* II/1, 2404, 39–51.

[113] *RIB* II/1, 2404, 7–10.

The mould-marks on ingots from the first century AD found on shipwrecks from the Baleares, Southern Gaul and Corsica first and foremost provide the name of individuals as owners of ingots.[114] The shipwreck Cabrera 6/Ses Salinas off Mallorca, for example, carried Dressel 20 amphoras and lead ingots, of which only few have survived. The ingots again carry mould-marks with the names of individuals like P. Aemilius Gallicus or Q. Cornutus, while the secondary stamps of Vespasian declare the ingots imperial property. Numbers indicating the weight are incised as well.[115] Further names of individuals are recorded on first-century AD lead ingots recovered from the shipwrecks Cabrera 4 and 5, Sud-Perduto 2, Sud-Lavezzi 2, and Port-Vendres 2. In contrast to the onomastic material on ingots from the Republican era associated with the Sierra de Cartagena, a geographic contextualization of the personal names provided by these ingots is not possible.[116] This is not the case with the ninety-nine lead ingots from Saintes-Maries-de-la-Mer: eight of them carry the mould-mark *[.]Flavi(i) Veruclae plumb(um) Germ(anicum)*, the 'German lead of [.] Flavius Verucla' and were stamped with IMP. CAES and incised with numerals; the other ninety-one ingots were merely stamped with L.FL.VE, IMP. CAES, EROTIS and incised with numerals.[117] Based on the lead isotopes and the rediscovery of a lead ingot with the mark *L. Fla*[-] at Bad Sassendorf-Heppen, it is possible that the lead originated from mines in the Sauerland, probably from the area near Brilon.[118] Given the mark IMP. CAES. on these ingots, they were likely produced during the period of Roman control of territories east of the Rhine during 8 BC to AD 9.[119]

The survey of inscribed lead (and copper) ingots of the Roman Principate reveals no significant changes in comparison to the Republican era, apart from an increase of the number of ingots owned by the emperor. This phenomenon has been understood as mirroring the growing control of

[114] Domergue 1990: 265; Domergue 1998: 201 f. On the shipwrecks, see Veny & Cerdá 1972; Parker 1974; Colls et al. 1975; Veny 1979; Liou & Domergue 1990; Bernard & Domergue 1991; Long & Domergue 1996.

[115] Veny 1969–70; Parker 1974; Parker 1992: no. 1017.

[116] The *nomen gentile* of C. Haterius Gallus, for example, a personal name attested only once in a mould-mark on a lead ingot from the shipwreck Cabrera 5, may appear in the epigraphic record of the Iberian peninsula (L. Hat[erius? . . .], Abascal Palazón 1994: 148). Nor do we find any attestation of the Mevii, recorded on lead ingots from the shipwreck Cabrera 6 in Southern Spain, see Abascal Palazón 1994: 185. For a list of names on 1st cent. AD lead ingots, cf. Domergue 1998: 214 f. (tables 1 & 2). For inscribed ingots and shipwrecks, cf. Veny 1969–70; Veny & Cerdá 1972: 302–10; Laubenheimer-Leenhardt 1973: 114–19; Parker 1974; Colls et al. 1977: 11–22; Veny 1979: 471–3, 483–8; Colls et al. 1986: 75–80; Liou & Domergue 1990: 74–90, 99–104; Bernard & Domergue 1991; Long & Domergue 1996: 837–51; Domergue 1998: 214, table 1a.

[117] Long & Domergue 1996: 818, 832–4.

[118] Rothenhöfer 2003; Eck 2004*b*: 19–21.

[119] Eck 2004*a*: 74–6.

mines by the emperor himself—a notion based on mould marks alone (cf. 3.3.5). The *opinio communis* holds that the names on the mould marks are those of individuals and companies mining on public land or on private property. The purpose of these mould-marks, however, must be seen in the context of further markings applied to the ingot. Most ingots carry a variety of markings, be they cast mould marks, stamps, holes, or incisions. Based on the superimposition of some of these markings on ingots found on shipwrecks from the first century AD (Cabrera 4,5,6; Port-Vendres II; Saintes-Maries-de-la-Mer 1; Lavezzi 1; Sud-Lavezzi 2; Sud-Perduto 2), Claude Domergue reconstructed their sequence of application.[120] Firstly, the lead was cast in the form of ingots, which carried a moulded mark. At a later stage the lead was stamped cold with different seals. The ingots were then probably nailed to a wooden surface (hence the nail holes) in order to prevent movement. Numerals were subsequently incised and occasionally a further cold stamp applied. Domergue offered a compelling explanation for this sequence of markings, based on the assumption that the ingots originated from Roman Baetica: the producer cast the lead into the form of ingots carrying his mould-mark or cast stamp. The lead ingots then might have been collected by *mercatores*; they stamped their own seals on the lead ingots. Nailing their cargo to the floor of the river boat to prevent movement, the *mercator* might have sailed down the Baetis, anchoring at every harbour on the northern bank of the river and collecting further lead ingots from the nearby mines of the Sierra Morena. At Hispalis, Baetica's large shipping port, the ingots were probably unloaded and stockpiled. They were then sold to a further trader, the ingots were weighed (incised numbers) and the stamp of the new owner applied. The second *mercator* then might have entrusted the lead cargo to a *navicularius* which transported them to Italy or Gaul. Evidently less complicated models are possible (Domergue offers three further possibilities), and one might imagine several of them being practised at the same time.[121] Interestingly, the incision of numbers signifying the weight of the ingots was performed after the ingots changed hands for the first time. Weighing the ingots appears therefore not to have been a necessary requirement during regular transactions. However, the exaction of trade taxes, such as the *portoria*, forced the

[120] The ships carrying the ingots, amongst other items, have been dated to the 1st cent. and, based on the amphora, may have originated from Roman Baetica. Domergue assumed that the ingots must also stem from southern Spain—an assumption refuted by the observations of Rothenhöfer on the origin of the Saintes-Maries-de-la-Mer lead ingots. Perhaps the freight has a more complex assemblage history. This evidence notwithstanding, Domergue's observations still provide a solid foundation for further deliberation, cf. Domergue 1998: 202 f., 206–9; Rothenhöfer 2003.

[121] Domergue 1998: 203–8.

traders to weigh and standardize their cargo. According to the *lex portorii Asiae*, the weight and number of any cargo being exported or imported by sea or land had to be declared to the customs office.[122] The tax on the ore was 4 asses per 100 pounds.[123] The numbers incised on the sides of our ingots indicated the number of pounds exceeding the normal weight, allowing the exact calculation of the charged tax-sum or of shipping costs.[124]

Whereas the application of cold stamps, holes, and numerals can be explained with some certainty, we are still uncertain about the role of the *socii* and individuals named in mould marks in the context of mining and metal trade. A list of the more informative mould marks on ingots from the Principate may provide some insights:

AE 1997: 1042	*Flavi(i) Veruclae plumb(um) Germ(anicum)*
RIB II/1, 2404.40	*L(uci) Aruconi Verecundi metal(li) Lutud(arensis)*
RIB II/1, 2404.51	*P(ubli) Rubrii Abascanti metalli Lutudare(n)s(is)*
RIB II/1, 2404.41.44–5	*Ti(berii) Cl(audii) Tr(ifernae?) Lut(udarense) Brit(annicum) ex arg(entariis)*
RIB II/1, 2404.46–50	*G(ai) Iul(i) Proti Brit(annicum) Lut(udarense) ex arg(entariis)*
RIB II/, 2404.53–5	*Soc(iorum) Lut(udarensium) Brit(annicum) ex arg(entariis)*

As the four ingots of Saintes-Maries-de-la-Mer (*AE* 1997: 1042) indicate, the genitive of the personal names (e.g. *Flavii Veruclae*) is a possessive, likely indicating ownership rather than origin of the lead ingot (*plumbum*).[125] The origin of the ingot is indicated either by a partitive genitive (e.g. *metalli Lutudarensis*), an adjective (e.g. *Britannicum, Lutudarense*), and/or by a preposition (e.g. *ex argentariis*).

The scholarly consensus is that the *societates* and individuals are named not only as owners of the ingots but as lessees, occupiers or private owners of mines as well. Even though it is plausible that in most cases the producers of the ingots put their name on their ingot, there is no evidence to date that satisfyingly corroborates this line of argument. In fact, the appearance of the same names either as mould-marks or cold stamps on ingots further calls this thesis into question, in particular when cold stamps are understood to have been applied by later owners or merchants (*mercatores/navicularii*):

122 Engelmann & Knibbe 1989: 80 f., ll. 45–6 (§18).

123 Engelmann & Knibbe 1989: 97 f., ll. 78–81 (§34). The weight norm of a 100 pounds for a wide range of transported goods also appears in the customs tarif of Zarai in North Africa (2nd cent. AD): *CIL* VIII 4508, cf. Domergue 1998: 209 fn. 41.

124 Domergue 1998: 203, 215 (table 2).

125 Domergue 2004: 225.

RIB II/1, 2404.41.44–5	*Ti(berii) Cl(audii) Tr(ifernae?) Lut(udarense) Brit(anicum) ex arg(entariis)*
RIB II/1, 2404.4–10.	Face (moulded): *Imp(eratoris) Vespasian(i) Aug(usti)* Front (moulded): *Brit(annicum) ex arg(entariis) Veb(-)* Stamp: *Ti(berii) Cl(audii) Trif(ernae?)*
Saintes-Maries-de-la-Mer 1	(*a*) Face (moulded)[.] FLAVI VERUCLAE PLUMB. GERM. Stamp: IMP.CAES. (*b*) no mould mark Stamps: IMP.CAES.; L.FL.VE.; EROTIS

Ti. Claudius Triferna thus appears as owner of lead ingots and—provided the interpretation of cold stamps is correct—as intermediary shipper of ingots owned by the emperor. Similarly, L. Flavius Verucla's name appears either as a mould-mark or as a stamp on ingots. This calls into question the idea that mould-marks only identified the individuals or companies mining the ore. To my knowledge, there is no indication in the sources which required the miners of ore to be named on the refined produce. In fact, numerous lead ingots carry no mould-marks at all, are 'anepigraphic', or carry moulded panels without lettering.[126] Unlike large-scale companies such as the *societas Sisaponensis*, a considerable number of individual miners in southern Spain are likely to have concentrated their efforts on the extraction of ore, whereas the trade in the produce was left to merchants. Some miners may have even yielded the task of smelting ore to another party, and thus saved on the costs of running their own furnace.[127] One could easily imagine the extracted ore being sold either to the operator of the furnace and/or a merchant prior to smelting. Only large-scale companies may have controlled the whole process from mining and smelting the ore down to shipping and trading the ingots. Consequently, the mould-mark of the ingot may provide the name either of a miner or mining company (provided they smelted the ore at their own cost), an owner of the furnace, and/or the merchant buying the ore from the miners.[128]

I therefore argue that the names displayed in the cast mould-marks on ingots at the most provide evidence of the initial owner of the ingot at the time it was cast, but do not conclusively identify the owner, occupant, or contractor of a mine where the ore originated. Hence, one cannot be certain

[126] E.g. Sancti Petri, Spain, cf. Parker 1992: 382, no. 1027; Cabrera 4, cf. Veny & Cerdá 1972: 302–10; Veny 1979: 471–3, 483–8.

[127] The Vipascan *lex metallis dicta*, for example, does not provide any information on who ran the *officinae*. The text implies, however, that the furnaces were not under the control of the mining authorities, cf. *LMD* ll. 1–4, 23–5.

[128] Poveda Navarro 2000.

that the production and/or trade of ingots was run by the same individuals that also exploited the mines.[129]

7.1.4. Conductores/mancipes/promagistri ferrariarum

Unlike the *conductores* known from the Vipasca tablets who leased auxiliary services within the mining district (e.g., shoemaker, barber, etc.), the function of *conductores ferrariarum* who are recorded in some of the mining provinces must have been of a different scale and nature altogether.[130]

The earliest contractor of iron mines may be documented in an inscription at Arelate/mod. Arles dated to the late first century AD: the title of *promagister f[errariarum] provincia[rum] Narbonens(is) Lu[gdun(ensis)] Aquitanic(ae) Belgi-[cae]* noted in the inscription of an unknown equestrian is perhaps the only indication of a *societas ferrariarum* operating on a superprovincial level within Gaul.[131] The term 'conductor' is often attested in Noricum: an inscription set up for Ti. Claudius Macro by his *vilicus* in Aquileia might be the oldest attestation of a *conductor ferrariarum Noricarum*, belonging to the early second century(?).[132] The inscriptions of other *conductores* found at Tiffen, Hohenstein, and Wieting are difficult to date, but according to Alföldy belong to the first half of the second century as well.[133] A further individual from Aquileia, an equestrian M. Trebius Alfius, *c(onductor) f(errariarum) N(oricarum), praefectus iure dicundo* at Aquileia, commissioned a votive inscription at Tiffen.[134] The Trebii are well attested in Aquileia and were probably part of the municipal elite.[135] Moreover, we find a Trebius Alfius as a high-ranking member (*pater*) of the Mithras cult at Virunum.[136] Another *conductor* is attested in Tiffen: [Pu]dens, the slave of a *c(onductor) f(errariarum) N(oricarum) [et as]ses⟨s⟩or [fe]rrariarum,*

[129] Andreau 1990: 105.

[130] For a similar view, cf. Täckholm 1937: 109.

[131] Pflaum 1960–1: 118–23, no. 52, 1053, based on the reading of *CIL* XII 671 by Benoit 1932: 138–41.

[132] *CIL* V 810 (Aquileia). On date, cf. Panciera 1957: 27 f.; Alföldy 1974: 115, 319 fn. 86. A further building inscription by a Sabi[nius], possibly a *conductor ferrariarum Noricarum*(?), who rebuilt the temple of Noreia at Hohenstein, names the presidial procurator of Noricum, Claudius Paternus Clementianus, attested around AD 120 (*CIL* III 14362, p. 2328, 197 + 14363 = *AE* 1968: 408). Besides providing a date for the inscription, the naming of Clementianus might also reflect a direct administrative connection between the *conductores* and the *procurator* at Virunum.

[133] Any attempt to date these inscriptions would have to rest on palaeographic or onomastic analysis. However, Ørsted 1985: 223–36, dates some of the discussed inscriptions to an earlier period.

[134] *CIL* III 4788 = ILS 1466 = *ILLPRON* 604, cf. Ørsted 1985: 223 f.

[135] *Inscr. Aqu.* III, p. 1310, Index.

[136] *AE* 1994: 1334 = *AE* 1996: 1189 = *AE* 1998: 1016. Piccottini 1994*b*.

appears in a fragmentary inscription.[137] Q. Septueius Clemens, *con(ductor) fer(rariarum)* and two *procuratores ferrariarum*, Tib. Claudius Heracla and Cn. Octavius Secundus, are named on a votive altar set up by a further *procurator ferrariarum*, Q. Septueius Valens, to Isis Noreia in Hohenstein.[138] The reading of the three letters N.P.D. following the words *con(ductor) fer(rariarum)* has been the subject of intense scholarly debate. Alföldy, Winkler, and Ørsted, based on Egger, read *conductor ferrariarum N(oricarum) p(artis) d(imidiae)*, while Hirschfeld, Rostovtzeff, Mócsy, Dušanić, and Andreau opted for *conductor ferrariarum N(oricarum) P(annoniarum) D(almaticarum).*[139] It has been rightly pointed out, in favour of the latter position, that there is no parallel for a *conductor (...) partis dimidiae.*[140] Moreover, as we shall see below, parallel examples of *conductores/mancipes* covering the iron mines of more than one province appear in the epigraphic record. The procurators honoured alongside the *conductor* Q. Septueis Clemens were his private procurators. This is suggested by the *nomen gentile* of the procurator commissioning the votive altar, Q. Septueius Valens.[141] Hirschfeld assumed that the three private procurators of Q. Septueius Clemens were each responsible for a province.[142] While Q. Septueius Clemens' personnel are well attested in Noricum, his own origins can again be traced to Aquileia.[143] A votive inscription to *Termunes* found at Friesach was set up for a Q. Calpurnius Phoebianus, *c(onductor) f(errariarum) N(oricarum)*, by his private procurator C. Iulius Hermes.[144] At Wieting a votive inscription records a Campilius Verus,

[137] *AE* 1957: 108 = *AE* 1969/70: 454; Alföldy 1969*b*: 23–5, no. 33; Alföldy 1970: 168 f. Contra Alföldy, cf. Ørsted 1985: 218. Alföldy believed the *assessor ferrariarum* had judicial competence within the Norican mining districts; the inscription, however, is fragmentary and does not support Alföldy's reading unequivocally. An *a(dsessor)* might be named in two inscriptions from Anicium/mod. Le Puy in Gallia Narbonensis, cf. *CIL* III 1576 = *ILA Vell.* 24. *CIL* III 1577 = *ILA Vell.* 25 with Aymard 1856: 344; Hirschfeld 1905: 158 f.; Wuilleumier 1948: 68 fn. 32; Sablayrolles 1989: 160 fn. 36; Rémy 1995: 76 f. They probably were connected with the *arca ferrariarum tres provinciae Galliae*, cf. Hirschfeld 1905: 158; Pflaum 1948: 18; Sablayrolles 1989: 160 with fn. 31; Rémy 1995: 77 f. A further funerary inscription from Mogontiacum/mod. Mainz, *CIL* XIII 11833 = *AE* 1913: 241, refers to an *adsessor fer(rariarum).*

[138] *CIL* III 4809 = *ILS* 1467 = *ILLPRON* 151.

[139] Rostovtzeff 1904: 448; Hirschfeld 1905: 152 fn. 3; Winkler 1969: 165 no. 2; Alföldy 1970: 176; Ørsted 1985: 225 f.; Andreau 1989: 101 f. Alföldy (2003: 226; 2004: 213 fn. 29) now reads N(oricarum) P(annoniarum) D(elmaticarum).

[140] Dušanić 1977: 82 fn. 199.

[141] Rostovtzeff 1904: 448; Winkler 1969: 164 f.

[142] Hirschfeld 1905: 152 fn. 3.

[143] Scherrer 2002: 19 with fn. 52; *CIL* V 8459 = *Inscr. Aqu.* I 1134; Inscr. Aqu. I 404, II 2139. The procurators aside, it has been pointed out that two further Quinti Septueii are attested within the province of Noricum: Q. Septueius Adiectus, named in an inscription from Weyer, perhaps was a freedmen of Clemens, cf. Alföldy 1970: 170 fn. 28; *CIL* III 5480 = *ILLPRON* 1482; for Teisendorf, cf. *CIL* III 5593 = *ILLPRON* 1552. A recently discovered inscription from Virunum/Zollfeld (*AE* 1986: 538) names a Thallus, a slave of Clemens.

[144] *CIL* III 5036 = *ILLPRON* 89.

conduct(or) ferrar(iarum), being honoured by his *vilicus* Fortunatus.[145] Calpurnii and Campilii do not appear in the epigraphic documents of the Norican municipal elite.[146] The Campilii, however, probably were based in Mediolanum/mod. Milano in the Transpadana.[147]

The preponderance of private individuals from Aquileia and northern Italy as *conductores* of Norican *ferrariae* is striking. Peter Scherrer pointed out that the municipal aristocracy of Norican towns was able to lease *portoria publica*, but was apparently unable to break into the market for mining contracts.[148] The exclusion of the municipal elite from such public contracts was not a phenomenon which can be observed in other parts of the empire. An inscribed votive altar from Roman Dacia, set up at Teliucu Inferior in the vicinity of iron mines, dates to the years AD 200–11.[149] Caius Gaurius Gaurianus, *sacerdos* of the *colonia Apulensis* and Flavius Sotericus, member of the *Augustales* at the *colonia Sarmizegetusa*, appear to both have been *conductores ferrariarum* at the beginning of the third century.[150] Unlike Noricum, the wealthy of Apulum and Sarmizegetusa certainly had access to these contracts. Whether these included all of the iron mines in Dacia or just the mines in the vicinity of Teliucu Inferior cannot be established, but it is reasonable to assume that the full title might have been *conductor ferrariarum Dacicarum*. This title is echoed in a barely readable inscription on a sarcophagus discovered in the garden of the priory St. Ulrich und Afra in Augusta Vindelicum/mod. Augsburg in Raetia.[151] The often overlooked inscription names P. Frontinius Decoratus as *manceps ferrariarum* for an unknown area, Raetia and the three Dacian provinces. Based on stylistic and epigraphic criteria, Hans Ulrich Nuber dated the sarcophagus and its inscription to the beginning of the third century AD.[152] A further contractor of iron mines in more than one province appears in two inscriptions from Pannonia. An altar found at Ljubija near the iron mines of the Japra valley names C. Iulius Agathopus as *conductor ferrariarum*. The inscription was set up by Callimorphus, a *vilicus*, on 21 April 201.[153] Agathopus' other altar at Mursa, commissioned by the *arkarius* Gamicus, renders his exact title as *c(onductor) f(errariarum) Pannoniar(um) itemque provinciar(um) transmarinarum*.[154]

The tasks of these conductores are rarely noted in our written sources. An inscription from Lugdunum/mod. Lyon, dated to AD 226, names a *societas* of

[145] *AE* 1995: 1195 = *ILLPRON* 720. [146] Scherrer 2002: 26, table 1.
[147] Lörincz 1999: 29. [148] Scherrer 2002: 29 with fn. 90, 91.
[149] Kienast 1996: 164. [150] *IDR* III/337 (Teliucu Inferior).
[151] Nuber 1985. [152] Nuber 1977: 233 f.; Nuber 1985: 131.
[153] *AE* 1973: 411 = *ILJug* 779. [154] Dušanić 1971*b*: 554; Bulat 1989: 36; Fitz 1993–5: 407 f., 740.

contractors collecting a *vectigal* from the iron mines of Memmia Sosandris.[155] A further *societas vectigalis ferrariarum* is mentioned in an inscription from Ostia; a Hilarus, a slave of this *societas*, erected a marble cippus to Silvanus in the year AD 102.[156] The named *societates* were apparently in charge of collecting *vectigal* from iron mines. Rostovtzeff and Hirschfeld thus assumed that all *conductores ferrariarum* simply collected the *vectigal ferrariarum* from miners.[157] Furthermore, the sheer geographical scope suggested by their titles supports the view of these *conductores ferrariarum* collecting the vectigal in one or more provinces. Just like the process of farming out the *portoria*, it has been suggested that their activities were monitored by the *procuratores metallorum* with 'similar regional circumscription' (cf. 6.1.3.4).[158] However, not all *conductores ferrariarum* would have dealt exclusively with vectigal: Peter Brunt argued that no proof has been provided for Hirschfeld's view that *conductores* simply collected a *vectigal* from small-scale operators. As regards the formation of corporations, the mid-second-century jurist Gaius remarked that *socii* (except those for public *vectigalia*) who ran gold, silver and salt mines received 'corpus habere', i.e. the right to have common property.[159] The *conductores*, *mancipes*, or *promagistri ferrariarum*, Brunt argued, were the heads of such corporations or partnerships.[160] Provided this is the case, the *conductores ferrariarum* could also be directly involved in the extraction of metal ore.[161]

This might be the case with a conductor named on a funerary stela from Narbo/mod. Narbonne. Perhaps dating to the second century AD, the stela was set up for his brother by Ti. Iunius Fadianus, *sevir Augustalis* of Narbo and *cond(uctor) ferrar(iarum) ripae dextrae.*[162] Belonging to the wealthy stratum (but not the municipal elite) of Narbo and being a *sevir Augustalis*, which was usually, but not exclusively, a lower grade priesthood for freedmen, Ti. Iunius Fadianus evidently gained a contract over mines on the *ripa dextra* i.e the right bank of a river.[163] The exact location of these *ferrariae*

[155] *CIL* XIII 1811 = *ILS* 8641 (Lugudunum).

[156] *CIL* XIV 4326. Despite the unsolved property issue—the question whether Memmia Sosandris owned the mines or simply had gained usufruct—it is likely that the *vectigal* was exacted by a *societas vectigalis*, cf. Hirschfeld 1905: 158 fn. 4; Sablayrolles 1989: 159 fn. 24.

[157] Rostovtzeff 1904: 448, 451; Hirschfeld 1905: 152; Täckholm 1937: 110; Alföldy 1970: 168.

[158] Brunt 1990*c*: 401 f.; Eck 1997*a*: 136.

[159] Gaius, *Dig.* 3.4.1pr.: 'Paucis admodum in causis concessa sunt huiusmodi corpora: ut ecce vectigalium publicorum sociis permissum est corpus habere vel aurifodinarum vel argentifodinarum et salinarum', cf. Brunt 1990*c*: 368 f., 401.

[160] Brunt 1990*c*: 398; France 2001: 384.

[161] See also Täckholm 1937: 109.

[162] *CIL* XII 4398 = *ILS* 6971. Hirschfeld 1905: 158, assumed Ti. Iunius Fadianus had contracted the iron mines belonging to the colony of Narbo.

[163] Based on Desjardins 1867–85, 414 f., Hirschfeld 1905: 158 with fn. 5, hesitantly located Fadianus' iron mines on the right bank of the river Aude. Sablayrolles 1989: 159 fn. 23, with further bibliography, locates the *ferrariae ripae dextrae* in the area of the Corbières. Given Tib.

remains unknown, but one may assume that they were located within Gallia Narbonensis.[164]

How these *conductores ferrariarum*, i.e. their companies, precisely functioned is not made clear by our written sources. The inscription from Hohenstein names three private procurators in the service of the *conductor ferrariarum N(oricarum) P(annoniarum) D(almatarum)* who, according to Hirschfeld, were put in charge of *ferrariae* in each (or rather the collection of *vectigal* from there).[165] Further insights are provided by the inscriptions at Mursa and Ljubija relating to C. Iulius Agathopus, the *conductor ferrariarum Pannoniarum itemque provinciarum transmarinarum.* The altar at Ljubija was set up by a *vilicus* named Callimorphus who in AD 201 may have functioned as the local agent of the contracting corporation at the iron mines. Perhaps he was in charge of the local collection of the *vectigal* or dealt with other management aspects of the mining operation.[166] The inscribed altar at Mursa was commissioned by Gamicus, an *arkarius*, obviously the local treasurer of Agathopus' company.[167] The personnel working for such companies may have been spread out over several provinces, in particular when their main responsibility was the collection of vectigal from the occupants of iron mines.

7.1.5. Summary

The majority of evidence for *conductores, mancipes,* or *socii* in the mining business is provided by epigraphic sources. However, the functional titles which appear in inscriptions, hardly allow us to reach an exact understanding of the contractual arrangements. The term 'conductor' implies that the basis of the arrangement was a 'locatio-conductio', a contract which could involve a wide array of different obligations.[168] The infrequent inclusion of the term 'vectigal' in the functional titles makes it likely that there was a collection of *vectigalia* or rents from the occupants of iron mines. In this context 'vectigal' is to be understood as

Iunius Fadianus' minor post of *sevir,* Hirschfeld believed it unlikely that he was a *conductor* of imperial mines, a position Sablayrolles 1989: 159, severely opposes.

164 Domergue & Leroy 2000. Whether a Gamicus documented in a Latin inscription from Haghio Pneuma near the iron mines of Monoekos, on the edge of the territory of Philippi in Macedonia, was a full-blown *conductor ferrariarum,* or simply a small-scale contractor of services within the iron mines there, is not clarified by his inscription. His socio-legal status as a freedman might suggest the latter: *AE* 1986: 629: *Gamicu[s] | conduct[or] | an(norum) X | lib(ertus) Pont[i] | Nov(i) SC [—] | [——],* cf. Samsaris 1985; Samsaris 1987: 156 f.

165 *CIL* III 4809 = *ILS* 1467 = *ILLPRON* 151, cf. Hirschfeld 1905: 152 fn. 3.

166 *AE* 1973: 411 = *IL Jug* 779. In AD 209 Callimorphus is named as *vilicus* in a further altar set up together with T. Flavius Verecundus, the *procurator (ferrariarum),* cf. *AE* 1958: 63 = *ILJug* 157.

167 Bulat 1989: 36.

168 Kaser & Knütel 2005: 219 f.

a revenue from the use of public property.[169] Usually, the titles bear no direct reference to the collection of a 'vectigal'; nonetheless, the *conductores* or *mancipes ferrariarum* as heads of companies and in charge of one or more provinces, are likely to have supervised the collection of *vectigal* from miners in public districts. Whether *conductores ferrariarum* in general exclusively collected 'vectigal' or were perhaps more closely involved in mining operations, remains unknown. Any clear evidence for the precise nature of these contracts remains elusive; they were apparently significant enough to merit monitoring by *procuratores ferrariarum*. The epigraphic evidence only provides particulars on *conductores* of iron mines; contractors of silver or gold mines are not directly identified in our sources.[170] This is in line with the special arrangements reported for the opening of *ferrariae* and the trade of iron and iron products (cf. 6.1.3.4).

The manner in which the *conductores ferrariarum* acquired the collection contracts for one or more provinces might be explained as follows. A conductor and his company may simply have agreed bilateral contracts with several provincial governors; or the contracts for the collection of *vectigalia* and the running of iron mines may have been concluded on a super-provincial level. Given the tentative evidence for centenarian *procuratores ferrariarum (Gallicarum)*, possibly in charge of up to ten provinces within the customs-district of the *quadragesima Galliarum*, and at least one further iron-mining procurator for the Danubian provinces, the latter explanation is to be favored. At an earlier stage contracts may have been concluded centrally with a *procurator ferrariarum* in Rome (6.1.3.4).

As regards the evidence from mould-marks and stamps on lead ingots, most inscriptions refer to the owner of the ingot, who occasionally might have been its producer as well. In contrast to scholarly opinion, it is not possible to certainly identify the function or status of the names of persons or companies named, whether they were the owner/contractor of a mine, of smelting furnaces, or a merchant. Likewise, the epigraphic evidence from mining sites does not provide information on the function of the civilian personnel present. Unless clear terminology is used (*coloni*, *occupatores*, *conductores*), it is virtually impossible to make any distinction between civilian workers in mines and the contractors or owners. Even the wooden tablets of Alburnus Maior do not offer much further information: apart from documenting the influx of foreigners to the mining district (together with the epigraphic evidence), the tablets provide insights only into the hiring of mine workers, without identifying the status of the employers within the operation. Only the

[169] *Dig.* 50.16.17.1; Pliny, *NH* 34.118; Mommsen 1887: 434 fn. 3; Pékary 1979; Burton 1999. On the sale/lease of public land, cf. Lintott 1993: 80–2; Rathbone 2003.

[170] Täckholm 1937: 108.

literary evidence sheds light on the use of *societates publicanorum* in the exploitation of rare metals. Even though this is mainly documented for the Republican era, the example of the *societas Sisaponensis* and the *metalla Antonianum* and *Samariense* suggest that the system continued at least into the early Principate.[171] One could thus argue tentatively that companies ran mines during the early Principate, perhaps even controlling the whole process from the extraction and smelting of ore to trading the final product.

If this interpretation is accurate, the *lex metallis dicta* of Vipasca shows a rather different approach to the organisation of extractive operations on public land. The mining administration in charge of the districts was not primarily interested in the extracted copper and silver per se; the authorities, however, were keen to sell mines off in return for an agreed sum. The 'sale' by the authorities of a mining plot within the Vipasca district to the *occupatores*, however, only occurred when ore was extracted. Only then did the authorities claim the *fiscus*' ownership of half of the ore and demanded compensation for the continued exploitation of ore by the occupiers of the plot. Hence, the mining administration at Vipasca was interested in seeing as many mining plots as possible sold to *occupatores/usurpatores* of unworked plots or orphaned mines. This helps explain the many restrictions on occupants. Nonetheless, after the 'sale' of mining plots the authorities strongly curtailed the ownership rights of the *colonus*: the resale of mines was limited to the other *coloni* at Vipasca, and if a silver-mine remained unworked for half a year, the *colonus* lost ownership. To acquire 'ownership' of one or more silver mining plots one had to buy an operational mine from the *fiscus* at an auction. Alternatively, one could occupy and then be assigned a plot or vacant mine which had to be operated until ore was discovered, for it it was only then that the administration would transfer ownership. Based on our current knowledge, it is uncertain whether this system documented at Vipasca was limited to southern Spain or whether it was employed in other provinces.

7.2. QUARRY LABELS AND QUARRYING CONTRACTORS

Over 300 quarried products with inscribed labels were unearthed from the marble yards of Rome during Visconti's excavations between 1868 and 1870 and published by Luigi Bruzza in 1870. A smaller number of blocks and columns were found at Portus as a result of the excavations from 1959 onwards. These have yielded the main bulk of inscriptions found on coloured and white

[171] Domergue 2004: 229.

marble.[172] A vast number of inscriptions on *pavonazetto* marble discovered in the quarries of Dokimeion, on *giallo antico* in Simitthus, and on granodiorite stone from Mons Claudianus, mostly from the years AD 22 to AD 236, have been published in the last two decades (see Appendix = App.). In spite of the highly abbreviated and formulaic nature of these label inscriptions, which make their interpretation a difficult task, they provide us with a limited understanding of internal work arrangements involving contractors.[173]

Based on the premise that the labels engraved on the marble were merely intended to convey relevant information as concisely as possible to a limited audience, the presence or absence of numerals and various abbreviations becomes highly significant in itself. In other words, every term or abbreviation utilized in these inscriptions delivered a distinct and unique message, often reflecting the inner organization of quarrying operations. Moreover, the fact that all surviving labels had been engraved on the stone and not merely painted on may suggest (apart from issues of practicality) that at least part of the information had to be readable for a specified length of time, possibly for the duration of storage at the quarry or, following shipment, to ensure it could be read at the destination.

7.2.1. Dokimeion and the Upper Tembris Valley

A vast amount of epigraphic evidence survives on *pavonazetto* and white marble products from the Bacakale quarries near Dokimeion/mod. Iscehisar (Turkey) and its vicinity, which dates mainly to the years between AD 73(?)/92–236.[174] Most inscribed labels on quarried marble blocks note provenance, responsibilities, amounts produced, as well as the year of 'extraction'(?). The exact significance and meaning of these terms have not, as yet, been satisfactorily resolved. In addition, due to the ongoing exploitation of marble, the initial locations of the inscribed blocks and columns within the Roman quarries at Dokimeion remain unknown.[175]

Significant changes in the epigraphic formula can be detected: the inscriptions on blocks discovered in the Bacakale quarries prior to the year AD 136 occasionally list different consular dates, numbers and abbreviations, e.g.

[172] Bruzza 1870; Baccini Leotardi 1979; Baccini Leotardi 1989; Pensabene 1994; Maischberger 1997; Fant 2001.

[173] The perhaps earliest inscription from the year 2 BC on a marble block so far was discovered in Rome and edited by Alföldy 1992 = App. no. 1281.

[174] On dates: Fant 1989*a*: 29. See App. nos. 1–437, for a listing of all label inscriptions from Dokimeion. For description of *pavonazetto* and other marbles, see Fant 1993*a*: 164 f.

[175] Fant 1989*a*: 12–16.

App. no. 61. (a) *loco IV b(racchio) III*
(b) *Sur(a) III co(n)s(ule) II | CCXXXVIII*
(c) *RMA Pal(ma) II co(n)s(ule)*
(d) *VFR Vop(isco) co(n)s(ule)*
(e) *b(racchio) tert(io)*

Around AD 136 the epigraphic formula appears to have been altered dramatically e.g.

App. no. 227. *Te[rt]ullo et Sacerdote co(n)s(ulibus)*
ex of(ficina) Andaev(i) caesura Alex(andri)
loco XCIX b(racchio)R

The abbreviations used in the formulaic inscriptions from Bacakale/Dokimeion appear in quarry labels from other imperial quarries as well, and therefore are worthwhile examining more closely. However, it remains to be shown whether the terms carry the same meaning in different imperial quarrying districts.

7.2.1.1. *Bracchia and Loci*

The term *bracchium* is not documented in full in any inscription at Dokimeion, but is abbreviated to BR or a simple B followed by the numbers I, II, III, IIII or by the abbreviated ordinal numbers PRIM, SEC/GUND, TERT, and QVART. The reading of the letters B or BR as *bracchium* was proposed by Theodor Mommsen and has never been questioned.[176] J. Clayton Fant pointed out that *bracchia* have always 'been understood as indicating the branch of the quarry from which a particular piece was extracted'. In his opinion, *bracchium* is a topographical term describing subdivisions within the quarries of Dokimeion.[177] This interpretation has been tested by the appearance of inscriptions with differing *bracchium*-numbers on the same stone.[178] As the same *bracchium*-number is quite often indicated twice or even three-times on one stone (usually connected with an inventory), it is tempting to explain the rare appearance of discrepancies as simple mistakes.[179] Alternatively, the different notations of *bracchia* might reflect the movement of quarried stones from one *bracchium* to the next, in order to clear the access to the quarry sections, or the rearrangement of stockpiling areas—an action necessitated by the spatial restrictions of the Bacakale

[176] Hirschfeld 1905: 163 fn. 4.
[177] Christol & Drew-Bear 1986: 84; Fant 1989*a*: 19 fn. 8.
[178] App. nos. 25, 39, 81, 275.
[179] Fant 1989*a*: 91.

quarries (cf. 2.1.6). All inscriptions with differing *bracchium*-numbers also carry more than one consular date. One therefore might conclude that the transfer of stones from one *bracchium* or section of the quarry to the next between inventories was duly noted in the quarry labels.[180]

Like *bracchium*, the term *locus*, accompanied by a number (henceforward: L-number), is only rarely found in the early inscriptions from the Dokimean quarry. The term is a regular feature in inscriptions from AD 136 onwards, while earlier appearances are attested for the first decade of the second century AD.[181] Usually, *loci* have been understood to be subdivisions of *bracchia*, but Fant has argued that 'locus indications were unique designations, first, of the site where a block was being freed and, later, of the block itself functioning as an annual serial number'. Fant's thesis that L-numbers were reissued annually is supported by the appearance of the same numbers in different years and by the fact that no number is repeated within one year. Furthermore, a sequence of L-numbers within a year can be assigned to different *bracchia, caesurae*, and *officinae*.[182] It thus seems probable that Fant's assumption of an annually assigned serial number, without any topographical significance, might be correct. If so, these numbers give us an idea of the annual production rate. L-numbers up to CCCCXXXIII (433) are documented (App. no. 165, AD 147). On average, however, L-numbers are rarely higher than 250.[183]

7.2.1.2. Caesura *and Officina in the Epigraphic Formula after* AD *136*

Caesurae and Caesura-Holders. The term 'caesura' once probably delimited an area of extraction or a quarry within a quarrying district.[184] It may have been introduced around AD 136 and, at first, appears only irregularly in quarry labels.[185] From AD 147 onwards, the term 'caesura' is a fixed element in the labels. After its introduction, 'caesura' is accompanied by a name in the genitive case.[186] The name associated has been understood as referring to

[180] On *bracchia*, cf. Christol & Drew-Bear 1987: 99; Fant 1989*a*: 21. For corrections of Fant's readings and additional material, see Christol & Drew-Bear 1986; Christol & Drew-Bear 1987; Christol & Drew-Bear 1991: 113 fn. 2, 173 f.; Drew-Bear 1994: 842 f.

[181] App. nos. 61, 108, 109, 110.

[182] Fant 1989*a*: 19 f., tables 1 and 2.

[183] Fant 1989*b*: 215 fn. 54.

[184] *TLL* III, p.115.

[185] Fant 1989*a*: 30 f.

[186] *Caesurae*: Tullius Saturninus, *centurio leg. XXII Primigenia*, AD 136/8: App. nos. 113–17. Ael(ius) Ant(oninus), centurio(?), AD 147/8: nos. 157, 162, 163, 165, 166, 168, 171–8, 180, 182, 183, 185, 188, 190; PERCLDAED, AD 149: no. 201; Alex(ander), AD 149, 157–60: nos. 203, 221, 223, 226–32, 234–49, 257; Ael(ius) Zos(imus), AD 153/4, 158/9: nos. 210–12, 215, 217, 218, 222, 224, 225, 233; Manl(ius) Alex(ander), AD 154: nos. 213, 216, 219–21; Claudian(us), AD 159–61: nos. 250–4, 256, 258, 261, 262, 264, 268; Efor(-), AD 160: no. 255; Fabi(us) AD 160/1: nos. 259,

the person directly responsible for the extraction procedure, who is henceforth identified as a '*caesura*-holder'.[187] Who these persons were is only partly revealed by the onomastic material provided by the quarry labels. Closely linked with the introduction of a new 'accounting' system (using terms such as *locus*-numbers, *bracchia, officinae,* and *caesurae*) in AD 136, is the appearance of Roman military personnel responsible for a *caesura*. Tullius Saturninus, *centurio* of the *legio XXII Primigenia*, is named in inscriptions found at Bacakale and on *pavonazetto* marble in Rome.[188] Perhaps a specialist in quarrying technology, he was put in charge of a *caesura*, possibly being responsible for the change of the label formula. A further *centurio* appears in charge of a *caesura* in AD 147/8 (cf. 5.1.1).

After the initial 'period' of military personnel in charge of a *caesura*, other names appear without any direct indication of social status. According to Thomas Drew-Bear, Aelius Zosimus, Ulpius Yacinthus(!), Aurelius Dionysus, Aurelius Theophilus, and Aurelius Demetrius might have been imperial freedmen manumitted by the emperors of the second century AD, but Iulius Neophytus, Iulius Gaurus, Manlius Alexander, and Ostillius Elpidephorus do not fall into this category. Fant seems to suggest that the latter four *caesura*-holders were 'private' entrepreneurs, an opinion Drew-Bear did not support. As the first two holders of a *caesura* were both *centuriones*, he finds it hard to imagine that private contractors could have had free access to *caesurae* afterwards.[189] However, Drew-Bear's assumption that all *caesura*-holders were linked to the *familia Caesaris* is not verifiable on the basis of the onomastic material.[190] The notation systems from other known imperial

263; Domest(icus), AD 161–4, 177: nos. 265, 267, 271–5, 281, 289, 384; Athenob(us), AD 175: nos. 278, 279; Tit(us), AD 177–80: nos. 282–8, 290–3, 295–301, 393; Ant(oninus?), AD 178: no. 293; Vetus, AD 179: no. 300; Aur(elius) Dionusi(us), AD 194, 197: nos. 302, 303, 307; Iul(ius) Neophyt(us), AD 197/8, 200, 209, 220, 229: nos. 306, 310, 328, 345, 365, 371, 373, 374; (H)ostili-(us) Elpidephor(us), AD 197–9, 206, 209, 210, 213: nos. 308, 311, 312, 315, 316, 318, 319, 321–3, 339, 343, 347, 348, 350; N, AD 197?: no. 309; Ulpi(us) Yacin(thus), AD 198: nos. 313, 317; Aur-(elius) Theophil(us), AD 198–203, 205, 209, 212: nos. 314, 320, 329–34, 336–8, 344, 349; Iul(ius) Gaur(us), AD 200: no. 325; Episco(-), AD 200: no. 326; Herculan(us), AD 206: nos. 340–2; (Nova, AD 217: nos. 352, 354); Phoenix, AD 218, 220: nos. 358, 366; Mar(-) Romae, AD 222: no. 368; Maryllin(us), AD 222, 229, 235: nos. 369, 370, 372, 375; Var(us), AD 236: no. 376.

187 Hirschfeld 1905: 164, 169; Dubois 1908: 37; Dworakowska 1983: 135; Fant 1989*a*: 27 f., 33. Nevertheless, one name following the term 'caesura' is not a personal name. See *caesura Nova* (App. nos. 352, 354), Christol & Drew-Bear 1991: 154.

188 Hirschfeld 1905: 170.

189 Fant 1989*a*: 29–31; Drew-Bear 1994: 806.

190 Drew-Bear (1994, 806 f. with fns. 188, 189) argues that the adjective *noster* or *dominus noster* in consular dates including the name of the emperor, is only used by members of the *familia Caesaris*. Hence, if *caesura*-holders are not freedmen or slaves of the emperor, *noster* does not appear. The epigraphic record at Bacakale, however, does not corroborate Drew-Bear's view. Cf. also Weaver 1972: 154–7. The adjective *noster* in consular dates appears with the names Claudianus, Domesticus, Fabius, Titus and Vetus, cf. App. nos. 261–5, 268, 282–90, 297–301.

quarries, such as Teos, Chios, or Luna, have the names of imperial officials regularly followed by an imperial status indication, such as *Aug(usti)* or *Caes(aris)* (cf. 7.2.4, 7.2.5). Freedmen and slaves of the *familia Caesaris* generally took much pride in making their status known.[191] Moreover, the names of the *centuriones* on the label inscriptions were followed by rank and unit. Thus, space limitations of the labels would not have prohibited an indication of social status. However, the names of the *caesura*-holders such as Aelius Zosimus, Ulpius Yacinthus, Aurelius Dionysus, Aurelius Theophilus, Aurelius Demetrius, Iulius Neophytus, Iulius Gaurus, Manlius Alexander, and Ostillius Elpidephorus are not provided with any indication of imperial status. Even less can be said for *caesura*-holders such as Alexander, Claudianus, Domesticus, Phoenix, Maryllinus, Titus, Fabius, Athenobus, Vetus, etc.[192] In consequence, we must conclude that at least some of these *caesura*-holders were not associated with the *familia Caesaris*, but were private entrepreneurs, most likely members of local communities in Phrygia. One might subscribe to Fant's view that they were perhaps not of a low social standing.[193] Whether the *caesura*-holders were present at the *caesura* themselves and in direct control of the quarrying procedure (as Fant believes), or whether they simply managed the extraction of stone while pursuing other economic interests, cannot be established on the basis of the available epigraphic material.[194]

A *caesura* at Bacakale probably included more than one extraction site: the *caesura* of the *centurio* Aelius Antoninus covered three *bracchia* in AD 147 (I, II, and IIII) and a fourth *bracchium* is mentioned in AD 148 (III). The quarrying operations of Alexander's *caesura* comprised the *bracchium* III in AD 149, whereas further *bracchia* were included in AD 157–60 (I, II, IIII). Other *caesurae*, such as the *caesura Claudiani* or the *caesura Domestici*, seem to have produced stones in one *bracchium* only. Hence we may conclude that 'caesura' was not a topographical term and could cover several *bracchia* within one year. *Caesurae* were held for varying periods of time. While some *caesura*-holders are only attested for one or two years, Iulius Neophytus seems to have run a *caesura* for more than two decades. Up to four *caesurae* of different contractors existed simultaneously within the Bacakale quarries (AD 198), extracting marble in the *bracchia* III and IIII. (This diminishes the possibility of the names in the genitive case after 'caesura' being those of

[191] There are a few attestations of slaves or freedmen omitting their imperial status indication, cf. Weaver 1972: 2, 80–6.

[192] Perhaps some of them were *servi Caesaris*, as their Latin slave names suggest. Fant 1989*a*: 34 f.; Solin 1996*a*: 6 (Titus), 13 (Claudianus), 18 (Fabius), 121 (Vetus) 130 (Domesticus).

[193] Fant 1989*a*: 34.

[194] Fant 1989*a*: 35.

procuratores).[195] The insertion of the numbers II and III between the words 'caesura' and *Athenob(ii)* in some quarry labels might indicate that Athenobus ran at least two or three *caesurae* within one year (AD 175).[196] Furthermore, a *caesura* can be named together with different *officinae*. The *caesura* of Claudianus appears on quarry labels together with the *officinae* PELAG, CONTRA PERS, and EPHES, whereas the *caesura Alexandri* is noted only in connection with the *officina* ANDAEV.[197] This shows that the *caesura*-holders were not necessarily subsections of an *officina* but, as Fant pointed out, probably operated independently from the latter.[198] The observation that Tullius Saturninus', Aelius Antoninus' or Manlius Alexander's *caesurae* are not combined with any *officina* in the quarry labels, confirms the view of both *officinae* and *caesurae* being separate and independent entities.

Three further inscriptions provide rare insight into accounting procedures. No photographs of the inscriptions are provided by Fant, preventing the verification of his readings. Following the consular dates in nos. 295 and 296, the labels display the words *recep̣ṭi a Tito nomene* (sic) *alicuius*, which one might translate as '(stones) received from Titus, in the name of someone(?)'. No. 300 reads *recepti ex officina Prusaen[si—] | in locqum*(sic) *lapidum quọṣ receperat promutuo Titus ex caesura Veteris*, or '(stones) received from the Prusaean *officina* [—] in place of the stones which Titus had received as a loan from the *caesura* of Vetus'.[199] Titus appears in various inscriptions of the same year (AD 179), running a *caesura*. This Titus may be the same Titus of the above-mentioned inscriptions; however, in inscription no. 295, Titus—not his *caesura*—is named as the person ***from whom*** the stones were received. In comparison to inscription no. 300 this seems rather peculiar. There, the preposition '*ex*' indicates the place ***from where*** the stones were received. The inscription nos. 295/296 and 300 probably do not refer to the same transaction, but to two different 'events'. In nos. 295 and 296 Titus delivers quarried blocks to an unknown location on behalf of an unknown person. In no. 300 stones are delivered from the *officina Prusaensis* to an unknown location (presumably the same location as in nos. 295/296) as a substitute for a delivery of stones from the *caesura Veteris*. The *caesura*-holder Vetus had given quarried blocks as a loan to Titus, who had not returned the loan in time. The missing stones were then delivered by the *officina Prusaensis*. Apart from this speculative reconstruction, two important insights may be gained: firstly, the need to designate clearly who delivered the stone indicates a strict

[195] App. nos. 311–15, 317–19. Fant 1989*a*: 34.
[196] App. nos. 278, 279.
[197] App. nos. 250–4, 256.
[198] Fant 1989*a*: 35.
[199] Fant 1989*a*: 31.

accounting policy. The person under whose name a product had been quarried had to be easily identified. Secondly, the option of receiving quarried stones 'as a loan' from a *caesura* (if 'promutuo' in label no. 300 has been read correctly) is only possible when one assumes that the *caesura*-holders had to deliver a previously agreed number of blocks quarried during a limited period of time. If this obligation could not be met, additional products could be supplied from an *officina*.

Based on these observations, the *caesura*-holders may be regarded as contractors of quarry work in a *locatio conductio operis*.[200] The contractor perhaps received a fixed payment for working a *caesura*, and in turn was obliged to extract a set amount of quarried products, presumably within a limited period of time. If the obligations of the contract were not met, the *caesura*-holder did not receive the agreed payment (*merces*). In order to prevent such a penalty, quarried products could be borrowed on an informal loan from other *caesura*-holders or *officinae*, on the condition that the origin of the products was properly designated. Another possibility might have been a *locatio conductio rei*, in which the *caesura*-holders could lease a quarry section and pay a rent for doing so. Some of the quarried products might have gone to the state/emperor directly, whereas the rest may have been sold privately. While the white marble of Dokimeion was in high demand as material for *sarcophagi* and for local Phrygian 'door-stones' (funerary monuments in the shape of doors), the exclusive use of the *pavonazetto* marble in imperial construction contexts appears to rule out any private trade of *pavonazetto* during the high empire.[201] Consequently, the contracting out of quarries at Bacakale along the lines of a *locatio conductio rei* may be less likely. Perhaps some of the white marble extracted at Dokimeion even remained under imperial control.

Officinae. The term *officina*, combined with a name in the genitive case, is also introduced in AD 136, although not regularly used until after AD 157.[202]

[200] Kaser & Knüttel 2005: 225 f.

[201] Waelkens 1985: 643; Fant 1985; Fant 1989*a*: 3.

[202] *Officinae*: *Papia*, AD 136–47: App. nos. 115, 116, 158; *Pelagon(i)*, AD 146–77: nos. 154, 155, 160, 161, 206, 222, 225, 250, 253, 254, 258, 260, 264, 268, 270, 278, 279, 288, 290; *Asiatic(i)*, AD 147–64: nos. 159, 180, 217, 233, 271, 273–5; *Ephes(ia)*, AD 149/50, 160: nos. 202, 203, 208, 251, 256; *Andaev(i)*, AD 149–62: nos. 221, 223, 226–8, 230–2, 234–49, 255, 257, 259, 263, 265, 266, 272; *contra Persis*, AD 158–212: nos. 224, 252, 267, 340, 341, 349; *Neicaens(is)*, AD 177/8: nos. 280, 292; *Veria(na)*, AD 177: nos. 282, 285; *Lucilli(-)*, AD 177–9: nos. 283, 298; *Commodiana*, AD 177–206: nos. 286, 287, 291, 293, 297, 299, 301, 325, 327, 342; *Prusaen(sis)*, AD 179: no. 300; *Smurnaiorum*, AD 194–7: nos. 302, 303, 307; *Pers(-)*, AD 197: no. 306; *Bassiana*, AD 197–213: nos. 308, 311, 319, 322, 323, 343, 347, 348, 350; *Urania*, AD 198: no. 310; *Herculi(ana)*, AD 198–206: nos. 312, 315, 316, 318, 321, 339; *Severi(ana)*, AD 198–235: nos. 314, 320, 329–33, 336–8, 344, 369, 370, 372, 375; *Mar(tis/tialis)*, AD 198: no. 317; *Horolog(-)*, AD 199: no. 324; *Antoniana*, AD

The first few attested *officinae* are followed by anthroponyms in the genitive form such as the *officinae Papia(ni), Pelagon(i), Asiatic(i), Andaev(i)* or *Crescent(i).*[203] It has been argued by Christol and Drew-Bear that after the reign of Marcus Aurelius these anthroponyms were increasingly replaced by names honouring members of the imperial family and/or toponyms and names of gods. This does not necessarily mean that the old *officinae* had been abandoned and replaced by new ones—some simply might have changed their names.[204] *Officinae* named after members of the imperial family or their protective deities are the *officinae Veria(na), Commodiana, Lucilli(-), Bassiana, Herculi(s/ana), Sever(iana), Antoni(ni)ana, Mar(tis/tialis),* and *Alex(andriana).*[205] Place names such as *(officina) Ephes(ia), Neicaens(is), Smurnaiorum,* or *Prusaens(is)* perhaps indicate a special relationship between these 'workshops' and important cities like Ephesos, Nicaea, Smyrna, or Prusa.[206] Christol and Drew-Bear argue that the emperor had permitted these cities to receive *pavonazetto.* The epigraphic evidence at Bacakale does not answer the question whether these towns in Asia and Bithynia received the marble as a gift through the emperor or whether they were allowed to quarry the *pavonazetto* columns or blocks.[207] Even so, an inscription from Smyrna records that Hadrian, through the agency of Antonius Polemo, provided seventy-two columns of Synnadian marble for a gymnasium(?).[208] A direct link between the quarries and the town of Smyrna certainly existed, and the *officina* might have been opened to deal with this large order of columns in particular. In the quarry labels, however, the *officina Smurnaiorum* is attested only from AD 194 to 197. A similar explanation is given for the abbreviation [OFF(ICINA)] HOROLOG (no. 324), which is understood to have been opened for the construction or restoration of a monumental *horologium.*[209]

In addition to these above-mentioned *officinae,* some carry toponyms referring to their location at the *Persis* ridge; *officinae iuxta Persidem, contra*

209–29: nos. 345, 364, 365, 374; *iuxta Persidem,* AD 218–22: nos. 358, 366, 368; *Erc(-),* AD 229: no. 371; *Crescent(i),* AD 229: no. 373; *Alex( ),* AD 236: no. 376.

[203] Christol & Drew-Bear 1987: 109; Christol & Drew-Bear 1991: 123 with fn. 45; Drew-Bear 1994: 776 fn.75, 791–6.

[204] Christol & Drew-Bear 1987: 149 f.; Christol & Drew-Bear 1991: 171.

[205] Christol & Drew-Bear 1991: 121 with fn. 37, 131 f.

[206] Christol & Drew-Bear 1991: 121 f. with fns. 36–8; Drew-Bear 1994: 800 fn. 157.

[207] Christol & Drew-Bear 1991: 122 with fn. 40.

[208] *CIG* 3148 = *IGR* IV 1431 = *IK* 24/4, 697 (Smyrna). Millar 1992: 421 f.; Fant 1993*a*: 155 f. with fn. 73; Boatwright 2000: table 6.2.

[209] Christol & Drew-Bear 1991: 137: 'les officines dont les noms dérivent d'ethnique de cités devaient produire du marbre pour des constructions dans les villes nommés'.

Pers(–), as well as *Pers(–)* are attested.[210] Assuming our interpretation of the *officina* followed by the toponym *Persis* is correct, it seems plausible that *officinae* were situated within the quarrying zone. The *officina Andaev(i)*, for example, is noted together with more than one *bracchium* in AD 159.[211] We therefore may assume that the *officinae* were 'mobile' units, with the task of dressing the rough shapes extracted from the bedrock into the basic column and block shapes required.[212] Christol and Drew-Bear have opposed Fant's view by pointing out that some blocks carrying a label with the full *officina-caesura* formula are not dressed but remain in a rough state.[213] Labels could be engraved on the rough-cut stone ***before*** they were dressed, probably to indicate which *caesura* cut the stone and which *officina* would transform the rough stone into the desired item. Hence, an *officina* might have not only refined the rough shapes, but accompanied the dressing process from the start.

Provided that omissions were made deliberately and were not simple mistakes, the absence of either *caesura* or *officina* from quarry labels is significant. Between AD 136 and 154 we do find inscriptions where *caesura* is omitted and only the *officina* mentioned or vice versa.[214] The omission of one or the other component, though, does not only confirm the independence of *officina* and *caesura* but supports the hypothesis of two entities, at least in the initial stages of the Antonine period, performing the tasks of cutting and/or dressing the extracted stone. The inscriptions nos. 295, 296, and 300 support this hypothesis; according to these inscriptions, an *officina* could lend stone blocks. Hence *officinae* with town names such as *Prusaensis* must have had the capacity to cut stones as opposed to only dressing them.[215] From AD 154 onwards *caesurae* and *officinae* are both mentioned in the same inscriptions. This is understood as reflecting a division of work in which the stone is cut by the *caesura* and then dressed in the *officina*. The actual division of work procedures might have been more complex: perhaps certain *officinae* cut stone in their own right under the direct control of the imperial officials. However, it is not possible to determine the exact procedure on the basis of the available evidence.

COMM, REPR, AND ℞. An additional mark which appears together with *caesura* and *officina* in the Dokimean inscriptions is COM(M), combined with an ordinal number (as indicated by the abbreviation PRIM in

[210] Christol & Drew-Bear 1991: 155 f. with fn. 169.
[211] Christol & Drew-Bear 1987: 101.
[212] Fant 1989*a*: 35.
[213] Christol & Drew-Bear 1991: 157 f.
[214] See App. nos. 154, 155, 158–60, 202, 206, 208, 270.
[215] Christol & Drew-Bear 1991: 137.

no. 329).[216] Pensabene thought of resolving COMM as *comm(issura)*. His emendation was hesitantly approved by Christol and Drew-Bear, but rejected by Fant.[217] The interpretation of COMM as a further sub-(or super?-)division of a quarry is supported by the use of ordinal numbers, as is the case for *bracchium*-numbers. The *bracchium*-numbers, though, seem independent from the three attested COMM-numbers (I, II). Moreover, there is no significant correlation of the COMM-numbers with the *caesurae* or *officinae*. After 198/9, COM(M) II is not used further, while COM III is only attested in AD 180.[218] Whatever its significance might be, the abbreviation was introduced in AD 180 and remained a common feature in Dokimeion quarry inscriptions until AD 236, the latest dated inscription on Synnadic marble.[219]

Another abbreviation appearing from AD 147 onwards is the mark REPR, which, based on an idea of Otto Hirschfeld, was understood by Thomas Drew-Bear and Werner Eck as *repr(obatum)*.[220] According to this interpretation, a 'probator' checked the quality of the stone and decided whether the specimen was worthy of export.[221] If not, the stone was marked with the letters REPR. The vast majority of stones left behind at Dokimeion do not carry this mark, which raises the question whether they were unfit for export and thus discarded or simply stacked in the quarry. Drew-Bear and Eck have suggested that the barred letter R̶ is to be resolved as *r(eprobatum)* as well.[222] With this, the number of rejected stones at Dokimeion would increase from 8 to roughly 40.[223] However, the resolution of the mark R̶ as *r(eprobatum)*, *r(ecensitum)*, or *r(ecepti sunt)* cannot be supported by any written evidence from the quarries, nor can any semantic content be discerned.[224]

Procuratores and sub cura-marks. Finally, the marks on *pavonazzetto* marble in Rome and Ostia, as well as on two quarried blocks discovered at Bacakale, name imperial procurators, such as Irenaeus in AD 136 and Maro in AD 194.

[216] Cf. App. nos. 301, 311, 312, 314–18, 320–3, 325, 327–34, 336, 337, 340, 342–56, 358, 360, 365, 366, 368–71, 373, 375, 376.

[217] Pensabene 1978: 115; Fant 1989*a*: 27; Christol & Drew-Bear 1991: 122 fn. 42, 127 fn. 53, 168.

[218] *COM* II, cf. nos. 315, 317, 322, 323. *COM* III, cf. no. 301.

[219] AD 180 (no. 301), AD 236 (no. 376).

[220] Hirschfeld 1905: 169; Drew-Bear & Eck 1976: 314 f. See App. nos. 159, 175, 191, 199, 200, 269, 270, 411.

[221] Hirschfeld 1905: 164 with fn. 3.

[222] Drew-Bear & Eck 1976: 315.

[223] For R̶ (in various contexts), see App. nos. 1, 10–12, 14, 18, 19, 21, 22, 28, 31, 32, 34, 35, 37, 38, 40, 42, 63, 74, 78, 89–91, 94, 96, 98, 100, 103, 105, 108, 110, 214, 227, 255, 257, 263, 266, 268, 278, 282, 287, 291, 407, 425, 430, 434.

[224] On the mark R̶, cf. Christol & Drew-Bear 1987: 104 f.; Fant 1989*a*: 22; Christol & Drew-Bear 1991: 816; Peña 1998: 128.

Hirschfeld and Fant claimed that the *sub cura*-formula was applied at a later stage of shipment, preferably in Synnada.[225] Christol and Drew-Bear discovered two inscriptions, which completely disproved Hirschfeld's assumption and confirmed the notion that the *sub cura*-formula was applied in the Dokimean quarries themself.[226] The first appearance of the *sub cura*-mark in AD 136 coincided with the introduction of the *caesura-officina* system, which again confirms our notion of an intentional change in the way quarrying procedures were to be organized. The fact that military personnel as well as an imperial procurator were named, strongly suggests a swift implementation of this new system by the administrative authorities of the Bacakale quarries.

7.2.1.3. *Ex r(atione) + PN*

Another abbreviated term used sporadically in inscriptions from Dokimeion is the mark 'ex r(atione)'. This mark is, with one exception, not found on blocks in Dokimeion but in Rome, Lepcis Magna, and Ostia.[227] Fant argues that the Dokimean marble exported to Rome, Lepcis Magna, etc. was marked with a different epigraphic formula using the term *n(umero)* followed by a number (henceforward: N-number), a consular date and the EX RAT-mark plus a personal name. Fant argues that this type of label inscription was meant for trade and might have been inscribed in Synnada, while the other types reflected the internal quarrying administration at Dokimeion.[228] This is not quite correct: the attestation of a *ratio Syr(i)* at Bacakale in AD 150 obviously shows that *ex ratione*-indications were inscribed at Dokimeion, and not in Synnada.[229] The exact significance of the term *ratio* is uncertain. The long-held view, proposed by Hirschfeld and based on the collection of inscriptions from the marble-yards in Rome and Ostia by Bruzza and others, is that the *ex ratione* formula designates the contractors of a quarrying lease.[230] However, the *ex ratione* formula is only used once in Bacakale, which hardly fits the idea that sections within a quarry were contracted out to private enterpreneurs on a regular basis—a semantic range the term *caesura* already covers. In the case of the Dokimean quarries, the term 'ex ratione' followed by a name in the genitive form presumably identified an activity outside of the actual

[225] Hirschfeld 1905: 169, and 164 (on *sub cura*-marks in general); Fant 1989*a*: 27.

[226] App. nos. 298, 299. Christol & Drew-Bear 1991: 120–4.

[227] *Ex r(atione) Oly(m)p(i) Caes(aris)* (AD 96, Ostia): App. nos. 28, 29, 33, 38, 424, 430; *ratione (—)* (AD 136?, Lepcis Magna): no. 111; *rationi Urbicae* (AD 136, Rome): nos. 115, 116; *ex rat(ione) Cl(audii) Zel(oti?)* (AD 142, 150, Ostia): nos. 140, 141; *ex rat(ione) Syr(i) III* (AD 173, Bacakale): no. 276.

[228] Fant 1989*a*: 17 f.

[229] App. no. 276.

[230] Hirschfeld 1905: 166 with fn. 1; Fant 1989*a*: 18, 19 f.

extraction process. One might think of contractors responsible for the import and/or transport of marble to Rome and other destinations (cf. 7.2.7).[231]

7.2.1.4. *The Notation System prior to* AD *136*

The *caesura-officina* system of AD 136 replaced an older notation system, which operated with a different set of abbreviations. Although *bracchia* and *loci* do appear in the earlier inscriptions, they do not surface regularly. The usual components of an early inscription are a consular date, a number and marks such as RMA, R, HE, VFR, ANT, etc. One very distinctive feature is the appearance of up to four or even five consular dates on one stone.[232] These dates provide evidence for the periodic inventories of stones stockpiled in the quarries. Most of the marks encountered in the Dokimeion inscriptions of the period prior to AD 136 are difficult to interpret.[233] Even so, they appear closely connected to inventories, which occurred in the years AD 100, 105, 107, 109, 114, 115, and 116.[234] If we follow Christol and Drew-Bear regarding the periodization of the inventory procedure and only count the major inventories in the Dokimean quarries beginning with 92/3, 100, 105, 109, 114/15, we might be close to a quinquennial cycle, a *lustrum*.[235] The marks RMA and VFR seem closely related to these inventory dates. R and HE are marked together with the annually distributed serial numbers.[236] Based on these periodically repeated inventory dates, Fant argues that the 'intense scrutiny of the Domitianic and Trajanic periods suggests an eager hunt for usable material'.[237]

7.2.1.5. *White Marble and Pavonazetto Quarries at Bacakale and Upper Tembris Valley*

The discovery of quarry labels on white marble similar to those on *pavonazetto* has raised the question whether or not the marble from the quarries at Bacakale and in the Upper Tembris Valley was also extracted under imperial

[231] Mommsen read the mark VEC as 'vectura'; hence the name after the mark belonged to a transport contractor(?). It is curious, however, that the only two reported attestations of a *vectura* are found exclusively on marble labels in Rome (see App. nos. 304, 437), cf. Bruzza 1870: no. 297; Hirschfeld 1905: 164 with fn. 4.

[232] App. nos. 1, 8, 9, 10–18, 21, 24–6, 31, 32, 35, 36, 38–47, 51, 53–9, 61, 63, 68, 70, 72, 74–6, 78–82.

[233] On the interpretation of these marks, cf. Christol & Drew-Bear 1987: 105 f.; Fant 1989*a*: 23 f.; Drew-Bear 1994: 815–37.

[234] Drew-Bear 1994: 837–41.

[235] Drew-Bear 1994: 841.

[236] Drew-Bear 1994: 837.

[237] Fant 1989*a*: 29.

control.[238] The white marble discovered at Bacakale was locally distributed in the shape of 'doorstones', and used for sarcophagi.[239] However, no inscriptions on white marble have been recorded at the Dokimeion quarries themselves. One inscription on a white marble block, recovered at Argithani near Tyriaion/mod. Ilgin, 100 km west of Synnada, seems to use the older epigraphic notation system (prior to AD 136), and could therefore originate from the quarries at Dokimeion or the Tembris Valley.[240] Another inscription found on statuary marble is believed to come from Dokimeion. Even so, in a slight deviation from the normal formula the inscription, dating to the year AD 206, lists an Epictetus, an *Augu[sti servus]* and *proactor pro*[-], and then goes on naming the *caesura*(?) of Aur(elius) Demetrius, and presumably the procurator Aur(elius) Epity(nchanus?).[241] This would suggest that at least part of the white marble also was controlled by imperial officials. The exact origin of the stone, though, cannot be established other than through the similarity of the epigraphic formulas.

The region of the Upper Tembris or Porsuk valley is rich in *pavonazetto* and white marble quarries. A quarry of *pavonazetto* was found near Çakirsaz, 50 km north-west of Dokimeion, and other white marble quarries have been discovered near Altintash, Alibey, and Kurt Köy.[242] The inscriptions from these *pavonazetto* and white marble quarries resemble the epigraphic evidence on quarried products known from Dokimeion. Not only is the older notation system used (consular date and number), but between AD 92 and 156 inventories were apparently made in the same years as at Bacakale.[243] However, the change of the epigraphic formula observed at Bacakale for the year AD 136 does not appear in the quarry labels of the Upper Tembris valley. Despite the slim quantitative basis of epigraphic evidence from Kurt Köy, the same inventory dates and use of abbrevations for consular dates suggest that the *pavonazetto* quarries in the Upper Tembris valley were subsidiary quarries of the imperial venture at Dokimeion.[244]

[238] Waelkens 1982: 124 f.; Waelkens 1985: 642.

[239] Fant 1985: 660.

[240] App. no. 408.

[241] Bruzza (1870: no. 279) and Dubois (1908: no. 207) believed the white marble to be from Paros or Greece: 'Epictetus Augus[—] | proactor pro[—] | [l]oc(o) IIII | b(racchio) sec(undo) COM I | [Al]bino et Aemilian[o co(n)s(ulibus)] | [caes(ura)?] Aur(elii) Demetri(i) b[—] | [sub cu]ra Aur(elii) Epity[nchani?] | VE CAVRT [—].' Cf. also Christol & Drew-Bear 1991: 119 fn. 31, 171 fn. 223.

[242] Waelkens 1985: 642.

[243] Waelkens 1985: 643 f.; Waelkens 1986: 88–93, believes that the exploitation of the white marble quarries continued after AD 156. The quarries were leased out to individuals, supplying a local workshop with the material for their votive and funerary sculpture.

[244] Fant 1989*a*: 9 f.

7.2.2. Simitthus

Whereas some of the terms known from quarry labels on *pavonazetto* also appear in the quarry labels from Simitthus, the epigraphic formulae of labels on *giallo antico* is markedly different from the formulae known from Bacakale. Although the earliest epigraphic evidence for regulated quarrying procedures at the *giallo antico* quarries near Simitthus is given by a quarry label dating to the year AD 64, it is from the AD 100s onwards that most quarried *giallo antico* blocks are marked with formulaic label inscriptions.[245] These labels on marble products discovered at Ostia, Rome, and Simitthus note a consular date, an N-number, EX RAT mark followed by a name in the genitive form, as well as a further name in the genitive form,[246] e.g.

App. no. 788. *Sura III et Senici(one) II co(n)s(ulibus)*
ex rat(ione) Felicis Aug(usti) ser(vi)
d(e) n(umero) DCXII XXX
(officina) Tiluris

As at Dokimeion the term *officina* on *giallo antico* is also often combined with personal names, the names of divinities or of members of the imperial family.[247] The term also appears to have a topographical meaning, as observed at Dokimeion.[248] Moreover, 'officina' and 'lapicaedina' appear to be interchangeable expressions. A quarry label in Rome is noted as being *[ex no]vis lapicaedinis Aurelianis* (AD 161), whereas another block in Simitthus is from the *officina nova Aureliana* (AD 199).[249] It is therefore conceivable that the words *officina* and *lapicaedina/lapidicina* covered a similar semantic range, although the term *lapicaedina* is used only once. Moreover, as observed for Bacakale, the term *officina* may not only identify a location where a stone was cut, but also where it was dressed and refined. Blocks inscribed with the name of the same *officina* have been found lying together in clusters—an observation which has led Dubois, Röder, and Kraus to try to locate their area of

[245] App. no. 787; Rakob 1997: 3 claims that a quarry administration at Simitthus is attested from the reign of Tiberius onwards.

[246] App. nos. 788–94.

[247] *Regia*: nos. 821, 823, 826; *nova Aurel[iana?]*: nos. 834, 837; *Genii Montis*: nos. 835; *nova Cael(estis?)*: nos. 794, 838; *Iunonis*: nos. 829. For *officina Iunonis*: *AE* 1994: 1876 = Kraus 1993*a*: Neu 29; for *officina Caelestis*, cf. *AE* 1994: 1884. Friedrich Rakob argues for an imperial takeover of the Simitthus quarries in 28 BC, perhaps with the involvement of M. Vipsanius Agrippa (hence the *officina Agrippae*, nos. 822, 824, 825), cf. Lassère 1980: 41; Fant 1993*b*: 75 with fn. 8; Rakob 1993: 7 fn. 29.

[248] This is supported by an inscription of the later 3rd cent. AD which was found on the rockface of a subterranean quarry (together with a christogram) and designated the *off(icina)* 'found/*inventa*' by Diotimus, *agens in rebus*, cf. no. 871.

[249] App. nos. 834, 837.

extraction. They pinpointed the *officinae Age(ntii)* and *Genii Montis* at the 'Gelber Berg'/Yellow Hill, and the *officinae Cael(estis), Nov(ati), Regia* and *Agrippae* at the 'Tempelberg'/Temple Hill.[250]

Besides the term *officina*, the *ex rat(ione)* mark found occasionally on *pavonazetto* blocks at Bacakale and Rome was also used in in the epigraphic formula on *giallo antico*. However, at Simitthus the term disappears during the major change in the epigraphic formula around AD 137. The personal names in the genitive form following the EX RAT mark, such as Felix, Callistus, Plurianus, Puteolanus, Puteo, belong to *servi* or *liberti Augusti*, while Lae-(tus?), [..]lutaris(?), and Agentius may be members of the *familia Caesaris* as well.[251] None of them, so far, has surfaced in funerary, votive or honorific inscriptions at Simitthus itself; besides the slave name 'Felix', none of the other names appear in the epigraphic documents of Africa Proconsularis.[252]

Theodor Kraus believed these 'rationarii' were in charge of an administrative subdivision of the Simitthus quarry, which included several *officinae*. This does not quite fit the evidence. First, no direct connection or indication of hierarchy between *ratio* and *officina* is recorded other than the fact that both terms appear in the same inscriptions. Secondly, 'ratio' is used exclusively on quarried material and is not attested in graffiti or inscriptions on the rockface of the quarries. Therefore the connection between 'ratio' and the actual extractive procedure cannot be confirmed. Thirdly, none of the personal names connected with the EX RAT mark are, as yet, documented elsewhere in the epigraphic record at Simitthus.[253] The term 'ratio', therefore seems not to be directly connected with the quarrying procedure, nor did it describe a subdivision of a quarry. The absence of the names of 'rationarii' (i.e. the persons named after EX RAT marks) from the epigraphic record at Simitthus might imply that these individuals were not present at the quarries but located somewhere else.

7.2.2.1. *First Change of Formula around AD 137*

The epigraphic formula for quarried products at Simitthus receives a substantial make-over at the end of Hadrian's reign. The emperor is now mentioned in the genitive clause, indicating his ownership of the stone. The EX

[250] Dubois 1908: 32 f.; Kraus 1993*a*: 61; Röder 1993: 31, 33, 36, 38. Kraus follows Röder in the assumption that *officinae* were subdivisions of *bracchia*. The latter do not appear regularly in inscriptions and are not followed by digits, but by abbreviated numerals such as *s(ecundum)* or *t(ertium)* (see App. nos. 820, 830), cf. Kraus 1993*a*: 62.

[251] App. nos. 787–94, 796, 797, 800, 839–46, 848–52.

[252] Felix: *AE* 1979: 656 (Hr. el-Messaouer; C. Iulius Felix, *libertus Augusti*). *ILAfr* 409 (Karthago; Felix, *Augusti servus*).

[253] Kraus 1993*a*: 62; Khanoussi 1996: 1013.

RAT mark is replaced by *sub cura* plus the name of a procurator, whereas the consular date, the N-number, and an indication of the *officina* remain,[254] e.g.

App. no. 816. *Imp (eratoris) Antonini Aug(usti) Pii d*
*n(umero)*vac. *of(ficina) Certi*
Stloga et Severo co(n)s(ulibus)
su(b) cura Agathae

The significance of this change is unknown, but the indication of imperial property raises the question of why this became necessary. The identification of blocks or columns belonging to Hadrian or Antoninus Pius suggests that blocks or columns which were not owned by the emperor, were quarried at Simitthus at the same time (cf. 3.3.4).[255]

The replacement of the EX RAT mark with the *sub cura* mark and the name of the procurator apparently mirrors a change in administrative procedures. Up to the year AD 137, stones were cut on behalf of an imperial official in charge of a *ratio*, a person not necessarily recorded at Simitthus or within the African province. Beginning with the year AD 137, the responsibility for extracting blocks belonging to the emperor seems to be laid squarely on the shoulders of imperial procurators. Agathas, *lib(ertus) Augusti*, Iulianus, Maximus, Athenodorus are clearly noted as *procuratores Augusti*, whereas Iulius Gallus, Nova(tus?), and Hymenaeus are documented in the genitive form following *sub cura* or *caesura* marks (at Simitthus, a more or less secure indicator of their procuratorial rank).[256]

A further change in the epigraphic formula took place at the latest under Septimius Severus but presumably already in AD 156,[257] e.g.

App. no. 835. *n(umero) CCCV of(ficina) Genii Montis*
Imp(eratore) Commodo Aug(usto) IIII
et Victorino [I]I co(n)s(ulibus)
caesura Maximi proc(uratoris)

The 'sub cura' mark is replaced by the term 'caesura', but still names *procuratores* as the responsible officials.[258] If 'caesura' precedes the name of a person in charge of the extraction procedure of stones, the use of this term could

[254] App. nos. 804–9, 811–20, 834.

[255] The emperor himself used *giallo antico* for his villae in Tivoli and Antium, cf. App. nos. 925–7, and presented Athens and Smyrna with *giallo antico* columns, cf. Pausanias 1.18.9. *CIG* 3148, l. 42. Dubois 1908: 34; Fant 1993*a*: 155 f. with fn. 73.

[256] App. nos. 804–9, 811–20, 834–8. Agathas is not noted as *procurator* in the label inscriptions on *giallo antico*, but is named in votive inscription at Simitthus: Kraus 1993*a*: 56; *CIL* VIII 14551 = Dubois 1908: no. 97.

[257] Kraus 1993*a*: 62.

[258] App. nos. 835–8, 859?, 863–6.

imply that the *procuratores* became more closely involved with the technicalities of quarrying.[259] The procuratorship at Simitthus may have required a basic knowledge of quarrying procedures. Against this view, one might argue that the term 'caesura' is a substitute for 'sub cura', covering the same semantic range. This opinion is supported by the fact that either term precedes the name of an imperial procurator. Yet, it would hardly make sense to replace *sub cura* with *caesura* unless the primary purpose was a change in meaning. Therefore one might argue that the procuratorial duties at Simitthus were altered to include the supervision of extraction techniques, in addition to other administrative responsibilities. Whether this resulted in specialization on the part of the procurators cannot be answered on current evidence.

7.2.3. Mons Claudianus

Unlike other quarries, the administration at Mons Claudianus used a simple set of abbreviations for their quarry labels. Most blocks are only marked with numbers. In other cases numbers are accompanied by the letter N for *n(umero)* and the abbreviation πο, as well as a limited set of abbreviations such as RACLP, PD/PDN/PRDN, CB, or CEP. No variation in the basic formula is discernible over time, since consular dates or regnal years are completely absent, e.g.

App. no. 988. (a) *RACLP*
(b) *P D XVII*

The low numbers on the blocks and columns seem to refer to the amounts of blocks extracted in one quarry, as the numbers are repeated on stones in other quarries at Mons Claudianus. Some blocks from the same quarry are inscribed with two or three different numbers.[260] It is not known whether this is a result of regular inventories or whether it reflects different counting systems, such as a serial number referring to the annual(?) production at Mons Claudianus and a number for the extracted products within one quarry.

A group of inscriptions beginning with the letters *Π°* for πό(δες)(?), indicate measurements in cubic feet.[261] These measurements apparently do not refer to the size of the unhewn blocks, but to the quarried block in its dressed state. The probable size of the blocks was noted by the quarrying

[259] Kraus 1993*a*: 62.
[260] App. nos. 942, 945, 952, 953, 957, 958, 960, 979, 980, 989, 998, 1009, 1013, 1017, 1022, 1023, 1025, 1072, 1077, 1080, 1088, 1089, 1102, 1106, 1135.
[261] App. nos. 936, 937, 960, 990, 996, 1079, 1096, 1097, 1099, 1120.

administration on the stone itself. As an ostracon proves, a note was made for the archive; it gives the dimensions, location, and the N-number of a stone as a reference.[262] Moreover, there is evidence from Wadi Hammamat to suggest that a similar administrative procedure was being followed in the Basanites quarries: an ostracon gives a list of block-measurements in Roman feet (πόδ(ες)).[263]

Presumably, each of the 130 quarries had its own name: the names of *latomia*, such as Myrismos, Cochlax or Apol(lon) and Epikomos, Harpocrates, Nikotyche, Hieronymos appear on some inscribed blocks.[264] Moreover, the water distribution list recently published by Hélène Cuvigny, apart from listing παγανοί and φαμελιάριοι in the quarries, provided a number of additional names of quarries such as Traiana, Mesêe, Chresmosarapis (the Oracle of Sarapis), Apis, Hera, Dioscurides, etc.[265] Furthermore, a hitherto unpublished and barely readable ostracon lists some of the mentioned quarry names.[266] The ligature EP, used in quarry labels and discovered on the rockface of quarries, has been read by Adam Bülow-Jacobsen as an abbreviation for Ep(aphroditos), the μισθωτής documented during Hadrian's reign in inscriptions at Mons Claudianus and Mons Porphyrites.[267] The ligature is often preceded by the letter C which most likely must be read *c(aesura)*.[268] This reading is supported by an ostracon, so far only published in translation. The text identifies two blocks by their N-numbers, lists their measurements, and gives the exact location by naming the quarry and the person responsible for the *caesura*.[269] Hence, the administrative system apparently kept track of the size and location of the stone and under whose responsibility it had been produced. Few stones give us such exact information: one inscription found on the rockface in quarry 109, C EP EX LAT ARPOCR, is emended to *c(aesura) Ep(aphroditi), ex lat(omia) (H)arpocrati*.[270] Quite often the C for *caesura* does not appear in a label inscription and only the EP ligature, followed by a number, is engraved in the stone.[271] The ligature EP appears in a

262 Bülow-Jacobsen 1996: 51 f.; Peacock & Maxfield 1997: 224.

263 Kayser 1993: 126 f. and no. 20. The measurements of blocks were important information and also processed by other quarrying organizations, cf. Monthel & Lambert 2002: 111 f.

264 App. nos. 935, 993, 1067, 1129, 1136, 1137.

265 Cuvigny 2005*b*: 314 f.

266 *O.Claud.*inv. 6969. Bülow-Jacobsen 1996: 51 f.; Peacock & Maxfield 1997: 223.

267 *I.Pan* 21, 42.

268 App. nos. 933, 934, 1068–70, 1103–6, 1119, 1129.

269 'Caesura of Epaphroditos; Dionysos quarry. N 57 revised(?) 3.3·1·0.9 m. N59 revised(?) 4.5·0.9·0.75 m', Bülow-Jacobsen 1996: 59.

270 App. no. 1129.

271 App. nos. 964, 1006–10, 1058. Peacock & Maxfield 1997: 6.61.

vast number of quarries. Quite obviously the term *caesura* does not identify a subdivision of a quarry.[272] The inscriptions suggest that the term 'caesura' has no specific topographical meaning. Hence, the use of this term at Mons Claudianus is comparable to Simitthus or Dokimeion. Similar observations can be made for the mark cb; Bülow-Jacobsen interprets the mark as a ligatured CIP and reads *c(aesura)* ⟨*H*⟩*i(meri) p(rocuratoris)*.[273] Whatever the exact reading of the ligature is, the ostraca do indicate that *procuratores* could be in charge of *caesurae*.[274] The use of the mark CIP in numerous quarries at Mons Claudianus (provided it is read correctly) supports our interpretation of the term *caesura* as an expression indicating the administrative responsibility for a series of quarries, rather than the subdivision of a quarry.[275] Moreover, an ostracon which refers to a procurator who ordered the *tabularius* Athenodoros to report on the amount of quarried blocks lying in the procurator's *caesura* corroborates that *caesura*-holders did not necessarily have to be present at the quarry itself.[276]

7.2.4. Quarry Labels on Marble from Karystos, Teos, Chios, and Paros

In other quarries, such as Luna, Karystos, Teos, Chios, and Paros, changes in the epigraphic formula of quarry labels are not as marked as at Simitthus or Dokimeion. There, the inscribed labels usually consist of a few basic elements: a consular date, an *ex rat(ione)* mark followed by a name, and the abbreviations *l(oco)*- or *n(umero)* followed by a number (henceforth L- and N-numbers). This minimalist practice makes it far more difficult to identify any alterations of the formulaic labels, as in some cases a further element is simply added or removed. In addition, the small quantities of dated label inscriptions hamper the identification of such changes. This makes it hard to discern whether the addition or omission of an abbreviation from a quarry label is a mistake by the inscriber or an actual change in the epigraphic formula.

[272] Quarries 6: nos. 933, 934; 11: no. 964; 55: nos. 1006–10; 62: no. 1058; 75: nos. 1068–70; 89: nos. 1103–5; 90: no. 1106; 96: no. 1119; 109: no. 1129.

[273] Maxfield and Rengen believe the lower case b to be a ligatured LD, cf. Bülow-Jacobsen 1996: 61; Peacock & Maxfield 1997: 223.

[274] Bülow-Jacobsen 1996: 60.

[275] Quarries 19: nos. 971–5; 23: no. 994; 36: no. 1005; 93: nos. 1113–15; 113: no. 1130.

[276] *O.Claud.*inv. 6483.

7.2.4.1. *Quarry Labels on Cipollino*

These inadequacies certainly apply to the quarry labels found on *cipollino*, a marble type said to originate from the quarries near Karystos. The scarcity of datable texts prevents any sound observations on changes in formula. Moreover, the main bulk of the 162 recorded inscriptions on *cipollino* discovered at Portus and Rome consist of little more than ligatured letters of an abbreviated name and an N-number. Despite these deficiencies, a probable change in epigraphic formula can be detected between AD 132 and 134. In AD 132 the *cipollino* blocks are marked with a consular date, the name of the imperial procurator under whose *cura* the stone was extracted and shipped, the mark PR/PROB followed by the name of a *libertus Augusti* (?), an N-number, and a L-number.[277] In AD 134/5, again during Hadrian's reign, the formula is altered; besides the consular date, the inscriptions found in Rome consist of an EX RAT-mark plus *nomen* (in the genitive form) and an N-number. The personal names following the *ex rat(ione)*-mark are Valens, Orbius Natalis, Caecilius Marcianus, Ω Zoes(imus?), and Cla[udianus?].[278] The only dated label inscription from Karystos itself, found on a *cipollino* block in its harbour, bears the name of Caecilius Marcianus.[279] The inscription does not indicate any direct connection to the *familia Caesaris*. Neither his name nor the *gentilicium* are recorded in the epigraphic evidence from Karystos or Euboea, nor do the names of the others appear in the local epigraphy.

The personal names coupled with *ex rat(ione)* marks set aside, other abbreviated names documented in undated labels do appear in the epigraphic record of Euboea. Hymenaeus and Thamyrus, names appearing on labeled *cipollino* marble at Rome, should probably be equated with Hymenaeus Thamyrianus, the *a lapicidinis Carystiis*, and Thamyrus Alexandrinus, *dispensator Augusti*, both of whom are known from two inscriptions found near Paleochora/Karystos and Rome (cf. 4.2.2.2).[280] Besides Minic(i)us Sanctus and C. Cerialis, perhaps 'patrimonial' procurators, and Sergius Longus, *centurio* of the *legio XXII Primigenia*, a unique PR/PROB-mark, followed by the *nomen* Crescens in the ablative form, also appears on *cipollino* blocks (cf. 4.1.3). If our reading of PROB as *prob(atus)*, 'checked', or *probator*, 'approver,

[277] App. nos. 603–10, 618.

[278] App. nos. 612–17, 619–22 (AD 134–61).

[279] App. no. 617.

[280] Hymenaeus: App. nos. 623–30. Thamyrus: nos. 631–47. Dubois 1908: 117 f. *CIL* III 563 + 12289 (Paleochora), cf. photo in Pensabene 1994, fig. 286. *CIL* VI 8486 (via Ostiense in S. Pauli).

inspector', is correct, the mark might indicate that the quarried product had been checked by an offcial, a *libertus (Augusti?)*.[281]

7.2.4.2. *Quarry Labels on Africano*

The quarry labels on *africano* and grey marble, both originating from the Karagöl quarry near Teos, provide a sufficient evidential basis for the detection of changes in the epigraphic formula.[282] Formulaic inscriptions on *africano* marble are documented on blocks found in Rome as early as AD 64.[283] The inscriptions usually name the consular year (in which the stone was cut), display the term *ex rat(ione)* followed by a name in the genitive form, and an N-number.[284] A change perhaps appears around AD 116 on the *africano* blocks, as the L-number is introduced.[285] They do not necessarily replace the N-numbers and seem to disappear from the dated inscriptions on *africano* again after AD 132.[286] Perhaps L-numbers were assigned only in AD 132, presumably during one general inventory of the marble products in stock. During the same period, from AD 116 and 132 the use of the ℞ mark emerges on *africano* marble found at the Marmorata in Rome and coincides with the use of the L-numbers.[287] The names in the genitive form following the *ex rat(ione)* mark seem to belong to members of the *familia Caesaris.*[288] Marks of imperial slaves on *africano* appear from AD 64 to AD 80. Soon after, the mark *Cae(saris)* disappears from the dated inscriptions. One inscription in AD 86, however, notes a Tychus, known as a *Cae(saris)* in AD 77, simply as a *ser(vus)*.[289] As both appearances are within five years of each other, it is probable that in both instances Tychus is one and the same person. Thus, the letters SER have to be read as *ser(vus Caesaris)*.[290] The name Laet(us) belongs to an imperial slave as well and in his case, the letters SER for *ser(vus)* seem again to be used alternatively to indicate a *ser(vus Caesaris)*.[291]

[281] App. nos. 603, 604, 618. Dubois 1908: 116. On 'probator', cf. Hirschfeld 1905: 164. Other names or rather abbreviations like Ti. Eup., Cλ.Pol., C. Apol., Cla. Phl., CIA, and Cla. B., cannot be emended easily, cf. Dubois 1908: 117.

[282] Both grey and *africano* marble are believed to have been quarried at the same spot and were handled by the same administration, cf. *ratio* of a Mamius Licinianus, App. nos. 489. 547.

[283] App. nos. 460, 461.

[284] Fant 1989*b*: 214 with fn. 52.

[285] App. nos. 486 ff.

[286] App. no. 486 carries both L- and N-numbers nos. 494 ff.

[287] App. nos. 486, 488, 490–2.

[288] App. nos. 463, 467, 470, 475, 476, 480, 483, 543–6.

[289] App. nos. 469, 470, 480.

[290] For Tychus, see App. nos. 469, 470, 480, 544–6.

[291] For Laetus, see App. nos. 462, 463, 467, 475, 476, 536–9, 543.

The strongly ligatured marks HER and PAED(–), followed by the letters CAE(*saris*), are difficult to interpret.[292] Both were abbreviations for names, presumably of 'rationarii'; this is the case with the abbreviation of HER, whereas PAE(D–) is never coupled with an EX RAT mark. Apart from identified members of the *familia Caesaris*, further names are linked with *ex rat(ione)* in the genitive clause. These personal names found in labels engraved on *africano*, however, are not tagged with any indication of status; Mul(–), Dac(–), Cl(audius) Zel(otus), Lucilius (H)erma or Lucilius Herma Sotienus(?), Claudius Sotericus, Sex(tus), Her(-ma?/-molaos?), and Mamius Licinianus are perhaps private individuals. Moreover, their names seem not to be attested at Teos or in Asia Minor.[293]

Dated label inscriptions on grey marble found at Karagöl emerge by AD 164, two years after the last consular date documented on *africano* blocks.[294] Hence, the grey Teian marble began to be quarried after the *africano* deposits had been exhausted. The labels recorded on these quarried products also register a slight change in formula; the mark 'loc(o)', followed by a number, appears and is now used simultaneously with 'ex r(atione)'-indications.[295] The names coupled with *ex r(atione)* are too abbreviated to be of any help, nor are any imperial status indications provided.[296]

7.2.4.3. *Quarry Labels on Portasanta (Chian Marble)*

All quarry inscriptions on *portasanta* marble were discovered in the marble yards of Rome and Ostia. The identification of the *portasanta* marble with the *marmor Chium* is based on the discovery of the *portasanta* quarries at Latomi north of the town of Chios.[297] The first dated inscription on *portasanta* appears in AD 67 and adheres to a simple epigraphic formula: the consular date is followed by a name (in the genitive form?) and an N-number.[298] The only name noted in this system is a Ian(uarius?), presumably a *servus*(?) *Caesaris*.[299]

[292] HER: App. nos. 460, 461, 527–32 (AD 64); PAED(–): nos. 533–5 (AD?).

[293] Mul(-): App. nos. 485, 486 (AD 116); Dac(–): no. 491 (AD 132); Cl(audius) Zel(otus), no. 495 (AD 150); Lucilius (H)erma or Lucilius Herma Sotienus(?): nos. 496, 497, 521–4 (AD 152); Claudius Sotericus: nos. 496, 497 (AD 152); Sex(tus): nos. 499–501 (AD 162); Her(ma?): no. 500 (AD 162); Mamius Licinianus: nos. 489, 547 (AD 124). Claudius Zelotus and Claudius Sotericus were perhaps members of the local population, cf. Claudii Soterichi at Ephesos (IK 7: 3221), Kibyra (IK 60: 347); for the epigraphic record of Teos, cf. McCabe 1985. Lucilii, cf. Tyana and Anazarbos in Cilicia (*IK* 55: 122. 123; *IK* 56: 592); Eastern Phrygia (Akshehir/Philomelion: *MAMA* VII 197, doorstone). Mamii, cf. Prusa ad Olympum (*IK* 39, 134).

[294] App. nos. 502 ff.

[295] App. nos. 503–11.

[296] Diod(–): App. nos. 503–5, 508–11; Pol(–): no. 507.

[297] Yalouris 1986: 144; Fant 1993*a*: 164.

[298] App. no. 565.

[299] App. nos. 565, 583.

This 'notation system' was probably changed in AD 96 to include an 'ex rat(ione)'-mark.[300] The names in the genitive case following the letters EX RAT all belong to members of the *familia Caesaris*: Successus, *servus*(?) *Caesaris*, Thras(–) Latianus, *servus Caesaris nostri*, and Rest(itutus), *servus Augusti*, who appears together with Hya(cinthus?), *servus Augusti*.[301] The only exception to this rule might be an Aurelius M[-], either an imperial freedman manumitted under Antoninus Pius or a provincial who had received the Roman citizenship.[302] None of the *servi Caesaris* is otherwise known from the epigraphic record of Chios. Whether these members of the imperial administration therefore were present on the island remains doubtful.

7.2.4.4. *Quarry Labels on Parian marble*

Even the inscriptions on Parian marble adhere to the regular formula of consular date, 'ex rat(ione)'-mark plus name, and a number. There is not enough epigraphic material to observe any changes in the formula.[303] Two inscriptions on Parian marble blocks found at Portus provide evidence for the 'rationarii' Sext(us) and Her(ma) in the years AD 163/4.[304] Another group of inscriptions on white marble, vaguely identified by Baccini Leotardi as 'marmo greco', seems also to originate from Paros.[305] These stones carry the 'ex r(atione)'-mark followed by the name Aur(elius) Geor(-). Again, the epigraphic formula consists of an N-number, consular date, 'ex rat(ione)' mark plus name, and a separately noted 'locus'-number. The inscriptions date to the years AD 153, AD 160, and AD 161 and seem to precede the Sextus and Herma inscriptions. As yet, no Aurelius Geor(gus) appears to be documented in the epigraphic record on Paros, nor does the Greek personal name '*Γεῶργος*' appear on Paros.[306]

[300] App. nos. 572–81, 584–9.

[301] Successus: App. nos. 572, 573 (AD 96); Thra(-) Latianus: nos. 574, 575, 584–6 (AD 101–4); Rest(itutus): nos. 576–8, 587, 588 (AD 108/18); Hya(cinthus?): nos. 576–8, 587, 588 (AD 108).

[302] Aurelius M[ ]: App. no. 581 (AD 150).

[303] On three exceptional labels on statues of Parian marble, cf. App. nos. 1230–2. Bruzza 1870: 193 nos. 177, 278.; Hirschfeld 1905: 163 fn.4.

[304] App. nos. 1243, 1244. Probably as a result of an inventory of older extracted stones, the two blocks in question were reinscribed during that time in the Parian quarries, replacing the 'Hermo(-)-plus-L-number'-inscriptions, cf. nos. 1243, 1244, 1246, 1247, 1250–2, 1255. The relative chronology of these inscriptions, though, is not at all secure; the Hermo(-)-formula might be of a later date.

[305] App. nos. 1235–44, 1255–9. See Hirschfeld 1905: 166 with fn. 4; Fant 2001: 184.

[306] See *IG* F.5 P.2, Indices. '*Γεῶργος*', *LGPN* I 107 (Andros, 2nd–3rd cent. AD). II 92 (1. Athens, *c.* AD 177/8). III/1, 98 (1. Misenum: 1st cent. AD).

7.2.5. Luna/Carrara—A special case?

This study has so far focused on imperial quarries located in the provinces of the Roman empire. The quarries at Luna/Carrara, however, are situated not in a province, but in the rather different administrative setting of Italy. Moreover, the quarry labels, as well as the epigraphic record of Luna display, organizational traits different to the imperial quarries in the provinces. They therefore merit separate treatment.

The quarry labels recorded at the Carrara quarries near Luna can hardly be described as 'complex'. They mostly consist of abbreviated and ligatured names; numbers and consular dates are mainly absent. Their simplicity is reminiscent of the quarry labels found at Mons Claudianus. Even so, when examined in connection with the epigraphic record of Luna, some quarry labels do allow a reconstruction of the unique quarrying history at Carrara. A slab of Luna marble carrying an inscription, known as the 'lapis Salvioni', discovered in the quarries (Valle di Colonnata) in 1810 or 1812 and kept in the house of Salverio Salvioni in Massa Carrara, provides a detailed list compiled by Hilario, a *vilicus magister*, who in AD 22 notes the names of *decuriones* of each year.[307] The names only provide vague indications of the social status of these men. As the usual *tria nomina* are missing and only one name is given, the listed *decuriones* were probably slaves.[308] Moreover, Hilario's title, *vilicus*, confirms the hypothesis that we are dealing with a group of *servi*.[309] What Hilario's specific task as a *vilicus* was is not stated. Despite the rather significant findspot, in the immediate vicinity of the Carrara quarries, no reference to the quarrying business is made in our inscription. Hilario, who compiled the list, is also called *mag(ister)*, which designates him as a presiding figure of a *collegium*.[310] Furthermore, other 'members' of a typical *collegium* are mentioned in the inscription as well. The inscription lists the four *decuriones* for each year and has the names of deceased *decuriones* marked with a *theta nigrum*.[311]

The *lapis Salvioni* raises a number of questions: what kind of *collegium* are we dealing with, and, more importantly, who was the owner of these slaves? Slaves and freedmen usually were accepted members in *collegia funeraticia*, amongst which Kornemann includes the *collegia*, collectives consisting of

[307] *CIL* XI 1356 = *ILS* 7228; for location, cf. map Banti 1931: 480, no. 24.

[308] Waltzing 1899: 484. Latin slave names, cf. Solin 1996*a*; for Greek slave names, cf. Solin 1996*b*.

[309] Fabian 1979.

[310] Royden 1988: 14.

[311] Waltzing 1899: 484. *Collegia* apparently copied the titulature used by municipial officials, cf. Royden 1988: 14; Stevenson & Lintott 1999.

slaves and freedmen either belonging to the emperor or a private person, and the *collegia familiae publicae*, which are *collegia domestica* consisting of *servi publici*.[312] Based on Kornemann's observations, we must assume that the slaves listed as *magister* and *decuriones* on the 'lapis Salvioni' probably constituted a *collegium funeraticium*.[313] Inscribed monuments of *collegia funeraticia*, and *collegia* in general, usually appear in graveyards or in towns. In the case of Luna, the colony in whose territory the Carrara quarries were situated, three other collegia set up their inscriptions within the town.[314] The findspot of the 'lapis Salvioni' in a remote area therefore can only be explained by the existence of a graveyard in the immediate vicinity of the dispersed Roman quarries. Funerary inscriptions were unearthed at Torano, confirming the notion of individuals being buried close to the quarries; as yet, no burials as such have been discovered there.[315]

As noted above, no link with the quarrying ventures is indicated by the inscription itself, except for its original location close to one of the Roman quarries. Charles Dubois observed that the names of some of the *decuriones* known from the 'lapis Salvioni' emerge in inscriptions on marble blocks from the Lunensian quarries, albeit in abbreviated and ligatured form. Dubois read the marks PHIL, HILAR, TB as Hilarus/Hilario, Philo, and Tiburtinus.[316] Other *decuriones* like Solumarus, Primus or Caloteimus are presumably also named in labels on quarried blocks.[317] Dubois' interpretation of the ligatured letters is not beyond doubt, but together with the findspot of the 'lapis Salvioni' it remains a powerful working hypothesis. Hence, the connection of our *collegium* with the quarrying business cannot easily be dismissed.

But who owned this group of slaves? Though the 'lapis Salvioni' does not give any indication of ownership, the quarry inscriptions on marble found in the Carrara quarries do provide limited information. Unlike the quarry labels naming imperial slaves or freedmen, the slave names known from the *lapis Salvioni* and documented in the quarry inscriptions are not followed by an imperial status indication.[318] This opens the field for candidates of ownership other than the emperor. Our slaves could have been owned by a private individual or a company involved in the quarrying business. So far, there is no direct evidence for the existence of a *societas* being active at the Lunensian

[312] Kornemann 1901: 387.

[313] Although slaves also could join a *collegium fabrum*, they were not legally able to form a *collegium fabrum* on their own without the consent of their owners nor would the formation of such a legal entity appear likely. Waltzing 1895: 346; Kornemann 1901: 417.

[314] *CIL* XI 1355, 1354. *AE* 1983: 390. See Waltzing 1899: 482 f.

[315] *CIL* XI 6994 with *CIL* XI 1355; *CIL* XI 1377, 1384, cf. map Banti 1931, nos. 47, 48.

[316] PHIL: App. nos. 1162–5; HILAR: nos. 1147, 1148, 1150, 1151, 1153; TB: nos. 1157–60.

[317] Solumarus: App. no. 1154. Primus: nos. 1180, 1181, 1184. Dubois 1908: 6.

[318] Weaver 1972: 2.

quarries. Private individuals apparently took part in the extraction of Carrara marble: votive inscriptions found in or near the Carrara quarries document a number of *vilici* owned by private entrepreneurs.[319] Moreover, the fact that these altars and other monuments are dedicated by the *vilici* to the gods Silvanus or Hercules—both divinities appear very often in quarrying or mining contexts—confirms our notion of an association with the quarrying business.

Two other inscriptions found at Luna, albeit not in the vicinity of the quarries, might confirm the involvement of a senatorial family in the extraction and/or trade of Lunensian marble. The fragmentary honorific inscription discovered near Luna (and now lost) mentions a Scribonius Proculus and his office as *curator operum publicorum*. Municipal *curatores* usually were either elected by the community or, in the case of direct financial participation in a municipal building project, appointed from amongst the members of the community by the emperor.[320] M. A. Speidel came to the conclusion that Scribonius Proculus probably was the senator P. Sulpicius Scribonius Proculus, who together with his brother held the governorship of both German provinces between AD 63 and 67.[321] Moreover, P. Sulpicius Scribonius Proculus was *curator aedium sacrarum et operum publicorum* in Rome and apparently had good contacts in Luna. This is confirmed by an inscription at Luna of a probable ancestor of his, the senator L. Scribonius Libo, whose *libertus*, Diogenes, set up a richly decorated altar in AD 2. It was made of Carrara marble and dedicated to Silvanus, a god often addressed by quarry workers. Scribonius Libo therefore might have had a hand in the extraction and trade of Luna marble.[322] The question arises why a *curator aedium sacrarum et operum publicorum* in Rome would receive an honorific inscription in Luna. Speidel assumes that Scribonius Proculus, as a *curator*, was able to contract out orders for Luna marble to private quarrying enterprises, in which he was probably involved himself as a consequence of his ancestor's business interests. The private entrepreneurs in return erected a monument for their senatorial business partner at Luna in gratitude. That *curatores* hired private contractors for building projects is well documented in Rome and there is no reason why this should have been limited to the city of Rome itself.[323] It therefore appears

[319] Carrara: *AE* 1985: 391 (Le Canalie/Bedizzano). *CIL* XI 1319 (Luna). 1320 (Cava della Carbonera). 1327 (Colonnata). 6947 (Cava del Polvaccio). *CIL* XI 6946 = Mentella 1990 (Luna), cf. map Banti 1931, nos. 2, 11, 25, 41; Dubois 1908: 6.

[320] Liebenam 1900: 385; Kolb 1993: 58.

[321] *CIL* XI 1340; Dio 63.17.1, 3, cf. Eck 1985: 125–8; Speidel 1994: 213.

[322] Speidel 1994, fn. 19: *CIL* XI 6948, cf. photo, see Angeli Bertinelli 1978: 15, fig. 6, with map Banti 1931: 482, no. 40. Silvanus appears quite often in inscriptions at Luna, see *CIL* XI 6947, 6949, 6945. *AE* 1985: 391.

[323] Speidel 1994: fn. 28. For orders to private enterpreneurs by *curatores operum publicorum*, see Kolb 1993: 115–17.

possible that the *colonia Lunensis*, or members of this community, quarried the marble in their own right or had contracted out work and sold the produce to the *curatores* in Rome. Cassius Dio records that P. Sulpicius Scribonius Proculus and his brother were summoned to Greece in AD 67 by Nero and driven to suicide; the emperor, having spent a considerable amount of his treasury on various building programmes, wanted access to the wealth of the brothers.[324] Speidel speculates that the Scribonii's business in Luna marble might have led Nero to get rid of the brothers. The demand for Luna marble increased significantly after the fire of Rome in AD 64 and the emperor perhaps wanted to gain direct access to parts of the quarries.

That private individuals controlled part of the extractive procedures in the Carrara quarries through slaves and freedmen is recorded in the epigraphic evidence from Luna. All the same, it appears less likely that one of these private entrepreneurs owned the slaves noted on the 'lapis Salvioni'. In fact, an observation by Dubois suggests a public ownership of these slaves. Some of the names from the 'lapis Salvioni' inscribed on marble products in the Luna quarries were accompanied by a number and the mark COL. This abbreviation was emended to *col(oniae)* and believed to refer to the town of Luna.[325] Thus, the names documented in connection with these COL or *col(oniae)*-marks were, in Dubois' opinion, *servi publici* of the *colonia Lunensis*. Moreover, a recently discovered, but undated grave inscription set up by a Chresimus, a (*servus*) *colonorum Lunensium*, provides evidence for the existence of public slaves in Luna.[326] If we assume Dubois' interpretation is correct, it is possible that the slaves noted on the 'lapis Salvioni' were public slaves of Luna and, therefore, owned by this municipality.[327]

The appearance of the mark CAES for *Caes(aris)* in some inscriptions on quarried blocks from Luna, has been used to argue for an early takeover of the quarries by Tiberius, at the latest in the year AD 27 (see no. 1145).[328] Dubois pointed out that some of the marble blocks, inscribed with the names of *decuriones* known from the 'lapis Salvioni', furthermore carry the names of imperial officials, which were obviously inscribed at a later date. The case of the *decurio* Philo helped Dubois in determining the *terminus post quem* of the inscriptions added by the imperial officials: on the 'lapis Salvioni' Philo's name is accompanied by a *theta nigrum*, marking his death after the year 22 (when the inscription was set up). The imperial officials therefore must have

[324] Dio 62.17.1.

[325] App. nos. 1147–51, 1154–6, 1164(?). Dubois 1908: 6, cf. also Eder 1980: 112 f.

[326] *AE* 1998: 435 (Luna).

[327] *AE* 1997: 504; *AE* 1998: 436c.

[328] App. nos. 1157–61, 1165(?), 1173–9. Hirschfeld 1905: 176 fn. 3; Dubois 1908: 7; Fant 1988: 150.

inscribed their names after AD 22. Based on a further inscription, which provides only the name of a consul for the year AD 27 and does not name any imperial official, Dubois concluded that the quarries near Luna were taken over by the emperor Tiberius during the years 22 to 27.[329] Although it is true that no consular dates are used in the known 'colonial' quarry inscriptions at Carrara, this is hardly a sound basis to argue for an imperial takeover, especially when consular dates do not appear in an imperial context either and the consular date of the year AD 27 is the sole attested example. Hence, an imperial takeover of the Luna quarries during the reign of Tiberius cannot be proven. The presence of imperial officials is attested at Luna from the Flavian period onwards, and the existence of *tabularii rationis marmorum Lunensium* for the same period suggests that part of the Luna quarries had perhaps come under imperial control. Even so, keeping in mind the very dispersed nature of the quarrying sites at Luna, private and municipal quarries might have easily continued to exist even after imperial control had been extended over the quarries.[330]

7.2.6. Caesura-Holders: Contractors of Quarry Work

The presence of private entrepreneurs in the quarrying business at Luna, as well as at the Bacakale quarries near Dokimeion, prompts a closer look at the legal basis of their involvement in the extractive procedures of imperial quarries. As the analysis of the quarry labels at Dokimeion and Simitthus has revealed, holders of a *caesura* could be army officers or members of the *familia Caesaris* as well as private individuals. Epaphroditos, a *servus Caesaris* appears on numerous quarry labels from Mons Claudianus as *caesura*-holder. Moreover, in two inscriptions found at Mons Claudianus and Mons Porphyrites he is also titled *μισθωτὴς τῶν μετάλλων*, a contractor in a *μίσθωσις* or *locatio conductio*.[331] Concerning the nature of *locatio conductio*, however, there are different scholarly opinions. Fitzler believed that the Roman state farmed out the extraction of granodiorite to 'Kleinpächter', small-scale lessees, who were allowed to keep part of the produce. The *μισθωτής* Epaphroditos was an 'Abgabenpächter', a contractor or *conductor* responsible for the collection of the quarried products owed by the 'Kleinpächter' to the Roman state.[332] Fitzler thus assumed that a *locatio conductio rei* was the basis of the

[329] Dubois 1908: 7.

[330] Fant 1993*b*: 76 with fn. 13; Speidel 1994: 213 fn. 28.

[331] *I.Pan* 21 (Mons Porphyrites), 42 (Mons Claudianus), with extensive commentary by A. Bernand, p. 102 f. On the *μίσθωσις*, cf. Taubenschlag 1955: 354 f.; Kaser 1971: 563.

[332] Fitzler 1910: 117.

μίσθωσις, with Epaphroditos simply collecting rent from the tenants.[333] However, none of these small-scale lessees or 'Kleinpächter' have so far surfaced in the ostraca. Moreover, the marble from Mons Claudianus and Porphyrites appears to have been quarried exclusively for use by the Roman emperor and was certainly not thought to be of lesser quality. Tenants were probably barred from selling their blocks of granodiorite stone privately within Roman Egypt.[334] Consequently, Hélène Cuvigny argued that Epaphroditos was a contractor of a *locatio conductio operis faciendi*, a contract for the completion of work. At Claudianus this was the extraction of a certain number of stone blocks from a quarry, for which the contractor perhaps received payment.[335] A similar arrangement can be seen at Bacakale/Dokimeion where holders of a *caesura* (who could extract stone from different sections of a quarry) produced a previously agreed number of blocks during one year (cf. 7.2.1). These contractors were named in the inscribed labels on marble blocks for two reasons: firstly, the conductor may have been held accountable for a faulty product;[336] secondly, the imperial administration needed to check the actual number of quarried products extracted during a year by one *caesura*-holder in order to ascertain whether or not he had met his legal obgligations.

Other branches of the Roman economy were partially organized along similar lines: three papyri provide a glimpse at the legal arrangements of pottery production in Roman Egypt. These contracts were drawn up between the owner of a pottery kiln and the work contractor: for a defined period of time the contractor was allowed to use storage room, pottery kilns, potter's wheels and further tools for production in order to manufacture an agreed number of pots. The contractor was to provide a number of workers to reach the set goal of production. In return, the contractor received a fixed sum of money per produced unit, payable in rates and provisions, as well as wood, water, and clay from the owner. The contractor could keep any additional pottery produced for himself.[337] These third-century work contracts of Roman Egypt might well apply to other branches of the economy. The stamps

[333] Rostovtzeff 1957: 327; Flach 1990: 88–100; Millar 1992: 180 f.; Cuvigny 2000*a*: 18 f.

[334] Pliny, *NH* 36.57. Klein 1988: 38; Cuvigny 2000*a*: 19.

[335] Hirschfeld 1905: 166 f. Cuvigny pointed out that giving a lease to Epaphroditos, a *servus Caesaris,* in the context of Roman private law is an unique event, if not a legal absurdity, cf. Brunt 1990*c*: 383, 400; Cuvigny 2000*a*: 20 f. The contract of Epaphroditos probably was not limited to Mons Claudianus, but covered Mons Porphyrites and Wadi Hammamat as well, more or less the administrative area the quarrying procurator was responsible for (Mons Claudianus: *I.Pan* 42; Mons Porphyrites: *I.Pan* 21. Wadi Hammamat(?): *I.Ko.Ko.* 54, 55. Cuvigny 1992*a*: 76; Kayser 1993: no. 19.

[336] Kaser 1971: 571.

[337] *P.Oxy.* 3595 (AD 245), 3596 (*c.* AD 250 s?), 3597 (AD 260); cf. Strobel 1987.

on roof tiles suggest similar procedures for the tile production in Rome: apart from naming the land where the clay was collected (e.g. *ex pr(aediis)* of PN) and the clay bed (e.g. *ex fig(linis)* of PN), the name of the person in charge of production could also be given (e.g. *op(us) dol(iare)* of PN, *ex of(ficina)* of PN, *ex conduc(tione)* of PN). Furthermore, a consular date and the type of product was occasionally indicated as well. The formula was altered over the course of the centuries, apparently becoming more elaborate in the second century. Even though no exact details are provided on the nature of the contract, the possibility of a work contract on the basis of a *locatio conductio operis faciendi* may be the most plausible explanation.[338] Given the probable proliferation of such organizational arrangements, it is not surprising to find a similar system in quarrying enterprises. In view of the possible contract arrangements, as documented for the pottery production in Roman Egypt, the Bacakale quarries might have seen similar agreements, in which a yearly quota of *pavonazetto* blocks was delivered in return for a fixed payment; any white marble quarried during this period the *caesura*-holders may have been allowed to keep.

By far the largest number of holders of a *caesura* appear on quarry labels at Bacakale/Dokimeion. Whereas the names associated with the term 'caesura' at Mons Claudianus or Simitthus are those of imperial officials or of members of the *familia Caesaris*, the *caesura*-holders at Bacakale apparently also include private individuals. The majority of the latter *caesura*-holders, however, do not appear otherwise in the epigraphic record from Dokimeion. Occasionally homonyms appear on inscribed honorary, funerary or votive monuments outside the quarries; the information provided is never sufficient enough nor the inscription precisely datable in order to identify the named persons securely with *caesura*-holders known from Bacakale. The problem is illustrated by the following examples. The name of the *caesura*-holder for the years AD 194 and 197, Aurelius Dionysios, is attested in a fragmentary and barely datable inscription in the vicinity of the quarry: an Aurelius Dionysios and his wife set up a funerary stela for their son at Sülümenli, an imperial estate immediately south of Dokimeion on the road to Synnada. Apart from the fact that they share the same name, there is no evidence that Aurelius Dionysios from Anossa and the Aurelius Dionysios from Dokimeion were one and the same person. The name itself is common in Asia Minor, particularly after the grant of Roman citizenship to all provincials in the *constitutio Antoniniana* of AD 212.[339] One might speculate that the findspot of the

[338] Steinby 1978: 1516 f.; Aubert 1994: 232 f.

[339] *IK* 52, 258 = *CIG* 3883 (and add. 1101); Zawadzki 1960; Strubbe 1975: 245–8; Gibson 1978: no. 3. Aur. Dionysii/Dionysidorii: Ephesos: *IK* 17, 1577, 3712a, 3804. Stratonikeia: *IK*

inscription at Sülümenli indicates that a connection with the administration of the imperial domain ultimately led Dionysios to gain access to a *caesura* at the imperial quarries of Dokimeion. That he had the required financial assets and had reaped financial profits is indicated by the richly ornated stela at Sülümenli.[340] A further homonym of a *caesura*-holder of white marble quarries in AD 206, Aurelius Demetrius, appears in the epigraphic record at Dokimeion.[341] An inscription, set up for Aelia Maximilla, wife of the *procurator provinciae Frygiae* Aurelius Marcio, names an ἀρχιερεὺς τῶν Σεβαστῶν Aurelius Demetrius.[342] From his priesthood we can infer that Aurelius Demetrius also was an important member of the local elite at Synnada at the end of the second and the beginning of the third century AD.[343]

The equation of *caesura*-holders with homonyms recorded in the inscribed monuments from the environs of the Dokimean quarry is not a convincing argument in and of itself. However, circumstantial evidence shows that part of the wealth articulated in highly ornate and inscribed stelae from the surrounds of Dokimeion or the quarries in the Upper Tembris Valley must derive from proceeds gained in the quarrying business. This is never pointed out by the inscriptions directly; the depiction of tools and scenes closely linked to quarrying activity on the monuments, however, strongly implies an economic correlation of this nature. A few examples must suffice: a so-called 'doorstone' (i.e. a stela crafted in the shape of a door) from Dokimeion, depicting a 'stonemasons'-hammer, is engraved with the inscription of Chysogonos for Babeis and his children Kyrila and Mousaios.[344] From Dokimeion an uninscribed stelae shows a bearded man with a chiton and a cloak. In his left hand he holds a scroll, and an adze or pick, a typical quarrying or stone working tool, appears beside his right foot. This unfinished stela had likely been intended for an administrative figure involved either in the quarries at Dokimeion or running a workshop.[345] Another

22: 319, 449, 457, 472. Smyrna: *IK* 23: 385, 258. *IK* 24: 793. Prusias ad Hypum: *IK* 27, 8 II 45. Kios: *IK* 29, 107. Mylasa: *IK* 35, 441. Laodikeia a. Lykos: *IK* 49, 94. Alexandria Troas: *IK* 53, 95. Arslani/Sevrihissar: *CIG* 4096. Laodicea Combusta/Kadyn Khan: *MAMA* I 154. Apollonia/Senirgent: *MAMA* IV 185. Eumeneia/Ishiklu: *MAMA* IV 338 A/B.

[340] Gibson 1978: no. 3, Taf. IIIa. Despite the style of depiction and artwork, common to a whole series of similar gravestones found in the Aka Çay Valley, Aurelius Dionysios probably had to pay a considerable sum to the stone-mason. Waelkens 1977: 288 f.; Gibson 1978: nos. 1–4; Strubbe 1997: 178 f., no. 258.

[341] Dokimeion: *MAMA* IV 4. Other Aurelii Demetrii: *MAMA* IV 117 (Lysias); 131 (Metropolis); 356 (Eumeneia); VII 470 (Bulduk); 537 (Insuyu); *IK* 22, 446–50, 454–6 (Stratonikeia); *IK* 28, 379(?) (Iasos); *IK* 31, 44 (Klaudiupolis); *IK* 49, 93 (Laodikeia a. Lykos); *TAM* IV 1. 185 (Nikomedia); *CIG* add. 4325d (Olympus/Deliktash).

[342] *MAMA* IV 66 (Synnada).

[343] On imperial cult and priesthoods, see Price 1984: 62–4.

[344] Waelkens 1986: 201 no. 501; Strubbe1997: no. 256.

[345] Waelkens 1977: 288; Waelkens 1986: 202, no. 502, Taf. 74.

doorstone at Afyon (third quarter of the third century AD) names a *lecticarius/λεκτεικάριος*, basically a 'litter bearer'; besides the depiction of a *lectica* with a marble block on it, a cart laden with three marble blocks and drawn by a pair of oxen is rendered.[346] Another doorstone from Afyon, now lost, showed a *regula*, a ruler used by stone masons. The inscription records Italian immigrants, L. Tullius Montanus and his wife Alennia Venusta, the former perhaps a stone-mason closely associated with either the imperial quarries or a workshop.[347] A further monument depicting objects such as a *regula*, a stone-mason's hammer, and two yard sticks was discovered in the vicinity of Altintash in the Upper Tembris Valley.[348] A 'doorstone' at Akmonia is illustrated with stone mason's tools, indicating that the person named in the inscription, a certain Zenon, pursued this line of work.[349] A funerary altar at Tiberioupolis/mod. Kirgil is engraved with an epigram for Euprepes, a stonemason, who claims to have learned his trade from his father Trophimos and his mother Asklepiodora.[350]

Given the involvement of certain elements of the populace in the quarrying business, it is likely that the *caesura*-holders recorded at Bacakale originate from the local surrounds. Quarrying activity was undoubtedly not limited to the *pavonazetto* outcrops under imperial control, but also covered white marble quarries, the produce of which was widely distributed. Inscriptions on white marble sarcophagi found throughout Asia Minor refer to 'Docimean marble' as a brand.[351] The existing expertise in the region provided a suitable pool from which the imperial officials could recruit eligible contractors for quarrying operations at Bacakale. Based on this hypothesis, one may assume that other *caesura*-holders named in the quarry-labels, and not otherwise recorded amongst the inscribed monuments of Phrygia, are likely to have a similar background. In addition, the term 'caesura' and its use in quarry labels on *pavonazetto* suggests a certain involvement of *caesura*-holders in everyday quarrying processes:[352] 'caesura' is found in overwhelming numbers on blocks at Bacakale (*c.*155) and rarely appears on blocks in the marble-yards of Ostia/Rome or the shipwreck of Punto Scifo (6). Although this observation is far from conclusive, it supports the tentative notion that the term 'caesura' is not

[346] Waelkens 1977: 288 = *MAMA* IV 10; Waelkens 1986: 196, no. 486, Taf. 75.

[347] Waelkens, 1977: 288; Waelkens 1986: 191, no. 471, Abb. 53. For inscription cf. Mendel, 1909; Gibson 1978.

[348] Waelkens 1977: 288.

[349] Waelkens 1986: 167 f., no. 417.

[350] *MAMA* IX, no.198a. cf. Reinach 1890: 55, no. 9; Merkelbach & Stauber 2001: 200, no. 16/22/05.

[351] *IK* 39, 39 (Prusa ad Olympum) cf. commentary to l. 6; Humann 1898, nos. 56, 158, 209, 213, 323 (Hierapolis).

[352] This is especially well documented in the inscribed labels App. nos. 295, 296, 300.

automatically linked to the handling of the produce outside the quarry. Consequently, the contractors of a *caesura*, e.g. of quarry work at Dokimeion, were most likely to be found in the vicinity of the quarry. An example is provided by the *μισθωτής* Epaphroditos, who is not only recorded on the inscribed architectural element of the main temple at Mons Claudianus, but also appears on the quarry labels and the ostraca as a 'caesura'-holder (cf. 7.2.3). The involvement of private individuals at Carrara can also be observed from the epigraphic evidence there, although it is not clear whether they owned or contracted out quarries of Luna marble from the colony of Luna or the Roman state. Hence one can argue that the *caesura*-holders at Dokimeion were mostly private entrepreneurs contracting out quarry work. They either supervised the extraction process in their *caesurae* directly or, as wealthy members of the local society, had a slave or freedman supervise the quarry work.[353]

7.2.7. *Ex ratione* Marks, *Redemptores Operis*, and the Marble Trade

Despite the differences in the use of the administrative 'vocabulary' (as recorded in the quarry labels), we have observed certain trans-regional uniformities such as the appearance of the 'ex rat(ione)' mark followed by a personal name (PN) in the genitive form, belonging to either private individuals or members of the *familia Caesaris*. Hitherto, the 'ex rat(ione)' mark has been understood to precede the names of *redemptores*, whom Otto Hirschfeld believed had an intermediary role(?) between lessees and the administrative officials at imperial quarries.[354] He based this observation on two arguments: (1) that some 'ex rat(ione)' marks are followed by two names of slaves, and (2) that in one case the 'ex rat(ione)' mark is followed by the term *redemptor* in the genitive form.[355] Hirschfeld used the first argument to discard the view that the 'ex rat(ione)' mark referred to the haulier of a quarried block. He thought the quarrying procurator was in charge of expediting the quarry's produce,

[353] The latter view is supported by an inscription on a block of Pentelic marble found at Rome (no. 1260, AD 166), which reads *caesura Cla(udii) Hier(odis) Attici et Apolloni Lupi (...)*. Based on this reading, Dubois believed the quarries of Pentelicon were the property of Herodes Atticus, who in the second century AD used Pentelic marble for different building projects, cf. Pausanias 1.19.6, 6.21.2, 10.32.1; Philostratus, *Vit.Soph.* 2.550. See also Ameling 1983*a*: 84–94; Korres 1995. It therefore is possible that Herodes Atticus and his partner(?), Apollonius Lupus, had leased or owned the quarries. Walter Ameling, however, believes the named Claudius Hierodes Atticus is that of his freedman, cf. Ameling 1983*b*: 216; Fant 1993*a*: 167. Herodes Atticus was undoubtedly not involved in the everyday work processes at the quarry.

[354] App. no. 548. Hirschfeld 1905: 166.

[355] Hirschfeld 1905: 166 fn. 1.

and found it difficult to explain why two slaves would be mentioned in such a context.[356] Furthermore, he interpreted the genitive 'redemptoris' in this inscription as replacing the name of a deceased or retired entrepreneur, as it had been inscribed over an erased word.[357] Hence, Hirschfeld believed that the personal names following the 'ex rat(ione)' mark were those of *redemptores* of quarry work or lessees of quarry sections. His observations were followed by Thomas Drew-Bear, J. Clayton Fant, and others.[358]

However, there are some problems concerning Hirschfeld's interpretation of the 'ex rat(ione)' mark. First and foremost, Fant noticed that some of the names following it appear in label inscriptions on different marbles; Laetus, *Caesaris servus*, is named on *africano* marble from Teos and on *giallo antico* from Simitthus.[359] Similarly, the names of Cl(audius) Zel(otus) or Sextus and Herm(-olaus/-a), following 'ex rat(ione)' marks, show up in label inscriptions on *africano* and *pavonazetto* marble or on *africano* and Parian marble.[360] According to Fant's interpretation, this would indicate that Claudius Zelotus was a contractor of quarry work at Teos and Dokimeion, while Sextus and Hermolaus did the same in Teos and Paros. Laetus, however, might have encountered some considerable difficulties supervising extraction procedures at Simitthus and Teos simultaneously. This led Fant to suggest that, 'personal supervision of operations was not necessary. This means that the contractors were much bigger men than was apparent earlier and probably had hundreds of men (slaves?) working for them'.[361] Baccini Leotardi explained the appearance of the same name on blocks from different quarries as the result of labels with the *ex rat(ione)* mark being engraved at a 'punto di raccolta del materiale', a collection point away from the quarries, or at Ostia itself. Hence, she believed that the officials named on blocks from different places probably coordinated their transport to Rome.[362] Maischberger pointed out that labels with the 'ex rat(ione)' mark also show up on blocks in the quarries, which makes Baccini Leotardi's idea of 'punti di raccolta', where the blocks were newly inscribed, untenable.[363] Moreover, Maischberger argues

[356] Hirschfeld 1905: 165 f.

[357] Hirschfeld 1905: 166 fn. 1.

[358] Täckholm 1937: 118 f.; Fant 1989*a*: 18–20; Fant 1993*a*: 157; Rostovtzeff 1904: 454, understood these *redemptores* to be 'redemptores operis' i.e. contractors in a *locatio conductio operis*.

[359] *Africano* (AD 65, 75–80): App. nos. 462, 463, 467, 475, 476, 536–9, 543; *giallo antico* (AD 64): no. 787.

[360] Claudius Zelotus: *pavonazetto* (AD 142, 150): App. nos. 140, 141, 207; *africano* (AD 150): no. 495. Sextus et Herm(a): *africano* (AD 162): nos. 499, 500; Parian marble (AD 164): nos. 1243, 1244.

[361] Fant 2001: 173, table 3.

[362] Baccini Leotardi 1989: 108 f., 116 ff.

[363] Maischberger 1997: 22.

that label inscriptions were not applied in the marble yards at Rome and Portus but in the imperial quarries.[364] However, none of the 'ex rat(ione)' names is documented in the epigraphy of the quarries and their vicinity; this can be observed in the case of distinctive names, where *nomina gentilia* and *cognomina* are legible. Although there are a number of reasons for this dearth of evidence, the fact that the same name can appear on marble from different quarries makes it likely that these 'ex rat(ione)'-men or 'rationarii' known from imperial quarries were not involved in the quarrying process.

This interpretation is in part confirmed by additional observations: firstly, the 'ex rat(ione)' marks do not appear often on marble blocks in the quarries. Of the 104 'ex rat(ione)'-marks counted on marble of 'known' provenience (Parian marble, *giallo antico, africano,* Teian grey, *portasanta, pavonazetto,* and *cipollino*), seventy-five were found on marble slabs and blocks in the marble yards of Portus and Rome and one on a *pavonazetto* slab in Lepcis Magna. Twenty-nine were found directly in the quarries of Teos (9), Simitthus (18), Dokimeion (1) and Karystos (1). Hence the main bulk of 'ex rat(ione)' inscriptions were discovered at Rome. Given the numbers of inscriptions involved, this is not mere coincidence. Moreover, the 'ex rat(ione)' marks are rarely documented in imperial quarries such as Dokimeion or Karystos, and seem to disappear from the *giallo antico* labels documented in Simitthus after AD 134.

The preponderance of 'ex rat(ione)' marks on stones found at the ports of destination, mainly Ostia and Rome, and the appearance of similar names on blocks of different origin, makes Hirschfeld's explanation that the 'ex rat(ione)' names, aka *redemptores,* were contractors of quarry sections less likely. The term *redemptor* perhaps refers to a different kind of contractor. Given the fact that the marble yards at Rome and Ostia yielded most 'ex rat(ione)' inscriptions, a connection of the 'rationarii' with the capital's marble trade is possible. The term *redemptor* can describe any kind of contractor. However, numerous occurrences in inscriptions dating from the first to third century AD suggest that the most common use of the term was to designate contractors of building projects and construction work.[365] This usage seems very

[364] Maischberger 1997: 23 with fn. 80; contra: Ward-Perkins 1971: 148 fn. 35; Baccini Leotardi 1979: 44 f.; Pensabene 1994: 17, 325.

[365] Building and construction work (see also Martin 1989, 52–62): Crawford 1996, no. 24. *CIL* XI 4127 = *ILS* 6027. *CIL* X 3821 = *ILS* 3662. *CIL* X 3703 with *CIL* X 1614 = *ILS* 7731a. *CIL* VI 9794 = *ILS* 7672. *CIL* VI 607, 9851, 33873. *CIL* IX 3650, 4694. *CIL* X 1549. *AE* 1971: 88 = *AE* 1993: 465 = *AE* 1995: 307. *CIL* XIV 2091 = *ILS* 3512. *CIL* XIV 3530 = *ILS* 3512 = *InscIt* 4/1, 611. *IL Tun* 732 = *AE* 1940: 16. *IRT* 275. *Supp.It.* 2/V, 25 = *AE* 1925: 87. Other *redemptores operis* (not of construction work): *CIL* II^2/5 1022, ll. LXIX 30 f. *CIL* VI 9854, 9852. Mateo 1999: 33–87.

much to reflect the meaning of the term in the Digest and in some of the literary sources.[366] As a working hypothesis, I propose that some of these 'rationarii' may have been building contractors for public or imperial buildings or had contracted the task of providing the required building material for Rome or for other towns in the empire.[367] The latter is possibly expressed by the term 'redemptor marmorarius', documented in inscriptions found at Rome, Puteoli, and Lepcis Magna.[368] Fant's observation that a number of 'rationarii' are recorded on marble blocks of different origin seems to corroborate this interpretation. Consequently, we are provided with a possible explanation for the 'ex rat(ione)' marks; public building contractors or contractors of the marble supply ordered the marble blocks directly from the imperial quarries, the stone blocks being earmarked with the words *ex ratione* PN or '(credited) on the account of PN'.[369] Hence, the quarries reacted directly to the demand of marble by producing custom-made products—in other words, a 'quarry-to-order' system.[370]

At Simitthus the EX RAT-mark is documented on *giallo antico* labels for a limited period of time (AD 64–134) only. One might speculate that the change in epigraphic formula on *giallo antico* around AD 134—when EX RAT marks seem to disappear—would reflect a departure from a 'quarry-to-order' to a 'quarry-to-stock' system. The latter system probably saw blocks being quarried automatically without directly reflecting actual demand and being stockpiled at a certain location in or near the quarries. The change of production system might be an explanation for the vast numbers of *giallo antico*, *pavonazetto*, or granodiorite left in the quarries at Simitthus, Dokimeion, and Mons Claudianus. In turn, this is confirmed by the observation that the numbers of 'ex rat(ione)'-marks on marble blocks found in the marble yards at Rome or Portus are by a 2.5:1-ratio significantly higher (75) than those found in quarries (29). The advantage of a 'quarry-to-order' system lies in limiting the output of surplus products from the quarry to a minimum; the disadvantage is that the individual who demanded the product may be

[366] Alf. *Dig.* 19.2.30.3, 19.2.60.3; Marcell. *Dig.* 45.1.98.1; Papir. *Dig.* 50.8.11; Ulp. *Dig.* 6.1.39, 11.6.7.3–4, 42.5.24.1, 50.8.3, 50.10.2.1; Cicero, *Ver.* 1.56.146, 1.57.150, 1.55.145, 3.7.16; Cicero, *Q.* 2.4.2, 2.5.3, 3.2.3; Cicero, *Att.* 4.25; Cicero, *Phil.* 9.7; Vitruvius, 7.5.8; Tacitus, *Hist.* 1.27; Festus, p. 270 M.; Sallust, *Hist.* 3.85; Horace, *Ep.* 2.2.72, *Carm.* 3.1.35.

[367] On building contractors in Rome, cf. Kolb 1993: 115 f. with further bibliography; Daguet-Gagey 1997: 223–30. On their wealth, cf. *CIL* X 3821 = *ILS* 3662. *CIL* VI 607 with Martin 1989: 57 fn. 57, 58 fn. 70 f., 60 f.; Adam 1999: 46 fig. 94. For *redemptores* in a *locatio conductio operis* providing building materials, see Alf. *Dig.* 19.2.30.3; Ulp. *Dig.* 6.1.39; cf. Kaser 1971: 570.

[368] *CIL* VI 33873. *CIL* X 1549. *IRT* 275.

[369] Fant 1993*b*: 84.

[370] We cannot be certain that all names following the 'ex rat(ione)'-mark belong to *redemptores*. Imperial slaves could appear as 'rationarii' which implies (though not conclusively) that *rationes*, accounts, were also opened for non-contractors.

stuck with an eventual surplus, which probably is the reason for the high number of blocks with the label 'ex rat(ione)' being stored at the marble yards in Rome or Portus. A further downside was that the person ordering pillars or blocks would have to wait a considerable time until they were ready.

The Karagöl quarries at Teos where *africano* and grey marble were extracted, apparently had a 'quarry-to-order'-system in place from AD 75 to AD 166. No basic change in the epigraphic formula can be detected there. This might reflect the sparse supply in *africano* marble, making a change to the perhaps more wasteful 'quarry-to-stock'-system unfeasible. Similarly, between AD 96 and AD 150 the quarries at Chios seem to have extracted marble to order, as did the quarries in the hinterland of Karystos between AD 132 to 161 and the Parian quarries between AD 153 and 164. At Dokimeion 'ex rat(ione)' marks (with one exception) do not appear to be used at all; a 'quarry-to-stock'-system could have been introduced early on in order to cope with the high demand for this exotic marble. Nevertheless, the existence of 'ex rat(ione)' marks on *pavonazetto* at Rome/Ostia or Lepcis Magna (11) and at Dokimeion (1) does prove that certain *pavonazetto* products still were quarried on the direct orders of 'rationarii'. The 'quarry-to-stock' system also permitted the standardisation of measurements. John Ward-Perkins's observations on the standardisation and prefabrication of quarried products seem to fit well into this model.[371] The archaeological and epigraphic evidence for standardized column lengths, though, is not as clear as Ward-Perkins suggests: Hazel Dodge lists a variety of different non-standardized pillar sizes, but it is not certain when and under what circumstances i.e. in a 'quarry-to-order' or 'quarry-to-stock'-system, these columns were produced.[372]

A number of personal names following the 'ex rat(ione)'-mark were not those of private individuals but of imperial slaves or freedmen. In the years AD 96–100 we find an imperial slave (?) responsible for direct orders of *pavonazetto* marble, while the names of imperial slaves follow the 'ex rat(ione)'-mark in labels on *giallo antico* from Simitthus in AD 64–134. Even in the case of quarries like Teos or Chios, where direct imperial control cannot be proven, members of the *familia Caesaris* are documented as 'rationarii' in AD 64–100 and AD 96–118. If our interpretation of the onomastic material is correct, 'rationarii' could be private individuals or members of the *familia Caesaris*. For the former we have argued that they were *redemptores*, not of quarry sections but of the marble import to Rome or other cities, and perhaps connected with specific public building projects. The latter, imperial slaves

[371] Dodge 1988: 72; Ward-Perkins 1992*b*: 25.

[372] Dodge & Ward-Perkins 1992: 25 fn. 18, with further bibliography, cf. also Fant 1993*b*: 74 fn. 6; Maischberger 1997: 24 f.

or freedmen, were to be found in a similar function in Rome, perhaps as members of an institutionalized marble bureau.[373]

7.2.8. Similarities, Differences, and Changes of Epigraphic Formula

The differences between the quarry labels from the large imperial quarries at Mons Claudianus and the inscriptions found on *pavonazetto* and *giallo antico* blocks at Dokimeion and Simitthus are significant. The labels from the latter two quarries exploit the 'administrative language' to the fullest: the terms *officina, caesura, ex ratione, sub cura, l(ocus), n(umerus),* etc., are used regularly and consular dates as well as N- or L-numbers are engraved on most blocks. At Mons Claudianus, on the other hand, a less complex system was employed: N- and L-numbers are used, but consular dates, are absent. Moreover, the terms *bracchium, caesura, sub cura, ex ratione,* or *officina* are completely missing. Strongly abbreviated and ligatured marks appear instead, and it is only through the ostraca that they become legible. The lack of any consular dates is a significant clue, and the inscribers at Mons Claudianus felt no need to display the year a block had been quarried on a label. I have argued that the brevity of the quarry labels at Mons Claudianus is a result of the geological and topographical conditions there, i.e. the outcrop size of the desired granodiorite stone (cf. 2.1.1). The 130 quarries (presumably opened in different years) were spread out over the vast area of *c.*9 square kilometers. Extracted blocks could be left in the quarries and did not have to be moved in order to make room for further exploitation. Rather than continuously exploiting old sites, new quarries were opened at different locations (preferably as close to the Wadi floor as possible). The authorities only had to note the *caesura*, the name of the 'contractor' (Epaphroditos) and the serial number of the extracted block on the quarryied blocks. Any further information could be easily deduced, as the blocks remained at the quarry until they were exported. Additional information may have been stored in a central

[373] A quarry label on *pavonazetto* naming a *ratio urbica* might indicate that marble for Rome's public building projects could also be ordered directly, perhaps by public magistrates(?). App. nos. 115, 116 (AD 136). Perhaps the *curatores operum publicorum* were in charge of the *ratio urbica* (?), cf. (*AE* 1945: 80 = 1947, p. 13, s.n. 20 = 1950: 83) with Hirschfeld 1905: 177 fn. 2; Kolb 1993: 131–6, 294–7, Nr. 6. Hirschfeld read the first line of a quarry label on Carystean marble (App. no. 618 = Bruzza 1870: no.1 = Dubois 1908: no.278 = *ILS* 8717) as *ex m(etallis) n(ovis) Caesaris n(ostri) r(ationis) d(omus) A(ugusti)....* Based on this reading he argued the *ratio domus Augusti* to be an account for marble delivered to embellish the imperial palace at Rome. The latter *ratio* as claimed by Hirschfeld and based on his reading of *CIL* VI 8531 was part of the *fiscus castrensis*, cf. Hirschfeld 1905: 177 f. with fns. 2, 3, 316 fn. 2.

archive at Mons Claudianus. Consequently, the degree of complexity in label inscriptions on granodiorite stone from Mons Claudianus was low, a result of the spacious layout and favourable geological conditions of the quarries.[374]

The geological and topographical constraints imposed on quarries like Simitthus and Dokimeion limited the available space significantly. The extraction zones did not exceed a size of 0.32 or 0.2 square kilometers (cf. 2.1.6). Old quarries had to be used continuously and extracted products (along with the accumulating rubble) could not be left in the quarries, but had to be moved. Furthermore, different contractors extracted blocks from the small outcrops. The high density of quarrying sections, the removal of blocks, and the changing of contractors in a quarry section necessitated a more complex accounting regime; consular dates, the contractors' names (*caesura*-holders), the workshops (*officinae*), the extraction zone before removal (*bracchium*), and a serial number had to be recorded on the marble blocks. The higher complexity of label inscriptions also allowed the quarrying authorities to check more easily whether or not the contractors had met their obligations. The quarries of Simitthus and Dokimeion possess more or less similar topographical and geological constraints. Even so, at Dokimeion/Bacakale a whole list of abbreviations (particularly ligatured marks like RMA, HE, VFR, ANT, etc.) are documented which do not occur at Simitthus. Furthermore, the elements of the administrative 'language' are combined differently at both quarries. The epigraphic formula at Simitthus for the years AD 64 to *c.* AD 135/6 consists of a consular date, an 'ex rat(ione)' mark plus the names of the 'rationarii', probably the name of the *officina* and an N-number. In Dokimeion the surviving evidence shows the use of a different formula. There, a consular date is accompanied by a ligatured mark such as RMA, HE, or VFR, a *b(racchium)* number as well as a (serial?) number. The appearance of numerous consular dates on one stone documents major inventories of stockpiled blocks over several years. This procedure is not recorded at Simitthus. Both quarries seem to alter their epigraphic formula at the end of Hadrian's reign. In Simitthus the 'ex rat(ione)' mark is replaced around AD 137 with a *sub cura* indication plus the name of *procuratores*, whereas the name of the emperor in the genitive form appears alongside the obligatory consular date. This change has been interpreted as an attempt to intensify the control of the imperial officials of the extraction process at Simitthus and, therefore, we must assume that the 'ex ratione' system was phased out. At Bacakale (but not in the Upper Tembris valley), the *caesura–officina* system seems to have been introduced around AD 136, replacing the older notation system.[375] The new system,

374 Comparable sites are the Carrara quarries near Luna or the *cipollino* quarries near Karystos.

375 Fant 1993*b*: 84 argues for a transitional stage in the evolution of labels, and therefore sees the change taking place less abruptly.

used at first for *caesurae* run by *centuriones* of the Roman army, allowed a more effective control of the contractors and workshops, under whose responsibility stones were extracted and/or refined into the desired products. The term *officina* designated the workshop where the extracted blocks were roughly hewn into the desired shape. A few inscriptions, though, seem to suggest that certain *officinae* may, from time to time, also have extracted stones in their own right. The *caesura* mark (followed by a name in the genitive form) presumably indicated the contractor of a *locatio conductio operis*, a work contract; a certain production quota had to be met by the *caesura*-holder in return for a money payment. The appearance of *caesura*-holders at Bacakale, most likely private individuals not related to the *familia Caesaris*, indicates that work in the *pavonazetto* quarries was increasingly contracted out. Whether the new epigraphic formula also bears witness to the introduction of contractors for quarry work and the refinement of the quarried products in AD 136, or whether quarry work already had been contracted out prior to that date, remains unknown.

No such system can be traced in Simitthus. There the imperial officials seem to have had a firm control of the extractive procedure. No work was contracted out to private individuals, perhaps a consequence of using convict labour in these quarries.[376] Whereas no significant changes are introduced in the epigraphic formula in Bacakale (apart maybe from the introduction of the COM(M) mark in AD 180), the *sub cura*-mark at Simitthus is replaced by the term 'caesura', followed by the name of a *procurator* during the reign of Antoninus Pius or later. Perhaps the synchronicity of the change in formula observed for AD 136/137 at Simitthus and Bacakale reflects a centrally issued administrative directive, but in itself does not provide enough evidence to argue for a centralized administration for the imperial quarries.[377] An increase in demand for quarried products might also have triggered the administrative changes in these quarries.[378]

[376] Mackensen 2000 with further bibliography.

[377] Cuvigny 2000*a*: 9 f., observed a change in administrative practice at Mons Claudianus, as from AD 136 onwards the receipts for advanced payments are supplied with a date. It is doubtful whether this small change should be seen in the same light as the changes in epigraphic formula at Simitthus and Dokimeion.

[378] On Dokimeion: Fant 1989*a*: 26–8; Maischberger 1997: 21 f. An abrupt change also can be detected at Karystos in AD 134/5, after which the *ex rat*-mark is used regularly. Hazel Dodge believes that, '(i)ncreased production is indicated by the opening of new quarries at Karystos under Hadrian . . . and at Chemtou under Antoninus Pius and Marcus Aurelius.' Her assumption is based on label inscriptions from Simitthus naming an *officina nova Aureliana/Augustea* and on a inscription from Karystos, emended by Dessau, cf. Dodge 1988: 72 no. 618 = *ILS* 8717 (Karystos); App.nos. 833, 834 (Simitthus). Dessau emended the beginning of the first line as: *ex m(etallis) n(ovis) Caesaris n(ostris) r(ationis) d(omus) A(ugusti)*. . . . Although his reading of EX MN is plausible, no direct parallel can be cited in support of his view, cf. Hirschfeld 1905: 177 fn. 3.

Based on the evaluation of the label inscriptions, the differences between these two imperial quarries are quite fundamental. While one gains the impression that private individuals were more or less 'excluded' from the extractive procedures at Simitthus, private contractors were employed on a regular basis to extract *pavonazetto* from Bacakale. Moreover, quarries like Mons Claudianus or Luna display distinctive quarry labels, suggesting fundamental differences from Simitthus or Dokimeion. For labels from quarries like Karystos, Teos, Chios or Paros (no imperial control is directly attested for the latter three) discrepancies or changes in formula are barely noticeable. All quarry labels from these quarries, as well as those from Dokimeion and Simitthus, display *ex rat*-marks followed by the name of the 'rationarii'. As none of the names of the 'rationarii' appear in the epigraphic records of these quarries, these specific abbreviations might not reflect organizational structures within the quarries but in the trade of marble (cf. below). Moreover, terms like *caesura* and *officina* appear to have the same or similar semantic content in every imperial quarry where these terms are used. Despite these similiarities, the significant discrepancies in organization, which cannot be explained by reference to different geological constraints, certainly underscore the 'autonomy' of the men-on-the-spot in organizing their own extractive procedures.

8

The Emperor and Imperial Extractive Operations

8.1. METALLA AND IMPERIAL INTERVENTIONS

There are various questions which still remain regarding the position of mining and quarrying operations within the overall administration of the Roman empire. These issues include the assignment of equestrian procurators, members of the *familia Caesaris,* and military specialists to imperial mining and quarrying ventures; the establishment of equestrian procuratorships for a certain category of mines on a provincial or even supra-provincial level; the synchronicity of changes observed in the epigraphic formula at Simitthus and Dokimeion; the documented (albeit limited) involvement of emperors in regulating mining ventures at Vipasca or the redirection of Numidian and Phrygian marble by Hadrian to cities in Asia Minor, etc. Furthermore, there is the role of the emperor and his Palatine bureaux in running mines and quarries during the Principate.

8.1.1. Human Resources

The emperor undoubtedly played a central role in the allocation of human resources to extractive operations. His involvement is particularly well attested in the case of the dispatch of military specialists to imperial quarries: the *centurio* Annius Rufus was ordered to Mons Claudianus *ab optimo Imperatore Traiano.* Similarily, the *centuriones* Tullius Saturninus, Aelius Antoninus, and Sergius Longus were transferred across the empire to Dokimeion and Karystos respectively, in order to put their knowledge of quarrying organization to use. Given the distances covered by these military specialists, it is likely that their assignment to imperial quarries came about at the initiative of the emperor and his Palatine bureaux. Similarly, the *centuriones frumentarii* seconded to Luna/Carrara and elsewhere must have acted on the orders of the emperor (cf. 5.1.1).

The workings of a central authority were probably not limited to military officers. Other personnel contributing to the organization of mining and quarrying operations might also have been allocated to their respective work places by a central authority: this appears to apply to convicts condemned *in metallum.* Ulpian in his ninth book on the 'Duties of a Proconsul' points out that governors whose provinces did not possess *metalla* were to deport their convicts to provinces with such precincts.[1] The patristic sources occasionally refer to transfers of condemned Christians from their provinces to 'foreign' *metalla*: according to Eusebius, ninety-seven Christian convicts were moved from a porphyry quarry in the Thebaid to the copper mines of Phaeno/mod. Wadi Faynan in Palaestina in AD 309/10; a further group of Egyptian Christians was deported to mines in Cilicia and Palaestina.[2] A group of Christian detainees in Phaeno were later deported to Cyprus, undoubtedly to work in the mines there as well.[3] During the reign of Commodus, a certain Callistus was condemned by the *praefectus urbi* to the *metalla* of Sardinia, for disturbing a ritual at a Jewish synagogue in Rome. He was later released, after a letter was received by the local procurator of Sardinia to set Christians free.[4] Ulpian's passage in the *Digest* and the deportation of Christian prisoners from one province to another offers circumstantial evidence that a central authority coordinated the assignment of prisoners to *metalla.* Occasionally, the Roman emperor is attested as being directly involved in the distribution of prisoners (albeit prisoners of war) to important construction projects: Titus had prisoners from the Jewish war sent from Iudaea to quarries in Egypt[5]; Nero allocated 6,000 Jewish captives to the commencement of work on the canal across the Corinthian isthmus[6]; he also ordered those in custody to be deported to Italy in order to work on canal projects between Misenum and Ostia and those convicted of a crime to be condemned *ad opus (publicum).*[7] In the late empire, Diocletian and Maximian answered a letter from Julian, proconsul of Africa, on the issue of how to deal with persons of higher status (*honorati*) who were guilty of 'thinking Persian thoughts', i.e. Manicheans. The rescript prescribes the confiscation of property and condemnation of the pursuer of 'Persian' teachings to the *metalla* of Phaeno and Proconnesus (marble quarries).[8] The available evidence strongly suggests that the emperor (i.e. his Palatine bureaux) directed the flow of convicts and prisoners to the *metalla* on a regular basis. Perhaps the distribution of prisoners clothing also involved Rome: it is at least questionable whether the 6,700 *στολαί* ordered by

[1] *Dig.* 48.19.8.4. [2] Eusebius, *MP* 8.1, 8.13, 9.1.
[3] Eusebius, *MP* 13.2. [4] *Haer.* IX, 11–12; Millar 1984: 140.
[5] Jos. *BJ* 6.418. [6] Jos. *BJ* 3.540.
[7] Suet. *Nero,* 31.3. [8] *Coll.* 15.3.7.

the *praefectus Aegypti* from the guilds of weavers in the Arsinoite nome in AD 153 were intended only for Egyptian convicts.[9]

Besides convicts and prisoners of war, the emperors directed newly acquired subjects to mining works. The resettlement of indigenous communities in north-western Spain confirms Florus' report that after the Cantabrian Wars Augustus ordered the Asturians to be moved from their hilltop villages to the valleys in order to work in the gold mines (cf. 6.1.3.2).[10] The edict(s) of Augustus regarding the *castellani Paemeiobrigenses* underlines his role in the reorganization of northwestern Spain after the Roman conquest. Apart from tribute payments and other burdens on the community of Paemeiobrigenses, Augustus' grant of *immunitas perpetua* may also have freed them of corvée labour in the mines.[11] Florus furthermore informs us that after the Pannonian mutiny Augustus' legate C. Vibius Postumus put the local Dalmatian population to work in the gold mines of the Vrbas valley.[12] The propagation of such measures in other conquered territories is not directly indicated by our sources. Yet, the available written evidence might imply that such measures were commonplace in newly annexed territories. In a speech to the Caledoni on the eve of the battle of Mt Graupius, Tacitus has the chieftain Calgacus declare that he would rather be killed in battle against the Roman invaders than face deportation to *metalla*.[13] Calgacus' purported words could reflect the experiences of some British tribes immediately after the Roman conquest. Lead ingots from Flintshire further provide insights on the issue: apart from a consular date on the face of the ingots, mould marks on the front render the adjective attributes *Deceangl(icum)* or *Brig(anticum)*.[14] Both adjectives derive from the ethnics Deceangli/Decangi and Brigantes, tribes located in northeastern Wales and in Yorkshire, respectively. The 'Deceangli' probably came under Roman rule after the final subjugation of north Wales around AD 73/4 and 77, as did the Brigantes by AD 78. The consular dates for the years AD 74, 76, and 84, on the 'Deceanglian' ingots and for the year AD 81 on the 'Brigantian' lead pigs illustrate the rapid pace with which mines were up and running after the Roman conquest. Thus, the use of indigenous populations in these local mines in the years immediately after the conquest seems plausible.[15] This might have also applied to lead mines opened in the 'provincia Germania' between 8 BC and AD 9. Recent studies on the lead ingots from Brilon in the Sauerland east of the Rhine imply that mines were

[9] *P. Graux* 30, Kol. VII, Kambitsis 1997.
[10] Florus, 2.32.59 f.
[11] *AE* 1999: 915 = *AE* 2000: 760, cf. Alföldy 2000*b*: 177.
[12] Florus 2.25.12, 2.32.59–60.
[13] Tacitus, *Agr.* 32.4; cf. Lassandro 1995.
[14] *RIB* II 2404.31–6. 61 f.
[15] Jones & Mattingly 1990: 66–77; Birley 2004.

opened there immediately after the Roman conquest. Perhaps some of the indigenous tribes were initially required to mine lead on behalf of the Roman state.[16] In Dacia the measures to allocate a work force to the deserted gold-mining ventures were different from the other provinces. Eutropius claims that in the aftermath of the Dacian wars, which deprived the country of its populace, Trajan had numerous subjects *ex toto orbe Romano* moved to Dacia.[17] The Pirustae, Baridustae, Sardiatae, and other Illyrians documented in the epigraphic record of Alburnus Maior in Roman Dacia have thus been regarded as deportees from their ancient tribal area in Dalmatia. The Pirustae in particular, located in the metalliferous region of the Upper Drina valley, were probably well acquainted with mining techniques.[18] Provided the Pirustae and other Illyrians were moved en bloc to Alburnus Maior as an initial measure to restart mining activites, the emperor himself must have initiated or at least sanctioned such a radical measure.

The most obvious aspect of the emperor's role in the administration of imperial mines and quarries was the appointment of equestrians and freedmen to procuratorial posts. Apart from career inscriptions, there is other evidence that members of the *familia Caesaris* were promoted from an unknown position to mining and quarrying posts: the commissioning of inscribed monuments first in Rome, and later in Karystos is clear evidence for the movement of Hymenaeus Thamyrianus from Rome to the cipollino quarries in Karystos (cf. 4.2.2). Dorotheus, *procurator massae Marianae*, likewise set up an inscribed monument to Silvanus at Ostia, likely on the eve of his departure to Baetica.[19] Apart from the assignment of individuals to administrative posts in the provinces, the initial allocation of imperial personnel en bloc to mining and quarrying districts was likely decided in Rome as well. Tiberius' directive, for example, to seize control over the mines of Sextus Marius in Baetica must have taken place simultaneously with the

[16] The tribe of the Sugambri (located in the region south of the river Lippe, within which the lead mines of Brilon lay) may, in part, have been deported to Gaul and settled on the Rhine, cf. Suet. *Aug.* 21.1, *Tib.* 9.2; *Epit.* 1.7; Orosius, *Hist. adv. pag.* 6.21.24 f.; *RGDA* 32; Tacitus, *Ann.* 2. 26. We have no direct indication that L. Flavius Verucla, named in the mould marks of the ingots, was anyone other than the owner of lead ingots from the mines at Brilon (cf. 7.1.3). On the 'provincia Germania', cf. Eck 2004*a*: 63–112; Eck 2004*b*.

[17] Eutropius 8.6.2.

[18] *CIL* III pp. 944 ff.,VIII; *CIL* III p. 937, VI. Hirschfeld, 1905: 154 f.; Mrozek 1969: 141–3; Wilkes 1969: 173 f.; Mrozek 1977: 99; Noeske 1977: 275 f; Wilkes 2004; Piso 2004: 292 f. Even though Dacian names do not occur in inscriptions at Alburnus Maior or Ampelum, one cannot completely exclude the presence of Dacians in the mines, cf. Gudea & Lobüscher 2006: 58–60, with further bibliography. The archaeological finds at Alburnus Maior imply pre-Roman mining activity (B. Cauuet et al., in Damian 2003: 467–526). This indicates indigenous knowledge of mining operations which may have been put to use by the Roman conquerors.

[19] *CIL* XIV 52 = ILS 1592; for 'massa' cf. *TLL* VIII, p. 429.

designation of a procurator and subaltern personnel to the site.[20] One might argue similarly in other instances. The argentiferous lead mines of the Ruteni in Gaul, noted by Strabo, must also have fallen under direct imperial control early in the Principate: an inscribed monument for a *vilicus* Zmaragdus was commissioned by the *familia Ti. Cae[sa]ris quae est in me[tal]lis.*[21] The allocation of imperial personnel to extractive operations in newly conquered regions such as Gaul, Egypt, Noricum, Dacia, and other Danubian provinces, must have been decided by the emperor and his Palatine bureaux. Although direct evidence for the process of imperial decision-making is extant, the inclusion of mines and quarries in public or patrimonial property either by conquest, by condemnation of individuals (*bona damnatorum*), by appropriation of unclaimed domains (*bona caduca*), or by private bequest, must have prompted the allocation of members of the *familia Caesaris* and equestrian procurators to these extractive operations, provided direct imperial control over these public or patrimonial assets was intended.[22]

The involvement of the empire's central authority was perhaps not only limited to the initial dispatch of parts of the *familia Caesaris* to newly acquired extractive operations, as their transferral (apart from single procurators and other officials) from other provinces to long-established imperial mining and quarrying districts may have been monitored by Rome as well. The onomastic evidence from Mons Claudianus suggests that besides Egyptians, a small group of φαμελιάριοι originated from Asia Minor, Syria, and Palestine. These 'foreign' φαμελιάριοι are attested as recipients of advance payments; they appear to have received wages for their work and, as their personal names partly suggest, were probably not imperial slaves.[23] Their relocation from Asia Minor, Syria, and Palestine to Mons Claudianus is likely to have involved the Palatine bureaux at some point—provided these 'Orientals' did not migrate to Egypt of their own volition.

8.1.2. Costs and Revenues

In his treatise on the water supply of Rome, Frontinus points out that '*Caesaris familia ex fisco accipit commoda*'. The members of the *familia Caesaris* assigned to the water supply of Rome alongside the public slaves

[20] Tacitus *Ann.* 6.19.

[21] *AE* 1892: 23 = *CIL* XIII 1550, cf. Hirschfeld 1905: 156 f. For the silver-mines of the Ruteni see Strabo 4.2.2.

[22] Brunt 1990*b*: 141–6; Millar 1992: 158–74.

[23] Cuvigny 1996*b*: 20; Cuvigny 2000*a*: 30–6, 287–309.

thus received a 'salary' from the *fiscus (Caesaris)*.[24] At Vipasca the *liberti et servi Caesaris* are also recorded as receiving *commoda*.[25] The ostraca from Mons Claudianus provide evidence of advance payments to φαμελιάριοι, the source of these payments likely being the *fiscus Caesaris*. The quarry workers in the Egyptian Eastern Desert received wheat and barley through the regular mechanisms of grain taxation and requisiton (cf. 6.1.2.4), and thus one may argue that the upkeep of φαμελιάριοι was provided at least in part by public revenues.

The grouping of the *familia* in a *numerus* of Alabastrine and a *numerus* of Porphyrites with its respective sub-divisions, such as the ἀριθμός of Mons Claudianus, is recorded in the receipts on advance payments to φαμελιάριοι at the earliest in AD 136/7. Before *c.* AD 140 the authors of these 'cheirographa' (i.e. the recipients of the advance payments) did not provide their affiliation to a *numerus* or ἀριθμός. The change of practice, bearing in mind that the precise dating of all receipts is difficult, has been tentatively suggested to mirror the inception of these organizational groupings under Antoninus Pius.[26] However, this notable extension of the formula might be interpreted slightly different. The indication of affiliation was obviously not a legal requirement of the 'cheirographa' at Mons Claudianus, otherwise the attestation of the recipient's precise association to one of the numeri would occur in the 'Hadrianic' receipts on advance payments as well. The indication of numerus and/or ἀριθμός was of relevance to the person making the payment (in other words, the recipient of the *cheirographon*) and may in fact reflect the accounting procedures of the κιβαριατής and his proxies; the κιβαριαταί must have compiled a list on papyrus of those φαμελιάριοι who had already received an advance payment. We can only assume that the indication of *numerus* and ἀριθμός must have been noted meticulously by them. The purpose of this is indicated by those receipts which stray from the 'numerus of Porphyrites / arithmos of Claudianus' formula: one Zambas ἐκ ν[ουμέρου] [Ἀλαβα]σ̣τ̣ρῶνος ἀριθμοῦ [...], obviously received his advance payment in AD 146 from the κιβαριατής at Mons Claudianus; likewise those of the ἀριθμός of Tiberiane also received theirs at Mons Claudianus; in AD 148/9 one Serenus received his advance from an official of the *numerus* of Alabastron.[27] The ostraca thus might be interpreted as reflecting an accounting regime in which the salaries were noted per *numerus* and *arithmos*. This information would have

[24] Frontinus, *aq.* 2.118, cf. Boulvert 1970: 103; Boulvert 1974: 114; Alpers 1995: 124 f. In this context 'fiscus' clearly identifies the *fiscus Caesaris*, as the *familia Caesaris* is juxtaposed to the *familia publica* paid by the *aerarium (Saturni)*.

[25] *LMV* ll. 23–4.

[26] Cuvigny 2000*a*: 81 f.

[27] *O.Claud.* 527, 528, 535, 536, 587, 597.

undoubtedly been of interest not only to the quarrying procurator but to his superiors within Egypt (perhaps the idios logos?) as well. Ultimately, a final account of the costs, arising amongst other things from the deployment of members of the *familia Caesaris* throughout the province, must have reached Rome.

The hiring of additional personnel to the quarry works in the Eastern Desert, be they *παγανοί* or freeborn *φαμελιάριοι*, undoubtedly affected the costs of these ventures considerably.[28] Similarly, the contracting out of work to private entrepreneurs at Dokimeion (cf. 7.2.6) or the probable use of convicts at Simitthus (cf. 2.1.6, 8.1.1) had an impact on the balance sheet. The local freedmen procurators for the quarries in the Eastern Desert and elsewhere undoubtedly formulated their personnel requirements based on their knowledge of topographical and geographical constraints and on the demand for marble. Yet, one wonders if it lay within the brief of the freedman procurator to hire additional personnel besides the imperial functionaries allocated to the quarries.[29] A significant increase in personnel either by hire of free labour, contracting out work to private entrepreneurs, or by the allocation of convicts, all of which would have resulted in an increase of costs, could (in theory) have required the emperor's approval.

In general, the opening of new quarries under imperial control may have necessitated the approval of the emperor. A building inscription commissioned by Epaphroditos Sigerianos at Mons Claudianus relates to (quarry) work ordered by Hadrian.[30] Furthermore, toponyms like Mons Claudianus or Tiberiane might imply the involvement of Claudius or Tiberius in their opening; and Pliny recalls Egyptian 'marble' being named 'Augustan' or 'Tiberian' because it was discovered during the reigns of Augustus and Tiberius.[31] This tentative interpretation might also pertain to *officina* in imperial quarries: the *officinae Veriana, Commodiana, Severiana, Antoniniana*

[28] The 'marble' quarried at Mons Claudianus was predominantly used in public construction or imperial buildings at Rome, cf. the list in Peacock & Maxfield 1997: 321; currently our evidence very much excludes the idea of these quarries being run commercially. In the case of red porphyry from Mons Porphyrites a precise reconstruction of the archaeological context is nearly impossible given that the material was sought after and thus reused in various contexts, cf. Maxfield & Peacock 2001*a*: 306–19. The distribution of polychrome marble from imperial quarries seems to be primarily restricted to public and imperial buildings. This observation by Fant (1993*a*: 153–5) virtually excludes free market trade of this marble during the Principate. Thus, the imperial quarrying ventures may not have generated any revenues for the imperial treasury.

[29] The legal sources on procurators of private principals do not directly address the issue of hiring personnel; yet, one may assume that a *procurator omnium rerum*, for example, was relatively free in deciding matters of human resources. The procurator, however, was obliged to hand in a financial report on a regular basis which allowed the principal to take action when necessary, cf. Schäfer 1998: 78–87.

[30] *I. Pan* 42, l. 2.

[31] Pliny, *NH* 36.55.

or *Alexandriana* at Dokimeion, *nova Aureliana* at Simitthus, or *Traiana* at Mons Claudianus, perhaps came into existence after specific orders of marble by the emperor.[32] The decision-making process which led to the opening of new quarries might be illuminated in part by an episode reported by Pliny the Elder. The presentation of red porphyry marble to the emperor Claudius by Vitrasius Pollio and the former's distaste for it might not be merely a story of ill-advised gift-giving: Pliny claims that after Vitrasius Pollio no one else presented specimens of marble to the emperor.[33] One might argue tentatively that prior to this episode it was not unheard of to make presentations of marble samples to the emperor. Apart from financial considerations, the emperor may have based his decision whether or not to open a quarry on these marble samples.

The emperors perhaps not only ordered the opening of new quarries or quarry sections but may have also become involved in the operative processes in quarries. The alteration of quarry labels on *giallo antico* from Simitthus and *pavonazetto* from Dokimeion echoes a central directive from AD 136/7 which was issued in regard to these quarries only (cf. 7.2.8). Even though the reason for this directive remains obscure, financial issues must be taken into consideration.[34] Whether directives for imperial quarries were promulgated on a regular basis remains unknown; yet, given the emperor's role as principal initiator of building projects at Rome and in the provinces and as 'distributor' of produce from imperial quarries, it is likely that he would have taken an interest in the organisation of quarrying procedures.[35]

The emperor Hadrian perhaps 'intervened' in pecuniary matters in the mining administration at Vipasca. The *lex metallis dicta* refers to Hadrian's generosity (*liberalitas*) in the context of the 'down payment' of 4,000 HS for the *pretium* owed to the fiscus.[36] The text does not provide any precise indication of the exact nature of the emperor's *liberalitas*; one thus can only assume that he reduced the initial downpayment (cf. 7.1.1). Given the geological similarity of Vipasca to other mining areas in southern Spain, Hadrian might have extended his *liberalitas* to other mining areas in this part of the

[32] App. nos. 282, 285, 286 f., 291, 293, 297, 299, 301, 314, 325, 320, 329–33, 327, 336–8, 342, 344, 340, 369, 372, 375, 345, 364 f., 374, 832, 836.

[33] Pliny, *NH* 36.57.

[34] Whether an administrative change from a 'quarry-to-order' to a 'quarry-to-stock' system, possibly reflected in the modification of quarry labels at these quarries, was triggered by shortcomings in the supply of 'Numidian' and 'Phrygian' marble for public building projects is not evidenced by other written sources (cf. 7.2.7).

[35] On Hadrian having *giallo antico* and *pavonazetto* shipped to Athens or Smyrna for public building projects, cf. Boatwright 2000: 153–7, 160 f. with table 6.2.

[36] *LMD*, ll. 4–7.

Iberian Peninsula as well.[37] At least on regulatory issues the emperor dealt with imperial mines in general or with certain mining areas in particular. Mateo argues that both Vipasca regulations likely originated with Vespasian; the well documented attempts to regiment public land strongly suggest his systematizing approach in dealing with the empire's assets.[38] Even though Mateo's argument is compelling, one ought to refrain from assigning such measures to one specific emperor without any definite indications. Nevertheless, the general orientation of the Vipasca tablets may suggest an attempt to increase revenues from the sale of mining rights and the contracting out of auxiliary services, and facilitate the permanent production of metal ore from the mines (cf. 6.1.3.1, 7.1.1).

Whether fiscal considerations otherwise played a role in administrative decisions regarding imperial mines is rarely demonstrated by our written sources. A *lex* limiting the output of lead in Gaul and Spain and increasing output in Britain instead is reported by Pliny to have been influenced by the abundance and relative shallowness of British galena deposits.[39] This decision to limit galena extraction to specific sites certainly identifies an awareness of extraction costs by the central authority. The productivity of mines not only varied significantly from venture to venture but could diminish over time, rendering the continuation of mining operations useless. The opening and closure of mining districts under imperial control thus must have followed successful prospection and an estimate of possible returns.

The imperial freedman and members of the *ordo equester* appointed to mining and quarrying procuratorships were, like the private *procurator cui mandatum est*, required to provide regular accounts of their tenure (*rationem reddere*).[40] The addressee of these 'financial reports' was, in principle, the emperor. Not all procuratorial reports may have reached Rome directly; some of them were transmitted to the equestrian officials responsible for the financial affairs of the respective province before being forwarded, perhaps in an abridged version(?), to the Palatine bureaux. Even though our sources are mostly mute on this issue, there is some circumstantial evidence. The *beneficiarii procuratoris Augusti* documented in the inscriptions of Villalís suggest a close relationship between the freedmen procurators present in the opencast mining area of the Duerna valley and the equestrian financial procurator for Asturia and Callaecia in nearby Asturica Augusta.[41] Similarily, the *beneficiarii procuratoris* documented at Ampelum may indicate the

[37] Domergue 1983: 151–66, esp. 185 f.

[38] Mateo 2001: 178–96 with sources; cf. also Levick 1999*b*: Domergue 2004: 229–31.

[39] Pliny, *NH* 34.164.

[40] Paulus, *Dig.* 2.13.9.pr. Aubert 1994: 183; Schäfer 1998: 74–6.

[41] *CIL* II 2552 =*EAstorga* 114; *AE* 1910: 1 = *AE* 1928: 176 = *IRLeón* 40.

involvement of the financial procurator of Dacia Apulensis with the gold-mining administration (cf. 4.1.9).[42] A close connection with the gold-mining administration is furthermore indicated by a votive monument commissioned by C. Sempronius Urbanus, financial procurator of Dacia Apulensis.[43] Based on these observations one could argue that the freedmen procurators at Ampelum and in the Duerna valley were perhaps subordinate to the financial procurator of Dacia Apulensis and Asturia and Callaecia, respectively. Thus, the *rationes* compiled by these freedmen procurators were likely forwarded to the financial procurators. The epigraphic evidence from Sardinia offers a further indication: the inscription commissioned by the *procurator metallorum et praediorum* at the beginning of the third century AD suggests that he was directly subordinate to the presidial procurator of the island (cf. 4.1.8).[44] Furthermore, Flach surmised that the *LMD* perhaps originated with the financial procurator of Lusitania.[45] As for the freedman procurators managing the quarries of the Egyptian Eastern Desert, any *rationes* were perhaps presented to the *praefectus Aegypti*: the inscriptions and ostraca from Mons Claudianus and Mons Porphyrites as well as occasional written documents from the Nile valley throw a sidelight on the role of the governor in running the quarries (cf. 6.1.2.6). There is however no firm evidence to suggest a direct referral of accounts to the *praefectus Aegypti*; other officials of the provincial administration within Egypt, such as the *idios logos* or perhaps the *procurator usiacus*, could have received accounting reports. A similar subordination might have been the case with the freedman *procurator 'provinciae' Phrygiae* who may have answered directly to the patrimonial procurator of Asia.[46] Again, positive evidence is extant, yet one might tentatively cite the quarry labels on Karystean marble as a plausible parallel: they mention *procuratores Augusti* such as Minicius Sanctus and C. C(a)erialis, both possibly patrimonial procurators of Achaia. This suggests that the Karystean quarries fell within the brief of these equestrian procurators.[47] Hence, in the light of the sparsity of evidence one can only argue cautiously for a subordination of freedman quarrying procurators under an equestrian provincial procurator within the administrative hierarchy. In the case of freedman mining procurators a subordination, for example, under the equestrian financial procurators

[42] *CIL* III 7833 = 1289 = *IDR* III/3, 300 = *CBFIR* 565. *CIL* III 1295 = *IDR* III/3, 310 = *CBFIR* 566. *AE* 1991: 1344. Nelis-Clément 2000: 370, no. 186.

[43] *CIL* III 1298 = *IDR* III/3, 316 with Noeske 1977: 356 (AMP 35). See also Pflaum 1960–1: 542 f., no. 200, 1065.

[44] *AE* 1998: 671.

[45] Flach 1979: 413 f.; Domergue1983: 170 f.; Lazzarini 2001: 87–98, with further bibliography.

[46] Eich 2005: 303–8, esp. 307, with bibliography.

[47] App. nos. 604, 618. PIR^2 M 628, C 670, cf. Pflaum 1960–1: 1001 f., no. 139 bis, 1071.

for Asturia et Callaecia or Dacia Apulensis is plausible. Nonetheless, one can not exclude a direct link between the freedman procurators and Rome. The equestrian *procuratores aurariarum* (*Dacicarum*), *argentariarum Pannonicarum* and/or *Dalmaticarum*, or *ferrariarum* for Gaul and the Danube provinces probably answered directly to Rome.[48] Be that as it may, the *rationes* from mines or quarries under imperial control must have reached Rome either directly or as part of a financial report submitted by the patrimonial or financial procurator.

8.2. PALATINE BUREAUX AND EXTRACTIVE OPERATIONS

A number of scholars have postulated the existence of a central authority in charge of mining and/or quarrying ventures under imperial control. Hélène Cuvigny argued that the monthly wages paid to free workers at Mons Claudianus were similar to those paid to miners in the gold mines of Alburnus Maior. She believed this observation of a fixed annual pay 'for all free workers in the *metalla* (mines and quarries) of the Empire' confirmed the hypothesis 'of a centralization of the administration of *metalla* throughout the Empire'.[49] Cuvigny had based her hypothesis on J. C. Fant's idea of a centralized 'marble bureau' controlling the production as well as the trade of marble.[50] The abbreviations, marks and formulae on various marbles, apparently part of the same administrative 'vocabulary', together with epigraphic evidence for a *statio marmorum* in Rome, led Fant and others to believe in a central marble authority, which controlled the exploitation and import of the desired coloured and white marbles to the capital.[51] Moreover, Slobodan Dušanić's hypothesis of a central bureau for imperial mines appeared to corroborate this picture. On the basis of the evidence from the Illyrian provinces he argued that it was possible to detect the outlines of a centralized administration of metal production. This centralization was, he believed, prompted by the constant need to supply the mint at Rome with gold, silver, and copper on a large scale. Thus, Dušanić assumed that there was a special *tabularium* attached to the bureau of the *a rationibus*. In support of this model, he

[48] Eich 2005: 289., cf. Although the subordination of equestrian officials under prefects is documented for Roman Egypt there is no evidence to suggest that this was also the case with equestrian mining procurators. Bowman 1996*a*: 152f.; Bowman 1996*b*: 66f.; Eck 1997*a*: 142.

[49] Cuvigny 1996*a*: 145.

[50] Fant 1993*a*: 145 f.

[51] For an overview of recent discussion, cf. Fant 1993*a*: 157; Maischberger 1997: 19 f.; contra Hirschfeld 1905: 176; Dubois 1908: XXXVIII.

cited the coinage especially intended for the imperial mining districts in Noricum, Pannonia, Dalmatia, and Moesia Superior. Moreover, he argued that the appearance of similar groups of migrants at mining districts in the Illyrian provinces reflected not only a coherent policy of deportation on Rome's part, but were unthinkable 'without the help of a mining *tabularium* in Rome'.[52]

Given the range of connections between emperor and individual extractive ventures as set out in the preceding chapters, there is no doubt that the Palatine bureaux dealt with issues as they arose, appointed members of the *familia Caesaris* to mines and quarries, decided on the distribution of convicts to *metalla* throughout the empire and received *rationes* from quarries or mines under imperial control. Yet, the identification of a particular Palatine bureau and its specific part in the activities described above remains a difficult issue.

8.2.1. The Palatine Bureaux and Imperial Mines

As regards the mines under imperial control one very obvious Palatine official involved with these operations is the *a rationibus*. In his funerary poem on the father of Claudius Etruscus, Statius notes the overall responsibility of the *a rationibus* for revenues and expenditure. Amongst the former, mining ventures figure prominently. The text alludes to mining ventures in Spain and Dalmatia (*quicquid ab auriferis eiectat Hiberia fossis, Dalmatico quod monte nitet...*).[53] A further connection is perhaps demonstrated by Beryllus' inscription at Vipasca, according to which *rationales* (the *procurator a rationibus* and the *procurator summarum rationum*) could have had a hand in the appointment of the *vicarius rationalium* Beryllus as procurator of the mining district there(?).[54] The mission of Beryllus, namely the restitution of the Vipascan mines, may suggest that the bureau of the *a rationibus* may have been directly informed of the situation at Vipasca. Undoubtedly the *rationes* from the mines under imperial control directly or indirectly (i.e. via the financial/patrimonial procurators) reached the *a rationibus*. Yet, it remains uncertain whether the allocation of freedman procurators was a regular task required of the *rationales*, i.e. the *a rationibus*. In fact, the few documents on the posting and promotion of imperial freedmen do not mention a bureau in charge of personnel matters. The letter of recommendation written by Fronto on behalf

[52] Dušanić 1989: 154 f.

[53] Statius, *Silv.* 3.3.87 f. Hirschfeld 1905: 30 fn. 4; Boulvert 1970: 383 f.; Weaver 1972: 284–94; Eck 1997*c*: 149–56; Eich 2005: 163 f.

[54] *IRPac* 121 = *AE* 1908: 233(?). The ultimate authority for promotions to administrative posts was the emperor, cf. Suet. *Vesp.* 23.2. Eck 1995*d*: 24.

of the imperial freedman Aridelus was addressed to Marcus Aurelius directly, as was the letter by Pliny on behalf of Maximus to Trajan.[55] Furthermore, two letters of promotion written by an unknown emperor are addressed to the imperial freedman Ianuarius.[56] Given the range of different procuratorial posts at hand, the number of incumbents, and the multitude of transfers or advancements not only of imperial freedmen, but of equestrians and military officers as well, it is likely that records on such issues were held at Rome.[57] Based on Statius, *Silv.* 5.1.94–100, it has been assumed that *ab epistulis* perhaps ran a 'human resources' department or an archive concerned especially with the selection of qualified equestrians and military officers for certain tasks.[58]

The Palatine bureau involved in the appointment of freedmen procurators to mining or quarrying ventures cannot as yet be identified. Moreover, it is not certain that the administrative issues arising from the imperial control of mining ventures only affected one specific Palatine bureau. Whereas the accounts from various mining and quarrying procurators must have reached the *a rationibus*, the 'pool' of equestrians and military specialists and the appointment of these to the *metalla* in the provinces was perhaps managed by the *ab epistulis*. Several tasks may have been distributed amongst further bureaux and Palatine officials. Under these circumstances, there would not have been a need for a specific mining *tabularium*.

8.2.2. An Imperial 'Marble Bureau' at Rome?

As regards the administration of quarrying operations throughout the provinces, the epigraphic evidence for a *statio marmorum*, *procuratores marmorum* and staff has been interpreted as the written remnants of a 'marble bureau' at Rome.

8.2.2.1. *Locating the 'Marble Bureau'*

The physical remains of a central quarrying or marble authority, i.e. an office building identified by an inscription, has yet to be discovered at Rome. Nonetheless, in the years 1868–70 a large wharf with quays, ramps, and embankment walls, the Marmorata, was excavated on the left bank of the Tiber west of the Aventine hill. More than 1,200 marble blocks, as well as

[55] Fronto, *Ad M. Caes.* 5.37, p. 87 Nab; Pliny, *Ep.* 10.85; Weaver 1972: 269; Boulvert 1974: 120, 162; Eck 1997*a*: 105.

[56] *CIL* VI 8619. Weaver 1972: 256; Boulvert 1974: 163.

[57] For a list of freedman procuratorial posts, cf. Boulvert 1970: *passim*.

[58] Birley 1963–4: 207; Birley 1992: 23 f., 41–54; Eck 1995*e*: 139–41.

10,000 slabs and fragments of different stones, white and coloured marble, and tons of stone chippings were found.[59] In 1886 five blocks of *cipollino* marble, two blocks of *africano* marble and twelve cubic metres of *giallo antico* chippings were discovered in the area of the so-called Porticus Aemilia. Moreover, Rodolfo Lanciani uncovered what he called an 'officina marmoraria' in the area of the Horrea Lolliana. In an $18 \times 15\,m^2$ large court, Lanciani noted eighteen column shafts and segments, column bases and capitals, as well as smaller fragments of coloured and white marble.[60] The wharf on the left Tiber bank and the Emporium were probably the major stockpiling and sawing areas for marble imported to Rome.[61] However, the preponderance of pre-Trajanic marble blocks recovered from the Emporium, has led Maischberger and Fant to the assumption that for a short period the marble was not stocked at the wharf.[62] During the restoration of the quays under Trajan, the marble yard was temporarily situated either at Portus or somewhere along the Tiber.[63] In the 1938 excavations in the area of the municipal gasworks (AGEA), a further deposit of twenty-six rough-hewn stones was discovered, of which there were fifteen *cipollino* blocks or column fragments, six *africano* column fragments, four *bigio* column fragments and one alabaster column.[64]

Maischberger showed that the imported marble was not stocked at the Campus Martius. The various workshops located there in connection with the ongoing building projects explain the finds of unfinished marble columns and roughed out statues, but there is no evidence for stockpiling at a central location. Given the archaeological evidence (or the lack thereof), Lanciani's idea of a central marble office located under the church of S. Apollinare is doubtful.[65] He had based his view on three inscriptions found in the Campus Martius area. An inscribed votive altar of an *optio tabellariorum stationis marmorum* was unearthed in the Parione quarter; and a funerary monument of a *a marmoribus* discovered near the Mausoleum of Augustus.[66] In particular, it was an inscription on a lead water pipe naming a *statio patrimonii*, located together with quarry-state columns under the S. Apollinare church, which led Lanciani to locate the *statio marmorum* there.[67] First, the method of locating the *statio marmorum* on the basis of a lead pipe

[59] Bruzza 1870.

[60] Maischberger 1997: 84 and fns. 341. 342, with bibliography.

[61] Maischberger 1997: 61–93.

[62] Bruzza 1870: nos. 138–40, 191, 220 (Nero); 3, 147–59 (Vespasian to Titus); 168, 170, 193, 194, 199 (Domitian); 200 (Trajan); 4, 6–12, 181, 257–9, 266 (Hadrian); 223 (Antoninus).

[63] Maischberger 1997: 80 ff.; Fant 2001: 194.

[64] Maischberger 1997: 58, with further bibliography.

[65] Maischberger 1997: 107, 110–37, 140 fn. 487 with further bibliography.

[66] *CIL* VI 410 = 30760. *CIL* VI 8483 = *ILS* 1598.

[67] *CIL* XV 7315: *Imp(eratoris) Antonini Aug(usti) Pii stationis / patrimoni sub cura Dioscori.*

inscription in the Campus Martius does not withstand closer scrutiny. Even if a *statio patrimonii* were to be located under the S. Apollinare church, no direct link can be established to a *statio marmorum*. The find of columns in this context is not significant, as numerous and dispersed findspots of columns located in the Campus Martius area do not bear witness of a large-scale marble yard. Secondly, the other two inscriptions hardly support the idea of a central marble office in the Campus Martius; the funerary inscription of the *a marmoribus* does not refer to a *statio marmorum*. Moreover, the findspot of the *optio tabellariorum* inscription (which does refer to a *statio marmorum*) is less than certain. It presumably was first unearthed on the Aventine hill close to S. Sabina. The Vigna Ciampolini, where the altar was found, became the property of Francesco Losca in the first half of the sixteenth century. Together with other sculptures it probably was moved to his house in the Via del Governo Vecchio in the Rione Parione on the Campus Martius, where it was fitted into a pillar on the ground floor. There it was re-discovered in 1888. In the year 1993 excavations in the Crypta Balbi yielded a further inscription naming a *statio marmorum*. A votive altar for the Bona Dea Apollinaris, commissioned by an M. Ulpius Restutus, a *libertus Augusti* and *praepositus ex statione marmorum*, had been reused in the construction of an early church in the fourth or fifth century AD. Whether this votive inscription first stood in the temple of the Bona Dea Subsaxana on the Aventine is doubtful, as Restutus might also have set up the altar in a *sacellum* close to his working place. Hence, none of these inscriptions provides satisfying evidence for the location of the *statio marmorum*. Moreover, the term *statio* itself poses a further problem. Maischberger, following Christer Bruun's discussion of the term *statio aquarum*, has argued that *statio* can be understood as an administrative authority in the abstract sense, or a real office building to be located within the city's topography.[68]

Inscriptions referring to marble trade and its administration appear to be concentrated in the Aventine/Emporium area. The votive altar of the *optio tabellariorum stationis marmorum* was first discovered in the sixteenth century on the Aventine.[69] An altar devoted to Hercules by Primigenius Iuvencianus, a *servus*(?) of emperor Vespasian and *tabularius a marmoribus*, was found in 1737 on the banks of the Tiber at the Vigna Sforza-Cesarini together with 'molti alabastri e marmi pregiati'.[70] Furthermore, the funerary inscription of C. Tullius Crescens, a *negotiator marmorarius*, found at the 'via

[68] Bruun 1991: 257 ff. with discussion of literary and epigraphic evidence; Maischberger 1997: 122, no. 35, 140–2.

[69] Maischberger 1997: 63.

[70] *CIL* VI 301 = 30731.

FIGURE 23. Rome (after Maischberger 1997: Abb. 17)

del Tritone Nuovo', seems to indicate that this marble trader was presumably located in or near the Horrea Galbana.[71] Another funerary inscription of a Marcus Aurelius Xenonianus Akylas documents that this person maintained the first(?) station in the Horrea Petroniana and was the 'finest' stone-trader.[72] The findspots of these inscriptions suggest that the main activity of marble trade in the capital was quite obviously located in the Emporium. This circumstantial evidence thus may allow us to tentatively locate the bureau of an imperial administrative authority for marble in this area.[73]

71 *CIL* VI 33886.
72 *SEG* IV 106 = *IG* XIV 2247 = *IGUR* II 413.
73 Maischberger 1997: 51 f.

8.2.2.2. *Defining the 'Marble Bureau'*

The location of a 'marble bureau' set aside, the description of its organizational outlines, development, and role is also problematic. The earliest attestation of an imperial official perhaps connected with the marble trade is provided by an inscribed statue base from Nepet/mod. Nepi north of Rome, commissioned by the *ab marmoribus* Hermeros Thyamidianus, a slave of emperor Claudius.[74] Three further inscriptions of Flavian date identify *tabularii (rationis) marmorum Lunensium*: an inscribed gravestone of T. Flavius Celadus was unearthed alongside the Via Ostiense; and two funerary inscriptions, discovered in the church of S. Croce in Gerusalemme and on the Via Praenestina, were commissioned by a T. Flavius Successus. Successus' position was that of *tabularius marmorum Lunensium* or *tabularius rationis marmorum Lunessium.* So far, this office is only attested under the Flavians, and there are no epigraphic indications that this *ratio* was continued afterwards, or had been introduced at an earlier date (cf. 7.2.5).[75] Other decorative stones imported were perhaps dealt with by *tabularii a marmoribus*; the votive monument of Primigenius Iuvencianus from Vespasian's reign documents the devotee in this position.[76] A funerary inscription found near Augustus' Mausoleum in Rome attests M. Ulpius Martialis as *a marmoribus.*[77] Furthermore, a *statio marmorum* in Rome is evidenced in a hitherto unpublished inscribed votive altar found in the Crypta Balbi in 1993 commissioned by an M. Ulpius Restutus, purportedly a *praepositus ex station(e) marmorum.*[78] Semnus, an *optio tabellariorum* of the *statio marmorum*, set up an altar to Iuppiter Maximus Dolichenus after AD 197.[79] The undated and fragmentary inscription discovered in S. Maria Maggiore renders the functional title of [-]us Antiochus, *libertus Augusti*, as *procurator marmorum.*[80] The same title is documented in the inscription of the *procurator provinciae Frygiae*, Aurelius Marcio in Kaimaz, who once probably held the office of *procurator marmorum*

[74] *CIL* XI 3199 = *ILS* 3481. As the title *ab marmoribus* is not followed by a topographical indication and no quarries of any importance are to be found in the vicinity of Nepet, Hermeros probably held this post in Rome. The inscription appears to have been set up at Nepet in commemoration of his role as *magister* of a *collegium Feroniae* for the name, cf. Weaver 1972: 212–23. On a *collegium Feroniae* in Furcona, cf. *AE* 1998: 406 and Buonocore 1998: 56 f.

[75] *CIL* VI 8484, 8485; Quilici 1974: 279, no.166 = *AE* 1974: 153.

[76] *CIL* VI 301 = 30731.

[77] *CIL* VI 8483.

[78] Saguì 1993: 413, fig. 6; Maischberger 1997: 141. The title *praepositus ex statione marmorum*—in analogy with the military usage of *ex* (cf. Speidel 1993)—might be understood to indicate a promotion of M. Ulpius Restutus from the *statio marmorum* to a different posting. This promotion obviously warranted the commissioning of a votive altar.

[79] *CIL* VI 410 = 30760 = *ILS* 1707.

[80] *CIL* VI 8482.

in Rome, perhaps sometime during the second half of the second century (cf. 4.1.3). Thus, the following chronology emerges:

Claudius		Hermeros Thymidianus *ab marmoribus*	
Vespasian			Primigenius Iuvencianus *tabularius a marmoribus*
Flavian			T. Flavius Celadus *tabularius marmorum Lunensium* T. Flavius Successus *tabularius (rationis) marmorum Lunensium*
Trajan	M. Ulpius Restutus *praepositus ex station(e) marmorum*		
		M. Ulpius Martialis *a marmoribus*	
Septimius Severus			Semnus *optio tabellariorum stationis marmorum*
Late 2nd/early 3rd c.	M. Aurelius Marcio *procurator marmorum*		
Undated	[-]us Antiochus *procurator marmorum*		

The arrangement of the evidence in chronological order may provide an outline of the gradual development of a 'marble bureau'. The office may have had its origins in a few imperial slaves acting as *ab marmoribus* under Claudius, and the evidence for *tabularii* for marble in general and a *ratio marmorum Lunensium* in particular for the Flavian period might mirror the increase of building activity in Rome.[81] The complexity of this administrative branch grew more and more elaborate during the second century: a *statio marmorum* is attested as early as Trajan's reign, whereas *tabellarii* might be a late second century addition—perhaps coupled with the creation of a freedman procuratorship. In view of the lacunose evidence, this model, which implies organic growth, might be misleading; the functions and administrative responsibilities could alternatively be seen as part of one administrative branch or 'marble bureau' headed by a procurator, which had already been established during the first century.

The purpose of the bureau is not revealed by the epigraphic sources. Whereas designations such as *a(b) marmoribus* are not particularly illuminating, the functional title of *tabularius rationis marmorum Lunensium* is rather precise. The administrative tasks of a *tabularius* are well known in general, and in the case of Mons Claudianus, the local *tabularius* is documented managing the marble stocks (cf. 6.2). One might well imagine the *tabularii a marmoribus* keeping track of numerous blocks and columns stocked at marble yards throughout Rome and Ostia. The duties of a *tabularius rationis marmorum Lunensium* were perhaps slightly different: the term *ratio* in this context may have described a specific registry for Lunensian, i.e. Carrara marble, arriving and being stored at Rome.[82] The use of *tabellarii* in the service of a *statio marmorum* raises the question of their purpose. As the *tabellarii stationis marmorum* are hitherto only attested in Rome, it appears less likely that they were responsible for the transfer of messages between Rome and the imperial quarries.[83] Given the fact that marble was also stockpiled at yards in Portus and Ostia *c.*30 km down the Tiber, the *statio marmorum* at Rome probably used the *tabellarii* at its disposal for the transfer of information between Rome and the marble yards, perhaps comparable to the fashion in which the *ἐπίτροπος τῶν μετάλλων* deployed *tabellarii* to the quarries in the Egyptian Eastern Desert (cf. 4.2.2). Perhaps some of the

[81] Perhaps these *tabularii rationis marmorum* stood under the direction of a *princeps* or *praepositus tabulariorum*, as is occasionally evidenced for other branches of the Roman administration, cf. Hirschfeld 1905: 460 f.; Sachers 1932: 1972.

[82] A parallel for the use of *ratio* in this context is provided by graffiti from the vaults of the archaeological complex at Barco Borghese in Monte Porzio Catione/Frascati, south-east of Rome, likely used as a logistics centre for construction work at the numerous imperial *villae* nearby. The letters *ratio marmoraria* painted on plaster, dated to the end of the 1st cent. AD., and other graffiti from one of its vaults, list *modii* of marble, a measurement normally used for liquids or grain. The constraints of the small vaulted room suggest that the marble stored in this room was pulverized marble in bags. The marble powder probably derived from rubble of marble work shops or directly from the quarries themselves as a useful by-product of the Roman marble industry, cf. Lega 2003.

[83] Hirschfeld 1905: 202 with fn. 1; Boulvert 1970: 186; Kolb 2000: 276.

numerous imperial officials named in the genitive after 'ex rat(ione)' on quarry labels were allocated to the *statio* (or perhaps even *stationes*?) at Rome (and Ostia or Portus?). The *procurator marmorum* is likely to have supervised this whole system. The responsibilities of the *procurator marmorum* or the role of the *statio marmorum* is perhaps comparable to the function of Ulpius Chresimus, the *procurator a marmoribus* known from Mylasa. Based on the geographical concentration of the epigraphic evidence, Chresimus was most likely stationed in the Maeander valley, if not at Miletos itself, probably overseeing either the marble transport and/or the supply of imperial marble for local public building projects in Asia Minor (cf. 4.1.4). While Ulpius Chresimus held this possibly temporary position in connection with particular building projects in cities of the Maeander valley, the constant building site which was the city of Rome certainly supported the creation of a more permanent office. The main purpose of this administrative branch was therefore to order marble blocks and columns for imperial building projects from various quarries and manage the stockpiles of marble at Portus, Ostia, and Rome. The names of private 'rationarii' on quarry labels may suggest that contractors of public building projects or contractors supplying building material may have ordered marble directly from the imperial quarries or via the 'marble bureau'. The question remains, however, whether this 'marble bureau' influenced operative procedures at the quarries such as Dokimeion, Simitthus, Mons Claudianus, or Karystos.

8.2.3. 'Marble Bureau' and Imperial Quarries

8.2.3.1. *Latin Quarrying 'Vocabulary'*

Based on the dissemination of Latin quarrying terminology documented in quarry labels throughout the empire, it has been argued that this central 'marble bureau' was in control of the exploitation of coloured and white marble and its import to the city of Rome.[84] The markedly diverse epigraphic formulae documented on quarry labels, however, record not only significant variations in geological and topographical constraints but also different approaches to organizing operative processes under similar geological conditions (cf. 7.2.8). While at Dokimeion quarry work was contracted out to private individuals, the quarry labels on *giallo antico* from Simitthus do not refer to contractors at all, but appear to reflect the running of extractive

[84] Fant 1993*a*: 157, contra Hirschfeld 1905: 176; Dubois 1908: XXXVIII.

procedures directly by imperial officials. Consequently, quarrying procurators might have been relatively independent in implementing the appropriate working procedures, and in deciding whether or not quarry work was to be contracted out.

Nonetheless, the abbreviations, marks, and formulae used in the imperial quarries of Dokimeion, Karystos, Luna, and Simitthus, and on marble from quarries such as Teos, Chios, Paros, Hymettos, or Pentelikon, are part of the same administrative 'vocabulary'.[85] The use of Latin terminology for accounting purposes within the quarries, let alone for the categorization of the work force (e.g. *pagani*, *familiarii*), in the context of a predominantly Greek-speaking population is significant. This suggests an intentional use of Latin terminology at least in quarry labels—a usage which needs to be explained. Assuming the above observations are correct, the information displayed by the quarry labels mostly concerns the internal organization of the quarries and does not refer to a central authority. The practice of detailing the contributors to the production of marble items in Latin seems not to have been associated with possible liability issues emerging from deficient produce being delivered to Rome. The quality of marble blocks is likely to have been examined prior to their export to Rome; this diminishes the relevance of quarry labels for the recipients of the marble produce. Only the formula 'ex rat(ione) PN' in quarry labels may relate to private contractors, 'redemptores', of building projects and imperial officials in Rome ordering marble from the respective quarries. This information would be of interest to a central authority coordinating the order of marble produce from the imperial quarries—perhaps this task fell to the 'marble bureau' at Rome and Ostia (cf. 8.2.2). If so, the use of Latin markings might be partly explained by the monoglot recipients of the marble exports to Rome. The observation, however, that the quarry labels reflect the internal organization of quarries calls for a different explanation: the appointment of imperial slaves and freedmen and of military personnel to the quarries partly explains the spread of Latin terminology to all imperial quarries, particularly Dokimeion or Karystos in the Greek East. Apart from inhabitants of Roman colonies (e.g. Alexandria Troas, Antiochia Pisidiae, or Berytus) and private individuals from the west, it is mostly members of the provincial administration and the Roman army who are recorded commissioning monuments inscribed in Latin during their service in the east. Furthermore, much of the internal military correspondence known from the papyri and ostraca of Roman Egypt and the Near East is composed in Latin.[86] These observations also apply to the staff documented at imperial quarries: the Latin inscriptions of *dispensatores* at Dokimeion and Krokeai, of

[85] For an overview of the recent discussion, cf. Maischberger 1997: 19 f.

[86] Millar 1994: 403–19.

tabularii at El-Ashmunein, of Aurelius Marcio at Kaimaz, of Chresimus at Mylasa and Tralles, and of Hymenaeus Thamyrianus at Karystos, certainly provide evidence that Latin was customarily used by members of the quarrying administration.[87] The spread of Latin as the language used in quarry labels is thus best explained by the appointment of Latin speaking officials to the quarries. Therefore the employment of a Latin quarrying 'vocabulary', seemingly common to quarry organizations under imperial control, must not be credited to a central marble bureau in Rome. The dissemination of this 'vocabulary' might as well be attributed to accounting practices shared by the military, imperial personnel, and the public domain. As for the involvement of a central marble bureau in every day quarrying operations the quarry labels do not provide any evidence in support of this view.

8.2.3.2. *Distribution of Organizational Tasks*

The use of Latin in quarry labels set aside, there is hardly any evidence for the influence of a 'marble bureau' on quarrying operations. Even though the emperor, i.e. his Palatine bureaux, can be observed in our sources opening and interfering in the running of quarries (cf. 8.1.2), there is nothing to suggest that this was a regular occurrence. In the light of their financial implications, some decisions, such as the expansion of existing quarries or the opening of new ones in order to meet the demand of public and imperial building projects in Rome, were probably ordered or at least sanctioned by the emperor. Further aid by the emperor, i.e. his Palatine bureaux, was required in the deployment of members of the *familia Caesaris*, military specialists, and perhaps convicts from other provinces to quarries (cf. 8.1.1). Only on one occasion is there any evidence for central intervention in ongoing quarrying operations: the change in epigraphic formula at Simitthus and Dokimeion may reflect a central directive most likely issued by the emperor (cf. 7.2.8). Most of the time Rome probably did not interfere. The varying geological and topographical constraints amongst imperial quarries, and the organizational implications arising from them, not to mention the lack of uniform traits in operative processes as expressed in the differing quarry labels, make it highly unlikely that there was centralized management of extractive operations in Rome. Consequently, it is doubtful whether the scale of the

[87] *AE* 1986: 674 = Christol & Drew-Bear 1986: no. 5 (Dokimeion); *CIL* III 493 = *IG* 5/1, 1569 (Krokeai); Cockle 1996: 23–8 (El-Ashmunein); *CIL* III 348 = *CIG* 3822 (Kaimaz); *CIL* III 7146 = *IK* 36/1, 148 (Tralles); *AE* 1988: 1028 = *SEG* XXXVIII 1073 = Herrmann 1988: 122–5 (Mylasa); *CIL* III 563, 12289 (Karystos).

emperor's, i.e. the Palatine officials', direct involvement in the organizational structures of imperial quarries merited the creation of a special administrative branch in Rome to cope with the information flow generated by the production ***and*** trade of marble. The *procuratores marmorum* and the *statio marmorum* in Rome were probably not responsible for the organization of the marble production, but were primarily concerned with the importing of marble from different imperial quarries for specific public building projects in Rome.[88]

The demand for marble produce from quarries under imperial control (Simitthus, Dokimeion, Mons Claudianus, Karystos) was predominantly formulated in the context of public and imperial building projects at Rome (and, infrequently, in other towns throughout the empire).[89] The conclusions gained from the analysis of quarry labels, if correct, imply that during an early phase of imperial control at quarries (such as Dokimeion, Simitthus, and others) contractors of imperial and public building projects and members of the *familia Caesaris* at Rome may have ordered marble elements (capitals, columns, bases, blocks, etc.) directly from the quarries. Perhaps the early forerunners of the *statio marmorum* in Rome, the *ab marmoribus* may also have been involved in relaying the orders to the quarries. The addressees were undoubtedly the freedmen procurators or other imperial officials placed in charge of the quarrying operations. They would have compiled and relayed these orders for marble to the individual quarries. It was probably up to the imperial official on the spot, the civilian and military engineers, and/or the foremen to organize the work gangs or distribute any incoming orders amongst the contractors. The produce was then extracted, dressed, and readied for transport. This 'quarry-to-order' system would have had some apparent disadvantages. The time elapsing from transmitting the order, the production of the required elements, to their shipment to their respective destination would have taken roughly one year or more. Moreover, the loss of

[88] On the role of the *statio marmorum* as provider of building materials, cf. Daguet-Gagey 1997: 207.

[89] The dissemination of polychrome marbles from imperial quarries in a private context appears to be limited mostly to small specimens found in the floors and veneers of villae in Italy and the provinces, cf. Fant 1993*a*: 153–5. How members of the Roman elite had come into possession of these samples is unclear. It is not known whether the marble blocks were presented as gifts by the emperor to meritorious senators or equestrians, or whether these coloured marbles were available to anyone who could afford such luxuries. Even though the Prices Edict of Diocletian of AD 301 (*CIL* III Suppl. pp. 2208–9, 'de marmoribus') does provide the maximum prices for specific coloured marbles, this may not necessarily be taken as evidence for free market availability of coloured marble in the preceding centuries.

customized marble elements, particularly columns, during transport (as a result of shipwrecks or other accidents), would have taken considerable time to replace. At Dokimeion and Simitthus this rather sluggish 'quarry-to-order' system may have been changed by AD 136/7. For Dokimeion the quarry labels could indicate a significant shift from a quarry-to-order to a quarry-to-stock system, perhaps in combination with a norming of measurements allowing for the prefabrication of capitals, bases, columns, etc. The contractors or work gangs produced certain marble elements in advance, stocking them at the quarries. This would have cut the response time considerably and would have speeded up construction at Rome (cf. 7.2.7). This 'new' system (albeit limited to Simitthus and Dokimeion) may have developed in step with the formation of the *statio marmorum* in Rome (provided one wishes to argue for the growth of this bureau) which possibly compiled the orders for marble at Rome and forwarded them to the quarries (cf. 8.2.2). The tons of marble shipped to Portus/Ostia and then to Rome could not be deposited in the immediate vicinity of the building sites given the spatial restrictions of the sprawling capital. Thus, central stocking areas at Portus, Ostia, and in Rome, such as the Emporium, were designated and the incoming marble distributed to the building sites throughout Rome. The administration of these stocks at Rome and at its harbours, as well as the transmission of orders to the quarries possibly lay with the *statio marmorum* in Rome.

How this demand for marble in specified shapes and sizes was to be met was basically up to quarrying authorities in the provinces. The aquaintance of quarrying procurators and the civilian or military specialists in charge of operations with the geological, topographical, and geographical constraints rule out constant interference from Rome. In the case of Dokimeion, however, the imperial officials in charge must have met considerable difficulties which prompted the dispatch of military specialists to the quarries (cf. 5.1.1). Most of the time, many of the organizational decisions are likely to have been taken by the procurator of a quarrying district. The recruitment and pay of the workforce (apart from convicts and members of the *familia Caesaris*), decisions on the general structure of working procedures (made together with the experienced military or civilian engineers), and the outsourcing of quarry work, all probably fell within the brief of the freedman quarrying procurator. The origin of the παγανοί from the Nile Valley makes it likely that the hire of free labour and their deployment was part of the procurator's responsibilities (contracted out to μισθωταί?), as was the formulation of the demand for φαμελιάριοι from Asia Minor, Syria, or Palestine, or for convicts in the case of Simitthus (cf. 8.1.1). Moreover, the procurator and his staff probably compiled the demands for provisions and logistical support reported to them by the military commanders and the civilian administrators on the spot (cf. 6.1.2.1). At Mons Claudianus the necessary supplies were then requested from the *praefectus Aegypti*—at least pro forma. The latter issued the relevant orders for the collection of grain or the requisition of animals from the provincial population (cf. 6.1.2.6). The responsibilities of the prefects of Roman Egypt, however, went beyond merely responding to the needs of the quarrying procurators. The military contingents detached from the *exercitus Aegypti* to the quarries stood under the general command of the *praefectus Aegypti*. Moreover, the army quarrying specialists from other provinces were likely seconded to the imperial quarries on the request of the prefect. The written evidence from the Egyptian Eastern Desert hints at occasional visits made by the *praefectus Aegypti* to the quarries at Mons Claudianus (cf. 6.1.2.6). The initiation of extractive operations as well as the presentation of marble samples to the emperor probably reflects the fact that the responsibility for the imperial quarries in Egypt was shared between the prefect and the freedman procurators. Provided the distribution of responsibilities between quarrying procurators and governors, as observed in Roman Egypt, was the same throughout the empire, one could assume that most decisions concerning the supply of quarries with provisions and logistical support, as well as the deployment of troops and military specialists, were taken by the governors and other equestrian officials of the provincial administration. The evidence for the support of imperial quarrying activities by other provincial governors is circumstantial: military contingents were probably detached to the quarries on the orders of the provincial governors and military specialists were requested by the governors from neighbouring provinces or from the emperor in Rome. Moreover, the gubernatorial authority of sentencing convicts to quarrying duty might imply a limited responsibility for part of the workforce in the *metalla* (cf. 6.1.2.6).

9

Imperial Mining and Quarrying Administration: A Conclusion

Arranging our written sources in a chronological order would allow one, in theory, to trace the respective origins and formation of quarrying and mining organizations within the framework of the empire's administration, perhaps even to identify a coherent 'policy' towards extractive operations under imperial control. The lacunose nature of our evidence, however, is a formidable obstacle to drawing definitive conclusions. For example, any proposal for a timeline of procuratorial posts established by Rome must inevitably rely on arguments *e silentio*; this prohibits any certainty or precision, for example, in pinpointing the date for the establishment of a 'new' procuratorship or in outlining the responsibilities of the office-holder. In consequence, any attempt at reconstructing the decision-making process at Rome which led to the establishment of these procuratorships remains hypothetical, and any essay on the evolution of mining and administration in the Roman provinces remains an educated guess.

The Augustan settlement of 27 BC saw the placing of 'unpacified' provinces of Spain, Gaul, Syria, Cilicia, Cyprus, and Egypt in the custody of Augustus.[1] Dio reports that the *princeps* seconded procurators to all provinces, to his own as well as those of the Roman populus. These officials would be guided by a set of rules and directives established by the emperor. The purported task of these officials was the collection of revenues and disbursements in accordance with their respective instructions.[2] In addition, Strabo highlights the presence of procurators in Hispania Citerior, who were in charge of supplying the soldiers with what they needed.[3] The administrative tasks bestowed on the financial procurator within a *provincia Caesaris*—the collection of taxes and revenue from public and imperial lands, the payment of the army stationed within the province and perhaps the expenditure for infrastructural construction work (e.g. roads, aqueducts)—would indicate the relative independence of financial procurators within the provincial administration.[4] As regards the opening and

[1] Dio 53.12.5 ff.; Kienast 1999: 86 f. [2] Dio 53.15.3 f. [3] Strabo 3.4.20.

[4] Brunt 1983: 52; Brunt 1990*b*: 165 f.; Burton 1993: 16; Eck 1997*a*: 135 f., 142 f.; Eck 1997*b*: 198 f.

running of mining operations on public or the emperor's land in these provinces, our knowledge is limited to the rare insight provided by Florus' testimonies on the early mining activities in north-western Spain or in Dalmatia: in the initial phases after the Roman conquest the local population was directed by force towards the exploitation of its own gold resources.[5] Similar processes may have been in play in other newly conquered territories such as the Augustan 'provincia Germania', or later, under Claudius and his successors, in Britain, where silver–lead and iron mines, respectively, fell into the hands of the Roman conquerors (cf. 8.1.1). The conspicuous location of legionary and auxiliary camps in the vicinity of mining regions, such as Legio/mod. León in north-western Spain, or the auxiliary forts at Dolaucothi, Charterhouse-on-Mendip, or Brough-on-Humber, strongly suggests that military forces were employed not only to provide security for these mining ventures, but were also perhaps there to enforce corvée-labour in the opencast or underground mining operations during the early phases of occupation (cf. 5.2). At what point, if ever, the exploitation regime in these regions changed from a forced approach to farming out or selling off mining plots, or whether potential mining areas were contracted out soon after 'pacification' of the occupied territories, cannot be determined based on the available evidence.[6] Even though military personnel may occasionally have been used as a workforce in the exploitation of iron or gold mines, their use for this purpose on a regular basis appears unlikely (cf. 5.3). Strabo's passage on the tasks of the procurators in Spain indicates that procurators in *provinciae Caesaris* were predominantly entrusted with the 'commissariat and pay of the troops'.[7] The build-up and onslaught of military expeditions, for example, into the territories east of the Rhine under Augustus and the early days of Tiberius' reign, and later the permanent garrison of Roman legions and auxiliary units along the Rhine frontier, must have considerably augmented the demand for metals and other supplies.[8] This demand was likely met by iron and lead mined in Gaul's provinces.[9] Similar arrangements must

[5] Florus 2.25.12, 2.32.59–60.

[6] The onomastic evidence for private individuals and *societates* named on British and 'German' lead ingots provides only an inadequate basis for identifying the organizational arrangements of individual mining operations. As argued above (cf. 7.1.3) the names in the genitive case in mould marks and stamps at most identify the (current) owners of the ingots, and not the owners or contractors of mining ventures.

[7] Strabo 3.4.20 with Brunt 1983: 52.

[8] The camp of Oberaden on the river Lippe, in use from 11–8 BC, yielded several tons of nails, which had been either manufactured there or imported. Nails, metal fittings, and lead must have been supplied in vast quantities for the construction of a great fleet by Drusus in 13/12 BC (Dio 54.32.2). The construction of 1,000 boats for Germanicus' last campaign east of the Rhine AD 15/16 must have required considerable quantities of metal (Tacitus, *Ann.* 2.6.2); Eck 2004*a*: 79.

[9] Strabo 4.2.2. For an overview, cf. Davies 1935: 77–93; Sablayrolles 2005: 416; Cauuet et al. 2005: 423–41.

have been in operation in other frontier regions and provinces of the Roman empire such as the Danube frontier. It thus is plausible, although there is no direct evidence, that the commissariat, if necessary, included the supply of the Roman army with iron and lead.

Setting aside the military requirements, the revenues and *vectigalia* from extractive operations must have been forwarded to the financial or presidial procurators and prefects in these provinces. As for the mining operations in the *patrimonium regni Norici*, their supervision was possibly the responsibility of the presidial procurator documented since Claudius' reign (cf. 4.1.9). Moreover, the control of opencast gold-mining ventures in first-century north-western Spain possibly fell within the mandate of the financial procurator of Hispania Citerior.[10] By the end of the first century, mining regions within provinces are documented as falling under the supervision of imperial freedmen. The mines of Mons Marianus, Rio Tinto in Baetica, or Vipasca in Lusitania were headed by freedmen procurators (cf. 4.1.6). The *metall(um) Alboc(rarense)* in Callaecia, mentioned by Pliny (*NH* 33.80), was likewise superintended by a freedman, the procurator M. Ulpius Eutyches.[11] If one takes the epigraphic record at face value, the evidence implies that mining districts were removed from the direct control of the financial or patrimonial procurators by the turn of the century. This measure—whether preconceived or in response to emerging needs—would have alleviated the administrative burden on the financial or patrimonial procurators. The assignment of a special representative of the emperor to the site allowed for an intensification of control over these districts. This result is likely to apply to Asturia and Callaecia as a mining region which was probably initially included in the mandate of the financial procurator for Hispania Citerior. The abundance of gold mines in northwestern Spain and the garrisoning of a legionary force and accompanying auxiliary units in the same area, however, may have prompted the emperor to install a procurator for Asturia and Callaecia by AD 79 (cf. 4.1.6) and, by the later years of Hadrian's reign, of *legati iuridici Asturiae et Callaeciae*.[12] The earlier measure would fit nicely with the administrative reorganization observed during the reign of Vespasian and his sons in pursuit of increasing fiscal revenues.[13]

[10] Pflaum 1960–1: 1048. Possibly in the early phases after conquest at the end of Augustus' reign and the succession of Tiberius, mining operations had already been under supervision of a *praefectus Asturiae* (*CIL* II 4616 = *IRC* I 101 = *IRC* V p. 23 = *ILS* 6948); a *praefectus Callaeciae* (*CIL* II 3271 = *CILA* III 92) is documented for the Flavian period who might also have dealt with mining operations within his district, cf. Zwicky 1944: 20; Brunt 1983: 55; Domergue 1990: 291 fn. 291.

[11] *CIL* II 2598.

[12] Alföldy 2000*a*: 32.

[13] According to Mateo (2001: 178–96), the Vipasca regulations must similarly be seen as product of Vespasian's fiscal policies (cf. 8.1.2).

This plausible reconstruction, nonetheless, rests on the premise that our evidence is not skewed by chance or by change in the epigraphic habit. One can not prove beyond doubt that freedman procurators did not head mining districts on the Iberian peninsula since the early days of the Principate. The estate of Sextus Marius in Baetica, for example, over which Tiberius established direct imperial control, may have been administered by a freedman procurator from the outset.[14] The occurrence of freedman *procuratores* at Ampelum after the founding of the new *provincia Dacia* lends support to the idea that the emperor placed particular mining ventures under the supervision of a special procurator when these ore deposits came into his sphere of control.[15] The evidence from Dacia, however, pertains to the second century AD and may not necessarily be indicative of the previous century or for other regions.

The presence of *beneficiarii procuratoris* in the later second century at Villalís, Luyego, and Ampelum highlights a further aspect (cf. 4.1.6, 4.1.8). The evidence implies the subordination of the freedman procurators to the respective financial procurators of their provinces—a notion corroborated partly by the inscribed monument commissioned by the Sardinian *procurator metallorum et praediorum* for the *procurator Augusti* and *praefectus provinciae Sardiniae* (cf. 4.1.7). This interpretation fits with the model of an evolving provincial administration, whereby particular financial tasks were gradually removed from the responsibilites of the financial procurator within a province and entrusted to special procurators instead. Moreover, the assumption that these branches of the administration evolved organically seems to be consistent with the increasing diversification and growth of the non-senatorial administration observed during the second century.[16] In the course of the second century, the supervision of the collection of certain taxes and those collected from a certain category of mines in one or more provinces was transferred to equestrian procurators. For the silver mines of Dalmatia and Pannonia, separate equestrian *procuratores argentariarum* appear already during the first half of the second century. The superintendence of these mines was apparently unified during the reign of Marcus Aurelius and perhaps was extended to include other metal resources such as copper and iron mines (cf. 4.1.10). At approximately the same time one can observe a change at Ampelum from imperial freedmen to equestrian *procuratores aurariarum*; the imperial freedmen were relegated to the post of *subprocurator aurariarum*. Perhaps this change in status went hand in hand with the extension of the geographical realm to include gold mines in all three Dacian

[14] Tacitus, *Ann.* 6.19.

[15] *CIL* III 1312 = *ILS* 1593 = *IDR* III/3, 366, cf. Noeske 1977: 347, AMP 1.

[16] Eck 1997*a*: 79–88, 135 f.

provinces (cf. 4.1.12). By the reign of Commodus the iron mines may have been singled out for separate management under an equestrian *procurator ferrariarum* whose administrative responsibilities might have covered super-provincial entities as vast as the customs district of the *publicum portorii Illyrici*. Likewise, by the early third century a *procurator (ad vectigal) ferrariarum* may have supervised iron mines in all of 'Gaul', that is, all the provinces within the customs district of the *quadragesima Galliarum*. As for the other equestrian *procuratores ferrariarum* of centenarian rank documented before the third century, they perhaps held office in Rome, contracting out the collection of *vectigal* from iron mines to *conductores* in different provinces (cf. 4.1.12; 6.1.3.3). As is the case with the tightening of imperial control over the *publicum portorii Illyrici* or the *quadragesima Galliarum*, there is scant evidence for an intensification of imperial control of vectigal collection from iron mines by the end of the second or the beginning of the third century (cf. 6.1.3.4). It appears plausible to assume similar developments for other mines under imperial control, although clear proof remains absent. The emergence of toll stations in the vicinity of mining areas in the Danubian provinces by the late second and early third century may not be a mere coincidence or a result of our lacunose records (3.2.3).

As for imperial quarries, the scarcity of epigraphic evidence does not allow for insights of similar detail; the evidence of ostraca and inscriptions from Roman Egypt, however, can be synthesized to provide an outline of the evolution of the quarrying administration in this province. The equestrian μεταλλάρχης, who supervised all extractive operations in Egypt at the beginning of the first century AD, appears to have been replaced at some point. As suggested by the ostraca at Mons Claudianus, which name the division of φαμελιάριοι into the *numeri* of Mons Porphyrites and of Alabastrine, the equestrian function of μεταλλάρχης might have been split into at least two entities. A freedmen procurator was appointed to one of these by the early second century AD at the latest. The magnitude of the tasks at hand may well have been too much for one equestrian official to handle, resulting in the abolition of the post of μεταλλάρχης. Apart from Roman Egypt, the epigraphic record of imperial quarries in other provinces does not display such obvious changes in officials. Some imperial quarries were perhaps not supervised by their own quarrying procurators: Karystos, Dokimeion, or Krokeai merely yield monuments inscribed by *dispensatores* (cf. 4.1.5, 6.2). There are rather vague indications as to the affiliation of these quarries to administrative districts: Dokimeion might have been supervised by a *procurator 'provinciae' Phrygiae* at Synnada (cf. 4.1.3), whereas the quarries at Karystos perhaps fell under the overall responsibilities of the *procurator provinciae Achaiae* (cf. 4.1.5). As for Simitthus, a sufficient number of inscriptions proves the presence of a freedman procurator there (cf. 4.1.6). The testimonies relating to the administration of imperial quarries date predominantly to the second century AD; for the first century there is hardly sufficient evidence to allow one to trace administrative developments with any precision.

Imperial freedmen and equestrians in the service of the emperor undoubtedly played a vital role in the administration of mines and quarries. In the function of 'procurator' they were placed in charge of a specific mining or quarrying district or mines within one or more provinces. Individual mines and quarries were defined as *metalla,* a term carrying the territorial notion of a district obviously distinct from other territorial entities within a province. Based on the *patrimonium regni Norici,* it had been assumed that other mining districts covered a vast area as well. Efforts to identify mining districts of this magnitude have rested solely on the assumption that the toll stations of the *publicum portorii Illyrici* marked the boundary of these districts—a notion refuted above. As for the legal nature of *metalla* under imperial control, the majority of them were most likely property of the Roman state, perhaps in rare instances property of the Roman emperor.

Most administrative responsibilities were common to all procuratorships and similar to those of private procurators—with the notable exception of the judicial authority in fiscal matters. Apart from the more general tasks (compilation of accounts, representing the 'principal' (i.e. the emperor) in a wide range of business deals based on a set of directives), it has proved somewhat difficult to allocate specific responsibilities within the organization of mining and quarrying operations to the procurator. As for the quarrying procurators responsible for Mons Claudianus (amongst other quarries), their responsibilities appear only in vague outline and are not easily contrasted with those of military and other officials connected with quarrying operations in the Egyptian Eastern Desert. Nevertheless, one can tentatively argue that procurators were generally in charge of recruiting, paying, and provisioning the workforce and transport animals, and relayed any orders for quarried products to Mons Claudianus. In their logistical efforts to supply Mons Claudianus or organize the removal of produce from the quarry, the procurators could fall back on the provincial administration. Particularly in Roman Egypt, the supply of the imperial quarries in the Eastern Desert profited from the ancient structures set in place for the collection of the grain tax, from which wheat and barley were regularly diverted to the soldiers and imperial personnel stationed in the desert. Camels and donkeys needed for various tasks in the quarries were requisitioned by similar means. Comparable arrangements may have been in place at Bacakale, Simitthus, or Karystos; the major difference of these quarries, however, was that they were not nearly as remote as Mons

Claudianus. This fact must have made the use of contractors far easier and alleviated the logistical burden on the procurator. There appears to have been no direct involvement of the procurators in the technicalities of quarrying—at least not at Mons Claudianus. There, civilian and military engineers may have organized extractive operations, as did the centurions at Bacakale.

The evidence for the responsibilities of mining procurators at Vipasca provides a similar picture: one of the procurator's responsibilities was to contract out the provision of auxiliary services (the collection of fees, the running of the baths, barbershops, etc.) to interested individuals. More importantly, the procurator was in charge of superintending mining operations. It appears that the procurator at Vipasca was responsible for the allotment and the sale of mining plots. He was required to keep track of changes in ownership, monitor the progress of mining operations and enforce compliance with the regulations. Should any occupant of mining plots at Vipasca be in breach of these mining laws, the procurator had the judicial power to fine or even evict culprits from the district. The theft of ore, the sabotage of mining infrastructure and the violation of security regulations were punished severely. This set of procuratorial responsibilities was probably unique to a certain type of mining venture located in southern Spain. Therefore, the obvious differences in organizational constraints which can be observed, for example, between individualized underground mining and large scale opencast operations, must have affected the responsibilities of the mining procurator. The operation of mines such as vast opencast operations in north-western Spain (Las Médulas), undoubtedly altered the responsibilities of the authorities. Like the procurators placed in charge of the Egyptian Eastern Desert, the imperial authorities at Las Médulas probably focused on coercing or recruiting a workforce and on related logistical issues.

The Vipasca tablets describe the duties of the freedman procurator responsible for a specific mining district within a province. In comparison, the titles of equestrian procurators (with the proviso that they are given in full), indicate that their responsibilities covered not one small mining district but numerous districts throughout one province or more. Given the sheer difference in size of the geographical area supervised by these equestrian procurators, their administrative responsibilities must have differed significantly from the comparably petty affairs overseen by the freedmen procurator at Vipasca. Outlining the duties of these provincial or 'super-provincial' mining procurators, however, remains difficult. Part of their responsibilities was likely the superintendence of contractors. These contractors either collected rents, 'vectigal', from occupants running mines on public land and/or they ran their own mining ventures or districts. This organizational approach is hitherto only fragmentarily recorded for iron mines, but one could imagine similar strategies for the *aurariae Dacicae* or the *argentariae Pannonicae et Delmaticae.* It seems plausible that the procurator in charge concluded the lease contracts with these mining companies directly. As a consequence of farming out mining operations and the collection of vectigal, the tasks of the procurator must have primarily been the monitoring of the activities of the contracting companies, that is, whether the *conductores* were in compliance with the contracts agreed and the Roman state received the expected revenues. If this did not prove to be the case, the procurators would have decided on possible sanctions. Like their colleagues in other administrative branches, they perhaps ruled on legal disputes regarding fiscal issues arising between the collectors of the *vectigal* and those running the mines. In the case of the *procuratores ferrariarum*, one inscription at Lyon may suggest they possessed the additional responsibility for supplying the Roman army with iron. Given the geographical expanse of procuratorial jurisdiction, and the judicial and monitoring responsibilities of equestrian procurators, it is reasonable to assume that they were required to occasionally visit the mining districts within their jurisdiction, much like governors of large provinces were required to visit the assizes of their province.

The usually short tenure of procuratorships by equestrians and imperial freedmen throughout the Roman empire and their common lack of expertise in the specifics of mining or quarrying management meant that the incumbents of these posts were heavily reliant on the support of an experienced staff. These *officiales* were based at the permanent headquarters of their procurators. Apart from providing a certain administrative continuity and expertise, the predominant task of the procuratorial staff was obviously to deal with the paperwork from the mining and quarrying procurators and handle any cash disbursements and receipts. Other members of the *familia Caesaris* in different functions (*tabularius*, *dispensator*, *arcarius*, *vilicus officinae ferrariae*), undoubtedly in contact with their respective procurator, were seconded to outlying *stationes* or to the mining and quarrying districts within the procuratorial jurisdiction. Further support was offered by military personnel in administrative and other tasks. The dispatch of military specialists from distant provinces to quarrying (and perhaps mining?) ventures in a specific organizational or engineering capacity certainly identifies the Roman army as one of the main providers of technical expertise. The predominant reason for a military presence at imperial mines and quarries, however, was to protect the operation, the export of produce and import of supplies against disruptive elements, and, furthermore, to guard convicts or direct corvée labour. The use of soldiers as a workforce in mining is little documented and perhaps was limited to rare occasions; quarrying stone, however, was seemingly part of regular army chores and soldiers were frequently deployed to construction work and related quarrying duties. This is where some of its officers acquired the experience needed to be seconded as specialists to imperial quarries.

Regarding the organization of quarries and mining ventures, the specifics were undoubtedly left to the men on the spot most of the time. Setting aside the sheer size of the Roman empire, this relative 'autonomy' of imperial mining and quarrying operations was mostly a consequence of varying organizational constraints. The analysis of the quarrying and mining ventures throughout the empire has shown that the geological, topographical, technological, and geographical ramifications could vary significantly, which had considerable implications for the organizational approach and the scale of logistical support. Further challenges were set by the 'customer'. In the case of quarries under imperial control, the demand for marble produce, predominantly for public and imperial building projects in Rome (and occasionally in other towns throughout the empire), meant that specified shapes (columns, capitals, blocks) had to be produced in the required numbers within a reasonable time. An increase in orders had an impact on the organization of work procedures. At Bacakale, the change in formula of quarry labels may perhaps reflect an alteration in the production system from quarry-to-order to quarry-to-stock by AD 136/7. The rationale behind this decision is undiscoverable. One can only assume that it was intended to increase the efficiency of the production cycle.

Certain major decisions may have been taken on a local level, rather than at Rome. On-site personnel would have decided whether the mine or quarry was to be directly exploited by a hired or forced workforce (convicts, corvée labour) under the general supervision of imperial officials and military personnel (cf. Mons Claudianus), or whether 'outsourcing' mining and quarrying operations to private companies and individuals was more desirable. In principle, the Roman state—in the Republic as well as during the Principate—was intent on delegating the exploitation of mines or quarries to private parties. This could mean that whole mining districts were either run by large scale companies in return for revenue or, like at Vipasca, were compartmentalized into plots sold off to its occupants. Furthermore, the collection of *vectigal* from occupants of iron mines on public grounds was farmed out to large scale companies. Occasionally, particular circumstances may have prevented the use of contractors either because of the topographical remoteness of the extractive operation or because the workforce operating the mines and quarries had been forced to do so. The former could have been the case with Mons Claudianus where contractors—with the exception of Epaphroditos Sigerianos—are virtually absent. Military personnel, φαμελιάριοι, and hired labour had to be brought in and were supplied and in part supervised by an ἐπίτροπος τῶν μετάλλων. The absolute remoteness of the quarries from the densely settled Nile Valley and the costs of supplying a workforce out in the desert may have rendered all quarrying operations unattractive for potential contractors. This was probably not the case with the Bacakale quarries near Dokimeion in a populated area of Phrygia, where private contractors were perhaps hired on the basis of a *locatio conductio operis*. At the quarries of Simitthus it was the possible use of convict labour which prohibited the employment of contractors.

The running of mining and quarrying districts under imperial control left a considerable footprint on the economic landscape of the province. For the inhabitants of Roman Egypt, the supply of the workforce and soldiers in the Eastern Desert with tax grain and requisitioned animals must have meant an additional burden. The population of other provinces with significant mining activity was perhaps pressed to corvée labour—at least in the early phases following conquest. Yet, mining and quarrying districts did not only have negative consequences for the populace: for a number of quarry workers, miners, and contractors these extractive operations were a source of income. The wealth resulting from the quarrying business in and around Bacakale near Dokimeion, for example, is reflected in the lavishly decorated stones of the area. Particularly in Spain and the Danubian provinces, mining districts emerge as focal points of migration. Whether or not the positive economic consequences for the populace formed part of the decision-making process of the Roman emperors is debatable. The emperors occasionally appear aware of the economic implications of their decisions.[17]

The importance of extractive ventures to the Roman state as source of revenue or as source of building materials is emphasized by the scores of imperial personnel, soldiers, officers, 'engineers', convicts, prisoners of war, coerced and hired workers, as well as contractors, who were directed to quarrying and mining operations, and the amount of provisions, tools, and materials supplied to these districts. Although the main reason for these measures was to produce revenue for the state, not all extractive operations under imperial control generated revenue: the distribution model argued for by Fant suggests that imperial quarries predominantly supplied visually distinctive marble elements for public and imperial building projects. These quarries (Simitthus, Mons Claudianus, Mons Porphyrites, Bacakale,

[17] This is implied by Suetonius, *Vesp.* 18: Vespasian reportedly did not implement a new mechanical device facilitating the transport of columns up to the Capitol. The reason given is that the *plebicula* were not to be deprived of their income. Given that the construction business sustained perhaps between 100,000 to 150,000 inhabitants at Rome (cf. Kolb 1995: 483–5), the use of a mechanical device would have lowered costs yet deprived scores of workers and their families of their daily income, cf. Drexhage et al. 2002: 32.

Karystos) were thus run primarily for non-economic reasons connected with the display of imperial power and its projection through the means of art and architecture.[18] As for mining districts under imperial control, the revenues generated by these were seen as an important asset for the Roman state: Statius' poem on the *a rationibus* highlights, amongst other things, the income from mining ventures, and numerous further references detail the sums acquired from specific mining districts for the Roman state. Apart from their importance as source of revenue, iron mines in particular appear to have had an additional strategic value emphasized by the obscure 'laws' on the production and distribution of iron.

The approach to controlling these public assets and ensuring the influx of revenue to the Roman state was flexible: mining on public land could seemingly be left to occupants in return for a *vectigal* payable to Rome which, in the case of iron mines in Gaul and the Danubian provinces, was handed to contractors of *vectigal* collection. Perhaps the mining of more precious metal ores was left to large-scale companies; mining plots of less precious or more abundant metals were sold to private entrepreneurs, as the Vipasca tablets seem to indicate. At first sight, Roman authorities appear hesitant to get directly involved in mining ventures. Written sources barely note the Roman state operating a mining district directly and the evidence for 'outsourcing' public mining operations to companies is far more abundant. The Vipasca tablets, however, indicate that Rome, represented by the local procurator, controlled nearly all aspects of mining activities even after mining plots had been sold to the occupants and thus were technically not public property anymore. This reluctance to give up control over the productivity of the mines at Vipasca may stem from the necessity to ensure the market was supplied with the desired metals in sufficient quantities. The Roman state may have further profited from the influx of metals to the economy by taxation of ores and metal ingots emitted from public mines—as is tentatively implied by the *lex portorii Asiae* (§34) and the occurrence of toll stations of the *publicum portorii Illyrici* near mining ventures in the Balkans. The control of mining ventures was increased mostly by seconding imperial personnel of lower clerical rank or freedmen procurators to the mining districts. Moreover, the creation of equestrian procuratorships for mines in one and more provinces during the second century was likely intended to keep the contractors of mines or the *conductores* of *vectigal* on a short leash. If our chronological outline is correct, administrative control over mining

[18] Peacock 1992; Fant 1993*a*: 146–51. This must not necessarily apply to all marble emitted from these quarries: the white marble from Bacakale/Dokimeion, for example, was worked into sarcophagi and found in private contexts throughout Asia Minor (cf. 7.2.6) and thus appears to have been available to the general public.

ventures was gradually intensified during the first and second century AD, either in conjunction with the growth of expenditure (particularly for the Roman army) towards the end of the second century and/or was precipitated by an increased strain on the administrative structures of the empire.

An overarching organizational system was probably never devised for the administration of mines and quarries. In fact, by the end of the second century AD in the Danubian provinces mining ventures for a specific metal were superintended by equestrian procurators covering one or more provinces, while in Spain or Britain there is hitherto no evidence for an equestrian post of such a magnitude. In combination with the organizational strategies put in place to deal with geological and topographical constraints for mining and quarrying ventures, one might argue that the Roman authorities were inclined to work out solutions in the local area in response to specific challenges. Consequently, decisions regarding the internal organization of quarries and mines were rarely made in Rome. The emperor and his Palatine bureaux were probably involved when mining and quarrying procurators were allocated and military specialists needed. Given the allocation of additional human and material resources resulted in an increase of costs, the opening of new ventures, particularly quarries, or the enlargement of existing operations could require imperial approval.

The guiding principle of the Roman mining and quarrying administration was to keep imperial involvement to a minimum without renouncing control of these ventures. Where possible, much of the work was entrusted to private entrepreneurs either in return for a fixed payment or ownership of the produce. Ideally the presence of military and imperial personnel would have been limited to a few officials receiving proceeds or making out payments to contractors and restricting any contract infringements by their private partners. Any aberration from this principle is primarily explained by unique organizational constraints of mining and quarrying operations which resulted in exploitation by a hired or coerced workforce under the direct supervision of imperial and military personnel. The authorities in charge were not primarily interested in acquiring the produce from mines *per se* (with the exception perhaps of gold and cinnabar) but in reaping the profits by contracting out the operation of mines. Besides pecuniary revenues, the Roman state had a vested interest in seeing the market supplied with sufficient quantities of metals—an interest implicit in the Vipasca regulations. The supply of Rome with marbles from different corners of its empire was, in comparison, a costly affair and seemingly not always subjected to an entrepreneurial rationale. The use of marble in the display of imperial wealth and power was the main reason for its extraction. In other words, the emperors had remote quarries worked not for the profit of the treasury, but simply because they had the capability to do so. Whatever the main impetus for the extraction and distribution of metals and marble from imperially controlled operations, the inception, growth, and differentiation of the administrative structures—centring on the procuratorial system in its local variations—ensured the continued supply of Rome and its economy with the desired materials. This elaborate organizational system remained vital to maintaining Rome's power throughout the Principate.

APPENDIX

Pavonazetto and White Marble: Dokimeion

1. a) *Ant?(onino) III co(n)s(ule)* (AD 140) | *[.]c[.]x*; b) *CLVIIII*; c) *Ant(onino) III XX*, d) *ÐCCLXXII*, e) *R̶*, f) *[.]GIE*; g) *RMA*; h) *RMA*; i) *C(aesare) XVI co(n)s(ule)* (AD 92); j) *Pal(ma) II c⟨o⟩(n)s(ule)* (AD 109); k) *VFR Verg(iliano) co(n)s(ule)* (AD 115); l) *Dom(itiano?) I̤I̤ co(n)s(ule)* | *[–]XX* (AD 73?)	Bacakale	Fant 1989: no. 11 + Drew-Bear 1994: 811 fn. 221.
2. a) *C(aesare) XVI co(n)s(ule)*; b) *GIΠA*	Bacakale, AD 92	Fant 1989: no. 6.
3. *C(aesare) XVI co(n)s(ule)* | *II*	Afyon, AD 92	Fant 1989: no. 1.
4. *C(aesare) XVI co(n)s(ule)* | *III*	Afyon, AD 92	Fant 1989: no. 2.
5. *C(aesare) XVI* | *IIII*	Afyon, AD 92	Fant 1989: no. 3.
6. *C(aesare) XVI co(n)s(ule)* | *XV*	AD 92	Fant 1989: no. 49^2 = *CIL* III 7009 = Ramsay 1882: no. 16.
7. a) *C(aesare) XVI co(n)s(ule)* | *XVIIII* (AD 92); b) *MA*; c) *GII[–]Π?Δ?* | *Traian(o) III co(n)s(ule)* (AD 100) *XXXVI*	Bacakale	Fant 1989: no. 8 + Drew-Bear 1994: 829, fn. 292.
8. a) *RMA ANT*; b) *[T]rai(ano) n(ostro) III co(n)s(ule)* (AD 100); c) *Ð CXII*; d) *C(aesare) XVI co(n)s(ule)* (AD 92)| *CXXXXV*; e) *NER Verg(iliano) co(n)[s(ule)]* (AD 115)	Afyon	Fant 1989: no. 16^2 = *CIL* III 7010, 7011.
9. *P Ailian(os) co(n)s(ule)* (AD 116) | *C(aesare) XVI co(n)s(ule)* (AD 92)| | [—] CC[–]	Iscehisar	Fant 1989: no. 62^2 = *JRS* 18: 1928: 22 f. no. 232 = *CIL* III 12228.
10. a) *DCII*; b) *R̶ Aelian(o) co(n)s(ule)* (AD 116)| *C(aesare) XVI co(n)[s(ule)]* (AD 92)| *XI*	Iscehisar	Fant 1989: no. 63^2 = *JRS* 18: 1928: 22 f. no. 233.
11. a) *C(aesare) XVI co(n)s(ule)* (AD 92) | *R Aelia(no) co(n)s(ule)*(AD 116) *CXXXVI*; b) *M D XIII* | *T DIIX co(n)s(ulibu)s*; c) *R Aelia(no) co(n)s(ule) C(aesare) XVI co(n)s(ule)* | *CXXXVI*; d) *M D XIII*	Iscehisar	Fant 1989: no. 33^2 = *CIL* III 12227.
12. a) *VFR Verg(iliano) co(n)s(ule)* (AD 115); b) *AA LX C(aesare) XVI co(n)s(ule)* | *V* (AD 92); c) *Traian(o) III co(n)s(ule)* (AD 100); d) *I R̶*	Bacakale	Fant 1989: no. 10.

13. a) *IIII b(racchio) IIII*; b) *C(aesare) XVI co(n)s(ule) VIII* (AD 92); c) *RM[A] Pal(ma) II co(n)s(ule)* (AD 109); d) *RMA*; e) *ANT*; f) *Traian(o) III co(n)s(ule) IIII* (AD 100); g) *VFR Verg(-iliano) | co(n)s(ule)* (AD 115); i) *GIΓ CNΓ*	Afyon	Fant 1989: no. 4.
14. a) *GIΓCNZ*; b) ₽; c) *RMA | Pal(ma) II co(n)s(ule)* (AD 109); d) ₽; e) *b(racchio) quart(o)*; f) *C(aesare) XVI co(n)s(ule)*(AD 92)| *VIIII*; g) *loco VI*; h) *Traian(o) III co(n)s(ule)* (AD 100); i) *b(racchio) IIII*; j) *VFR Verg(iliano) co(n)s(ule)* (AD 115); k) *ANT*	Afyon	Fant 1989: no. 5 + Christol & Drew-Bear 1991: 117 f. fn. 25.
15. a) *[T?]r(aiano) III co(n)s(ule)[—]* (AD 100?); b) *RMA ANT*; c) *b(racchio) IIII*; d) *Pal(ma) II co(n)s(ule)*(AD 109); e) *ĐCCXXI*; f) *VFR Verg(iliano) co(n)s(ule)* (AD 115); g) *C(aesare) XVI co(n)s(ule)*(AD 92) | *XX*; h) *b(racchio) quar[t(o)]*; i) *GIIPΛΓ*	Bacakale	Fant 1989: no. 12.
16. a) *VFR Verg(iliano) co(n)s(ule)*(AD 115); b) *C(aesare) XVI co(n)s(ule)* (AD 92); c) *Trai(ano) n(ostro) III co(n)s(ule)*(AD 100)| *XXXVII*; d) *Pal(ma) II co(n)s(ule)*(AD 109); e) *GII[.]XZ*; f) *RMA*; g) *XLVII*; h) *ANT*; i) *XXV*; j)*RMA*	Afyon	Fant 1989: no. 9 + Drew-Bear 1994: 808 fn. 196.
17. a) *VFR Verg(iliano) co(n)s(ule)*| *DCCLXIII* (AD 115); b) *GIΓΓNE?*; c) *b(racchio) tert(io)*; d) *XXXXV*; e) *RMA Pal(ma) II co(n)s(ule)* (AD 109); f) *DCCLXVII*; g) *C(aesare) XVI co(n)s(ule)* (AD 92); h) *Sura III co(n)s(ule)* (AD 107), *HER CCCXII*, *b(racchio) ter(tio)*	Bacakale	Fant 1989: no. 13.
18. *C(aesare) XVI co(n)s(ule)*(AD 92) | *L* | *XX* |₽ *Aelia[n(o)]* | *co(n)s(ule)* (AD 116)	Altintas	Fant 1989: no. 14.
19. *Coll(ega) co(n)s(ule)*| *XIIII* ₽ (AD 93)	Afyon	Fant 1989: no. 102[2] = Christol & Drew-Bear 1986: no. 6.
20. *Coll(ega) co(n)s(ule) XIX* (AD 93)	Aizanoi?	Fant 1989: no. 37b[2] = R. Naumann, Der Zeustempel zu Aizanoi, Berlin 1979: 80.
21. a) *VI* ₽ *III* | *Coll(ega) co(n)s(ule)* (AD 93) ; b) *VFR Verg(iliano)* | *co(n)s(ule)* (AD 115)	Afyon	Fant 1989: no. 103[2] = Christol & Drew-Bear 1986: no. 7.
22. a) *Coll(ega) co(n)s(ule)* (AD 93) *VIII* ₽; b) *VER Verg(iliano)* | *co(n)s(ule)* (AD 115); c) *RMA*	Afyon	Fant 1989: no. 104[2] = Christol & Drew-Bear 1986: no. 8.

23. *Co[l] ḷ(ega?) co(n)[s(ule)—]X̣* (AD 93)	Bacakale	Fant 1989: no. 15.
24. a) *Ṛ*; b) *Coll(ega) co(n)s(ule) LXXIX* (AD 93); c) *Pal(ma) II co(n)s(ule)* (AD 109); d) *Tra(iano) n(ostro) III co(n)s(ule)| XCVIII* (AD 100); e) *DCCCL*; f) *R ANT*; g) *RMA*, h) *CR RMA*, i) *VFR Verg(iliano) co(n)s(ule)* (AD 115)	Bacakale	Fant 1989: no. 16 + Drew-Bear 1994: 808 fn. 196.
25. a) *DCCCLXIX b(racchio) IIII*; b) *Trai(ano) n(ostro) III co(n)s(ule)| XCVI* (AD 100); c) *RMA*; d) *Coll(ega) co(n)s(ule) CLXXV | b(racchio) III* (AD 93); e) *CLXIX*; f) *CHR XCVI*; g) *ÐCCCLXIX Pal(ma) II co(n)s(ule)* (AD 109); i) *VFR Verg(iliano) co(n)s(ule)* (AD 115)	Bacakale	Fant 1989: no. 17 + Drew-Bear 1994: 824.
26. a) *C(aesare) XVII co(n)s(ule) XXXV* (AD 95); b) *(mille) (mille) LXX, b(racchio) IIII*; c) *Traian(o) III co(n)s(ule) CCXVI* (AD 100)	Bacakale	Fant 1989: no. 18.
27. *C + IXXXXVIII | C(aesare) XVII co(n)s(ule) XXXI*	Iscehisar, AD 95	Fant 1989: no. 50[2] = *CIL* III 7012 = Ramsay 1882: no. 17.
28. a) *L ∞ C℞)* ; b) *ex r(atione) Olyp(i?) Caes(aris)*; c) *Ve(tere) e(t) [Val(ente)] co(n)s(ulibus*	Ostia, AD 96	Baccini Leotardi 1989: no. 40
29. a) *ex r(atione) Ol(ym)p(i) Caes(aris?)*; b) *n(umero) CXXXXII || II*; c) *Ve(tere) e(t)Val(ente) co(n)s(ulibus)*	Ostia,AD 96	Fant 1989: no. 83[2] = Baccini Leotardi 1979: no. 35 = Pensabene 1994: 52, fig. 55
30. a) *CXLVI*; b) *VETER COS XCXC*	Iscehisar, AD 96?	Fant 1989: no. 55[2] = *CIL* III 7014 = Ramsay 1882: no. 22.
31. a) *Veter(e et Gallicano) co(n)s(ulibus) ∞ LXXV[-]*; b) *CLXX[—]*; c) *℞*	Bacakale, AD 96?	Drew-Bear 1994: no. 1
32. a) *Ve(tere) e(t) Val(ente) co(n)s(ulibus)*; b) *n(umero) DXXXII*; c) *L ∞ C ℞*	Ostia, AD 96	Baccini Leotardi 1989: no. 39 = Fant 1989: no. 92[2].
33. a) *Pa(lma) et Sen(ecione) co(n)s(ulibus)*, b) *ex r(atione) Olyp(i?) Caes(aris)*	Ostia, AD 99	Fant 1989: no. 84[2] = Baccini Leotardi 1979: no. 36 = Pensabene 1994: 74, no. 20
34. a) *X | CXXI*; b) *DXC*; c) *℞*; d) *VFR Verg(iliano) | co(n)s(ule)* (AD 115); e) *MAR*; f) *Traian(o) III co(n)s(ule)| VIII* (AD 100); g) *ANT*; h) *T[.] XLVII*	Bacakale	Fant 1989: no. 23.
35. a) *Tra(ia)n(o) III co(n)s(ule)* (AD 100) *CXLIIX | CHR RMA*(?); b) *∞ CLX T℞*	Bacakale	Fant 1989: no. 18[2] = *CIL* III 7015 = III

Pal(ma) co(n)s(ule) (AD 109?); c) *VE(!)-R Verg(iliano) co(n)s(ule)* (AD 115)		13653 = Fant 1989: no. 30[2].
36. *Traian(o) III co(n)s(ule) CLXXVI*	Bacakale, AD 100	Fant 1989: no. 19.
37. a) *MAR*; b) ∞*(mille) CDLIX R*; c) *b(racchio) II[—]*; d) *CHR RMA*; e) *Traian(o) III co(n)s(ule) CCCX[—]*	Bacakale, AD 100	Fant 1989: no. 20 + Drew-Bear 1994: 825 fn. 280.
38. a) *LDCIX* | *R*; b) *Trai(a)n(o) III co(n)s(ule)*; c) *[e]x r(atione) Olyp(i?) Caes(aris) N CXIIII*	Ostia, AD 100	Fant 1989: no. 85[2] = Baccini Leotardi 1979: no. 37.
39. a) *[P]ịsone et Bolanọ [cos]* | *XXXIIX* (AD 111); b) *b(racchio) II*; c) ∞*(mille) DCCLXI b(racchio) III*; d) *b(racchio) tert(io)*; e) *Tr(aiano) [I]II et F[ro]n[tino]* | *[I]II co(n)s(ule)* (AD 100)	Bacakale	Fant 1989: no. 21.
40. a) *[—]o*, b) *[T]ra(iano) III co(n)s(ule)* (AD 100) *HE C* | *XLVII R*, c) *VF[R—]* (AD 114/115)	Iscehisar, AD 100	Fant 1989: no. 17[2] = *CIL* III 7018, 7019 = Christol & Drew-Bear 1986: no. 9.
41. a) *RMA*; b) *C[and(ido) II] co(n)s(ulibu)s XLIII*	Bacakale, AD 102?	Fant 1989: no. 27
42. a) *Cand(ido) II co(n)s(ule) HE* (AD 105) | *CCLXXV*; b) *R*; c) *VFR Verg-(iliano) co(n)s(ule)* (AD 115) | *DCCCIX*; d) *Sur(a) II⟨I⟩ co(n)s(ule)* (AD 102) | *HE CCCLXXXIV*; e) *RMA Pal(ma) II co(n)s(ule)* (AD 109); f) *DCCLV*	Bacakale	Fant 1989: no. 28 + Drew-Bear 1994: 833 fn. 315.
43. a) *XXXIV*; b) *DLXIII b(racchio) tert(io)*; c) *RMA Pal(ma) II co(n)s(ule)* (AD 109); d) *VFR Vop(isco) co(n)s(ule) CCXL* | *IIII* (AD 114); e) *b(racchio) III*; f) *Cand(ido) II co(n)s(ule) HE DLXIII* (AD 105); g) *Trai(ano) V co(n)ṣ(ule) HE X* (AD 103)	Bacakale	Fant 1989: no. 24 + Drew-Bear 1994: 829 fn. 289.
44. *Cand(ido) I[I] c̣ọ(n)[s(ule)]* (AD 105) | *VFR Vop(isco) co[s]* (AD 114)	Bacakale	Drew-Bear 1994: no. 6.
45. a) *Pal(ma) [I]I co(n)s(ule)* (AD 109) *HE CXCIX* | *[C]anḍ(ido) [II] c[os]* (AD 105) ; b) *[VF]R Vop(isco) co(n)s(ule) DCXC[—]* (AD 114)	Bacakale	Drew-Bear 1994: no. 5.
46. a) *RMA Cand(ido) II co(n)s(ule)* (AD 105) | *HE CCCXXVII*; b) *Pal(ma) II co(n)s(ule)* (AD 109)	Bacakale	Drew-Bear 1994: no. 3.
47. *RMA Pal(ma) II co(n)s(ule)* (AD 109) | *Cand(ido) II co(n)s(ule)* (AD 105) *HE CCCXXXV*	Afyonkarahisar	Drew-Bear 1994: no. 2.
48. a) *[—]o cos CC* (AD *?*); b) *[P]a[l(ma)] II co(n)s(ule)* (AD 109); c) *RMA*;	Bacakale	Fant 1989: no. 29 + Drew-Bear 1994: 823 fn. 276, 831 fn. 300.

d) *Cand(ido) II co(n)s(ule)* | *HE XV* (AD 105); e) *DXCVI b(racchio) III*		
49. a) *Cand(ido) II co(n)s(ule)*(AD 105) *Pal(ma) II* (AD 109) *co(n)s(ule)*; b) *[—]R Vop(isco) co(n)s(ule)*(AD 114) *CXLIV*; c) *Cand(ido) II co(n)s(ule)* | *HE VII* ; d) *RMA*; e) *Cand(ido) II co(n)s(ule)*; f) *XXX*	Iscehisar	Fant 1989: no. 3[2] =AE 1984: 846 = Christol & Drew-Bear 1986: no. 35 + Drew-Bear 1994: 831 fn. 298.
50. a) *b(racchio) tert(io) XXXV*; b) *Cand(-ido) II co(n)s(ule) [—]* (AD 105); c) *V-[FR] V[e]r(giliano) co(n)s(ule)* (AD 115)	Bacakale	Fant 1989: no. 41.
51. a) *RMA Pa(lma) II co(n)s(ule)* (AD 109) | *Cand(ido) II co(n)s(ule)* (AD 105) *HEC XXC* | *VFR [Vo]p(isco) co(n)s(ule)* (AD 114) *CLXIV*; b) *Ɖ XXXI*	Afyon	Fant 1989: no. 30 = Christol & Drew-Bear 1986: no. 17. + Drew-Bear 1994: 830 fn. 295.
52. *Cand(ido) II co(n)s(ulibu)s XL [—]*	Afyon, AD 105	Fant 1989: no. 106[2]= Christol & Drew-Bear 1986: no. 11.
53. a) *Cand(ido) II co(n)s(ulibu)s XLII* (AD 105); b) *RMA*; c) *R VIIII*	Afyon, AD 105	Fant 1989: no. 26 = Christol & Drew-Bear 1986: no. 10 + Drew-Bear 1994: 751 fn. 13, 827 f., fn. 286.
54. *Cand(ido) II co(n)s(ule)* (AD 105) | *HE CCXXCII*	Bacakale, AD 105	Fant 1989: no. 25.
55. a) *CXX[. . .]* | *RM(A?) P[. . .]* | *VFR Vo[p(isco)co(n)s(ule)]* (AD 114?)| *Ɖ C[—]*; b) *Sur(a) III* (AD 107)	Iscehisar	Fant 1989: no. 46[2] = *CIL* III 7007 = Ramsay 1882: no. 12.
56. a) *Sura III co(n)s(ule) IL* | *DECCXXIII*; b) *ƉEDCCCXXX* | *R III*	Iscehisar, AD 107	Fant 1989: no. 52[2] = *CIL* III 7021 = Ramsay 1882: no. 19.
57. a) *VFR Vop(isco) co(n)s(ule)* (AD 114) *ƉCLVI* | *RM(A?) Pal(ma) II* (AD 109) *co(n)s(ule)*; b) *Sur(a) III co(n)s(ulibu)s* (AD 107) | *CXXV*	Iscehisar, AD 107	Fant 1989: no. 45[2] = *CIL* III 7006 = Ramsay 1882: no. 11.
58. a) *Sur(a) II co(n)s(ule)* (AD 107) | *CXCIIII* | *Palma II co(n)s(ule)* (AD 109); b) *caes(ura) b(racchio) III* | *∞DCLXVI*	Iscehisar	Fant 1989: no. 19[2] = *CIL* III 7022.
59. a) *Sur(a) III co(n)s(ule) HE CCXXII* (AD 107); b) *RMA Pal(ma) II co(n)s(ule)* (AD 109)	Afyon	Fant 1989: no. 31.
60. *RMA Pal(ma) II co(n)s(ule)* (AD 109) | *Sur(a) III co(n)s(ule)* (AD 107) *HE* | *CCXXXVII*	Bacakale	Drew-Bear 1994: no. 4.
61. a) *loco IV b(racchio) III*; b) *Sur(a) III co(n)s(ule) II* (AD 107) | *CCXXXVIII* ; c) *RMA Pal(ma) II co(n)s(ule)* (AD 109); d) *VFR Vop(isco) co(n)s(ule)* (AD 114); e) *b(racchio) tert(io)*	Bacakale	Fant 1989: no. 40.

62. a) *Sur(a) III co(n)s(ule) HE* (AD 107) | *CCCLXXXIII*; b) *RMA Pal(ma) II co(n)s(ule)* (AD 109); c) *DCCCIV*; d) *b(racchio) III*; e) *VFR Verg(iliano)* | *co(n)s(ule) DCC* | *LX* (AD 115)	Bacakale	Fant 1989: no. 32 + Drew-Bear 1994: 833 fn. 311.
63. a) *VFR Verg(iliano) co(n)s(ule)* (AD 115); b) *DCCLIII*; c) *Sur(a) III co(n)s(ule)* (AD 107); d) *HE CCCXXC*; e) *RMA Pal(ma) II co(n)s(ule)* (AD 109); f) *DCCCII* ₽	Bacakale	Fant 1989: no. 33 + Drew-Bear 1994: 834 fn. 316.
64. *Brad(ua) co(n)s(ule)* (AD 108)	Afyon, AD 108	Fant 1989: no. 43 + Drew-Bear 1994: 834 fn. 318.
65. a) *P[a]l(ma) II co(n)s(ule), HE CCXVIIII* (AD 109); b) *Brad(ua) co(n)s(ule)* (AD 108)| *HE XIIỊ*	Afyon	Fant 1989: no. 42 + Drew-Bear 1994: 834 fn. 317.
66. *Pal(ma) II co(n)s(ule)* (AD 109)	Bacakale, AD 109	Fant 1989: no. 44
67. *VFR V[—]* | *RMA Pa[l(ma) II co(n)s(ule)?* (AD 109) *—]*	Iscehisar	Fant 1989: no. 28[2] = *CIL* III 7042.
68. *[. . .]r() co(n)s(ule)* | *[. . .]XIV* || *P(alma) II co(n)s(ule)* || *P(alma) II co(n)s(ule)RMA*	Iscehisar, AD 109	Fant 1989: no. 6[2] = AE 1984: 848.
69. *Pal(ma) II co(n)s(ule)*| *HE XXXIII T*	Afyon, AD 109	Fant 1989: no. 47 = Drew-Bear 1994: 835 fn. 321.
70. *Pal(ma) II co(n)s(ule) HE XXVI*	Afyon, AD 109	Fant 1989: no. 36
71. *Pal(ma) II co(n)s(ulibu)s XLIX*	Afyon, AD 109	Fant 1989: no. 108[2] = Christol & Drew-Bear 1986: no. 13.
72. a) *VFR Vop(isco) co(n)s(ule)* (AD 114) | *DCXCII*; b) *Pal(ma) II co(n)s(ule)* (AD 109) | *HE CXVII*	Afyon	Fant 1989: no. 46.
73. *Pal(ma) II co(n)s(ule) HE CXLI*	Bacakale, AD 109	Fant 1989: no. 49.
74. a) *c (mille) VI* ₽; b) *Pal(ma) II co(n)s(ule)* (AD 109) | *HE CLXXIII*; c) *VFR Vop(isco) co(n)s(ule)* (AD 114) | *CCCXXXV*	Bacakale	Fant 1989: no. 45.
75. a) *(mille) (mille) CIX b(racchio) III*; b) *Pal(ma) II co(n)s(ule) HE CLXXIIX T*	Bacakale, AD 109	Fant 1989: no. 48.
76. a) *Pal(ma) II co(n)s(ule)* | *HE CCLIX*; b) *VFR Verg(iliano) co(n)s(ule)* (AD 115)| *DCCCCXI*	Bacakale, AD 109	Fant 1989: no. 37.
77. a) *RMA Pal(ma) II co(n)s(ule)* (AD 109); b) *DCCLVII*	Bacakale, AD 109	Fant 1989: no. 35.
78. a) *RMA ANT*; b) *ANT CLX*; c) *RMA*; d) *VFR Verg(iliano) co(n)s(ule)* (AD 115); e) *Pal(ma) II co(n)s(ule)* (AD 109); f) *DCCXXC* ₽	Bacakale	Fant 1989: no. 34.

No.	Text	Findspot	Reference
79.	a) *BOΔARTSLI* | *ĐCCCXL* | *b(racchio) III C* | *XIIX* | *A Pal(ma) II co(n)s(ule)* (AD 109); b) *NER Ve[rg(iliano)] co(n)s(ule)* (AD 115)	Iscehisar	Fant 1989: no. 38^2 = *CIL* III 7017 = Ramsay 1882: no. 4.
80.	a) *ĐCCCLXIX Pal(ma) II co(n)s(ule)* (AD 109) ; b) *VFR Verg(iliano) co(n)s(ule)* (AD 114) ; c) *CHR XCVI*	Bacakale	Drew-Bear 1994: no. 7.
81.	a) *CCCLXX*; b) *b(racchio) III XIIXXS*; c) *b(racchio) sec(undo)* ; d) *Pal(ma) II co(n)s(ule) XVI*	Iscehisar, AD 109	Fant 1989: no. 20^2 = *CIL* III 7023.
82.	a) *Orfito co(n)s(ule)* (AD 110) *HE XV*; b) *Orfito co(n)s(ule)* (AD 110) *HE XV*; c) *VI b(racchio) III*; d) *VFR Verg-(iliano) [c]o[(n)s(ule)]* (AD 115) | *COM[-] CXCIV* | *co(n)s(ule) XIIX [-]*	Bacakale	Fant 1989: no. 50.
83.	a) *VFR Verg(iliano) co(n)s(ule)* (AD 115); b) *(mille) CI*; c) *CDXI b(racchio) II*; d) *b(racchio) secund(o)* | *Orf(ito) co(n)s(ule)* (AD 110) | *[—]XVIII*	Bacakale	Fant 1989: no. 38 + Drew-Bear 1994: 835 fn. 330.
84.	a) *VFR Vop(isco)* | *co(n)s(ule)* (AD 114); b) *(mille) DC[C]LXṾ*; c) *[—]E?[—] Cels(o)* | *II co(n)s(ule)* (AD 113?) *DCXXXIX*; d) *b(racchio) III*; e) *Pi[sone] et Boḷ(ano)* | *co(n)s(ule)* (AD 111)	Bacakale	Fant 1989: no. 51.
85.	a) *Pisone co(n)s(ule)* (AD 111) | *DSN*; b) *VFR Vop(isco)*(AD 114)	Bacakale	Fant 1989: no. 52.
86.	a) *RIVLEIIIACELS*; b) *VFR Vop(isco) co(n)s(ule)* (AD 114) ; c) *Pisone co(n)s(ule)* (AD 111) | *II co(n)s(ule)*; d) *DSN*	Bacakale, AD 111	Fant 1989: no. 79^2 = Röder 1971: J1.
87.	*Traiano* | *n(ostro) VI co(n)s(ule) DI*	Bacakale, AD 112	Fant 1989: no. 22 + Drew-Bear 1994: 840 fn. 349.
88.	*VFR Vop(isco) co(n)s(ule)*	Bacakale, AD 114	Fant 1989: no. 53
89.	*Ɍ Vop(isco) co(n)s(ule) C[—]* | *V[—]*	Iscehisar, AD 114	Fant 1989: no. 111^2 = Christol & Drew-Bear 1986: no. 16.
90.	*VFɌ Vop(isco) co(n)s(ule) XXIIX*	Iscehisar, AD 114	Fant 1989: no. 109^2 = Christol & Drew-Bear 1986: no. 14.
91.	a) *VFɌ Vop(isco) co(n)s(ule)* | *CIII*; b) ∞*(mille) LXXIV* | *b(racchio) III*	Iscehisar, AD 114	Fant 1989: no. 110^2 = Christol & Drew-Bear 1986: no. 15.
92.	*VFR Vop(isco) co(n)s(ule)* | *CCI* | *Λ I co(n)s(ule?)*	Iscehisar, AD 114	Fant 1989: no. 47^2 = *CIL* III 7008 = Ramsay 1882: no. 13.

93.	a) *Vopisco ⟨co(n)⟩s(ule) | ĐLXXXVII*; b) *Vop(isco) co(n)s(ule) ĐLXXXVII*	Bacakale, AD 114	Fant 1989: no. 54.
94.	a) *Verg(iliano) co(n)s(ule) CXXC | Ꞧ*; b) *CCXCỊ b(racchio) III*	Bacakale, AD 115	Drew-Bear 1994: no. 8.
95.	*VFR Verg(iliano) co(n)s(ule)| CHR | RMA*	Bacakale, AD 115	Drew-Bear 1994: no. 9.
96.	a) *ĐCCCXXVII Ꞧ*; b) *Ꞧ Verg(iliano) co(n)s(ule)*; c) *CLXIIX*	Afyon, AD 115	Fant 1989: no. 39 + Drew-Bear 1994: 841 fn. 350.
97.	[. . .]*co(n)s(ule?)| C | VFR Verg(iliano) co(n)s(ule)| III b(racchio)*	Bacakale, AD 115	Fant 1989: no. 80^2 = Röder 1971: J2.
98.	*Ꞧ Verg(iliano) co(n)s(ule)*	Iscehisar, AD 115	Fant 1989: no. 15^2 = *CIL* III 7005A.
99.	a) *Verg(iliano) co(n)s(ule) ĐCCXXXIV Ꞧ*; b) *ĐC[C?]XXXIV*	Afyon, AD 115	Fant 1989: no. 55.
100.	*Hadriano III co(n)s(ule)| XIIII | Ꞧ*	Bacakale, AD 119	Fant 1989: no. 58.
101.	*Hadr(iano) III [co(n)s(ule)] | (mille) CCCCXX[—]*	Bacakale, AD 119	Fant 1989: no. 57.
102.	*Hadr(iano) III co(n)s(ule)| (mille) CCCCLXIIII*	Bacakale, AD 119	Fant 1989: no. 56.
103.	a) *Hadr(iano) III co(n)ṣ(ule)| de XIV*; b) *XXIX Ꞧ*	Bacakale, AD 119	Fant 1989: no. 59.
104.	*Hadr(iano) III co(n)s(ule) de LIV*	Bacakale, AD 119	Fant 1989: no. 60.
105.	a) *Asiatico II co(n)s(ule)| XXVI*; b) *VII Ꞧ*; c) *X*	Afyon, AD 125	Fant 1989: no. 61.
106.	a) *CLIX | b(racchio) III*; b) *Ver(o) III co(n)s(ule) CCCXCVI*	Bacakale, AD 126	Fant 1989: no. 63.
107.	a) *Ver(o) III co(n)s(ule)CCCXXX*; b) *b(racchio) quart(o)*; c) *CLXVI | b(racchio) IIII*	Bacakale, AD 126	Fant 1989: no. 62 = Drew-Bear 1994: no. 10.
108.	*loc(o) V Ꞧ | Rufin[o co(n)s(ule)]*	Bacakale, AD 131?	Fant 1989: no. 64.
109.	*Augurin(o) cos | l(oco) C ĐVI R*	Roma, AD 132	Fant 1989: no. 9^2 = Bruzza 1870: no. 257.
110.	a) *l(oco)ĐCCCXV Ꞧ | Augurin(o) co(n)s(ule)*; b) *XV*	Ostia, AD 132	Fant 1989: no. 86^2 = Baccini Leotardi 1979: no. 38.
111.	*[Imp(eratore) Ca]es[a]re IIII et Balbino co(n)s(ulibus) ratione [—]*	Lepcis Magna AD 136?	Fant 1989: no. 75^2 = *IRT* 794b = *ZPE* 75, 1988: 291–4.
112.	a) *b(racchio) tert(io)*; b) *b(racchio) III*; c) *Comm(odo) e⟨t⟩ Civic(a) co(n)ṣ(ulibus) b(racchio) III | loco XXVI Tul(lii?)*	Iscehisar, AD 136	Christol & Drew-Bear 1987: no. 1.
113.	*Comm(odo) et Civic(a) co(n)s(ulibus) loco VII, b(racchio) III | caesur(a) Tulli 7(centurionis)*	Bacakale, AD 136	Fant 1989: no. 66.

114.	*Commodo et Civic(a) co(n)s(ulibus)| caesur(a) Tulli 7(centurionis) loco XV | b(racchio) III*	Bacakale, AD 136	Fant 1989: no. 67.
115.	a) *L. Aelio | [C]aesare n(ostro) III et Bal|bino co(n)s(ulibus) rationi | urbicae sub cur(a) Irenaei | Aug(usti) lib(erti) proc(uratoris) caesura Tulli | Saturnini 7(centurionis) leg(ionis) XXII Prim(igeniae)*; b) *off(icina) Pa(pia) | n(umero) LXXVI*; c) *locus NII(?)CIA | loc(o) XVI b(racchio)*	Roma, AD 136	Fant 1989: no. 10[2]/ p. 251, 1a–c = Bruzza 1870: no. 258 = Dubois 1908: no. 199.
116.	a) *A[elio etc.* dot | *Tu[lli etc.]*; b) *L. Aelio Caesare n(ostro) III | et Balbino co(n)s(ulibus) r⟨a⟩tioni | urbicae sub cur(a) Iren|[a]ei ; Aug(usti) lib(erti) proc(uratoris), caesura | [Tu]lli Saturnini 7(centurionis) leg(ionis) | XXII Prim(igeniae)*; c) *off(icina) Papi(a) | n(umero) XCIV | loco XX*	Roma, AD 136	Fant 1989: no. 11[2]/ p.252 f. no. 2a–c = Bruzza 1870: no. 259 = Dubois 1908: no. 200.
117.	a) *Comm(odo) et Civic(a) co(n)s(ulibus) b(racchio) III | loco XXV Tul(li)*; b) *b(racchio) III*; c) *b(racchio) tert(io)*	Bacakale, AD 136	Fant 1989: no. 68.
118.	a) *L(ucio) Aelio Caes(are) n(ostro) II et Balbino | co(n)s(ulibus) loco XXII b(racchio) IIII*; b) *b(racchio) quart(o)*	Afyon, AD 137	Drew-Bear 1994: no. 11.
119.	*L. Aelio Caes(are) n(ostro) II| e⟨t⟩ Balbino co(n)s(ulibus)| loc(o) XLI[—]*	Bacakale, AD 137	Fant 1989: no. 69.
120.	*L. Caes(are) n(ostro) II co(n)s(ule)| loco CXXIII b(racchio) IIII*	Bacakale, AD 137	Fant 1989: no. 65 + Christol & Drew-Bear 1991: 123 fn. 46 + Drew-Bear 1994: 808 fn. 194.
121.	a) *Camer(ino) co(n)s(ule)| loco XXVIIII*; b) *VE[—]*	Bacakale, AD 138	Fant 1989: no. 72.
122.	*Cam(erino) et Nig(ro) loco [—] | b(racchio) III b(racchio) tert(io)*	Bacakale, AD 138	Fant 1989: no. 77.
123.	*[Nig]ro et Cam(erino) co(n)s(ulibus) caes(ura) Tulli | loco XIIX b(racchio) IIII*	Bacakale, AD 138	Fant 1989: no. 76.
124.	*Cam(erino) et Nig(ro) co(n)s(ulibus) loco XXXX | [b(racchio)] IIII*	Bacakale, AD 138	Fant 1989: no. 78.
125.	*Cam(erino) et [Nig(ro)] | [loco –] ẊXXXXIIX | CVSPOL*	Bacakale, AD 138	Drew-Bear 1994: no. 12.
126.	*b(racchio) II [—] | Camerino c[o(n)s(ule)] | loco LXXX*	Bacakale, AD 138	Fant 1989: no. 71.
127.	*Nigrino et Camer(ino) | de loco LXXXV de b(racchio) ter(tio)*	Iscehisar, AD 138	Christol & Drew-Bear 1987: no. 4.

128. *Nigr⟨in⟩o et Camer(ino) | de loco LXXXXV de br(acchio) ter(tio)*	Bacakale, AD 138	Fant 1989: no. 73 + Drew-Bear 1994: 767 fn. 49.
129. a) *Cam(erino) et Nig(ro) loc(o) CXII | b(racchio) IIII*; b) *b(racchio) quart(o)*	Bacakale, AD 138	Fant 1989: no. 74 + Drew-Bear 1994: 767 fn. 49.
130. *Cam(erino) et Nic(rino) | loc(o) XXCIII b(racchio) IIII*	Bacakale, AD 138	Fant 1989: no. 75 = Christol & Drew-Bear 1987: no. 3.
131. a) *b(racchio) tert(io) | Cam(erino) et Niç(rino)*; b) *ḷoco CXXXVII b(racchio) III*	Afyon, AD 138	Christol & Drew-Bear 1987: no. 2 = Fant 1989: no. 70?
132. *Imp(eratore) Ant(onino) n(ostro) II co(n)s(ule)| b(racchio) IIII loco LXII*	Bacakale, AD 139	Drew-Bear 1994: no. 13.
133. *Imp(eratore) Ant(onino) n(ostro) II co(n)s(ule) | loco XCII b(racchio) IIII*	Iscehisar, AD 139	Christol & Drew-Bear 1987: no. 5.
134. *[Im]p(eratore) Ant(onino) III co(n)s(ule)*	Iscehisar, AD 140	Fant 1989: no. 120[2] = Christol & Drew-Bear 1986: no. 30.
135. a) *Imp(eratore) Ant(onino) III co(n)s-(ule) | loco XXXXVII | b(racchio) III* ; b) *b(racchio) tert(io)*	Bacakale, AD 140	Fant 1989: no. 216 = Christol & Drew-Bear 1987: no. 6.
136. *Imp(eratore) Ant(onino) n(ostro) III co(n)s(ule) | loco LXVI*	Dokimeion, AD 140	Fant 1989: no. 119[2] = Christol & Drew-Bear 1986: no. 29.
137. *Prisc(ino) et Sever(o) co(n)s(ulibus) loc(o) XXIIII b(racchio) IIII*	Afyon, AD 141	Fant 1989: no. 79.
138. *Prisci(no) et Seve(ro) co(n)s(ulibus) | loco XLVI[I]I*	Iscehisar, AD 141	Fant 1989: no. 112[2] = Christol & Drew-Bear 1986: no. 18.
139. *b(racchio) tert(io) | Prisci(no) et Sever(o) co(n)s(ulibus) | loco CVIII b(racchio) III*	Nuri Bey, AD 141	Fant 1989: no. 61[2] = *MAMA* IV 6 = Drew-Bear 1994: no. 14.
140. a) *NOΦNVB*; b) *Claro II co(n)s(ule) ex r(atione) | no CIII | Cl(audii) Zel[o(ti)]*	Ostia, AD 142	Fant 1989: no. 91[2] = Baccini Leotardi 1989: no. 42.
141. *Claro II co(n)s(ule) | ex r(atione) Cl(audii) Zel(oti) | No CXXXV*	Ostia, AD 142	Fant 1989: no. 91b[2] = Baccini Leotardi 1989: no. 43.
142. *Flacco et Herod(e) co(n)s(ulibus) | loco XI b(racchio) IIII*	Iscehisar, AD 143	Fant 1989: no. 113[2] = Christol & Drew-Bear 1986: no. 19.
143. a) *b(racchio) quart(o) | Flacco et Ḥerod(e) co(n)s(ulibus) | loco CLXVI b(racchio) IIII*; b) *IA*	Iscehisar, AD 143	Christol & Drew-Bear 1987: no. 9.

144. a) *b(racchio) tert(io)*; b) *Flacco et Herod(e) co(n)s(ulibus) | loco CXXI b(racchio) III*	Bacakale, AD 143	Fant 1989: no. 82.
145. *Flacco et Herod(e) co(n)s(ulibus) | loco CXXII b(racchio) III | b(racchio) tert(io)*	Bacakale, AD 143	Fant 1989: no. 81.
146. a) *Flac(co) et Hero(de) co(n)s(ulibus) | [l]oco CCXI[–b(racchio) III]*; b) *b(racchio) tert(io)*	Iscehisar, AD 143	Christol & Drew-Bear 1987: no. 8.
147. a) *Flac(co) et Hero(de) co(n)s(ulibus) | loco CCXLIX b(racchio) III*; b) *b(racchio) tert(io)*	Bacakale, AD 143	Fant 1989: no. 80 = Christol & Drew-Bear 1987: no. 7.
148. *Maxi(mo) et Avi(to) co(n)s(ulibus) | loco IX b(racchio) [–]*	Iscehisar, AD 144	Christol & Drew-Bear 1987: no. 13 = Fant 1984: no. 2.
149. *Maxi(mo) et Aivt(o)(!) co(n)s(ulibus) | b(racchio) quart(o) | loco XIX b(racchio) IIII*	Iscehisar, AD 144	Christol & Drew-Bear 1987: no. 12.
150. a) *Maxim(o) et Avit(o) co(n)s(ulibus)* (AD 144) | *loco XXV, b(racchio)* vac.; b) *b(racchio) quart(o)*	Bacakale, AD 144	Fant 1989: no. 83.
151. a) *Maxi(mo) et Aivt(o)(!) co(n)s(ulibus) | loco XXXIX | b(racchio) IIII* |; b) *b(racchio) quart(o)*	Ischeisar, AD 144	Christol & Drew-Bear 1987: no. 11.
152. *Maxi(mo) et Avi(to) co(n)s(ulibus) | loco CXIX b(racchio) III | b(racchio) tert(io)*	Iscehisar, AD 144	Christol & Drew-Bear 1987: no. 10.
153. *Ma[x(imo) et] Avit(o) co(n)s(ulibus) | loco CCXLV | b(racchio) quar(to)*	Iscehisar, AD 144	Fant 1989: no. 40^2 = *CIL* III 7013 = Ramsay 1882: no. 6 = Christol & Drew-Bear 1987: p.85.
154. *Claro II et Sever(o) co(n)s(ulibus) | loco CCXXXI | b(racchio) tert(io) | off(icina) Pela(goni)*	Iscehisar, AD 146	Fant 1989: no. 21^2 = *CIL* III 7025 = Christol & Drew-Bear 1986: no. 20.
155. *Clar(o) II et Sever(o) co(n)s(ulibus) | loco CCCX | b(racchio) tert(io) | off(icina) Pelag(oni)*	Bacakale, AD 146	Fant 1989: no. 84.
156. *Clar(o) II e[t Severo co(n)s(ulibus) | loc(o) LX[—] | o(f)fi[cina—]*	Iscehisar, AD 146	Fant 1989: no. 56^2 = *CIL* III 7026 = Ramsay 1882: no. 23.
157. *Larg(o) et Mess(alino)[co(n)s(ulibus) | loco CXC | caesura Ael(ii) Ant[onini] | b(racchio) quar[t(o)]*	Iscehisar, AD 147	Fant 1989: no. 32^2 = *CIL* III 14192.1 = Christol & Drew-Bear 1987: 87 f.

158. *Largo et Messall(ino) co(n)s(ulibus) | loco XVI b(racchio) secund(o) | off(icina) Papia*	Afyon, AD 147	Fant 1989: no. 85 = Christol & Drew-Bear 1991: 123 fn. 45 = Christol & Drew-Bear 1986: no. 21.
159. *Larg(o) et Messal(lino) co(n)s(ulibus)| loco XXXVII repr(obatum) | off(icina) Asiat(ici)*	Iscehisar, AD 147	Fant 1989: no. 22[2] = *CIL* III 7027.
160. *Largo et Messall(ino) co(n)s(ulibus) | loco CCXCVI b(racchio) secund(o) off(icina) Pelag(oni)*	Bacakale, AD 147	Fant 1989: no. 86= Christol & Drew-Bear1987: no. 14 + Drew-Bear 1994: 772 fn. 66.
161. *Largo et Messal(ino) co(n)s(ulibus) | loco CCXCVII R(=b(racchio)?) quart(o) | off(icina) Pelag(oni)*	Iscehisar, AD 147	Christol & Drew-Bear 1987: no. 16.
162. *Largo et Messal(lino) co(n)s(ulibus) | loco CCCXXIII b(racchio) prim(o) | caesu(ra) Ael(ii) Antonini 7(centurionis)*	Bacakale, AD 147	Fant 1989: no. 87.
163. *Largo et Messal(lino) co(n)s(ulibus) | loco CCCCXXIX b(racchio) secund(o) | caesu(ra) Ael(ii) [A]ntonini 7(centurionis)*	Bacakale, AD 147	Fant 1989: no. 89.
164. *Largo et Messal(ino) co(n)s(ulibus) | loco CCCCXXXII b(racchio) quart(o)*	Iscehisar, AD 147	Christol & Drew-Bear 1987: no. 15.
165. *Largo et Messal(lino) co(n)s(ulibus) | loco CCCCXXXIII b(racchio) quart(o) | caesu(ra) Ael(ii) Ant(onini) | 7(centurionis)*	Bacakale, AD 147	Fant 1989: no. 90.
166. *Largo et Messal(lino) co(n)s(ulibus) | loco CCC[.]XIV b(racchio) secund(o) | caesu(ra) [Ael(ii) Ant]onini 7(centurionis)*	Bacakale, AD 147	Fant 1989: no. 88.
167. *Torqua(to) et Iulian(o) co(n)s(ulibus) | loco*	Iscehisar, AD 148	Christol & Drew-Bear 1987: no. 25 = Fant 1989: no. 111?
168. *Torqu[a(to) et Iulian(o) co(n)s(ulibus)] | loco [—b(racchio)—] | caesu[r(a) —]*	Bacakale, AD 148	Fant 1989: no. 112 + Drew-Bear 1994: 779 fn. 84.
169. *Torqua(to) et Iulian(o) co(n)s(ulibus) | loco Ç[—] | [——]*	Iscehisar, AD 148	Christol & Drew-Bear 1987: no. 19.
170. *[Tor]qua(to) et Iulia̤ṇ(o) co(n)s(ulibus)| [loc]o XIIII | [——]*	Bacakale, AD 148?	Drew-Bear 1994: no. 18.
171. *Torqua(to) et Iul[ian(o) co(n)s(ulibus)] | loco XVII | caesur(a) Ael(ii) [Ant(onini) centurionis]*	Bacakale, AD 148	Fant 1989: no. 91.

172. a) *Torqua(to) et Iuli(a)n(o) co(n)s(ulibus) | loco XXI b(racchio) quart(o) | caesur(a) Ael(ii) Antonini 7(centurionis)* ⊕; b) ⊕	Afyon, AD 148	Drew-Bear 1994: no. 16.
173. a) *Torqua(to) et Iulian(o) co(n)s(ulibus) | loco XIIX b(racchio) quart(o) | caesur(a) Ael(ii) Antonini*; b) ⊕	Bacakale, AD 148	Fant 1989: no. 92 + Drew-Bear 1994: 775 fn. 72 = Christol & Drew-Bear 1987: no. 21.
174. *Torqua(to) et Iulian(o) co(n)s(ulibus) | loco XXXII b(racchio) tert(io) | caesur(a) Ael(ii) Antonini*	Synnada, AD 148	Fant 1989: no. 101[2] = *CIL* III 14402i = Monceaux, BullSocAnt 1900: no. 3.
175. a) *Torqua(to) et Iulian(o) co(n)s(ulibus) | loco XLVI repr(obatum) | caesur(a) Ael(ii) Antonini 7(centurionis)*; b) ⊕	Bacakale, AD 148	Fant 1989: no. 93 + Drew-Bear 1994: 775 fn. 71.
176. *Torqua(to) et Iulian(o) co(n)s(ulibus) | loco LIII b(racchio) ṭẹrt(io) | caesur(a) Ael(ii) Antonini 7(centurionis)*	Bacakale, AD 148	Christol & Drew-Bear 1987: no. 23 = Fant 1989: no. 94.
177. a) *Torqua(to) et Iulian(o) co(n)s(ulibus) | loco LV b(racchio) quart(o) | caesur(a) Ael(ii) Antonini 7(centurionis)* ; b) ⊕	Bacakale, AD 148	Christol & Drew-Bear 1987: no. 24 = Fant 1989: no. 95? + Drew-Bear 1994: 775 fn. 71.
178. *Torqua(to) et Iulian(o) co(n)s(ulibus) | loco LVI b(racchio) quart(o) | caesur(a) Ael(ii) Antonini 7(centurionis)*	Iscehisar, AD 148	Christol & Drew-Bear 1987: no. 28.
179. *Torqua(to) et Iulian(o) co(n)s(ulibus) | loco LXI b(racchio) tert(io)*	Bacakale, AD 148	Fant 1989: no. 96.
180. *Torqua(to) et Iulian(o) co(n)s(ulibus) | loco LXV b(racchio) secund(o) | caesur(a) Ael(ii) Antonini 7 (centurionis) off(icina) Asia(tici)*	Bacakale, AD 148	Drew-Bear 1994: no. 17.
181. *Torqua(to) et Iulian(o) co(n)s(ulibus) | loco LXXXV b(racchio) tert(io)*	Bacakale, AD 148	Fant 1989: no. 97.
182. *br(acchio) quart(o) | Torqua(to) et Iulian(o) co(n)s(ulibus) | loco XCIII | caesur(a) Ael(ii) Antoni(ni) 7(centurionis)*	Bacakale, AD 148	Fant 1989: no. 98 + Drew-Bear 1994: 779 fn. 82, 802 fn. 165.
183. *Torqua(to) et Iu[li]an(o) [co(n)s(ulibus)] | loco XCIIII [b(racchio) –] | caesur(a) Ael(ii) Ạ[ntonini]*	Iscehisar, AD 148	Fant 1989: no. 114[2] = Christol & Drew-Bear 1986: no. 22.
184. *Torqua(to) et Iulian(o) co(n)s(ulibus) | loco CIII b(racchio) quart(o)*	Bacakale, AD 148	Fant 1989: no. 99.
185. *Torqua(to) et Iulian(o) co(n)s(ulibus) | loco CXIIII b(racchio) tert(io) | caesur(a) Ael(ii) Antonini 7(centurionis)*	Bacakale, AD 148	Fant 1989: no. 100 + Drew-Bear 1994: 774 fn. 68.

186. *Torqua(to) et Iulian(o) co(n)s(ulibus) | loco CXXIX b(racchio) seg(undo)*	Synnada, AD 148	Fant 1989: no. 100[2] = *CIL* III 14402h = Monceaux, *BullSocAnt* 1900: no. 2.
187. *Torqua(to) et Iulian(o) co(n)s(ulibus) | loco CXXIIX b(racchio) segun(do)*	Synnada, AD 148	Fant 1989: no. 99[2] = *CIL* III 14402g = Monceaux, *BullSocAnt* 1900: no. 1.
188. *Torqua(to) et Iulian(o) co(n)s(ulibus) | loco CLXIII b(racchio) quart(o) | caesu(ra) Ael(ii) Antonini 7(centurionis)*	Bacakale, AD 148	Drew-Bear 1994: no. 15.
189. a) *Torqua(to) et Iulian(o) co(n)s(ulibus) | loco CLXXVI b(racchio) quart(o)*; b) ⊕	Afyon, AD 148	Fant 1989: no. 101 = Drew-Bear 1994: 774 fn. 71 = Christol & Drew-Bear 1986: no. 23.
190. *Torqua(to) et Iulian(o) co(n)s(ulibus) | loco CCXỊIX | caesur(a) Ael(ii) Antonini 7(centurionis) | b(racchio) secun(do)*	Afyon, AD 148	Fant 1989: no. 102 = Christol & Drew-Bear 1987: no. 27 = Drew-Bear 1994: no. 19.
191. *Torqua(to) et Iulian(o) co(n)s(ulibus) | loco CCXXIIII repr(obatum)*	Bacakale, AD 148	Fant 1989: no. 103 = Christol & Drew-Bear 1987: no. 18.
192. *[Tor]qua(to) et Iulian(o) co(n)s(ulibus) | loco CCXXV, b(racchio) quart(o)*	Bacakale, AD 148	Fant 1989: no. 104
193. *Torqua(to) et Iulian(o) co(n)s(ulibus) | loco CCXXXIIII | b(racchio) quart(o)*	Iscehisar, AD 148	Fant 1989: no. 105 = Christol & Drew-Bear 1987: no. 26.
194. a) *Torqua(to) et Iulian(o) co(n)[s(ulibus)] | loco CCXXXVI [—] | [b(racchio) quart(o)*; b) ⊕	Bacakale, AD 148	Christol & Drew-Bear 1987: no. 20 = Fant 1989: no. 106 + Drew-Bear 1994: 774 fn. 71.
195. a) *b(racchio) qua(rto)*; b) *loco | CCXXX̣IIX*; c)⊕; d) *Torqua(to) et Iuli[an(o) co(n)s(ulibus)]*	Bacakale, AD 148	Fant 1989: no. 107 + Drew-Bear 1994: 774 fn. 71.
196. *Torqua(to) et Iulian(o) co(n)s(ulibus) | loco CCLXXV | b(racchio) quart(o)*	Bacakale, AD 148	Fant 1989: no. 108.
197. a) *Torqua(to) et Iulian(o) co(n)s(ulibus) | loco CCLXXVI | b(racchio) quart(o)*; b) ⊕	Bacakale, AD 148	Christol & Drew-Bear 1987: no. 22 = Fant 1989: no. 109.
198. *Torqua(to) et Iulian(o) co(n)s(ulibus) | loco CCLXX[X]X | b(racchio) quart(o)*	Iscehisar, AD 148	Christol & Drew-Bear 1987: no. 29.
199. *Torqua(to) et Iulịạn(o) [co(n)s(ulibus)] | loco CCXVI [—] | r[e]pr(obatum)*	Iscehisar, AD 148	Christol & Drew-Bear 1987: no. 30.

200. a) *Torqua(to) et Iulian(o) co(n)s(ulibus) | loco CCXXCIX | repr(obatum)*; b) ⊕	Bacakale, AD 148	Fant 1989: no. 110 = Christol & Drew-Bear 1987: no. 17 + Drew-Bear 1994: 774 fn. 71.
201. a) *Orfito et Prisco co(n)s(ulibus)| loco XII b(racchio) III | caes(ura) PERCḶDAED* (?); b) *XII*	Bacakale, AD 149	Fant 1989: no. 114 = Christol & Drew-Bear 1987: no. 31 + Drew-Bear 1994: 783 fn. 102.
202. *Orfito (et Pudente) co(n)s(ulibus) | off(icina) Ephe(sia) loco LXXIV*	Dokimeion, AD 149	Fant 1989: no. 115² = Christol & Drew-Bear 1986: no. 24 + Drew-Bear 1994: 800 fn. 157.
203. *Orfito et Pudente | co(n)s(ulibus) offi(cina) Ephes(ia) caes(ura) | Alex(andri) | loco LV b(racchio) IIII*	Bacakale, AD 149	Fant 1989: no. 157 + Drew-Bear 1994: 800 fn. 157.
204. *Gallicano et Vetere | co(n)s(ulibus)*	Bacakale, AD 150	Fant 1989: no. 116
205. *Gallicano et V|etere co(n)s(ulibus)*	Bacakale, AD 150	Drew-Bear 1994: no. 20.
206. a) *Gallicano et Vetere | co(n)s(ulibus)*; b) *off(icina) Pelag(oni) | loco XXIX b(racchio) III*	Bacakale, AD 150	Fant 1989: no. 115.
207. a) *Gallicano et Ve⟨te⟩r(e) co(n)s(ulibus)*; b) *ex ra(tione) Cl(audi) Zel(oti) | no LXXVIII*	Ostia, AD 150	Baccini Leotardi 1989: no. 44.
208. a) *off(icina) Ephe(sia) loco LXV[—]| Callicano et Vete[re] | co(n)s(ulibus)*; b) *b(racchio) II*	Iscehisar, AD 150	Christol & Drew-Bear 1987: no. 32.
209. *Maxi et Sil(o) co(n)s(ulibus) | loco IX ḅ(racchio) [—]*	Iscehisar, AD 152	Fant 1989: no. 2² = *AE* 1984: 845 = Christol & Drew-Bear 1986: no. 25 + Drew-Bear 1994: 810 fn. 213.
210. *Praesente et Rufino co(n)s(ulibus) | ca⟨e⟩sura Ael(ii) Zosim[i] | loco L b(racchio) II*	Bacakale, AD 153	Fant 1989: no. 117.
211. *Praesente e(t) Rufino co(n)s(ulibus) | caesura Aeli(i) Zosimi | loco* vac. *| b(racchio) III*	Iscehisar, AD 153	Christol & Drew-Bear 1987: no. 33.
212. *Commodi et Laterani co(n)s(ulibus) | caes(ura) Zos(imi) | loco LIIII b(racchio)* vac.	Iscehisar, AD 154	Christol & Drew-Bear 1987: no. 34.
213. *L(ucio) Ael(io) Aurel(io) Aug(usto) fil(io) Commodo | et Laterano co(n)s(ulibus) | caes(ura) Manli(i) Alexandr(i)*	Bacakale, AD 154	Drew-Bear 1994: no. 23.
214. *L(ucio) Ael(io) Aur(elio) Aug(usto) ⟨f⟩il(io) Commodo | et Laterano co(n)s(ulibus) | caes(ura) Manli(i) Ale xandri | loco CI ₽*	Bacakale, AD 154	Drew-Bear 1994: no. 24.

215. *Commodo et Laterano | co(n)s(ulibus) caes(ura) Zosimi | loco XXXIII br(acchio) II*	Bacakale, AD 154	Fant 1989: no. 118.
216. *loco XIII* ⅄ | *Commodo et Lateran[o] | co(n)s(ulibus) caesur(a) Ma⟨n⟩li Alex-(andri)*	Bacakale, AD 154	Fant 1989: no. 119 = Christol & Drew-Bear 1987: no. 35 + Christol & Drew-Bear 1991: 139 fn. 122, 161 fn. 189 + Drew-Bear 1994: 783 fn. 103.
217. *loco V b(racchio) secu(ndo) | Silvano et Au|gurino co(n)s(ulibus) ex off(icina) | Asiatici ex cae(sura) | Zosim(i)*	Bacakale, AD 154?	Fant 1989: no. 122.
218. *loco | XXVI | b(racchio) II | Commodo | et Laterano co(n)s(ulibus) | caes(ura) Ael(ii) Zos(imi)*	Bacakale, AD 154	Fant 1989: no. 121.
219. *Commodi et La⟨t⟩erani co(n)s(ulibus) | caes(ura) Manl(ii) Alexand(ri) | loco XXXX*	Afyon, AD 154	Drew-Bear 1994: no. 22.
220. *Commodi et Laterani co(n)s(ulibus) | caes(ura) Manl(ii) Alexand(ri) | loco LV b(racchio) IIIỊ*	Afyon, AD 154	Fant 1989: no. 120 = Drew-Bear 1994: no. 21.
221. *loco* [vac.] *b(racchio)* [vac.] | *Barbaro et Regulo co(n)s(ulibus)* (AD 157) | *ex of(f)i(cina) Andae(vi) ex cae(sura) Alex(andri)*	Bacakale, AD 157	Fant 1989: no. 123 = Christol & Drew-Bear 1987: no. 36 + Christol & Drew-Bear 1991: 161 fn. 189.
222. a) *loco IIII b(racchio) III | Barbaro et Regulo co(n)s(ulibus) ex off(icina) Pela(goni) | ex cae(sura) Zosi(mi)* b) *PAL*	Bacakale, AD 157	Fant 1989: no. 124.
223. *[lo]co XI b(racchio) pṛ[im(o)] | [Ba]rbaro et Reg[u]|[lo c]ọ(n)s(ulibus) ex of(ficina) And[ae(vi)] | [ex ca]e-(sura) Ạḷẹx(andri)*	Bacakale, AD 157	Fant 1989: no. 125 = Drew-Bear 1994: no. 25.
224. *Tertullo et Sacerdote co(n)s(ulibus) | ex off(icina) contra caesura Zosi(mi) | loco* [vac.] *b(racchio) quar(to)*	Bacakale, AD 158	Fant 1989: no. 128.
225. a) *Tertullo et Sacerdote co(n)s(ulibus) | ex off(icina) Pelagon(i) caes(ura) | Zosim(i) loco IIII b(racchio) quar(to)*; b) *VLI*	Bacakale, AD 158	Fant 1989: no. 126.
226. *[Te]rtullo et Sacẹr[dote] co(n)s(ulibus) | [ex] off(icina) Andaev(i) caes(ura) Alex(andri) | loco LIIII b(racchio) quar(to)*	Iscehisar, AD 158	Christol & Drew-Bear 1987: no. 37.
227. *Ṭẹ[rt]ullo et Sacerdote co(n)s(ulibus) | ex of(ficina) Andaev(i) caesura | Alex(andri) | loco XCIX b(racchio)* Ꝛ	Bacakale, AD 158	Fant 1989: no. 127 = Christol & Drew-Bear 1987: no. 38 + Christol & Drew-Bear 1991: 162 fn. 189 + Drew-Bear 1994: 779 fn. 82, 783 fn. 100, 795 fn. 142.

228. *[Q]uintillo et Prisco co(n)s(ulibus) | ex of(ficina) Andae(vi) caes(ura) | Alex(andri) loco XV b(racchio) q[ua]rt(o)*	Iscehisar, AD 159	Christol & Drew-Bear 1987: no. 41.
229. *Quintilio et | Prisco co(n)s(ulibus) | loco XVII | caes(ura) Alex(andri) | b(racchio) quar(to)*	Iscehisar, AD 159	Christol & Drew-Bear 1987: no. 46 = Fant 1989: no. 130.
230. *Quintillo et Prisco co(n)s(ulibus) | ex off(icina) Andae(vi) caesur(a) | Alex(andri) loco XVIII b(racchio) tert(io)*	Bacakale, AD 159	Fant 1989: no. 129.
231. *Quintillo e(t) Prisco co(n)s(ulibus) | ex of(ficina) Andae(vi) caesur(a) | Alex(andri) | loco XVIIII b(racchio) tert(io)*	Iscehisar, AD 159	Christol & Drew-Bear 1987: no. 44.
232. *Quintillo et Prisco co(n)s(ulibus) | ex of(ficina)Andae(vi) caesu(ra) Alex(andri) | loco CII b(racchio) ter(tio)*	Iscehisar, AD 159	Fant 1989: no. 1[2] = *AE* 1984: 844 = Christol & Drew-Bear 1986: no. 27: + Drew-Bear 1994: 773 fn. 67, 793 f. fn. 137.
233. *[loco—] XXXI b(racchio)* vac. *| [Quintillo et] Prisco co(n)s(ulibus) | [ex off(icina) Asiat]ịc(i) caes(ura) Zos(imi) | b(racchio) secund(o)*	Bacakale, AD 159	Fant 1989: no. 141 = Drew-Bear 1994: no. 29.
234. *[Q]u[i]nt[i]llo et Prisco co(n)s(ulibus) | ex of(ficina) Andae(vi) caesur(a) | Alex(andri) loco XXXIIII b(racchio) tert(io)*	Bacakale, AD 159	Fant 1989: no. 131.
235. *Quintillo et Prisco co(n)s(ulibus) | ex of(ficina) Andae(vi) ca⟨e⟩sura | Alex(andri) loco XXXXI b(racchio) tert(io)*	Bacakale, AD 159	Fant 1989: no. 132.
236. *Quịnṭil[lo et Pris]c̣ọ co(n)s(ulibus) | ex of(ficina) Andae(vi) ca⟨e⟩sura | Alex(andri) loco XXXXII | b(racchio) qu(arto)*	Bacakale, AD 159	Drew-Bear 1994: no. 27.
237. *Quin[t]iḷlo et Prisco [c]o(n)s(ulibus) | ex of(ficina) Andae(vi) caesur(a) | Alex(andri) loco XCVI b(racchio) tert(io)*	Bacakale, AD 159	Fant 1989: no. 133.
238. *Π[.]EIA | Quintillo et Pris[co] | co(n)s(ulibus) ex of(ficina) Andae(vi) | caesu(ra) Alex(andri) loco c[..] | b(racchio) quart(o)*	Bacakale, AD 159	Fant 1989: no. 142.
239. *Quintillo et P[ri]sco co(n)s(ulibus) | ex of(ficina) Andae(vi) caesur(a) Alex(andri) | loco CXXV b(racchio) tert(io)*	Bacakale, AD 159	Fant 1989: no. 134.

240. *[Qui]ntillo et | [Pri]sco co(n)s(ulibus) ex | of(ficina) Andae(vi) caes(ura) | Alex(andri) | loco CXXXXIII | b(racchio) quart(o)*	Bacakale, AD 159	Drew-Bear 1994: no. 26.
241. *Quintillo et P[risco] | co(n)s(ulibus) ex of(ficina) Andae(vi) | caes(ura) Alex(andri) | loco CXXXXVI | b(racchio) terti(o)*	Iscehisar, AD 159	Fant 1989: no. 117[2] = Christol & Drew-Bear 1986: no. 26.
242. *[Qui]ntillo et Prisco co(n)s(ulibus) ex of(ficina) Andae(vi) | [c]aes(ura) Alex(andri) | loco CXXXXIX b(racchio) terti(o)*	Bacakale, AD 159	Christol & Drew-Bear 1987: no. 45 = Fant 1989: no. 135
243. *Quintillo et Prisco co(n)s(ulibus) | ex of(ficina) Andae(vi) caes(ura) Alex(andri) | loco CLXIII b(racchio) tert(io)*	Iscehisar, AD 159	Christol & Drew-Bear 1987: no. 39.
244. *Quintill[o et] Prisco co[s] | ex of(ficina) Andae(vi) caes(ura) Alex(andri) | loco CLXXVIII | b(racchio) quart(o)*	Bacakale, AD 159	Fant 1989: no. 136 = Christol & Drew-Bear 1987: no. 42.
245. *Quintillo et Prisco co(n)s(ulibus) | ex of(ficina) Andae(vi) caes(ura) Alex(andri) || loco CLXXXII | b(racchio) quar(to)*	Bacakale, AD 159	Fant 1989: no. 137 + Drew-Bear 1994: no. 30, 804 fn. 173.
246. *Quintillo et Prisco co(n)s(ulibus) | ex of(ficina) Andae(vi), caesu(ra) | Alex(andri), de loco CXCVI, b(racchio) qua[r(to)]*	Bacakale, AD 159	Fant 1989: no. 138.
247. *[Quint]illo et P[ris]|co co(n)s(ulibus) ex off(icina) | Andaev(i) caes(ura) A[lex(andri)] | loco CCLXXXIII | b(racchio) tee(!)r(tio.)*	Iscehisar, AD 159	Drew-Bear 1994: no. 28.
248. *Quintillo et Prisco co(n)s(ulibus) | ex off(icina) Andae(vi) caes(ura) Alex(andri) | loco CCCLXXIIII | b(racchio) secun(do)*	Bacakale, AD 159	Fant 1989: no. 139 = Christol & Drew-Bear 1987: no. 43.
249. *[Quintillo] et Prisco co(n)s(ulibus) | off(icina) Andae(vi), caes(ura) Alex(andri) | loco CCCLXXXXV[–] | b(racchio) quar(to)*	Afyon, AD 159	Fant 1989: no. 140 = Christol & Drew-Bear 1987: no. 40 + Christol & Drew-Bear 1991: 162 fn. 189 + Drew-Bear 1994: 802 fn. 162.
250. *Bradua et Varo co(n)s(ulibus) | off(icina) Pelag(oni) caesur(a) | Claudiani loco* vac. *| b(racchio) quar(to)*	Iscehisar, AD 160	Christol & Drew-Bear 1987: no. 49.
251. *Bradua et Varo co(n)s(ulibus) off(icina) Eph[esia] | caes(ura) Claud(iani) loco VIIII b(racchio) [—]*	Afyon, AD 160	Fant 1989: no. 118[2] = Christol & Drew-Bear 1986: no. 28.
252. *Bradua et Varo co(n)s(ulibus) off(icina) Contra Pers(is) | caes(ura) Claud(iani) | loco XIIII b(racchio) quar(to)*	Bacakale, AD 160	Fant 1989: no. 143.

253.	*Bradua et Varo co(n)s(ulibus) | off(icina) Pelag(oni) caes(ura) | Claudian(i) b(racchio) qua(rto) | loco XVIII | b*	Iscehisar, AD 160	Fant 1989: no. 23[2] = *CIL* III 7029 = *MAMA* IV 7 = *JRS* 2, 1912: 251 no. 5.
254.	*Bradua et Varo co(n)s(ulibus) | off(icina) Pelag(oni), caesur(a) | Claudian(i) | loco b(racchio) quar(to) XXII*	Bacakale, AD 160	Fant 1989: no. 147.
255.	*Bradua et Varo co(n)s(ulibus) | offi(cina) Andae(vi) caes(ura) | Efor( ) | loco XXVI Ɍ*	Iscehisar, AD 160	Christol & Drew-Bear 1987: no. 47.
256.	*Bradua et Varo co(n)s(ulibus) | [o]ff(icina) Ephes(ia) caesur(a) | [Clau]diani loco LIX b(racchio) qu[ar(to)]*	Bacakale, AD 160	Fant 1989: no. 144 + Christol & Drew-Bear 1991: 121 f., fns. 37, 40 + Drew-Bear 1994: 800 fn. 157.
257.	*Bradua et Varo co(n)s(ulibus) | off(icina) Andae(vi) caesu(ra) Alex(andri) | loco LX K | Ɍ*	Bacakale, AD 160	Fant 1989: no. 145.
258.	*Bradua et Varo co(n)s(ulibus) | off(icina) Pelag(oni) caesur(a) | Claudian(i) | loco LXVI b(racchio) quar(to)*	Bacakale, AD 160	Fant 1989: no. 146.
259.	*Bradua et Varo co(n)s(ulibus) | off(icina) Andae(vi) caes(ura) | Fabi loco LXVII | b(racchio) quart(o)*	Iscehisar, AD 160	Christol & Drew-Bear 1987: no. 48.
260.	*Imp(eratore) Anto|nino n(ostro) III et | Imp(eratore) Vero n(ostro) I | co(n)s(ulibus) off(icina) Pe|lag(oni) caes(ura) [—]*	Iscehisar, AD 161	Fant 1989: no. 24[2] = *CIL* III 7030 = *MAMA* IV 8.
261.	*Imp(eratore) Antonino n(ostro) III et | Imp(eratore) Ve[ro] n(ostro) [II] co(n)s(ulibus) off(icina) | caes(ura) Claudiani loco | DT*	Iscehisar, AD 161	Fant 1989: no. 25[2] = *CIL* III 7031 = III 12233.
262.	*Imp(eratore) An[to]ni(no) no(stro) I[II] | Imp(eratore) Vero II co(n)s(ulibus) | caes[ura Cla]udiani*	Roma, AD 161	Fant 1989: no. 12[2] = Bruzza no. 261.
263.	a) *[I]mp(eratore) Antonino n(ostro) III [e]t I[mp(eratore)] | Vero n(ostro) II co(n)s(ulibus) off(icina) Anda[e(vi)] | caes(ura) Fabi loco X | Ɍ*; b) *loco X*	Bacakale, AD 161	Fant 1989: no. 148 = Christol & Drew-Bear 1987: no. 52 + Christol & Drew-Bear 1991: 154 fn. 163: + Drew-Bear 1994: 807 fn. 189.
264.	*Imp(eratore) Antonino n(ostro) III et | Imp(eratore) Vero n(ostro) II co(n)s(ulibus) | off(icina) Pelag(oni) caes(ura) | Claudia(ni) loco XII | b(racchio)quart(o)*	Iscehisar, AD 161	Fant 1989: no. 122[2] = Christol & Drew-Bear 1986: no. 32.
265.	*Imp(eratore) Antonino | n(ostro) III et Imp(eratore) Vero | n(ostro) II*	Bacakale, AD 161	Fant 1989: no. 149 = Christol & Drew-Bear

	co(n)s(ulibus) off(icina) Andae(vi) | caes(ura) Domes(tici) | loco XIII b(racchio) quar(to)		1987: no. 50 + Christol & Drew-Bear 1991: 154 fn. 163.
266.	*Imp(eratore) Antonino n(ostro) | III et Imp(eratore) Vero n(ostro) II | co(n)s(ulibus) off(icina) Andaev(i) | caes(ura)* vac. *| loco XVIII R̶*	Iscehisar, AD 161	Fant 1989: no. 121^2 = Christol & Drew-Bear 1986: no. 31.
267.	a) *Imp(eratore) Antoni(no) III et | Vero II co(n)s(ulibus) off(icina) | contr(a) caes(ura) Dom(estici)*; b) *b(racchio) quar(to)*; c) *loco XXVIII*	Bacakale, AD 161	Fant 1989: no. 150.
268.	a) *Imp(eratore) Antonino n(ostro) III | et Imp(eratore) Vero n(ostro) II cos | offi(cina) Pelag(oni) caes(ura) Claud(iani)*; b) R̶; c) *loco XXXXV [—]*	Bacakale, AD 161	Fant 1989: no. 151 = Christol & Drew-Bear 1987: no. 51 + Drew-Bear 1994: 806 fn. 189.
269.	*Imp(eratore) Anto(nino) [II]I et A[.]A caes(ura?) | loco LIII repr(obatum)*	Iscehisar, AD 161?	Fant 1989: no. 31^2 = *CIL* III 14192.
270.	*Imp(eratore) Anto(nino) IIII et Arel | caes II cos | loco CCCXXXIII | repr(obatum) off(icina) Pela(goni)*	Iscehisar, AD 161?	Fant 1989: no. 41^2 = *CIL* III 7024 = Ramsay 1882: no. 7.
271.	*b(racchio) quarṭ(o) | Rustico II et Aquilino co(n)[s(ulibus)] | off(icina) Asiatic(i) caesur(a) | Domes(tici) loco LXII*	Bacakale, AD 162	Christol & Drew-Bear 1987: no. 55 = Fant 1989: no. 152.
272.	*Rustico II et Aquilino co(n)s(ulibus) | off(icina) And(aevi) caes(ura) Dom(estici) | b(racchio) IIII loco CVI |*vac. *AN | ~~loco LXXXX~~XI | ~~b(racchio) quar(to)~~*	Bacakale, AD 162	Fant 1989: no. 153 = Christol & Drew-Bear 1987: no. 54.
273.	*[L]aeḷ[iano et] Pastore co(n)[s(ulibus)] | offi(cina) Asiat(ici) caes(ura) Dom(estici) | loco LVIIII b(racchio) quar(to) | M Ẋ*	Bacakale, AD 163	Christol & Drew-Bear 1987: no. 56 = Fant 1989: no. 154.
274.	a) *Macrino et Celso co(n)s(ulibus) | off(icina) Asiat(ici) caes(ura) | Dom(estici) | loco XXXIII b(racchio) quar(to)*; b) *Δ*	Bacakale, AD 164	Fant 1989: no. 155.
275.	*Macrino et Celso co(n)s(ulibus)| off(icina) Asiatic(i) caes(ura) Dom(estici) | b(racchio) [q]var(to) b(racchio) III Δ | [l]oco XXXXIIII*	Iscehisar, AD 164	Fant 1989: no. 37^2 = *CIL* III 7032 = Ramsay 1882: no. 3.
276.	a) *Severo II co(n)s(ule) | ex rat(ione) Syr(i) | III*; b) *VII b(racchio) III*; c) *b(racchio) tert(io)*	Bacakale, AD 173	Fant 1989: no. 158.
277.	*Sever(o) II co(n)s(ule)| VIII*	Bacakale, AD 173	Fant 1989: no. 159.
278.	a) *II R̶ | CIII*; b) *LVIII MS*; c) *A II*; d) *Iuliano et Peison(e) co(n)s(ulibus) | offic(ina) Pelag(oni), caes(ura) III Athenob(i) | loco II*	Bacakale, AD 175	Fant 1989: no. 160.

279. *Iuliano et Peisone co(n)s(ulibus) | of(ficina) Pelag(oni), caes(ura) II Athen[ob(i)] | loci XII b(racchio) II*	Bacakale, AD 175	Fant 1989: no. 161.
280. *Commodo dom(ino) [n(ostro) et Quintillo] | co(n)s(ulibus) off(icina) Neicae(nsis) [—]*	Iscehisar, AD 177?	Christol & Drew-Bear 1987: no. 63.
281. *loco XX[—b(racchio?)—] | Imp(eratore) Aug(usti) filio C[ommodo] | et Quintillo co(n)s(ulibus) [off(icina)?—]| ca[es(ura)] Dom(estici)*	Bacakale, AD 177	Fant 1989: no. 165 + Christol & Drew-Bear 1991: 140 fn. 125.
282. a) *of(fi)c(ina) Veria(na) caes(ura) Titị | loc(o) II Ɍ Com(m)odo dom(ino) n(ostro) et Qv|intillo co(n)s(ulibus)*; b) *T ∼* ; c) *LXV*	Bacakale, AD 177	Fant 1989: no. 167 = Christol & Drew-Bear 1987: no. 59 + Christol & Drew-Bear 1991: 121 fn. 37.
283. *[Co]mmodo [d]om(ino) n(ostro) II et Mạrtio Ve[ro II] | co(n)s(ulibus) off(icina) Lucilli(ana?) caes(ura) Titi loc(o) II | b(racchio) tert(io)*	Iscehisar, AD 177	Christol & Drew-Bear 1987: no. 66.
284. *[Commod]o do[m(ino) n(ostro)] II et Martio V[e]ro II co(n)s(ulibus) ọf[f(icina)] | [—] d(e) caes(ura) Titi loc(o) IIII b(racchio) q[u]art(o)*	Iscehisar, AD 177	Christol & Drew-Bear 1987: no. 65.
285. a) *Com(m)odo d[om(ino) n(ostro) et Qu]intill[o] | co(n)s(ulibus) off(icina) Veṛ[i(ana)] c̣aes(ura) Titi | loc(o) IIII b(racchio) quar(to)*; b) *b(racchio) quar(to) | T ∼*	Bacakale, AD 177	Fant 1989: no. 168 = Christol & Drew-Bear 1987: no. 58 + Christol & Drew-Bear 1991: 121 fn. 37.
286. *Com(m)odo do[m(ino)] | n(ostro) et Quintillo | co(n)s(ulibus) of(ficina) Com(m)od(iana) | caes(ura) Titi | loc(o) X b(racchio) ter(tio)*	Iscehisar, AD 177	Christol & Drew-Bear 1987: no. 60.
287. *C[o]m(m)odo [dom(ino) n(ostro)] et Q[u]intillo | co(n)s(ulibus) off(icina) Com(m)odiana caes(ura) Titi | loc(o) XI Ɍ*	Bacakale, AD 177	Fant 1989: no. 169 = Christol & Drew-Bear 1991: 128 fn. 57.
288. *Com(m)odo dom(ino) n(ostro) et Quiṇ[til]ḷo co(n)s(ulibus) o⟨fi⟩c(ina) Pelag(oni) caes(ura) Titi loc(o) XXIII | b(racchio) quar(to)*	Bacakale, AD 177	Fant 1989: no. 164 = Christol & Drew-Bear 1987: no. 61.
289. *loco XXIX ḅ(racchio) [—] | Imp(eratore) Aug(usto) filio Com(modo) d[om(ino) n(ostro)] | et Quintillo co(n)s(ulibus) of(f)ic(ina) [—] | c[aes(ura)] Dom(estici)*	Iscehisar, AD 177	Christol & Drew-Bear 1987: no. 62. Fant 1989: no. 166.
290. *LXVIII | b(racchio) o(!)uar(to)| Comodo dom(ino) n(ostro) et Qu|intillo cq(!)(n)s(ulibus) of⟨f⟩ic(ina) Pelạ(goni) | caes(ura) Titi*	Iscehisar, AD 177	Christol & Drew-Bear 1987: no. 57.
291. *Orfito et Ṛ[ufo co(n)s(ulibus) o]ff(icina) Com(m)ọḍịan[a] |*	Bacakale, AD 178	Fant 1989: no. 172.

caes(ura) Titi loc(o) [—]i[-] ℞ IX | ADN ex k(alendario) nov(o)		
292. *Ọrf[it]o et Rufo | co(n)s(ulibu)s offi(cina) Neicaens(is) | caes(ura) Titi loc(o) III b(racchio) IIII*	Bacakale, AD 178	Fant 1989: no. 170 = Christol & Drew-Bear 1987: no. 64 + Christol & Drew-Bear 1991: 121 fn. 37.
293. a) *Oṛfito et Rufo co(n)s(ulibus) off(icina) Comṃ(odiana) caes(ura) Titi loc(o) III b(racchio) quar(to)*; b) *C ANT loc(o) IIII | ANT[.]o[.]*; c) *ΕΦ*	Bacakale, AD 178	Fant 1989: no. 171.
294. *[—] Vero II co(n)s(ulibus) |* vac. *LXIII*	Iscehisar, AD 179?	Christol & Drew-Bear 1987: no. 53.
295. *Commodo dom(ino) n(ostro) II et Martio | Vero II co(n)s(ulibus) recepṭi a Tito | n[o]men[e] ZNT IOVI[.] | [. . .] b(racchio) quar(to)*	Bacakale, AD 179	Fant 1989: no. 178 + Drew-Bear 1994: 809 fn. 200.
296. *[Commodo, dom(ino)] n(ostro) II et Martio Ver(o) | [II co(n)s(ulibus) recep]ti a Tito nomene | [—] b(racchio) quar(to)*	Bacakale, AD 179	Fant 1989: no. 176.
297. *Commodo dom(ino) n(ostro) II | et Martio Vero II co(n)s(ulibus) off(icina) | Comod(iana) caes(ura) Titi loc(o) I b(racchio) quar(to)*	Bacakale, AD 179	Fant 1989: no. 175.
298. *Commodo dom(ino) n(ostro) II et Martio Ver[o] | II co(n)s(ulibus) off(icina) Lucilli caes(ura) Titi loc(o) II | b(racchio) tert(io)*	Bacakale, AD 179	Fant 1989: no. 174.
299. *[Comm]odo [dom(ino) n(ostro)] II et Martio Vero II co(n)s(ulibus) of(ficina) | [Como]d(iana?), caes(ura) Titi loc(o) IIII b(racchio) quart(o)*	Bacakale, AD 179	Fant 1989: no. 173 + Drew-Bear 1994: 793 fn. 133.
300. *Commodo dom(ino) n(ostro) II et [Martio Vero, co(n)s(ulibus) II] | recepti ex off(icina) Prusaen[(si)—] | in locqum lapidum quọṣ receper(at) | promutuo Titus ex caesu(ra) Veteris | b(racchio) quar(to), loco XXV*	Bacakale, AD 179	Fant 1989: no. 177 + Christol & Drew-Bear 1991: 122 fn. 38 + Drew-Bear 1994: 808 f. fns. 199, 200.
301. *Commo[do dom(ino) n(ostro) II et Ma]ṛṭio Ṿẹṛọ Ị̣Ị̣ c̣[o(n)s(ulibus)] | off(icina) Com(m)[o(diana) ca]es(ura) Ṭịṭị loc(o) XI b(racchio) quar(to) | C̣OM III*	Bacakale, AD 180	Christol & Drew-Bear 1991: no. 5.
302. a) *In(!)p(eratore) L(ucio) Septiṃio Severo II et Clodio Albino | Caes(are) II co(n)s(ulibus) off(icina) Smurnaiorum caesura Dio|nusi sub cura Maronis Aug(usti) lib(erti)*; b) *loc(o) IIII b(racchio) Ị̣II*	Bacakale, AD 194	Christol & Drew-Bear 1991: no. 1.

303.	*loc(o) XXXI b(racchio) IIII* | *In(!)p(eratore) L(ucio) Septim(i)o Sev(e)ro II et Clodio* | *Albino S(!)aes(are) II co(n)s(ulibus) off(icina) Smurnaio*|*rum caes(ura) Aur(elii) Dionusi sub cura* | *Maronis Aug(usti) lib(erti) proc(uratoris)*	Bacakale, AD 194	Christol & Drew-Bear 1991: no. 2.
304.	a) *VECYACIN* | *[—]EGNLO* | *[—]PH [—]*; b) *LOC TER* | *SEC* | *Laterano* | *[et R]ufino* | *[—]*	Ostia, AD 197	Orsi, *NotSc* 1921: no. 7.
305.	*Laterano et Ruphino co(n)s(ulibus)*| *[—]* | *[—] sub c[ura]*| *[—]proc(uratoris)*	Ostia, AD 197	Fant 1989: no. 97^2 = Orsi, *NotSc* 1921: 493 f. no. 5.
306.	*[l]oc(o) III b(racchio) ter(tio)* | *Laterano et Ru*|*fino co(n)s(ulibus) off(icina) Pers(is)* | *caes(ura) Neophyti*	Bacakale, AD 197	Fant 1989: no. 179.
307.	*loc(o) VIII b(racchio) sec(undo)* | *Laterano et Ruphi*|*no co(n)s(ulibus) off(icina) Smur(naiorum)* | *caes(ura)* vac. *Di(onusi?)* vac.	Bacakale, AD 197	Christol & Drew-Bear 1991: no. 3.
308.	*loc(o) XXIII b(racchio) quar(to)* | *Laterano et Rufino* | *co(n)s(ulibus) off(icina) Bass(iana) caes(ura) Os*|*till(i) Elpidefo(ri)*	Iscehisar, AD 197	Christol & Drew-Bear 1987: no. 67.
309.	a) *sub c(ura)* | *lib(erti) (p)roc(uratoris)* | *[..]s[..]* |; b) *loc(o) XXX b(racchio) I* |*[Latera]no et Ruf[ino]* | *caes(ura) N*	Punto Scifo, 197?	Fant 1989: no. 93^2 = Orsi, *NotSc* 1921: 493 f., no. 1 + 2 = Pensabene 1978: no. 6.
310.	*loc(o) [. . .]* | *Saturnino et Gal(lo)* | *co(n)s(ulibus) off(icina) Urania cae(sura)* | *Iul(ii) Neophyti*	Iscehisar, AD 198	Fant 1989: no. 82^2 = Röder 1971: J4.
311.	*loc(o) I b(racchio) quar(to) COMM I* | *Sa{n}(tu)rnino et Gallo* | *co(n)s(ulibus) off(icina) Bass(iana) caes(ura) Os*|*ta(!)illi Elpidepho(ri)*	Iscehisar, AD 198	Christol & Drew-Bear 1987: no. 68.
312.	*loco II b(racchio) q[ua]r(to)COMM I Saturnino* | *et Gallo co(n)s(ulibus) off(icina) He(r)culi(ana?)* | *cae(sura) Ostilli Elpide*|*pho(ri)*	Bacakale, AD 198	Fant 1989: no. 182 = Christol & Drew-Bear 1987: no. 69 + Christol & Drew-Bear 1991: 131 fn. 75, 139 fn. 104, 155 fn. 167.
313.	*loc(o) III b(racchio) quar(to)* | *Latera [no et Rufi]*|*no co(n)s(ulibus) of[f(icina)—]* | *caes(ura) Yac[inthi]*	Bacakale, AD 198	Christol & Drew-Bear 1991: no. 4.
314.	*loc(o) VI b(racchio) quar(to) COM I* | *Saturnino et Gallo c[o(n)s(ulibus) off(icina)]* | *Sever(iana) caes(ura) Aur(elii) The[ophili]*	Bacakale, AD 198	Fant 1989: no. 180 = Christol & Drew-Bear 1987: no. 71 + Christol & Drew-Bear 1991: 132 fn. 78.

315. *ḷọc̣(o) Ṿị b(racchio) quar(to) COM II Saturnino et | Gallo co(n)s(ulibus) off(icina) Hercul(iana?) caes(ura) Ostil (li) | Elpide(phori)*	Bacakale, AD 198	Fant 1989: no. 184 = Christol & Drew-Bear 1991: no. 6, 130 fn. 72.
316. *loc(o) VIIII* vac. *COMM I | Saturnino et Gallo | co(n)s(ulibus) off(icina) Herculi(ana?) caes(ura) | Ostilli Elpidepho(ri)*	Bacakale, AD 198	Fant 1989: no. 181 = Christol & Drew-Bear 1987: no. 70 + Christol & Drew-Bear 1991: 132 fn. 80.
317. *[loc(o)] X̣ b(racchio) quar(to) COM II Saturnino e(t) Gallo co(n)s(ulibus) | [of]f(icina) Mar(tis/-tialis?) caes(ura) Ulpi(i) Yacin(thi)* vac.	Bacakale, AD 198	Christol & Drew-Bear 1991: no. 7.
318. *loc(o) XIIII b(racchio) quar(to) COMM I Satur|nino et Gallo co(n)s(ulibus) off(icina) Hercu|li-(ana?) caes(ura) Ostilli Elpidepho(ri)*	Bacakale, AD 198	Fant 1989: no. 183.
319. *loco XVIIII b(racchio) quar(to)(COM I?) | Saturnino et Gal[lo] | co(n)s(ulibus) off(icina) Bass(iana) caes(ura) Os|till(i) Helpid(ephori)*	Bacakale, AD 198	Fant 1989: no. 185 + Christol & Drew-Bear 1991: 138 fn. 108, 155 fn. 167.
320. *loc(o) II b(racchio) tert(io) COM I | Anullino II et Fron|ṭone co(n)s(ulibus) off(icina) Severi(ana?) | caes(ura) Aur(elii) Theophilli*	Bacakale, AD 199	Fant 1989: no. 187 = Christol & Drew-Bear 1991: no. 8, 122 fn. 41, 133 fns. 91, 92, 137 fn. 98.
321. *loc(o) II[I? b(racchio) qu]ar(to), CO[M I] | Anullino II et Frontone | co(n)s(ulibus) off(icina) Hercul(iana?) caes(ura) | Ostilli Helpidephor(i)*	Bacakale, AD 199	Fant 1989: no. 186.
322. *l[o]c(o) III[I] b(racchio) teṛ[t](io) COM II | [A]nul[l]ino II e(t) Fron|ṭone co(n)s(ulibus) off(icina) Bassia(na) | c̣[aes](ura) Ostil(l)i Elpi|dephor[i]*	Bacakale, AD 199	Christol & Drew-Bear 1991: no. 9.
323. *loc(o) VIIII b(racchio) quar(to) COM II | Anullino II et Frontone co(n)s(ulibus) | off(icina) Bassia(na) caes(ura) Ostill(i) Help[i]|dephor(i)*	Bacakale, AD 199	Christol & Drew-Bear 1991: no. 10.
324. *loc(o) LIII [b(racchio)—COM –] | Anullino II et Fro[ntone co(n)s(ulibus) off(icina)] | Horolog caes(ura) E[—]| Saturnino iuss[—]*	Bacakale, AD 199	Christol & Drew-Bear 1991: no. 11.
325. a) *[. . . .] l(oco) b(racchio) se(cundo) COM ⊥ Severo et(?) Vi|ctorino co(n)s(ulibus) off(icina) Comodiana | caes(ura) Iuli(i) Gayri sub cura*; b) *[. . . .]gres[. . .]| avgger [..] | tor Iulii | gayri[.. . . .] | oph | eepictet[..] Eλ*	Punto Scifo, 200	Fant 1989: no. 95^2 = Orsi, *NotSc* 1921: 494-5: no. 3 + 4 = Pensabene 1978: no. 10.

<table>
<tr><td>326.</td><td>loc(o) T b(racchio) pri(mo)[—] Vic⟨to⟩|rin o et Severo co(n)s(ulibus) off(icina) | [.. . . . . .]episco</td><td>Punto Scifo, 200</td><td>Fant 1989: no. 96^2 = Orsi, NotSc 1921: 493 f. no 6 = Pensabene 1978: no. 11.</td></tr>
<tr><td>327.</td><td>loc(o) III | b(racchio) II COM I | Severo et Victo|rino co(n)s(ulibus) off(icina) | ⟨Co⟩modiana | Iuli(i) ss</td><td>Punto Scifo, 200</td><td>Fant 1989: no. 98^2 = Pensabene 1978: no. 5.</td></tr>
<tr><td>328.</td><td>loc(o) VI b(racchio) ter(tio) COM I Sẹvero | et Victorino co(n)s(ulibus) [off(icina)—] | caes(ura) Nẹoph[yti]</td><td>Bacakale, AD 200</td><td>Fant 1989: no. 188 = Christol & Drew-Bear 1991: no. 12; 140 fn. 125.</td></tr>
<tr><td>329.</td><td>loc(o) XXVIIII b(racchio) quarṭ(o) | COM prim( ) Severo [[—]] | et Victorino co(n)s(ulibus) off(icina) | Sever(iana) vac. caes(ura) Aur(elii) The|ophill(i)</td><td>Bacakale, AD 200</td><td>Fant 1989: no. 189 = Christol & Drew-Bear 1987: no. 72 + Christol & Drew-Bear 1991: 122 fn. 41, 139 f. fn. 123.</td></tr>
<tr><td>330.</td><td>[loc]o XXXX b(racchio) quar(to) COM I | [Mu]ciano et Fabiano co(n)s(ulibus) off(icina) | Seve(riana) ca[e]s(ura) Aur(elii) Theo(phili)</td><td>Bacakale, AD 201</td><td>Fant 1989: no. 190.</td></tr>
<tr><td>331.</td><td>loc(o) XXI b(racchio) ter(tio) COM I | dominis nn(ostris) invictis | piis L(ucio) S(eptimio) Severo III et Anto|nino co(n)s(ulibus) off(icina) Sever(iana) | caes(ura) Aur(elii) Theoṗ[hili]</td><td>Bacakale, AD 202</td><td>Christol & Drew-Bear 1991: no. 13.</td></tr>
<tr><td>332.</td><td>loc(o) XXXI [b(racchio) . . . CO]M I | P(ublio) S(eptimio) Geta II et Plautiano et P | II co(n)s(ulibus) off(icina) Sever(iana) caes(ura) Aur(elii) Theophili</td><td>Bacakale, AD 203</td><td>Fant 1989: no. 191.</td></tr>
<tr><td>333.</td><td>[loc(o)–] b(racchio) tert(io)CO[M I] | M(arco) Ạ[y]ṛ(elio) Antonino [II] | Aug(usto) et Sep(timio) Geta Caes(are) | co(n)s(ulibus) off(icina) Sever(iana) caes(ura) A[ur(elii)] Theophili</td><td>Bacakale, AD 205</td><td>Christol & Drew-Bear 1991: no. 15.</td></tr>
<tr><td>334.</td><td>[loc(o)—] b(racchio) quar(to) COM I | [M(arco) Aur(elio) An]ṭonino II et Sep(timio) Geta | [co(n)s(ulibus) off(icina)—] caes(ura) Ayr(elii) Theop(hili)</td><td>Bacakale, AD 205</td><td>Christol & Drew-Bear 1991: no. 16.</td></tr>
<tr><td>335.</td><td>lo[c(o)—b(racchio)—COM I] | M Ay[r(elio) Antonino II Aug(usto) et Sep-(timio) Geta] | Caẹ[s(are) co(n)s(ulibus) off(icina)—caes(ura)—]</td><td>Bacakale, AD 205</td><td>Christol & Drew-Bear 1991: no. 30.</td></tr>
<tr><td>336.</td><td>loc(o) VIII b(racchio) tert(io) COM I | M(arco) Ayr(elio) Antonino II et | Sep(timio) Geta co(n)s(ulibus) off(icina) Seveb(iana) caes(ura) | Ayr(elii) Theophil(i)</td><td>Bacakale, AD 205</td><td>Fant 1989: no. 192 = Christol & Drew-Bear 1991: no. 14, 143 fn. 139.</td></tr>
</table>

337. *[loc(o)] X̣VIII b(racchio) quart(o) COM I | [M(arco) Au]ṛ(elio) Antonino II Ayg(usto) et Sep(timio) | [Geta-C]aesar(e) co(n)s(ulibus) off(icina) Sever(iana) caes(ura) | Aur(elii) Theophil(i)*	Bacakale, AD 205	Christol & Drew-Bear 1991: no. 17.
338. *[loc(o)—b(racchio)—COM-] | Albino et Aẹm[iliano] | co(n)s(ulibus) off(icina) Sever(iana) caeṣ(ura) | Ayr(elii) Theophili*	Bacakale, AD 206	Fant 1989: no. 196 = Christol & Drew-Bear 1991: no. 18, 147 fn. 146.
339. *loc(o) I b(racchio) sec(undo) C[OM—] | Albino e(t) Aemili[ano co(n)s(ulibus)] | off(icina) Hercul(iana?) caes(ura) [Ostilli] | Helpid(ephori)*	Iscehisar, AD 206	Christol & Drew-Bear 1987: no. 73 = Fant 1989: no. 195 + Christol & Drew-Bear 1991: 138 fn. 113, 146 fn. 145.
340. *loc(o) III b(racchio) quar(to) COM I | Albino et Aemiliano co(n)s(ulibus) off(icina) | Contra Persis caes(ura) Herc⟨u⟩lan(i)*	Bacakale, AD 206	Fant 1989: no. 194.
341. *loc(o) VI [b(racchio) —] | Albino et [Aemi]liano | off(icina) Contra Persis, caes(ura) | Herculan(ii)*	Bacakale, AD 206	Fant 1989: no. 193 + Christol & Drew-Bear 1991: 146 fn. 145.
342. *loc(o) XXXIỊ [b]r(acchio) t(ertio) COM I | Albino et Aem(i)liano co(n)s(ulibus) | off(icina) Comm(odiana) caes(ura) Herc(u)-lan(a)*	Iscehisar, AD 206	Christol & Drew-Bear 1987: no. 74.
343. *[——] | COM I Pompeiano et VA(!)ito co(n)s(ulibus) off(icina) Bassi(ana) | caes(ura) Ostil(li) Helpid(ephori)*	Iscehisar, AD 209	Christol & Drew-Bear 1987: no. 75.
344. *[loc(o)–]I b(racchio) teṛ[t(io)] | COM I Poṃ[p]ei|{a}ano et Avito co(n)s(uli-bus) off(icina) | Sever(iana) caes(ura) Theophil*	Bacakale, AD 209	Fant 1989: no. 197 = Christol & Drew-Bear 1991: no. 19, 149 fn. 149.
345. *loc(o) XXXV b(racchio) tebt(io) | COM I Pompeiano | et Avito co(n)s(ulibus) off(icina) | Antoniana | caes(ura) Neophyti*	Bacakale, AD 209	Christol & Drew-Bear 1991: no. 20.
346. *loc(o) XI b(racchio) quar(to) COM I K | Faust(ino)* vac.	Bacakale, AD 210	Fant 1989: no. 198 = Christol & Drew-Bear 1987: no. 76 + Christol & Drew-Bear 1991: 129 fn. 62.
347. a) *loc(o) XVI b(racchio) quar(to) C ỌṂ Ị | Faustino et Rufino co(n)s(ulibus) | off(icina) Bas(siana) caes(ura) Ostil(ii) Helpid(ephori)*; b) *loc(o) XVI b(racchio) quar(to)*	Bacakale, AD 210	Christol & Drew-Bear 1991: no. 22.

348. *loc(o) XXVII b(racchio) ter(tio) COM I | Faustino et Rufino cos | off(icina) Bassiana caes(ura) Hostil(ii) | Elpidephori*	Bacakale, AD 210	Fant 1989: no. 199 = Christol & Drew-Bear 1991: no. 21, 150 fn. 155.
349. *loc(o) XC b(racchio) quar(to) COM I duabus Aspris co(n)s(ulibus) | off(icina) Contra Pers(is) caes(ura) Aur(elii) Theophili*	Bacakale, AD 212	Fant 1989: no. 200.
350. *loc(o) XXVII b(racchio) qua(rto) COM I | I[m]p(eratore) Antonino Aug(usto) | IIII et Balbino II co(n)s(ulibus) | off(-icina) Bas(siana) caes(ura) Helpidephori*	Bacakale, AD 213	Fant 1989: no. 201 = Christol & Drew-Bear 1991: no. 23.
351. *loc(o) III b(racchio) ter(tio) COM I | Sabino co(n)s(ulibus)*	Bacakale, AD 214	Fant 1989: no. 202.
352. a) *loco b(racchio) quart(o) COM I | Praesente*; b) *caes(ura) Nova*	Bacakale, AD 217	Fant 1989: no. 204 = Christol & Drew-Bear 1991: no. 24.
353. *loc(o) VII b(racchio) quart(o) COM I | Prae(se)nte* vac.	Iscehisar, AD 217	Christol & Drew-Bear 1987: no. 77.
354. a) *loc(o) XI b(racchio) quart(o) COM I | Praesente* ; b) *caẹs(ura) Nova*	Bacakale, AD 217?	Fant 1989: no. 205.
355. *loc(o) XV b(racchio) tert(io) COM I | Praesente* vac.	Iscehisar, AD 217	Christol & Drew-Bear 1987: no. 78.
356. a) *loc(o) XL b(racchio) quart(o) COM I | Praesente co(n)s(ulibus)*; b) *caesura*	Bacakale, AD 217?	Fant 1989: no. 203.
357. a) *loc(o) XXCII b(racchio) qu(arto)*; b) *caesuṛa*	Bacakale, AD 217?	Fant 1989: no. 206.
358. *loc(o) XXVII b(racchio) quar(to) COM I | Antonino Aug(usto) et Advento co(n)s(ulibus) | off(icina) Iuxta Persidem | caes(ura) Phoenicem*	Bacakale, AD 218	Fant 1989: no. 207.
359. *Imp(eratore) Ant(onino) loc(o) IIX*	Bacakale, AD 218	Fant 1989: no. 209.
360. *loc(o) XV b(racchio) quart(o) COM I | Anton(i)no Aug(usto)*	Bacakale, AD 218	Fant 1989: no. 208.
361. *Imp(eratore) Ant(onino) n(ostro) II co(n)s(ulibus) | loco XCII*	Afyon, AD 219	Fant 1989: no. 210.
362. *b(racchio) prim(o) | Imp(eratore) Ant(onino) n(ostro) II co(n)s(ulibus) | loco XCIIII*	Bacakale, AD 219	Fant 1989: no. 211.
363. *Imp(eratore) Ant(onino) n(ostro) II co(n)s(ulibus) | loco XCV*	Afyon, AD 219	Fant 1989: no. 212.
364. *loc(o) VII b(racchio) teṛ[t(io) COM I] | Imp(eratore) Anto[nino Aug(usto) III] | et Comaz[on(e) co(n)s(ulibus)] | ọff(icina) Anṭon[iana caes(ura)—]*	Bacakale, AD 220	Christol & Drew-Bear 1991: no. 26.

365. *[l]oc XIX b(racchio) quar(to) COM I | Imp(eratore) Antonino Aug(usto) III | et Comazo(ne) co(n)s(ulibus) off(icina) Antoni|ana, caer(!)(sura) Iuli(i) Neophyti*	Bacakale, AD 220	Fant 1989: no. 215 = Christol & Drew-Bear 1987: no. 79 + Christol & Drew-Bear 1991: 154 fn. 163.
366. *loc(o) XVII b(racchio) quar(to) COM I| Imp(eratore) Ant[o]nino Aug(usto) III | et Comaz[o]n(e) co(n)s(ulibus) | off(icina) Iux(ta) Pe[r]s(idem), cae(sura) Phoenic(is)*	Bacakale, AD 220	Fant 1989: no. 214 = Christol & Drew-Bear 1991: no. 25.
367. *l[oc(o)] XXXVIII b(racchio) quar(to) | Imp(eratore) Antonino Aug(usto) III*	Bacakale, AD 220	Fant 1989: no. 213.
368. *loc(o) II b(racchio) te(rtio), COM [I] | Alexandro Aug(usto) co(n)[s(ule)] | off(icina) Iux(ta) Pers(idem) caes(ura) Mar(yllini) | Romae*	Bacakale, AD 222	Fant 1989: no. 217.
369. *loc(o) V b(racchio) quar(to) COM I | Alexandro Aug(usto) co(n)s(ulibus) | off(icina) Sever(iana) caes(ura) Mary|-(l)lin(i)*	Bacakale, AD 222	Christol & Drew-Bear 1991: no. 27.
370. *[loc(o)-] b(racchio) qvar COM I | A[l] eIII et Dione II co(n)s(ulibus) | off(icina) Sever(iana) caes(ura) Maryllin(i)*	Bacakale, AD 229	Fant 1989: no. 221 = Christol & Drew-Bear 1991: no. 28.
371. *loc(o) II b(racchio) quar(to) COM I | Alexandro Aug(usto) II[I] | et Dion(e) II co(n)s(ulibus) of(ficina) Erc(uliana) | [c]aes(ura) Iuli Ne[ophyti]*	Bacakale, AD 229	Fant 1989: no. 219.
372. *[loc]o V b(racchio) quar(to) [COM I?] | Imp(eratore) [Alexa]ndro Aug(usto) III et Dione II co(n)s(ulibus) | off(icina) Sever(iana) [caes(ura) M]aryllin(i)*	Bacakale, AD 229	Fant 1989: no. 220 + Christol & Drew-Bear 1991: 159 f.
373. *loc(o) V b(racchio) pr(imo) T COM I | Alexanord (!) Aug(usto) III | et Dion(e) TI co(n)s(ulibus) off(icina) Cres|cent(is) caes(ura) Iul(ii) Neo⟨p⟩hy(ti)*	Iscehisar, AD 229	Christol & Drew-Bear 1987: no. 80.
374. *loc(o) VIII [b(racchio) quar(to?) COM I] | [Alexandro Aug(usto)] | III et Dion(e) II co(n)s(ulibus) of(ficina) | Antoni(niana) caes(ura) [I]uli Neo|phyti*	Bacakale, AD 229	Fant 1989: no. 218.
375. *loc(o) [–] COM I Se[vero et Quintiano] | [–]co(n)s(ulibus) off(icina) Se[ver-(iana)—] | [–caes(ura)] Mary[llin(i)—]*	Afyon, AD 235	Christol & Drew-Bear 1991: no. 33.
376. *P loc(o) LXXIII b(racchio) quart(o) COM I | Imp(eratore) Maximino Aug(usto) | et Africano co(n)s(ulibus) off(icina) Alex(andriana) | caes(ura) Vari*	Bacakale, AD 236	Fant 1989: no. 222 = Christol & Drew-Bear 1991: 132 fn. 81.

377. *off(icina) Pel[ag( )] | loco vii, b(racchio) ii*	Bacakale	Fant 1989: no. 162.
378. a) *iussu Aur(elii) [.]r[..]n | FRANS*; b) *D*; c) *repr(obatum)*	Bacakale	Fant 1989: no. 163.
379. *Loc(o) XXXX b(racchio) quar(to) COM I*	Bacakale	Christol & Drew-Bear 1991: no. 29.
380. *caes(ura) Nova*	Bacakale	Christol & Drew-Bear 1991: no. 31.
381. *[—]COM I | [—c]o(n)ṣ(ulibus) | [—caes(ura) Maryl]ḷịṇ*	Bacakale	Christol & Drew-Bear 1991: no. 32.
382. *[L]oc(o) XIII b(racchio) IIII || de XXVI || d(e) XXVI || RMA || [. . .]ci r(ecognitum?)*	Iscehisar	Fant 1989: no. 4^2 = *AE* 1984: 847.
383. *[. . .]ippae*	Roma	Fant 1989: no. 8^2 = Bruzza 1870: no. 255.
384. *[. . . cos, off(icina)] | [- c.4–5 -] N, caes-(ura) Domes(tici) | loco xiiii, b(racchio) quart(o)*	Bacakale	Fant 1989: no. 156.
385. *EX RAP LIB | n(umero) CCCCXXI*	Roma	Fant 1989: no. $13a^2$ = Bruzza 1870: no. 262.
386. *[. . . C]AES*	Roma	Fant 1989: no. $13b^2$ = Bruzza 1870: no. 263.
387. *[——]| sub cu[ra–]*	Roma	Fant 1989: no. $13c^2$ = Bruzza 1870: no. 264.
388. *G I P*	Roma	Fant 1989: no. 14^2 = Bruzza 1870: no. 265.
389. ⊕	Bacakale	Fant 1989: no. 113 + Drew-Bear 1994: 774 fn. 71.
390. *loc(o) XXXX, b(racchio) quar(to), COM* i	Bacakale	Fant 1989: no. 223.
391. *[—c.12—]|[—c.17—]|[—c.18—]*	Bacakale	Fant 1989: no. 224.
392. *b(racchio) secund(o)*	Bacakale	Fant 1989: no. 225.
393. a) *off(icina)* [.c.3..], *b(racchio) IIII, | caes(ura) Titi, loc(o) III*; b) *ΤΩ*	Bacakale	Fant 1989: no. 226.
394. a) *b(racchio) tert(io)*; b) *[. . .]NIOṚ c̣ọṣ | DCCCCX*; c) *DCCCX | XV | b(racchio) III*	Bacakale	Fant 1989: no. 227.
395. *ΤΩ*	Bacakale	Fant 1989: no. 228.
396. a) *loco XVII*; b) *b(racchio) I*	Bacakale	Fant 1989: no. 229.
397. *loco CXXXIII | b(racchio) III | O[–] VIOR*	Bacakale	Fant 1989: no. 230.
398. *de loco CV*	Bacakale	Fant 1989: no. 231.
399. *loc(o) II, b(racchio) tert(io), COM I*	Bacakale	Fant 1989: no. 232.
400. *loco | b(racchio) III, b(racchio) ter(tio)*	Bacakale	Fant 1989: no. 233.
401. a) *loc(o) XVIII*; b) *loco X, b(racchio) II*	Bacakale	Fant 1989: no. 234.
402. *loco CLXIV | b(racchio) III*	Bacakale	Fant 1989: no. 235.

403. a) *E*; b)*AY*; c) *C* (or *Γ*); d) *E*; e) *EΦ*	Bacakale	Fant 1989: no. 236.
404. *VVV*	Bacakale	Fant 1989: no. 237.
405. *B TERT*	Bacakale	Fant 1989: no. 26² = *CIL* III 7039.
406. *RMA PλR CLX[—]| [—] C B*	Bacakale	Fant 1989: no. 27² = *CIL* III 7040.
407. *RMA PA R̶ CLX | [—]C R*	Bacakale	Fant 1989: no. 29² = *CIL* III 12235.
408. a) *RV COSS CXXXIX | CAD[I]I COSS CDXLVI,* ; b) *RNA PA B OART | II COS | VIR VOP COS BII | DCCXVI CCXI*	?	Fant 1989: no. 34² = *MAMA* VII 134 = *IK* 62: 360.
409. a) *GXVICOI | IXXVIII | B TERT XXVII* ; b) *XXVII*	Iscehisar	Fant 1989: no. 35² = *CIL* III 12229.
410. a) *FRVEPC COS | ƉCCCCLXXV*; b) *Λ III CO RMA | B III | CAD II COSSCS IV*; c) *RV COSS | XLVLI*; d) *ƉCCCCLXXXVI*; e) *BQVART*	Iscehisar	Fant 1989: no. 39² = *CIL* III 7016 = Ramsay 1882: no. 5.
411. a) *EMES |XLIII | REPR*; b) *III*	Iscehisar	Fant 1989: no. 43² = *CIL* III 7028 = Ramsay 1882: no. 9.
412. a) *VLVII RIII*; b) *B TERT[—]*; c) *[—]I (?)APRO*	Iscehisar	Fant 1989: no. 44² = *CIL* III 7035 = Ramsay 1882: no. 10.
413. *[—]ANO COS ANDAN IV*	Iscehisar	Fant 1989: no. 51² = *CIL* III 7033 = Ramsay 1882: no. 18.
414. *[—]M(?)R ANTONINO AV | EST-FEOR III*	Iscehisar	Fant 1989: no. 53² = *CIL* III 7034 = Ramsay 1882: no. 20.
415. *CCC | R III | IALIC COS*	Iscehisar	Fant 1989: no. 57² =*CIL* III 7005 = Ramsay 1882: no. 24.
416. *DE ƉCCXXIX*	Iscehisar	Fant 1989: no. 58² = *CIL* III 7037 = Ramsay 1882: no. 25.
417. *[L]OCO XV Λ*	Iscehisar	Fant 1989: no. 59² = *CIL* III 7036 = Ramsay 1882: no. 26.
418. *∞∞CX | TAΩ*	Iscehisar	Fant 1989: no. 60² = *CIL* III 7038 = Ramsay 1882: no. 27.
419. *[—]Cels[—]*	Lepcis Magna	Fant 1989: no. 74² = *IRT* 794a.
420. *[—]FRV[—]*	Lepcis Magna	Fant 1989: no. 76² = *IRT* 794c.

<table>
<tr><td>421. a) [—]i sub cura [—] | [—s]ub cura Hesp[erii—] | [—]; b) [—] lib(erti) proc(uratoris) [—] | [—] proc(uratoris) ex [—]</td><td>Lepcis Magna</td><td>Fant 1989: no. 77^2 = IRT 794d.</td></tr>
<tr><td>422. a) [—] III co(n)s(ule) [—] ; b) [—Au]g(usto) co(n)s(ule) [—] | [—]I</td><td>Lepcis Magna</td><td>Fant 1989: no. 78^2 = IRT 794e.</td></tr>
<tr><td>423. L | Cae CCXCVIIIS</td><td>Lepcis Magna</td><td>Fant 1989: no. 81^2 = Röder 1971: J3.</td></tr>
<tr><td>424. a) ex r(atione) Olyp(i?) Caes(aris); b) n(umero) CXVI</td><td>Ostia</td><td>Fant 1989: no. 88^2 = Baccini Leotardi 1989: no. 41.</td></tr>
<tr><td>425. a) <s>D</s>; b) l(oco) C ∞XCIII | <s>R</s></td><td>Ostia</td><td>Baccini Leotardi 1989: no. 45.</td></tr>
<tr><td>426. CEL COS ex r(atione)</td><td>Ostia</td><td>Baccini Leotardi 1989: no. 46 = Pensabene 1994: 73, no. 19.</td></tr>
<tr><td>427. XXVII</td><td>Ostia</td><td>Baccini Leotardi 1989: no. 47 + 127 = Pensabene 1994: 73, no. 17.</td></tr>
<tr><td>428. Hyac(inthi) Ca(esaris?)</td><td>Ostia</td><td>Fant 1989: no. 87^2 = Baccini Leotardi 1979: no. 40.</td></tr>
<tr><td>429. IMP</td><td>Ostia</td><td>Fant 1989: no. 90^2 = Baccini Leotardi 1989: no. 129 = Pensabene 1994: 73, no. 18.</td></tr>
<tr><td>430. a) ex r(atione) O(lym)p(i) (?) Caes(aris) | n(umero) CDXXXIII; b) CCCI | L" DCCCLVI <s>R</s>(?)</td><td>Ostia</td><td>Fant 1989: no. 92b^2 = Bruzza no. 302 = Lanciani, AnnInst 1868: 180.</td></tr>
<tr><td>431. sub c[—]| lib proc | [—]s[—]</td><td>Ostia</td><td>Fant 1989: no. 94a,b^2 = Orsi, NotSc 1921: 493 f. no 2 = Pensabene 1978: no. 7?</td></tr>
<tr><td>432. RMA</td><td>Afyon</td><td>Fant 1989: no. 107^2 = Christol & Drew-Bear 1986: no. 12.</td></tr>
<tr><td>433. loco | X</td><td>Dokimeion</td><td>Fant 1989: no. 123^2 = Christol & Drew-Bear 1986: no. 33.</td></tr>
<tr><td>434. a) D XXΛI | [—]CI <s>R</s> | [—l]oc(o) XIII b(racchio) IIII | de XXVI ; b) RMA</td><td>Bacakale</td><td>Christol & Drew-Bear 1986: no. 34.</td></tr>
<tr><td>435. a) loco XXVIII; b) HER; c) LXXI[—]</td><td>Iscehisar</td><td>Fant 1989: no. 7.</td></tr>
<tr><td>436. S d ATTAII</td><td>Ostia</td><td>Pensabene 1994: 74, no. 21.</td></tr>
</table>

437. *Epictetus Augus[—] | proactor pro[—] | [l]oc(o) IIII | b(racchio) sec(undo) COM I | [Al]bino et Aemilian[o co(n)s(ulibus)] | [caes(ura)?] Aur(elii) Demetri(i) b[—] | [sub cu]ra Aur(elii) Epity[nchani?] | VE CAVRT [—]*	Rome, AD 206	Bruzza 1870: no. 279 = Dubois 1908: no. 207; Christol& Drew-Bear 1991: 119 fn. 31, 171 fn. 223.

Pavonazetto and White Marble: Upper Tembris Valley

438. a) *n(umero) XXXIIII*; b) *C(aesare) XVI co(n)s(ule)*; c) *CIΓΞH*	Kurt Köy, 92	'pavonazetto': *MAMA* X 122.
439. a) *[—] XXI | [—]o cos*; b) *co(n)s(ule) C(aesare) X[VI?]*	Alibey Köy, 92?	'white marble'?: Drew-Bear & Eck 1976: no. 16; Fant 1989: no. 64[2].
440. a) *COS*; b) *M | XIΓ*; c) *AN*; d) *RΛA (=RMA?) P(alma II co(n)s(ule?))*	Altintash, 109?	'Docimian marble (with blue patches) *MAMA* X 72. Christol & Drew-Bear 1986: 75–80.
441. a) ℞ *Sil(vano) co(n)s(ule)* (AD 156)| *n(umero) CXIII*; b) *Vop(isco) co(n)s(ule)*(AD 114) | *XXXIV*; c) *Aeliano co(n)s(ule)* (AD 116)	Kurt Köy,	'pavonazetto': *MAMA* X 124.
442. ℞ *Aeliano co(n)s(ule)* (AD 116) | *[—] XII* ℞ *S[i]lvan(o) c⟨o(n)⟩s(ule)* (AD 156) | *Vop(isco) co(n)s(ule)* (AD 114) *n(umero) CCXLVI | XXIX*	Kurt Köy	'pavonazetto': *MAMA* X 125.
443. = | X | *XI* X | ℞ *Aelian(o) co(n)s(ule)*	Çakırsaz (Soa), 116	'white marble'?: Drew-Bear & Eck 1976: no. 25; Fant 1989: no. 73[2].
444. *[—A]elian(us) co(n)s(ule)* (AD 116)	Altintash Köy, 116	'pavonazetto': *MAMA* X 71; Drew-Bear & Eck 1976: 315; Waelkens 1985: 643.
445. ℞ *Aelian(o) co(n)s(ule)*	Alibey Köy, 116	'white marble'?: Drew-Bear & Eck 1976: no. 17; Fant 1989: no. 65[2].
446. a) *R AEL|MA CLO*; b) *IEOE | LXVIII*	Alibey Köy, 116?	'white marble'?: Drew-Bear & Eck 1976: no. 18; Fant 1989: no. 66[2].
447. *Silvan(o)*	Kurt Köy	'pavonazetto': *MAMA* X 123.

448. *[—] co(n)s(ule) | n(umero) IX*	Kurt Köy	‘pavonazetto’: *MAMA* X 126.
449. *Rhọd( ) NXIV*	Kurt Köy	‘pavonazetto’: *MAMA* X 127.
450. *DXC̣*	Kurt Köy	‘pavonazetto’: *MAMA* X 128.
451. *n(umero) IX*	Kurt Köy	‘pavonazetto’: *MAMA* X 129.
452. *+ X////*	KurtKöy	‘pavonazetto’: *MAMA* X 130.
453. *n(umero) LXXXII*	Alibey Köy	‘white marble’: *MAMA* X 94; Drew-Bear & Eck 1976: 315 f.
454. οζ’ | *LXXVII*	Alibey Köy	‘white marble’?: Drew-Bear & Eck 1976: no. 19; Fant 1989: no. 67^{2}.
455. *CXIIX* | ριη’	Alibey Köy	‘white marble’?: Drew-Bear & Eck 1976: no. 20; Fant 1989: no. 68^{2}.
456. *C̣ES C AE IIII | COS LXIII*	Çakırsaz (Soa)	‘white marble’?: Drew-Bear & Eck 1976: no. 21; Fant 1989: no. 69^{2}.
457. *POPR XVII*	Çakırsaz (Soa)	‘white marble’: Drew-Bear & Eck 1976: no. 22; Fant 1989: no. 70^{2}.
458. *POPR XVIII*	Çakırsaz (Soa)	‘white marble’: Drew-Bear & Eck 1976: no. 23; Fant 1989: no. 71^{2}.
459. *HII | CVIII*	Çakırsaz (Soa)	‘white marble’?: Drew-Bear & Eck 1976: no. 24; Fant 1989: no. 72^{2}.

Africano and Grey Marble: Teos

460. a) *Frugi et Basso co(n)s(ulibus)*; b) *HER Cae(saris?) | n(umero) XXXIII*	Rome, 64	‘africano’; Bruzza 1870: no. 138 = Dubois 1908: no. 430.
461. a) *Frugi et Basso co(n)s(ulibus)*; b) *HER Cae(saris) | n(umero) LI*	Rome, 64	‘africano’; Bruzza 1870: no. 139 = Dubois 1908: no. 431.

462. a) *Rufo et Capitone c[o(n)s(ulibus)];* b) *Lae(ti?) Cae(saris?) | n(umero) LX*	Rome, 67	'africano'; Bruzza 1870: no. 140 = Dubois 1908: no. 432.
463. a) *Imp(eratore) Vespasiano VI | Tito Caes(are) IIII co(n)s(ulibus),* ; b) *ex rat(ione) Laet(i) ser(vi) | n(umero) LXXXV*	Rome, 75	'africano'; Bruzza 1870: no. 147 = Dubois 1908: no. 439.
464. *Imp(eratore) Vespasiano [VI] | Tito Caes(are) IIII co(n)s(ulibus)*	Rome, 75	'africano'; Bruzza 1870: no. 148 = Dubois 1908: no. 440.
465. *[Imp(eratore) Vespasia]no VI | [Tito Caes(are)] IIII co(n)s(ulibus)*	Rome, 75	'africano'; Bruzza 1870: no. 149 = Dubois 1908: no. 441.
466. *Imp(eratore) Vespasiano VII | Tito Caesar(e) V co(n)s(ulibus)*	Rome, 76	'africano'; Bruzza 1870: no. 150 = Dubois 1908: no. 442.
467. a) *[Imp(eratore)] Vespasiano VII | [Tit]o Caesar(e) co(n)s(ulibus)*; b) *[e]x rat(ione) Laet(i) ser(vi) | n(umero) XXXXII*	Rome, 76	'africano'; Bruzza 1870: no. 151 = Dubois 1908: no. 443.
468. *Imp(eratore) Vesp[—] | Tito Ca[esare —]*	Rome	'africano'; Bruzza 1870: no. 152 = Dubois 1908: no. 444.
469. a) *Imp (eratore) Vesp(asiano) Caes(are) Aug(usto) VIII | Domit(iano) Caes(are) V co(n)s(ulibus)*; b) *ex rat(ione) Tyc(hi?) Cae(saris) | n(umero) LXXIII*	Rome, 77	'africano'; Bruzza 1870: no. 153 = Dubois 1908: no. 445.
470. a) *Imp(eratore) Vesp(asiano) Cae(sare) Aug(usto) VI[II] | Domitiano V co(n)s(ulibus)*; b) *ex rat(ione) Tyc(hi?) Cae(saris) | n(umero) XVIII*	Rome, 77	'africano'; Bruzza 1870: no. 154 = Dubois 1908: no. 446.
471. *Imp(eratore) Vesp(asiano) Aug(usto) VIII | Domit(iano) Cae(sare) V co(n)s(ulibus)*	Rome, 77	'africano'; Bruzza 1870: no. 155 = Dubois 1908: no. 448.
472. *Imp(eratore) Vesp(asiano) Au(gusto) VIII | Domit(iano) Cae(sare) V co(n)s(ulibus)*	Rome, 77	'africano'; Bruzza 1870: no. 156 = Dubois 1908: no. 449.
473. *Imp(eratore) T(ito) Cae(sare) VIII co(n)s(ulibus) | Domitiano Cae(sare) VII*	Rome, 80	'africano'; Bruzza 1870: no. 157 = Dubois 1908: no. 450.
474. *Imp(eratore) T(ito) Cae(sare) Aug(usto) VIII co(n)s(ulibus) | VIII*	Rome, 80	'africano'; Bruzza 1870: no. 158 = Dubois 1908: no. 451.
475. a) *Imp(eratore) T(ito) Cae(sare) VIII co(n)s(ulibus) Domi(t)i(ano) C(aesare) VI | ex rat(ione) Laeti Caes(aris);* b) *ex rat(ione) Laeti Caes(aris) | n(umero) XVIIII*	Rome, 80	'africano'; Bruzza 1870: no. 159 = Dubois 1908: no. 452.

<table>
<tr><td>476.</td><td>a) ex rat(io)n(e) Laeti Caes(aris) | n(umero) X; b) Imp(eratore) T(ito) Cae(sare) VIII co(n)s(ule) | Domitiano Cae(sare) VI</td><td>Ostia, 80</td><td>'africano bigio'; Pensabene 1994: 197 fig.1.</td></tr>
<tr><td>477.</td><td>Imp(eratore) T(ito) Cae(sare) VIII | co(n)s(ulibus) | Domitiano Cae(sare) VI</td><td>Ostia, 80</td><td>'africano'; Pensabene 1994: 198 fig.4.</td></tr>
<tr><td>478.</td><td>a) [I]mp(eratore) T(ito) C(aesare) A(ugusto) VIII | [D]omitiano Cae(sare) VI co(n)s(ulibus); b) ∞ CCC [.]XXIII</td><td>Ostia, 80</td><td>'africano'; Baccini Leotardi 1989: no. 1 = Pensabene 1994: no. 6.</td></tr>
<tr><td>479.</td><td>?</td><td>Ostia, 82</td><td>'africano'; Bruzza 1870: no. 167 = Dubois 1908: no. 460.</td></tr>
<tr><td>480.</td><td>a) Imp(eratore) Domitian[o] Aug(usto) XII c[o(n)]s(ule); b) ex rat(ione) Tyc(hi?) ser(vi)</td><td>Rome, 86</td><td>'africano'; Bruzza 1870: no. 168 = Dubois 1908: no. 461.</td></tr>
<tr><td>481.</td><td>Imp(eratore) D[o]mitian[o] Aug(usto) XII c[o(n)]s(ule)</td><td>Rome, 86</td><td>'africano'; Bruzza 1870: no. 170 = Dubois 1908: no. 463.</td></tr>
<tr><td>482.</td><td>Domitiano Aug[(usto) —] XII</td><td>Rome, 86</td><td>'africano'; Bruzza 1870: no. 171 = Dubois 1908: no. 464.</td></tr>
<tr><td>483.</td><td>Imp(eratore) Traiano Aug(usto) n(ostro) III co(n)s(e) | ex r(atione) Spend(-) ser(vi) PME | XXXIX</td><td>Ostia, 100</td><td>'africano'; Baccini Leotardi 1979: no. 3.</td></tr>
<tr><td>484.</td><td>[V]etere et Ae[li]|ano co(n)s(ulibus)</td><td>Ostia, 116</td><td>'africano'; Baccini Leotardi 1989: no. 2.</td></tr>
<tr><td>485.</td><td>a) Vetere et A(eliano) | [..]CCONK V; b) ex rat(ione) Mul( ) | n(umero) CLII</td><td>Ostia, 116</td><td>'africano'; Baccini Leotardi 1989: no. 3.</td></tr>
<tr><td>486.</td><td>a) [e]x rat(ione) Mul( ) | n(umero) CLVII; b) [V]ete[re et A]elia|no co(n)-s(ulibus); c) l(oco) CCCXXVII Ꞧ</td><td>Ostia, 116</td><td>'africano'; Baccini Leotardi 1989: no. 4.</td></tr>
<tr><td>487.</td><td>Hadriano n(ostro) III Q(uinto) Dasumio Rustico co(n)s(ulibus)</td><td>Ostia, 119</td><td>'africano'; Baccini Leotardi 1989: no. 5.</td></tr>
<tr><td>488.</td><td>a) Augurin(o) co(n)s(ule) | l(oco) CCXXIX | Ꞧ (AD 132) b) Hadriano n(ostro) III Q(uinto) Dasumio Rustico c[o(n)s(ulibus)] (AD 119)</td><td>Ostia</td><td>'africano'; Baccini Leotardi 1989: no. 6.</td></tr>
<tr><td>489.</td><td>ex r(atione) Mami(i) Liciniani n(umero) I | Glabrione et Tebaniano co(n)s(ulibus)</td><td>Rome, 124</td><td>'africano'; Bruzza 1870: no. 181 = Dubois 1908: no. 474.</td></tr>
<tr><td>490.</td><td>[A]ugurin(o) co(n)s(ule) | l(oco) ÐCLXXIII Ꞧ</td><td>Ostia, 132</td><td>'africano'; Baccini Leotardi 1979: no. 6.</td></tr>
<tr><td>491.</td><td>a) Augurin(o) co(n)s(ule) | l(oco) CCXXXIIX Ꞧ; b) ex r(atione) Dac( ) | n(umero) XVII</td><td>Ostia, 132</td><td>'africano'; Baccini Leotardi 1979: no. 8.</td></tr>
</table>

492.	*co(n)s(ule) | Augurin(o) | l(oco) CCXCVI ℞*	Ostia, 132	'africano'; Baccini Leotardi 1979: no. 9.
493.	a) *Augurin(o) | co(n)s(ule) | l(oco) CC XIIX*; b) [-c.3 -]*N | n(umero) LVIIII | E*	Ostia, 132	'africano'; Baccini Leotardi 1989: no. 7.
494.	*Augurin(o) co(n)s(ule) | lo(c)o CVII*	Ostia, 132	'africano'; Pensabene 1994: 56, fig. 56 = Ostia Inv. 36760.
495.	*IH | Claro II co(n)s(ule) ex | ra(tione) Cl(audi) Zel(oti)*	Ostia, 150	'africano'; Baccini Leotardi 1989: no. 8.
496.	*ex ratione Lucili Er|ma(e) et Claudi Soterici | Glabrioni (!) et Homul|lo con(sulibus) [—]*	Ostia, 152	'africano'; Baccini Leotardi 1989: no. 9.
497.	*ex ratione Lucili | Erma(e) et Claudi | Soterici | Glabrion(e) et Horm[ul]|lo consuli(bus)*	Ostia, 152	'africano'; Baccini Leotardi 1989: no. 10.
498.	*Rust(ico) II et Aq(uilino) co(n)s(ulibus) | n(umero) LXIIII*	Ostia, 162	'africano'; Baccini Leotardi 1989: no. 11 = Pensabene 1994: 56, no. 16.
499.	*n(umero) LXIII | Ru[s]t(ico) II et A[q(uilino) co(n)s(ulibus)] | ex r(atione) Sext(i?) et [—]*	Ostia, 162	'africano'; Baccini Leotardi 1989: no. 12.
500.	*Rust(ico) II et Aq(uilino) | co(n)s(ulibus) | ex r(atione) Sext(i?) et | Her(mae?)*	Ostia, 162	'africano'; Baccini Leotardi 1989: no. 13.
501.	*Rust(ico) II et | Aq(uilino) co(n)s(ulibus) ex r(atione) Sext(i?)*	Ostia, 162	'africano'; Baccini Leotardi 1979: no. 10.
502.	*Laeliano et Pasto(re) | co(n)s(ulibus) Aur(elii?) Corn(eliano?) R | ORG | loco IIII || IE*	Karagöl, 164	grey marble?; *CIL* III 419 a = Bruzza 1870: no. 243 = Dubois 1908: no. 241 = Béguignon 1928: no. 1 = McCabe 1985: 266.1.
503.	*Orfito et Pud(ente) | co(n)s(ulibus) loco | CXLVI ex r(atione) Diod(-)*	Karagöl, 165	grey marble; *CIL* III 419 b = Bruzza 1870: no. 244 = Dubois 1908: no. 242.
504.	*Orfito et [Pud(ente)] | co(n)s(ulibus) loc(o) CXLIX | ex r(atione) Dio(d)*	Karagöl, 165	grey marble; *CIL* III 419 c = Bruzza 1870: no. 245 = Dubois 1908: no. 243.
505.	*Puden(te) et P[ollione?] | [c]o(n)s(ule) loc[o] CXX | ex r(atione) Dio(-)*	Karagöl, 166	grey marble; *CIL* III 419 g, (reads POL and CXXXIIII COS) = Bruzza 1870: no. 246 = Dubois 1908: no. 244 = Béguignon 1928: no. 2, fig.1 = McCabe 1985: 266.2.

506.	*[Pude]nte et Pol(lione) | [loc]o L | Pu[dente?]*	Karagöl, 166	grey marble; Béguignon 1928: no. 8 fig.3 = McCabe 1985: 266.8.
507.	*Puden(te) et Pol(lione) co(n)s(ulibus) | loco XXII ex r(atione) Pol(-)*	Karagöl, 166	grey marble; *CIL* III 419 d = Bruzza 1870: no. 249 = Dubois 1908: no. 247.
508.	*Puden(te) et Pol(lione) co(n)s(ulibus) | loco XL ex r(atione) Di(-)*	Karagöl, 166	grey marble; *CIL* III 419 e = Bruzza 1870: no. 251 = Dubois 1908: no. 249.
509.	*Puden(te) et Pol(lione) co(n)s(ulibus)| loco LXV ex r(atione) Dio(-)*	Karagöl, 166	grey marble; *CIL* III 419 f = Bruzza 1870: no. 250 = Dubois 1908: no. 248.
510.	*l(oco) XI | Puden(te) et Pol(lione) co(n)s(ulibus) | loco CXXXVIII | ex r(atione) Dio(-)*	Karagöl, 166	grey marble; *CIL* III 419 h = Bruzza 1870: no. 247 = Dubois 1908: no. 245.
511.	*Pudente ⟨et⟩ Pol[lione] co(n)[s(ulibus)]| loco CLXXX | ex r(atione) Dio(-)*	Karagöl, 166	grey marble; *CIL* III 419 i = Bruzza 1870: no. 248 = Dubois 1908: no. 246 = McCabe 1985: 266.3 = Béguignon 1928: no. 3.
512.	*n(umero) XXI*	Karagöl	grey marble; *CIL* III 419 l = Dubois 1908: no. 253.
513.	*loco IIII*	Karagöl	grey marble; *CIL* III 419 k =Bruzza 1870: no. 252 = Dubois 1908: no. 250 = Béguignon 1928: no. 4 = McCabe 1985: 266.4.
514.	*loc(o) XXVIII*	Karagöl	grey marble: *CIL* III 419 m = Bruzza 1870: no. 253 = Dubois 1908: no. 251 = Béguignon 1928: no. 1, McCabe 1985: 266.5.
515.	*loco XXX*	Karagöl	grey marble?; *CIL* III 419 n = Bruzza 1870: no. 254 = Dubois 1908: no. 252 = Béguignon 1928: no. 6 = McCabe 1985: 266.6.
516.	*ex r[at(ione)—] | loco [—]*	Karagöl	grey marble; Béguignon 1928: no. 7, fig. 2 = McCabe 1985: 266.7.

517. *P n(umero) XXXI | [Au]r(elius) Cor(nelianus) b(racchio) III* — Karagöl — grey marble; Béguignon 1928: no. 9, fig. 4 = McCabe 1985: 266.9.

518. *Cor(nelius?) | [—]* — Karagöl — grey marble; Béguignon 1928: no. 10 = McCabe 1985: 266.10.

519. *L ∞ CCXCL[.] | T* — Ostia — 'africano'; Baccini Leotardi 1979: no. 4.

520. *[.]III* — Ostia — 'africano'; Baccini Leotardi 1979: no. 5.

521. *n(umer)o CXXXXVI | ET (!)[.r]atione Lucili Her|mae soti[.]eni* — Ostia — 'africano'; Baccini Leotardi 1989: no. 14.

522. a) *n(umero) CXXXXIII;* b) *ETE [r]atione Lucili Herma() [s]oti[.]eni* — Ostia — 'africano'; Baccini Leotardi 1989: no. 15.

523. a) *n(umer)o CXXXIX;* b) *ETE [r]atione Lucili He[r]ma[e] | [soti.e]ni* — Ostia — 'africano'; Baccini Leotardi 1989: no. 16.

524. a) *n(umer)o CXLI;* b) *ETE [r]atione Lucili Hermae | soti[.]ni* — Ostia — 'africano'; Baccini Leotardi 1989: no. 17.

525. *NELVC | T | RIONIET | ON* — Ostia — 'africano'; Baccini Leotardi 1989: no. 18.

526. *VII | COS* — Ostia — 'africano'; Baccini Leotardi 1989: no. 19.

527. a) *AP;* b) *HER Cae(saris?) n(umero) XXII* — Rome — 'africano'; Bruzza 1870: no. 141 = Dubois 1908: no. 434.

528. *HER Cae(saris?) n(umero) I* — Rome — 'africano'; Bruzza 1870: no. 142 = Dubois 1908: no. 435.

529. *HER Cae(saris?) n(umero) IIII* — Rome — 'africano'; Bruzza 1870: no. 143 = Dubois 1908: no. 436.

530. *HER Cae(saris?) | n(umero) VIII* — Rome — 'africano'; Bruzza 1870: no. 144 = Dubois 1908: no. 437.

531. *HER Cae(saris?)* — Rome — 'africano'; Bruzza 1870: no. 145 = Dubois 1908: no. 438.

532. *HE[—]* — Rome — 'africano'; Bruzza 1870: no. 146 = Dubois 1908: no. 439.

533. *Pae cae | X V* — Rome — 'africano'; Bruzza 1870: no. 175 = Dubois 1908: no. 468.

534. *Pae cae | n(umero) XIIX* — Rome — 'africano'; Bruzza 1870: no. 176 = Dubois 1908: no. 469.

535. *Paed cae | n(umero) III* — Rome — 'africano'; Bruzza 1870: no. 177 = Dubois 1908: no. 470.

536. *Lae(ti) Cae(saris) n(umero) V* — Rome — 'africano'; Bruzza 1870: no. 160 = Dubois 1908: no. 453.

537. *Lae(ti) Cae(saris) n(umero) XIV* — Rome — 'africano'; Bruzza 1870: no. 161 = Dubois 1908: no. 454.

538. *Lae(ti) Cae(saris) n(umero) LIX* — Rome — 'africano'; Bruzza 1870: no. 162 = Dubois 1908: no. 455.

539. *Lae(ti) Cae(saris)* — Rome — 'africano'; Bruzza 1870: no. 163 = Dubois 1908: no. 456.

540. *EDAIEC Cae* — Rome — 'africano'; Bruzza 1870: no. 183 = Dubois 1908: no. 476.

541. *[. . . .]iser | [. .]XXX* — Rome — 'africano'; Bruzza 1870: no. 165 = Dubois 1908: no. 458.

542. *[. . .]aet ser(vi) | [. .]VIII* — Rome — 'africano'; Bruzza 1870: no. 164 = Dubois 1908: no. 457.

543. *ex rat(ione) Laet(i) ser(vi) | n(umero) LXXX* — Rome — 'africano'; Bruzza 1870: no. 166 = Dubois 1908: no. 459.

544. *ex rat(ione) Tyc(hi?) ser(vi) n(umero) LXVIII* — Rome — 'africano'; Bruzza 1870: no. 169 = Dubois 1908: no. 462.

545. *ex rat(ione) Tyc(hi?) ser(vi) n(umero) V* — Ostia — 'africano'; *CIL* XIV 2016 = Dubois 1908: no. 447.

546. *[ex] rat(ione) Tyc(hi?) ser(vi) Cae-(saris) | n(umero) III* — Ostia — 'africano'; Pensabene 1994: 55, no. 10.

547. *ex rat(ione) Mami(i) Licin(iani?)* — Rome — grey marble; Bruzza 1870: no. 239 = Dubois 1908: no. 237.

548. *Pontiano et Atiliano co(n)s(ulibus) ex ratione | redemptoris n(umero) * XXVIIII* — Rome — 'africano'; Bruzza 1870: no. 182 = Dubois 1908: no. 475.

549. *ex r(atione) Ces[—] | [—] co(n)s(ulibus)* — Rome — 'africano'; Bruzza 1870: no. 185 = Dubois 1908: no. 478.

550. *[—]sar Dom[—]* — Rome — 'africano'; Bruzza 1870: no. 172 = Dubois 1908: no. 465.

551. *[—]o Aug(usto)*	Rome	'africano'; Bruzza 1870: no. 173 = Dubois 1908: no. 466.
552. *Cae(sare) Aug(usto) | n(umero) V*	Rome	'africano'; Bruzza 1870: no. 174 = Dubois 1908: no. 467.
553. *[—]ae | [—]XII*	Rome	'africano'; Bruzza 1870: no. 178 = Dubois 1908: no. 471.
554. *Cae[—] | VI[—]*	Rome	'africano'; Bruzza 1870: no. 179 = Dubois 1908: no. 472.
555. *[—]Cae | [—]XXXXI*	Rome	'africano'; Bruzza 1870: no. 180 = Dubois 1908: no. 473.
556. *Ner[—]*	Rome	'africano'; Bruzza 1870: no. 184 = Dubois 1908: no. 477.
557. *Cae() n(umero) CLIX | CIA*	Rome	'africano'; Bruzza 1870: no. 186 = Dubois 1908: no. 479.
558. *Cn(aeo) Pompei(o)*	Rome	'africano'; Bruzza 1870: no. 187 = Dubois 1908: no. 480.
559. *[—]CM[—]*	Rome	'africano'; Bruzza 1870: no. 188 = Dubois 1908: no. 481.
560. *CX LIII*	Rome	'africano'; Bruzza 1870: no. 189 = Dubois 1908: no. 482.
561.  *N*	Rome	'africano'; Bruzza 1870: no. 190 = Dubois 1908: no. 483.
562. B-ΠAY	Rome	grey marble; Bruzza 1870: no. 240 = Dubois 1908: no. 238.
563. *TA*	Rome	grey marble; Bruzza 1870: no. 241 = Dubois 1908: no. 239.
564. *V*	Rome	grey marble; Bruzza 1870: no. 242 = Dubois 1908: no. 240.

Portasanta: Chios

565. a) *Rufo et Ca[pitone]*; b) *Ian(uarii) Caes(aris) n(umero) CCL*	Rome, 67	Bruzza 1870: no. 191 = Dubois 1908: no. 486.

566. *Imp(eratore) Domit(iano) Cae(sare) Aug(usto) VIIII | Petil(io) Rufo c[o(n)s(ulibus)]*	Rome, 83	Bruzza 1870: no. 193 = Dubois 1908: no. 488.
567. *Imp(eratore) Domit(iano) Cae(sare) Aug(usto) X | [C. Oppio Sa]bino co(n)s(ulibus)*	Rome, 84	Bruzza 1870: no. 194 = Dubois 1908: no. 489.
568. a) *Imp(eratoris) Caesaris Domitian(i) | Augusti Germanici | n(umero) III*; b) *LIXIIII*	Rome, 83–93	Bruzza 1870: no. 195 = Dubois 1908: no. 490.
569. *Coll(ega) et Priscino co(n)s(ulibus) | CC∞XXXII* R	Rome, 93	Bruzza 1870: no. 196 = Dubois 1908: no. 491.
570. *Colleg(a) et Priscino*	Rome, 93	Bruzza 1870: no. 197 = Dubois 1908: no. 492.
571. a) *Collega et [P]riscino co(n)s(ulibus)*; b) *SVC III*	Ostia, 93	Baccini Leotardi 1989: no. 64.
572. *Veter(e) et Vale[nte co(n)s(ulibus)] | ex r(atione) Suc(cessi) Caes(aris) VIIII*	Rome, 96	Bruzza 1870: no. 198 = Dubois 1908: no. 493.
573. *[Vetere] et Vale[nte co(n)s(ulibus)] | [ex r(atione) Su]cces(i) Cae[s(aris)—]*	Rome, 96	Bruzza 1870: no. 199 = Dubois 1908: no. 494.
574. *Caes(are) Traiano n(ostro) IIII co(n)s(ulibus) | ex rat(ione) | Thras Caes(aris) n(ostri) s(ervi) Laetiani*	Rome, 101–104	Bruzza 1870: no. 200 = Dubois 1908: no. 495.
575. *Caes(are) Traiano n(ostro) IIII co(n)s(ule) | ex rat(ione) | Thras Caes(aris) n(ostri) ser(vi) Laetiani*	Rome, 101–104	Bruzza 1870: no. 201 = Dubois 1908: no. 496.
576. *Gallo et Bradua co(n)s(ulibus) | ex rat(ione) | Rest(ituti) et Hya(cinthi) Caes(aris) n(ostri) | n(umero) XI*	Rome, 108	Bruzza 1870: no. 205 = Dubois 1908: no. 500.
577. *Gallo et Bradua co(n)s(ulibus) | ex rat(ione) | [Re]st(ituti) et Hy[a(cinthi) Caes(aris) n(ostri) ser(vorum)]*	Rome, 108	Bruzza 1870: no. 207 = Dubois 1908: no. 501.
578. *Gallo et Bradua co(n)s(ulibus) | ex rat(ione) | Rest(ituti) et Hya(cinthi) Caes(aris) n(ostri) ser(vorum) | n(umero) CXLIX*	Rome, 108	Bruzza 1870: no. 209 = Dubois 1908: no. 504.
579. a) *[Pal]ma II et Tullo co(n)s(ulibus) | ex rat(ione) | S* vac. *SER | n(umero) CXLVI*, b) *l(oco?)*ĐCCXIIXR	Ostia, 109	Baccini Leotardi 1989: no. 65.
580. *Imp(eratore) Hadriano n(ostro) II co(n)s(ule) | ex rat(ione) | Rest(ituti) Caesaris n(ostri) ser(vi) | n(umero) CXXIII*	Ostia, 118	Baccini Leotardi 1989: no. 66.
581. *Galli(cano) et Ve|te(re) co(n)s(ulibus) ex r(atione) Aure|li M[—]*	Ostia, 150	Baccini Leotardi 1989: no. 67 = Pensabene 1994: 99, no. 37.
582. *Commodo Caes(are) n(ostro) II co(n)s(ulibus) | SEHTT [—]PXƎ*	Rome, 179	Bruzza 1870: no. 210 = Dubois 1908: no. 505.

583. *Ian(uarii) Cae(saris) n(umero) XXXXV*	Rome	Bruzza 1870: no. 192 = Dubois 1908: no. 487.
584. *ex rat(ione) | Thras Caes(arum) nn(ostrorum) | ∞CDLIV*	Rome	Bruzza 1870: no. 202 = Dubois 1908: no. 497.
585. *ex rat(ione) | Thras Cae(saris) n(ostri) CΓXVI*	Rome	Bruzza 1870: no. 203 = Dubois 1908: no. 498.
586. *ex rat(ione) | Thras Cae(saris) n(umero) CLXIIX*	Rome	Bruzza 1870: no. 204 = Dubois 1908: no. 499.
587. *ex rat(ione) | Rest(ituti) et Hya(cinthi) Cae(saris) n(ostri) ser(vorum) n(umero) CXLII*	Rome	Bruzza 1870: no. 206 = Dubois 1908: no. 502.
588. *[—]cos | ex rat(ione) | Rest(ituti) Caes(aris) n(ostri) ser(vi) n(umero) XXI*	Rome	Bruzza 1870: no. 208 = Dubois 1908: no. 503.
589. *AA I V III | ex rat(ione)*	Ostia	Pensabene 1994: 99, no. 38.
590. *[——] | [—] Caes(aris?) n(ostri?) | CXXIII*	Rome	Bruzza 1870: no. 211 = Dubois 1908: no. 506.
591. *[—]VA | [—]S | [—]XIII*	Rome	Bruzza 1870: no. 212 = Dubois 1908: no. 507.
592. *[—]ILA | [——]*	Rome	Bruzza 1870: no. 213 = Dubois 1908: no. 508.
593. *FRON*	Rome	Bruzza 1870: no. 214 = Dubois 1908: no. 509.
594. *LXXIV*	Rome	Bruzza 1870: no. 215 = Dubois 1908: no. 510.
595. *[——-] | [—]CCCLXIX ꝶ*	Rome	Bruzza 1870: no. 216 = Dubois 1908: no. 511.
596. *[—] CƉII ꝶ*	Rome	Bruzza 1870: no. 217 = Dubois 1908: no. 512.
597. *l(oco) ƉCCCXL ꝶ*	Rome	Bruzza 1870: no. 218 = Dubois 1908: no. 513.
598. *l(oco) ∞ XCV ꝶ*	Rome	Bruzza 1870: no. 219 = Dubois 1908: no. 514.
599. a) *No CXXXIIX*, b) *V*, c) *PI∈*, d) V	Ostia	Baccini Leotardi 1989: no. 68.
600. *l(oco) ƉCCXIX | ꝶ*	Ostia	Baccini Leotardi 1989: no. 69 = Pensabene 1994: 44, fig.40.

Cipollino: Karystos

601. a) *cae n II. CIA | CAE[.]IRVFO | co(n)s(ulibus)*; b) *cae n(umero) CXXIIX | CIA*	Rome, 17?	Bruzza 1870: no. 2 = Dubois 1908: no. 281.
602. *[D]om(itiano) Caes(are) II co(n)s(ule)*	Rome, 73	Bruzza 1870: no. 3 = Dubois 1908: no. 282.

603. a) *sub cura Minici* | *Sancti proc(uratoris) Aug(usti)* | *pr(ob.) Crescente lib(erto)* | *n(umero) XLII*; b) *Augurin(o) co(n)s(ule)* | *l(oco) CXXCVI* ℞	Rome, 132	Bruzza 1870: no. 4 = Dubois 1908: no. 283.
604. *sub cura Minici Sa[ncti—]* | *pr(ob.) Crescente lib(erto) n(umero) I[—]*	Rome, 132?	Bruzza 1870: no. 5 = Dubois 1908: no. 284.
605. a) *Augurin(o) co(n)s(ule)* | *l(oco) LXXX* ℞; b) *l(oco) LXXX* ℞; c) *n(umero) LXV* | *Cae(?) N*	Rome, 132	Bruzza 1870: no. 6 = Dubois 1908: no. 285.
606. a) *Augur(ino)*; b) *l(oco) XXVII R*; c) *n(umero) LIV Cae⟨?⟩s N*	Rome, 132	Bruzza 1870: no. 7 = Dubois 1908: no. 286.
607. a) *Augurin(o) co(n)s(ule)*; b) *n(umero) XXXIV* | *Cae(?)s N*	Rome, 132	Bruzza 1870: no. 8 = Dubois 1908: no. 287.
608. a) *Augurin(o) co(n)s(ule)* | *l(oco) CCCCXXXVI* ℞; b) *CA n(umero) XC* | *CIA*	Ostia, 132	Baccini Leotardi 1989: no. 50 = Pensabene 1994: 86, no. 32.
609. *Augurin(o) co(n)s(ule)* | *C*	Ostia, 132	Baccini Leotardi 1989: no. 51.
610. *Augurin(o) co(n)s(ule)* | *l(oco) ĐXCV* ℞	Ostia, 132	Baccini Leotardi 1979: no. 43 = Pensabene 1994: 86, no. 31.
611. *Augurin(o) co(n)s(ule)*	Ostia, 132	Pensabene 1994: 177, no. 118 = *CIL* XIV 5242.
612. *Serviano III co(n)s(ulibus)* | *ex rat(ione)* | *Valentis* | *n(umero) VII*	Rome, 134	Bruzza 1870: no. 9 = Dubois 1908: no. 288.
613. *Serviano III co(n)s(ule)* | *ex rat(ione)* | *Valentis* | *n(umero) CXXXII*	Rome, 134	Bruzza 1870: no. 10 = Dubois 1908: no. 289.
614. *[S]erviano III co(n)s(ule)* | *ex rat(ione)* | *Valentis* | *LXXXIIII*	Rome, 134	Bruzza 1870: no. 11 = Dubois 1908: no. 290.
615. *Pontiano co(n)s(ule)* | *ex r(atione)* | *VRBEI Natalis* | *n(umero) II*	Rome, 135	Bruzza 1870: no. 12 = Dubois 1908: no. 291. cf. Hirschfeld, 1905: 177, with fn. 2.
616. *Pontiano co(n)s(ule)* | *ex r(atione) Valentis* | *n(umero) CL XII*	Rome, 135	Bruzza 1870: no. 13 = Dubois 1908: no. 292.
617. *L(ucio) Aelio Caesare n(ostro) II c(onsule)* | *ex rat(ione)* | *Caecili Marciani* | *n(umero) X VII.*	Karystos, 161	CIL III 12288.
618. *ex M N Caesaris N R D A sub cur(a) C(aii) Caerialis pr(ocuratoris)* | *subseq(uente) Sergio Longo 7(centurioni) leg(ionis) XXII Primig(eniae) prob.* | *Crescente lib(erto)* vac. *n(umero) VIIII* || *n(umero) VIIII*; in red paint: *CXXX n(umero) VIIII*	Rome	Bruzza 1870: no. 1 = Dubois 1908: no. 278 = *ILS* 8717.

619. *ex r(atione) | Ω ZOES*
Ostia
Baccini Leotardi 1989: no. 56.

620. *ex r(atione) Cla[—] | XXC*
Ostia
Baccini Leotardi 1979: no. 55 = Pensabene 1994: 39, fig. 31.

621. *ex r(atione) | Orbi Natales | n(umero) XXIIII*
Roma
Bruzza 1870: no. 15 = Dubois 1908: no. 295.

622. a) *Orbi Na[tales—]*; b) *n(umero) XXXVI | V*
Roma
Bruzza 1870: no. 16 = Dubois 1908: no. 296.

623. *Hyme(naei?) Cae | dom(-) n(umero) IX*
Roma
Bruzza 1870: no. 17 = Dubois 1908: no. 298.

624. *Hyme(naei?) Cae*
Roma
Bruzza 1870: no. 18 = Dubois 1908: no. 299a.

625. *Hyme(naei?) Cae*
Roma
Bruzza 1870: no. 19 = Dubois 1908: no. 299b.

626. *n(umero) CL [—] | Hyme(naei?) [—]*
Roma
Bruzza 1870: no. 20 = Dubois 1908: no. 300.

627. a) *n(umero) CC | Hyme(naei?) C[ae?]* ?; b) Å
Roma
Bruzza 1870: no. 21 = Dubois 1908: no. 301.

628. a) *CLXXXIII*; b) *n(umero) VI | Hyme(naei?) Cae | A H*
Roma
Bruzza 1870: no. 22 = Dubois 1908: no. 302.

629. *n(umero) CCCXXX | Hyme(naei?) | Z*
Roma
Bruzza 1870: no. 23 = Dubois 1908: no. 303.

630. *Hy(menaei?)*
Roma
Bruzza 1870: no. 24 = Dubois 1908: no. 304.

631. *Tha(myri?) Cae(saris) | Minui[..]til | co(n)s(ule)*
Roma
Bruzza 1870: no. 14 = Dubois 1908: no. 294.

632. *T(h)amyri*
Roma
Bruzza 1870: no. 25 = Dubois 1908: no. 305.

633. *Th(a)myr(i) Cae*
Roma
Bruzza 1870: no. 26 = Dubois 1908: no. 306.

634. *VII T(ha)my(ri) Cae*
Roma
Bruzza 1870: no. 27 = Dubois 1908: no. 307.

635. *Th(a)my(ri) C*
Roma
Bruzza 1870: no. 28 = Dubois 1908: no. 308.

636. *Thamy(ri) Cae*
Roma
Bruzza 1870: no. 29 = Dubois 1908: no. 309.

637. *Thamy(ri) C*
Roma
Bruzza 1870: no. 30 = Dubois 1908: no. 310.

638. *Th(a)my(ri) Cae*
Roma
Bruzza 1870: no. 31 = Dubois 1908: no. 311.

639. *n(umero) XLIII | Thamy(ri) C*
Roma
Bruzza 1870: no. 32 = Dubois 1908: no. 312.

640. a) *n(umero) CV*; b) *Thamy(ri) | Cae*
Roma
Bruzza 1870: no. 33 = Dubois 1908: no. 313.

641. *Th(a)my(ri) Cae*	Roma	Bruzza 1870: no. 34 = Dubois 1908: no. 314.
642. *L[—] | [T]hamy(ri) C[—]*	Roma	Bruzza 1870: no. 35 = Dubois 1908: no. 315.
643. a) *Tha(myri);* b) *n(umero) LVI*	Roma	Bruzza 1870: no. 36 = Dubois 1908: no. 316.
644. *n(umero) XIV | Th(a)my(r)e ZO*	Roma	Bruzza 1870: no. 39 = Dubois 1908: no. 319.
645. a) *Th(a)my(r)e | n(umero) CLXVI* ; b) *K H*	Roma	Bruzza 1870: no. 40 = Dubois 1908: no. 320.
646. *n(umero) CCXII | Th(a)my(ri) C*	Roma	Bruzza 1870: no. 41 = Dubois 1908: no. 321.
647. a) *n(umero) CCCXLI T(ha)my(ri) C;* b) *I [—] CXI*	Roma	Bruzza 1870: no. 42 = Dubois 1908: no. 322.
648. *Ti Eup | CCXL*	Roma	Bruzza 1870: no. 51 = Dubois 1908: no. 331.
649. *Cae n(u)m(ero) IX | C λ POL*	Roma	Bruzza 1870: no. 52 = Dubois 1908: no. 332.
650. *CL PH*	Roma	Bruzza 1870: no. 71 = Dubois 1908: no. 351.
651. *CL PH | n(umero) XIII*	Roma	Bruzza 1870: no. 72 = Dubois 1908: no. 352.
652. *L PLI*	Roma	Bruzza 1870: no. 73 = Dubois 1908: no. 353.
653. *CL PH*	Roma	Bruzza 1870: no. 74 = Dubois 1908: no. 354.
654. *n(umero) CXXII | PH*	Roma	Bruzza 1870: no. 75 = Dubois 1908: no. 355.
655. *n(umero) LXXX | PH*	Roma	Bruzza 1870: no. 76 = Dubois 1908: no. 356.
656. *n(umero) CXLV | CL PH*	Roma	Bruzza 1870: no. 77 = Dubois 1908: no. 357.
657. *n(umero) CXLV | CLA PHL*	Roma	Bruzza 1870: no. 79 = Dubois 1908: no. 359.
658. *n(umero) CCXXXV | CL PH*	Roma	Bruzza 1870: no. 78 = Dubois 1908: no. 358.
659. *Cae n(umero) I | CIA*	Roma	Bruzza 1870: no. 81 = Dubois 1908: no. 361.
660. *Cae n(umero) I [—] | CIẠ[—]*	Roma	Bruzza 1870: no. 82 = Dubois 1908: no. 362.
661. *Cae n(umero) XIV | CIA*	Roma	Bruzza 1870: no. 83 = Dubois 1908: no. 363.
662. *Cae | n(umero) XVI | CIA*	Roma	Bruzza 1870: no. 84 = Dubois 1908: no. 364.
663. *Cae | n(umero) XXVI | CIA*	Roma	Bruzza 1870: no. 85 = Dubois 1908: no. 365.

664. *C ⧺ | C III | CIA*	Roma	Bruzza 1870: no. 86 = Dubois 1908: no. 366.
665. *n(umero) XX | CIA*	Roma	Bruzza 1870: no. 87 = Dubois 1908: no. 367.
666. *A | Cae n(umero) XXVIII | CIA*	Roma	Bruzza 1870: no. 88 = Dubois 1908: no. 368.
667. *Cae n(umero) LIX | XIII CIA*	Roma	Bruzza 1870: no. 89 = Dubois 1908: no. 369.
668. *Cae n(umero) LX | II CIA*	Roma	Bruzza 1870: no. 90 = Dubois 1908: no. 370.
669. *Cae n(umero) LXI | CIA*	Roma	Bruzza 1870: no. 91 = Dubois 1908: no. 371.
670. a) *Cae n(umero) LXXI*, b) *Cae n(umero) | CIA*	Roma	Bruzza 1870: no. 92 = Dubois 1908: no. 372.
671. *Cae n(umero) CCXIII | CIA*	Roma	Bruzza 1870: no. 93 = Dubois 1908: no. 373.
672. *DN III | CIA*	Roma	Bruzza 1870: no. 94 = Dubois 1908: no. 374.
673. a) *C*; b) *l(oco) ∞XXXVII | ℞*; c) *n(umero) XXXI | CLA B*	Roma	Bruzza 1870: no. 98 = Dubois 1908: no. 378.
674. a) *℞ | L∞XL*; b) *n(umero) IX | CL*	Roma	Bruzza 1870: no. 99 = Dubois 1908: no. 379.
675. a) *l(oco) ∞XI III ℞*; b) *n(umero) XVI CLA B | AP | IIIↃ*	Roma	Bruzza 1870: no. 100 = Dubois 1908: no. 380.
676. *T*	Roma	Bruzza 1870: no. 37 = Dubois 1908: no. 307.
677. *Λ*	Roma	Bruzza 1870: no. 38 = Dubois 1908: no. 308.
678. *[—]omy[—]*	Roma	Bruzza 1870: no. 43 = Dubois 1908: no. 323.
679. *ΘΥΡ*	Roma	Bruzza 1870: no. 44 = Dubois 1908: no. 324.
680. a) *ΡΟΥ*; b) *caes n(umero) CXVI*	Roma	Bruzza 1870: no. 45 = Dubois 1908: no. 325.
681. ⊟ *n(umero) XXI*	Roma	Bruzza 1870: no. 46 = Dubois 1908: no. 326.
682. *Φ*	Roma	Bruzza 1870: no. 47 = Dubois 1908: no. 327.
683. *K*	Roma	Bruzza 1870: no. 48 = Dubois 1908: no. 328.
684. ψ	Roma	Bruzza 1870: no. 49 = Dubois 1908: no. 329.
685. *AP P | I N*	Roma	Bruzza 1870: no. 50 = Dubois 1908: no. 330.
686. *n(umero) IX | caes n(umero)*	Roma	Bruzza 1870: no. 53 = Dubois 1908: no. 333.

687. *Cae n(umero) | CLXI* Roma Bruzza 1870: no. 54 = Dubois 1908: no. 334.

688. *n(umero) XXIIII | cae* Roma Bruzza 1870: no. 55 = Dubois 1908: no. 335.

689. *n(umero) XXVI | caes n* Roma Bruzza 1870: no. 56 = Dubois 1908: no. 336.

690. *Cae n(umero) L* Roma Bruzza 1870: no. 57 = Dubois 1908: no. 337.

691. *Cae n(umero) LX* Roma Bruzza 1870: no. 58 = Dubois 1908: no. 338.

692. *Cae n(umero) LXII* Roma Bruzza 1870: no. 59 = Dubois 1908: no. 339.

693. *n(umero) LXV | caes n* Roma Bruzza 1870: no. 60 = Dubois 1908: no. 340.

694. *Cae n(umero) CLIII* Roma Bruzza 1870: no. 61 = Dubois 1908: no. 341.

695. *Cae n(umero) CCX* Roma Bruzza 1870: no. 62 = Dubois 1908: no. 342.

696. *Cae n(umero) CCL* Roma Bruzza 1870: no. 63 = Dubois 1908: no. 343.

697. *n(umero) CL | II Cae* Roma Bruzza 1870: no. 64 = Dubois 1908: no. 344.

698. *DN CXXXII* Roma Bruzza 1870: no. 65 = Dubois 1908: no. 345.

699. *DN CLX* Roma Bruzza 1870: no. 66 = Dubois 1908: no. 346.

700. a) *DN CLXXXII*; b) *n(umero) LXXXII* Roma Bruzza 1870: no. 67 = Dubois 1908: no. 347.

701. *n(umero) CXXXVI | M D | A N* Roma Bruzza 1870: no. 68 = Dubois 1908: no. 348.

702. *N* Roma Bruzza 1870: no. 69 = Dubois 1908: no. 349.

703. *No* Roma Bruzza 1870: no. 70 = Dubois 1908: no. 350.

704. *HP* Roma Bruzza 1870: no. 80 = Dubois 1908: no. 360.

705. *CL* Roma Bruzza 1870: no. 95 = Dubois 1908: no. 375.

706. *CLA* Roma Bruzza 1870: no. 96 = Dubois 1908: no. 376.

707. *n(umero) CX[—] | CL [—]* Roma Bruzza 1870: no. 97 = Dubois 1908: no. 377.

708. *BI | n(umero) CXXIIX* Roma Bruzza 1870: no. 101 = Dubois 1908: no. 381

709. *n(umero) XVI | XIIXXX N* Roma Bruzza 1870: no. 102 = Dubois 1908: no. 382.

710. *XXVI [—]*	Roma	Bruzza 1870: no. 103 = Dubois 1908: no. 383.
711. *n(umero) LI*	Roma	Bruzza 1870: no. 104 = Dubois 1908: no. 384.
712. *n(umero) LVI*	Roma	Bruzza 1870: no. 105 = Dubois 1908: no. 385.
713. *LIX*	Roma	Bruzza 1870: no. 106 = Dubois 1908: no. 386.
714. *NL+V*	Roma	Bruzza 1870: no. 107 = Dubois 1908: no. 387.
715. *LXXXII*	Roma	Bruzza 1870: no. 108 = Dubois 1908: no. 388.
716. *XXCVI*	Roma	Bruzza 1870: no. 109 = Dubois 1908: no. 389.
717. *IXC*	Roma	Bruzza 1870: no. 110 = Dubois 1908: no. 390.
718. *XCVII*	Roma	Bruzza 1870: no. 111 = Dubois 1908: no. 391.
719. *n(umero) LXXXXII*	Roma	Bruzza 1870: no. 112 = Dubois 1908: no. 392.
720. *n(umero) XCVI*	Roma	Bruzza 1870: no. 113 = Dubois 1908: no. 393.
721. *n(umero) CXVIIIIX*	Roma	Bruzza 1870: no. 114 = Dubois 1908: no. 394.
722. *n(umero) CXX [—]*	Roma	Bruzza 1870: no. 115 = Dubois 1908: no. 395.
723. *CXXXX*	Roma	Bruzza 1870: no. 116 = Dubois 1908: no. 396.
724. *CLXXIII*	Roma	Bruzza 1870: no. 117 = Dubois 1908: no. 397.
725. *n(umero) CXC*	Roma	Bruzza 1870: no. 118 = Dubois 1908: no. 398.
726. *n(umero) CXCVI*	Roma	Bruzza 1870: no. 119 = Dubois 1908: no. 399.
727. *n(umero) ƆƆ | CCXL*	Roma	Bruzza 1870: no. 120 = Dubois 1908: no. 400.
728. *n(umero) CCXIII*	Roma	Bruzza 1870: no. 121 = Dubois 1908: no. 401.
729. *n(umero) CCXX | VXXI*	Roma	Bruzza 1870: no. 122 = Dubois 1908: no. 402.
730. *n(umero) CCXXXI*	Roma	Bruzza 1870: no. 123 = Dubois 1908: no. 403.
731. *CCXLIX*	Roma	Bruzza 1870: no. 124 = Dubois 1908: no. 404.
732. *n(umero) CCCI*	Roma	Bruzza 1870: no. 125 = Dubois 1908: no. 405.

733. a) *l(oco) CCCXLIII Io CCCII*; b) *A | CLXXVII*	Roma	Bruzza 1870: no. 126 = Dubois 1908: no. 406.
734. *n(umero) CCCXLVII*	Roma	Bruzza 1870: no. 127 = Dubois 1908: no. 407.
735. *l(oco) ÐXC Ɍ*	Roma	Bruzza 1870: no. 128 = Dubois 1908: no. 408.
736. *C∞LXX | IIX7*	Roma	Bruzza 1870: no. 129 = Dubois 1908: no. 409.
737. *MIX | ∞CLXX*	Roma	Bruzza 1870: no. 130 = Dubois 1908: no. 410.
738. *ÐƆ ∞∞ | H | CXII*	Roma	Bruzza 1870: no. 131 = Dubois 1908: no. 411.
739. *VIII*	Roma	Bruzza 1870: no. 132 = Dubois 1908: no. 412.
740. *[—]VII | [—]OI*	Roma	Bruzza 1870: no. 133 = Dubois 1908: no. 413.
741. *V*	Roma	Bruzza 1870: no. 134 = Dubois 1908: no. 414.
742. *XII*	Roma	Bruzza 1870: no. 135 = Dubois 1908: no. 415.
743. *TA*	Roma	Bruzza 1870: no. 136 = Dubois 1908: no. 416.
744. *T A*	Roma	Bruzza 1870: no. 137 = Dubois 1908: no. 417.
745. *LÐXI | F*	Ostia	Baccini Leotardi 1989: no. 52 = Pensabene 1994: no. 34.
746. *NΘI*	Ostia	Baccini Leotardi 1989: no. 53.
747. *∞ CCCXXX n(umero) IIII | C*	Ostia	Baccini Leotardi 1989: no. 54.
748. *n(umero) LIII*	Ostia	Baccini Leotardi 1989: no. 55.
749. XXI	Ostia	Baccini Leotardi 1979: no. 44.
750. a) *l(oco) CXLIII*, b) *n(umero) CLXX*, c) *CL*	Ostia	Baccini Leotardi 1979: no. 45.
751. a) *l(oco) ∞ CLXVI | Ɍ*; b) *NP*; c) *PH*	Ostia	Baccini Leotardi 1979: no. 47.
752. *XVX*	Ostia	Baccini Leotardi 1979: no. 48.
753. *ΘYR*	Ostia	Baccini Leotardi 1979: no. 49.
754. *CCCCX[—]*	Ostia	Baccini Leotardi 1979: no. 50.
755. a) *n(umero) CCXLII*; b) *n(umero) CCXLII* vac. *CXXXIII | Ɍ*	Ostia	Baccini Leotardi 1979: no. 52.

756. *CLV* | ₽	Ostia	Baccini Leotardi 1979: no. 56 = Pensabene 1994: 40, fig. 34.
757. *D* ∞	Ostia	Baccini Leotardi 1979: no. 57.
758. *CXXV[—]*	Ostia	Baccini Leotardi 1979: no. 58 = Pensabene 1994: 86, no. 29.
759. *CCCX[—]*	Ostia	Baccini Leotardi 1979: no. 59 = Pensabene 1994: 40, fig. 33.
760. *D* | *N*	Ostia	Baccini Leotardi 1979: no. 60 = Pensabene 1994: 41, fig. 35.
761. *LXXIIII* ₽	Ostia	Baccini Leotardi 1979: no. 61.
762. *l(oco)* ∞ *CXL [—]*	Ostia	Baccini Leotardi 1979: no. 62.
763. *Fl(avii) Claudiani*	Styra	Vanhove 1989.
764. ∞ *CCCXXXIII n(umero) III*	Ostia	Baccini Leotardi 1989: no. 54.
765. *Cae(saris) n(ostri)* | *XIX* | *C M? P O L*	Ostia	Pellegrino 1990: 219 fig. 6.7 = Pensabene 1994: 131, no. 67.
766. a) *Cae(saris) n(ostri) V I* | *C. Apol(-)*; b) *DXXXIIX* | *LXIV*₽; c) *CLXXXI*	Ostia	Pellegrino 1990: 219 fig.8 = Pensabene 1994: 131, no. 68.
767. *I X CL XXXV*	Ostia	Pensabene 1994: 131, no. 69.
768. a) *l(oco)* ∞ *CCCXR* | *I*; b) *X C I I* | *X C A*	Ostia	Pensabene 1994: 131, no. 70.
769. *XPTP*	Ostia	Pellegrino 1990: 219, fig.9= Pensabene 1994: 131, no. 71.
770. a) *Dom[*; b) *CXXIV*	Ostia	Pellegrino 1990: 219 fig.9 = Pensabene 1994: 132, no. 72.
771. *LXI*	Ostia	Pensabene 1994: 132, no. 73.
772. *X*	Ostia	Pensabene 1994: 132, no. 74.
773. *n(umero) XXXIX*	Ostia	Pensabene 1994: 132, no. 75.

<table>
<tr><td>774. a) CLXIX; b) DXXI; c) XLIIX | H</td><td>Ostia</td><td>Pensabene 1994: 132, no. 76.</td></tr>
<tr><td>775. HP</td><td>Ostia</td><td>Pensabene 1994: 132, no. 77.</td></tr>
<tr><td>776. CCXII</td><td>Ostia</td><td>Pensabene 1994: 132, no. 78.</td></tr>
<tr><td>777. DNGF</td><td>Ostia</td><td>Lugli & Filibeck 1935: 80 = Pensabene 1994: 188, no. 180.</td></tr>
<tr><td>778. a) Fl(avii) St(i)l(i)c(onis); b) DXIX | D.LIANI</td><td>Ostia</td><td>Pensabene 1994: 194, no. 187 = CIL XIV 165a = Thylander 1952: 365.</td></tr>
<tr><td>779. CCCXLR</td><td>Ostia</td><td>Pensabene 1994: 198, no. 7.</td></tr>
<tr><td>780. VE</td><td>Ostia</td><td>Pensabene 1994: 198 f., no. 8.</td></tr>
<tr><td>781. CE</td><td>Rome</td><td>Pensabene 1994: 214, no. 27.</td></tr>
<tr><td>782. EP XLIIX</td><td>Rome</td><td>Pensabene 1994: 214, no. 29.</td></tr>
<tr><td>783. CE</td><td>Rome</td><td>Pensabene 1994: 215, no. 30.</td></tr>
<tr><td>784. a) EP XCVI; b) n(umero) XXIIX</td><td>Rome</td><td>Pensabene 1994: 215, no. 31.</td></tr>
<tr><td>785. a) DL'XXXIV; b) VE</td><td>Rome</td><td>Pensabene 1994: 216, no. 33.</td></tr>
<tr><td>786. VE</td><td>Rome</td><td>Pensabene 1994: 216, no. 35.</td></tr>
</table>

Giallo Antico: Simitthus

<table>
<tr><td>787. a) M(arco) Crasso c[o(n)]s(ule) | ex r(atione) Lae; b) n(umero) CCCXCIIX</td><td>Roma; AD 64</td><td>Bruzza 1870: no. 220 = Dubois 1908: no. 67.</td></tr>
<tr><td>788. Sura III et Senici(one) II co(n)s(ulibus) | ex rat(ione) Felicis Aug(usti) ser(vi) | d(e) n(umero) DCXII XXX | (officina) Tiluris</td><td>Chemtou, 107</td><td>CIL VIII 14560 = Dubois 1908: no. 68 = Kraus 1993: 56f.</td></tr>
<tr><td>789. a) Orfito et Priscino f co(n)s(ulibus) | ex rat(ione) Callist(i) | d(e) n(umero) LXI (officina) Cer(ti); b) Orfito et Priscino f co(n)s(ulibus) | ex rat(ione) Callist(i) | d(e) n(umero) L d(e) n(umero) L</td><td>Chemtou, 110</td><td>CIL VIII 25637 = Dubois 1908: no. 73 = Kraus 1993: 55.</td></tr>
</table>

790.	*Orfito et Priscino co(n)s(ulibus) | ex rat(ione) Callist(i) | n(umero) DX (officina) Cer(ti)*	Chemtou, 110	*AE* 1994: 1848 = Kraus 1993: Neu 1.
791.	*Orfito et Priscino co(n)s(ulibus) | ex rat(ione) Callist(i) | [n(umero)] LIII (officina) Cer(ti)*	Chemtou, 110	*CIL* VIII 14561 = Dubois 1908: no. 71
792.	a) *Orfito et Priscino co(n)s(ulibus) | ex rat(ione) Callist|ti*(!) *n(umero) XCIX (officina) Cer(ti)*	Chemtou, 110	*CIL* VIII 14562 = Dubois 1908: no. 72.
793.	*Orfito et Priscino f(?) co(n)s(ulibus) | ex rat(ione) Callist(i) | n(umero) CII (officina) Cer(ti)*	Chemtou, 110	*CIL* VIII 14563 = Dubois 1908: no. 70.
794.	*Pisone et Bolano co(n)s(ulibus) | ex rat(ione) Felicis Aug(usti) ser(vi) | d(e) n(umero) XXX of(ficina) no(va) Cael(?)*	Chemtou, 111	*AE* 1994: 1874 = Kraus 1993: Neu 27.
795.	*Hadriani Aug(usti) co(n)s(ule) n(umero) NCLXXII*	Roma, 118	Bruzza 1870: no. 221 = Dubois 1908: no. 78.
796.	a) *Domitiani Aug(usti) | ex rat(ione) | Felicis ser(vi)*; b) *Augurin(o) co(n)s(ule) Ꞧ | L ĐCCLXVII | FCL | DINIR*; c) *NN CCXXCIII*	Ostia, 132	Baccini Leotardi 1979: no. 76.
797.	a) *Domitiani Aug(usto) | ex rat(ione) | Felicis ser(vi)*; b) *Augurin(o) co(n)s(ule) | L ĐCCLXXIIX | Ꞧ | DINTR FCL*	Ostia, 132	Baccini Leotardi 1979: no. 77.
798.	a) *Augurin(o) co(n)s(ule) | L ĐXXIIX | DINTR [F]CL | Ꞧ*; b) *NTLVIII*	Ostia, 132	Baccini Leotardi 1979: no. 78 = Pensabene 1994: 107, no. 44.
799.	*Hadriani Aug(usti) d(omini) | n(umero) C[—] agr | Hibero co(n)s(ule) | SA*	Chemtou, 133	*CIL* VIII 14564 = Dubois 1908: no. 79.
800.	*(ex r)at(ione) Pluriani Aug(usti) l(iberti) | Iun (?) et Serviano III (consulibus) n(umero)*	Chemtou, 134	*AE* 1994: 1855 = Kraus 1993: Neu 8.
801.	*Hadriani Aug(usti) | N(?) Iun(?) et Serviano III*	Chemtou, 134	*AE* 1994: 1856 = Kraus 1993: Neu 9.
802.	*Hadriani Aug(usti) d(omini) P | n(umero)* vac. *Cal | Serviano co(n)s(ule) | SA*	Chemtou, 134	*CIL* VIII 14565= Dubois 1908: no. 81.
803.	*Hadriani Aug(usti) | CV | (—et) Serviano III (consulibus)*	Chemtou, 134	*AE* 1994: 1857 = Kraus 1993: Neu 10.
804.	*Hadriani Aug(usti) | Aelio Caesare II co(n)s(ule) | sub cura Agathae lib(erti Augusti)*	Chemtou, 137	*AE* 1994: 1868 = Kraus 1993: Neu 21.
805.	*Hadriani Aug(usti) | loc(o) XXXV | Camerino et Nig[ro co(n)s(ulibus) | su(b) cura Iul(ii) Galli[—].*	Chemtou, 138	*CIL* VIII 14566 = Dubois 1908: no. 82.

<table>
<tr><td>806. Hadriani Aug(usti) d(omini) loc(o) | Camerino et Nigro co(n)s(ulibus) | sub cura Agathae lib(erti)</td><td>Chemtou, 138</td><td>AE 1994: 1869 = Kraus 1993: Neu 22.</td></tr>
<tr><td>807. Hadriani Aug(usti) d(omini) | loc(o) | Camerino et Nigro co(n)s(ulibus) | sub cura Agathae lib(erti)</td><td>Chemtou, 138</td><td>CIL VIII 25634 = Dubois 1908: no. 84.</td></tr>
<tr><td>808. Hadrian(i) Aug(usti) d(omini) loc(o) | Camerino et Nigro co(n)s(ulibus) | sub cura (—) Galli</td><td>Chemtou, 138</td><td>AE 1994: 1866 = Kraus 1993: Neu 19.</td></tr>
<tr><td>809. [Antonini A]u(gusti) Pii d(omini) | [Camerino et] Nigro co(n)s(ulibus) | [sub c]ura Galli</td><td>Chemtou, 138</td><td>AE 1994: 1870 = Kraus 1993: Neu 23.</td></tr>
<tr><td>810. Hadriani Aug(usti) d(omini) | IND[— lo]c(o) EGN | Camer(ino et) Nigro co(n)s(ulibus) | [—]</td><td>Chemtou, 138</td><td>Dubois 1908: no. 83.</td></tr>
<tr><td>811. Hadriani Aug(usti) d(omini) | loc(o) | Camerino et Nigro co(n)s(ulibus) | sub cura Agathae lib(erti)</td><td>Chemtou, 138</td><td>Dubois 1908: no. 84.</td></tr>
<tr><td>812. [Antonini Augusti] Pii d(omini) | AL | [Camerino et] Nigro co(n)s(ulibus) | [sub cura] Agathae li(berti Augusti)</td><td>Chemtou, 138</td><td>AE 1994: 1871 = Kraus 1993: Neu 24.</td></tr>
<tr><td>813. Imp(eratoris) Antonini Aug [usti—] | n(umero) DCCLI of[f(icina). —] | Imperatori Antonino et [—] | sub cura Ag[athae—]</td><td>Chemtou, 139 or 140/145</td><td>CIL VIII 25636 = Dubois 1908: no. 96 = Kraus 1993: 55.</td></tr>
<tr><td>814. Imp (eratoris) Antonini Aug[(usti) pii d] | n(umero) vac. of(ficina) Cer[ti] | Stloga et Severo co(n)s(ulibus) | sub cura Agatha[e][lib] | ΛXXƆ|</td><td>Chemtou, 141</td><td>CIL VIII 14571 = Dubois 1908: no. 89.</td></tr>
<tr><td>815. Imp (eratoris) Antonini Aug (usti) p[ii d] | n(umero) vac. of(ficina) Cert[i] | Stloga et Severo co(n)s(ulibus) | sub cura Agathae | LIR</td><td>Chemtou, 141</td><td>CIL VIII 14572 = Dubois 1908: no. 90.</td></tr>
<tr><td>816. Imp (eratoris) Antonini Aug(usti) pii d(omini) | n(umero) vac. of(ficina) Cerii | Stloga et Severo co(n)s(ulibus) | su(b) cura Agathae[—]</td><td>Chemtou, 141</td><td>CIL VIII 14573 = Dubois 1908: no. 91.</td></tr>
<tr><td>817. Imp (eratoris) Antonini Aug(usti) pii d(omini) | n(umero) vac. of(ficina) Cer-(ti) | Stloga et Severo co(n)s(ulibus) | sub cura Agathae l</td><td>Chemtou, 141</td><td>CIL VIII 14574 = Dubois 1908: no. 92.</td></tr>
<tr><td>818. Imp (eratoris) Antonini Aug(usti) || d(omini) | n(umero) vac. of(ficina) Cer(ti) | Stloga et Severo co(n)s(ulibus) | sub cura Agath[ae]</td><td>Chemtou, 141</td><td>CIL VIII 14575 = Dubois 1908: no. 93.</td></tr>
<tr><td>819. Imp (eratoris) Antonini Aug (usti) pi[i d] | n(umero) vac. of(ficina) Cert[i] | Stloga et Severo co(n)s(ulibus) | sub cura Agathae l</td><td>Chemtou, 141</td><td>CIL VIII 14576 = Dubois 1908: no. 94.</td></tr>
</table>

820.	a) *Imp(eratoris) Antonini Aug(usti) P(ii) d(omini) | [n(umero)] LXIV of (ficina) al | Stloga et Severo co(n)s(ulibus) | [s]ub cura Agatiae (!) lib(ertus) | b(racchio?) s(ecundo?)* b) *RO*	Chemtou, 141	*CIL* VIII 14577 = Dubois 1908: no. 95.
821.	*Imp(eratoris) Antonini Aug(usti) Pii | n(umero) DCVI of(ficina) Regia | Orfito et Prisco co(n)s(ulibus) | FPC*	Chemtou, 149	*CIL* VIII 14578 = Dubois 1908: no. 98.
822.	*[Imp(eratoris) Antonini Aug(usti)] Pii d(omini) | [n(umero)—of(ficina) A]gripp(ae) | [Gallicano et Vete]re co(n)s(ulibus)*	Chemtou, 150	*CIL* VIII 14582 = Dubois 1908: no. 102.
823.	*Imp(eratoris) Antonini Aug(usti) d(omini) | n(umero) CCCXLIII (officina) Reg(ia) | Gallicano et Vet(ere) | FPC*	Chemtou, 150	*CIL* VIII 14579 = Dubois 1908: no. 99.
824.	*Imp(eratoris) Antonini Aug(usti) Pii d(omini) | n(umero) D[—] of(ficina) Agrippae | Gallicano et Vetere co(n)s(ulibus) | FPC*	Chemtou, 150	*CIL* VIII 14580 = Dubois 1908: no. 100.
825.	*Imp(eratories) Antonini Aug(usti) Pii d(omini) | n(umero) DXVII of(ficina) Agrippae | Gallicano et Vetere co(n)s(ulibus) | FPC*	Chemtou, 150	*CIL* VIII 14581 = Dubois 1908: no. 101.
826.	*Imp(eratoris) Antonini Aug(usti) Pii d(omini) | n(umero) CXXCIX of(ficina) Regia | Condiano et Maximo | FPC*	Chemtou, 151	*CIL* VIII 14583 = Dubois 1908: no. 103.
827.	*(I)mp(eratoris) Antonini Aug(usti) Pii d(omini) | n(umero) of(ficina) | (C)ommodo et Lateran(o consulibus) BAB*	Chemtou, 154	*AE* 1994: 1873 = Kraus 1993: Neu 26.
828.	*Imp(eratoris) Antonini Aug(usti) Pii | n(umero)* vac. *of(ficina) | Silvano et A(ugurino consulibus) | V*	Chemtou, 156	*AE* 1994: 1867 = Kraus 1993: Neu 20.
829.	*(I)m(peratoris) Antonini Aug(usti) Pi(i) d(omini) | n(umero) [—] off(icina) Iun(onis) | Barbaro et Regulo (consulibus) | P C*	Chemtou, 157	*CIL* VIII 14584 = Dubois 1908: no. 104 = Khanoussi 1996: 1003 f.
830.	*[Impe]ratorum Caesarum | [Antoni]ni et Veria Augustorum d(ominorum) | [—]n(umero) IXX of(ficina) Aur(elii?) [Ant]onino III et Vero II co(n)s(ulibus) | PC | b(racchio) t(ertio)*	Roma, 161	Bruzza 1870: no. 222 = Dubois 1908: no. 108.
831.	*Imp(eratoris) Antonini A[ug(usti)] | n(umero) XXI | Antonino III [et Vero] II | [——]*	Roma, 161	Bruzza 1870: no. 223 = Dubois 1908: no. 107.
832.	*Imp(eratoris) Antonini Aug(usti) Pii n(umero) LX of(ficina) [—] | Aurelio Vero C[ae]s(are) III et Commodo [II] | FPC*	Chemtou, 161	*CIL* VIII 14585 = Dubois 1908: no. 106.

833.	*Imp(eratore) L(ucio) Aurelio Vero III et Q[u]adrato co(n)[s(ulibus)] | n(umero) XIIX of(ficina) nova Augustea | TAVT*	Chemtou, 167	*CIL* VIII 14587 = Dubois 1908: no. 109.
834.	*[Orfit]o et Maximo co(n)[s(ulibus)] | [ex no]vis lapicaedinis | Aurelianis | [—]ASU[—]MO Caesare | [sub] cura Iuliani proc(uratoris) Aug(usti)*	Roma, 172	Bruzza 1870: no. 224 = Dubois 1908: no. 110.
835.	*n(umero) CCCV of(ficina) Genii Montis | Imp(eratore) Commodo Aug(usto) IIII et Victorino [I]I co(n)s(ulibus) | caesura Maximi proc(uratoris)*	Chemtou, 199	*CIL* VIII 14588 = Dubois 1908: no. 111.
836.	*n(umero) XII of(ficina) | Anullino II et Frontone co(n)s(ulibus) | caesura Athenodori proc(uratoris)*	Chemtou, 199	*AE* 1994: 1860 = Kraus 1993: Neu 13.
837.	*n(umero) LIII of(ficina) Nova Aurel-[iana] | [A]nullino II et Frontone co(n)s-(ulibus) | [ca]esura Athenodori proc(u-ratoris)*	Chemtou, 199	*CIL* VIII 14589 = Dubois 1908: no. 112.
838.	*n(umero) CCXXX.. of(ficina) Cael(i) | Luciano et Fabi(a)no co(n)s(ulibus) | caesura Athenodo(ri) pro(curatoris)*	Chemtou, 201	*AE* 1994: 1861 = Kraus 1993: Neu 14.
839.	a) *ex rat(ione) | of(ficina) Age(ntii ?) Calist(i) lib(erti) Aug(usti)*; b) *n(umero) X*	Chemtou	*AE* 1994: 1852 = Kraus 1993: Neu 5.
840.	a) *ex rat(ione) | of(ficina) Age(ntii ?) Calist(i) lib(erti) Aug(usti)*; b) *n(umero) CCCLXXIX*	Chemtou	*AE* 1994: 1853 = Kraus 1993: Neu 6.
841.	a) *ex rat(ione) | of(ficina) Aga(thae) Callist(i) li(berti)*; b) *CCCCXIII*	Chemtou	*CIL* VIII 25693 = Dubois 1908: no. 74 = Kraus 1993: 56.
842.	a) *ex rat(ione) [of(ficina)] | Age(ntii ?) Callis(i) lib(erti) Aug(usti)*; b) *n(umero) CDXLIV*	Chemtou	*CIL* VIII 25639 = Dubois 1908: no. 75 = Kraus 1993: 56.
843.	*[ex ratione?] Callist(i) d(e) (?) n(umero) CX de(?) co|l(umna) p(edum) XXX*	?	*AE* 1994: 1877a, b = Kraus 1993: Neu 30.
844.	a) *ex rat(ione) | Puteolani Ca(esaris?) | C*; b) *N∞XXCVI*	Ostia	Baccini Leotardi 1979: no. 79.
845.	a) *l(oco?) ∞ĐCCL | DINT(R) FCL [R?]*; b) *LDF (?) n(umero) CCCXLIIX*; c) *Cai | Puteolani | b*	Ostia	Baccini Leotardi 1979: no. 80.
846.	*Puteolani Cae(saris?) | A[—]*	Chemtou	*CIL* VIII 14593 = Dubois 1908: no. 117.
847.	∞ *Đ III | ĐINTF*	Ostia	Baccini Leotardi 1979: no. 81.
848.	a) *[ex] rat(ione) Felicis Au[g(usti)] s[e]r(vi)*; b) *[—]lv*	Ostia	Baccini Leotardi 1979: no. 82= Pensabene 1994: 107, no. 45.

No.	Text	Findspot	Reference
849.	*ex rat(ione) | Felicis [—]i Aug(ustorum) | nn(ostrorum) DCCXXIIII*	Chemtou	*CIL* VIII 25641 = Dubois 1908: no. 69.
850.	*ex rat(ione) | Puteon(is?) l(iberti) A(ugusti)*	Chemtou	*AE* 1994: 1858 = Kraus 1993: Neu 11.
851.	*ex[ratione] | [Sa]lutaris [—]*	Chemtou	*AE* 1994: 1859 = Kraus 1993: Neu 12.
852.	*ex rat(ione) | of(ficina) Aga(thae) Callist(i) li(berti?)*	Henchir Frouri	Dubois 1908: no. 74.
853.	*Hadriani D [—] | n(umero)* vac. *[—]*	Chemtou	*CIL* VIII 14567 = Dubois 1908: no. 85.
854.	*TA | Hadriani Aug(usti) d(omini) | n(umero) CXXCIII Fort[.]VQVAIVOI*	Chemtou	*CIL* VIII 14569 = Dubois 1908: no. 87.
855.	*Hadriani Aug(usti) d(omini) n(umero) CCLXIII | N[..]IR O NE cos | EP*	Chemtou	*CIL* VIII 14568 = Dubois 1908: no. 86.
856.	*[—]es sub c[ura—] | [—p]roc(uratoris) f[—]*	Leptis Magna	*IRT* 794g.
857.	*[—]ER[—]] [—]lib(erti) pr[oc(uratoris)—]*	Leptis Magna	*IRT* 794j.
858.	*[—]nodori proc(uratoris)*	Chemtou	*AE* 1994: 1862 = Kraus 1993: Neu 15.
859.	*[—An]tonin[i—] | XXXI |[caesura No]vat[i?—]*	Chemtou	*AE* 1994: 1863 = Kraus 1993: Neu 16.
860.	*[Imperat]orum Caesa[rum] | [—]et Veri Aug(u)sti | [numero] LIV of[ficina | [Laelia]no et Pastor(e)*	Chemtou	*AE* 1994: 1865 = Kraus 1993: Neu 18.
861.	*Antonini Au[gusti Pii] | [—]reta et Vero c(onsulibus)*	Chemtou	*AE* 1994: 1872 = Kraus 1993: Neu 25.
862.	*Antonino III co(n)s(ule) | DCC*	Chemtou	*CIL* VIII 14570 = Dubois 1908: no. 88.
863.	a) *[—]AILIS [caes]ura Nova(ti?) | [—]∞ CXXX loc(o) XII | Tellur(is)*; b) *[Imp(eratoris) A]ntonini Aug(usti) | [—]atianam fecit*	Chemtou	*CIL* VIII 14586 = Dubois 1908: no. 105.
864.	*[—]et[—] | caesur[a] | praef[—]*	Chemtou	*AE* 1994: 1864 = Kraus 1993: Neu 17.
865.	*caes(ura?) N(ovati?) | n(umero) MCXXΛIIII | Tellur(is?) d*	Chemtou	*AE* 1994: 1875 = Kraus 1993: Neu 28.
866.	*caesura Hymenaei*	Chemtou	*AE* 1994: 1854 = Kraus 1993: Neu 7.
867.	a) *n(umero) ~~D~~CCXX*; b) *Corinth Caes(aris servi) | C*	Chemtou	*CIL* VIII 14594 = Dubois 1908: no. 119a.
868.	a) [—]; b) *Corinth Caes(aris servi) | C*	Chemtou	*CIL* VIII 14595 = Dubois 1908: no. 119d.
869.	a) *n(umero) CCXXI* ; b) *Corinth Caes(aris servi)*	Chemtou	*CIL* VIII 14596 = Dubois 1908: no. 119b.

870. a) *n(umero) ƉCCXXI;* b) *Corinth Caes(aris servi) | C*	Chemtou	*CIL* VIII 14597 = Dubois 1908: no. 119c.
871. (christogram) | *off(icina) inven|ta a Dio|timo* (palm twig) | *agen(te) in r(ebus)*	Chemtou	*CIL* VIII 14600 = Dubois 1908: no. 113.
872. *lat(omia) IX Hiron(ymi?)*	Chemtou	*AE* 1994: 1849 = Kraus 1993: Neu 2.
873. *n(umero) DCCXV*	Chemtou	*AE* 1994: 1850 = Kraus 1993: Neu 3.
874. *X Abas[—]*	Chemtou	*AE* 1994: 1851 = Kraus 1993: Neu 4.
875. *Abas(canti) Cae[s(aris servi)]*	Chemtou	Dubois 1908: no. 115.
876. *P[o]R Cae | n(umero) (mille) (mille) XLIII*	Roma	Bruzza 1870: no. 225 = Dubois 1908: no. 120.
877. *n(umero) CƉXXCII | Glaeb*	Roma	Bruzza 1870: no. 226 = Dubois 1908: no. 121.
878. *Cae*	Roma	Bruzza 1870: no. 227 = Dubois 1908: no. 122.
879. *Sap Cae*	Roma	Bruzza 1870: no. 228 = Dubois 1908: no. 80.
880. *Q Canusi Praenestini | N II*	Roma	Bruzza 1870: no. 236 = Dubois 1908: no. 118.
881. *Corni | Cas | n(umero) DCC XX*	Chemtou	*CIL* VIII 14591 = Dubois 1908: no. 114.
882. *Eutychi*	Chemtou	*CIL* VIII 14592 = Dubois 1908: no. 116.
883. *(mille) (mille) (mille)CCXCIIX*	Roma	Bruzza 1870: no. 229 = Dubois 1908: no. 123, 1.
884. *n(umero) (mille)CLXXV*	Roma	Bruzza 1870: no. 230 = Dubois 1908: no. 123, 2.
885. *DCCCXII*	Roma	Bruzza 1870: no. 231 = Dubois 1908: no. 123, 3.
886. *DCCCXXII*	Roma	Bruzza 1870: no. 232 = Dubois 1908: no. 123, 4.
887. *n(umero)CCCXCIIX*	Roma	Bruzza 1870: no. 233 = Dubois 1908: no. 123, 5.
888. XCIV	Ostia	Baccini Leotardi 1989, no. 74 = Pensabene 1994: 107, no. 42.
889. |||	Ostia	Baccini Leotardi 1989, no. 75.
890. a) *CΣA*, b) *CΣA*, c) *CAE*, d) ⋈ | *n(umero) XXIX*, e) *CCIII*, f) *DCCLIII*	Chemtou	*CIL* VIII 14598.
891. *n(umero) CƉXCII*	Chemtou	*CIL* VIII 14599a = Dubois 1908: no. 123,14.

892. *n(umero) ĐCIII*	Chemtou	*CIL* VIII 14599b = Dubois 1908: no. 123, 10.
893. *n(umero) ĐCCXV*	Chemtou	*CIL* VIII 14599c = Dubois 1908: no. 123, 13.
894. *[n(umero)] ĐCCXVIII*	Chemtou	*CIL* VIII 14599d = Dubois 1908: no. 123, 15.
895. *n(umero) ĐCCXIX*	Chemtou	*CIL* VIII 14599e = Dubois 1908: no. 123, 8.
896. *n(umero) ĐCCXX*	Chemtou	*CIL* VIII 14599 f = Dubois 1908: no. 123, 11.
897. *n(umero) ĐCCLXVI*	Chemtou	*CIL* VIII 14598g = Dubois 1908: no. 123, 12.
898. *n(umero) DCCCXXX*	Chemtou	*CIL* VIII 14598h = Dubois 1908: no. 123, 9.
899. *[—]loc(o) II p[—]*	Leptis Magna	*IRT* 794 f.
900. *[—]ER[—] | [—]VA[—]*	Leptis Magna	*IRT* 794 h.
901. *[—]NEQVADOC[—] | [—]XXXVI[—]*	Leptis Magna	*IRT* 794 k.
902. *[—]ITIANA[—]*	Leptis Magna	*IRT* 794 l.
903. |Y|	Rome	Bruzza 1870: no. 234 = Dubois 1908: no. 123, 6.
904. *Δ*	Rome	Bruzza 1870: no. 235 = Dubois 1908: no. 123, 7.
905. *n(umero) XII*	Rome	Dubois 1908: no. 123, 16.
906. *XXXIIX*	Rome	Dubois 1908: no. 123, 17.
907. *n(umero) CXXI*	Rome	Dubois 1908: no. 123, 18.
908. *n(umero) IICCCXVIII*	Rome	Dubois 1908: no. 123, 19.
909. *n(umero) CCCXXIII*	Rome	Dubois 1908: no. 123, 20.
910. *n(umero) (mille) DDLLXIII*	Rome	Dubois 1908: no. 123, 21.
911. *n(umero) (mille) (mille) CCXIIV*	Rome	Dubois 1908: no. 123, 22.
912. *n(umero) (mille) (mille) CCLXXII*	Rome	Dubois 1908: no. 123, 23.
913. *n(umero) (mille) [—]*	Rome	Dubois 1908: no. 123, 24.
914. *n(umero) X*	Rome	Dubois 1908: no. 123, 25.
915. *CCIII*	Rome	Dubois 1908: no. 123, 26.
916. *DCCLIII*	Rome	Dubois 1908: no. 123, 27.
917. ⋈	Rome	Dubois 1908: no. 123, 28.
918. *CZA*	Rome	Dubois 1908: no. 123, 29.
919. *CAE*	Rome	Dubois 1908: no. 123, 30.

920. ⊕	Rome	Dubois 1908: no. 123, 31.
921. a) *Hil(?) Cae(saris)*, b) *n(umero) LVI* | *A*	Rome	Dubois 1908: no. 123, 32.
922. a) *Hil(?) Cae(saris)*, b) *n(umero) LVII* | *A*	Rome	Dubois 1908: no. 123, 33.
923. [—]*ARA*	Rome	Dubois 1908: no. 123, 34.
924. *HSE*	Chemtou	*CIL* VIII 14679 = Dubois 1908: no. 124.
925. *Hadrianus Augustus*	Antium	Dubois 1908: no. 126; Two plombs found in 1697 in a floor pavement of yellow marble. *CIL* XV 7956.
926. *L.Coc(cei?)*	Antium	Dubois 1908: no. 127; *CIL* XV 7926.
927. *Hadrianus Augustus*	Tivoli	Dubois 1908: no. 128 = *CIL* XV 7927, 3. Circular plombe found in piece of yellow marble in Hadrians villa, Tivoli.
928. *[—]Αυ[—]Σεβ(αστοῦ)*	Rome	Dubois 1908: no. 129 = *CIL* XV 7937a. Circular plombe, inserted into a piece of yellow marble.
929. Anepigraphic token showing picture of Hadrian	?	Dubois 1908: no. 130.

Granodiorite: Mons Claudianus

930. *n(umero) II*	Mons Claudianus	Peacock & Maxfield 1997: 217, quarry 1, no. 1.
931. *IIII*	Mons Claudianus	Peacock & Maxfield 1997: 217, quarry 2, no. 1.
932. *V I*	Mons Claudianus	Peacock & Maxfield 1997: 217, quarry 2, no. 2.
933. *CEP n(umero) II*	Mons Claudianus	Peacock & Maxfield 1997: 217, quarry 6, no. 1.
934. *CEP n(umero) III*	Mons Claudianus	Peacock & Maxfield 1997: 217, quarry 6, no. 2.
935. APOL | EPIKωMOC	Mons Claudianus	Peacock & Maxfield 1997: 217, quarry 7, no. 1.
936. a. *AP9P*; b. *Π̊ I B L*	Mons Claudianus	Peacock & Maxfield 1997: 217, quarry 8/9, no. 1.
937. a. *Π̊ I E*; b. *RACLP*; c. *LVI*	Mons Claudianus	Peacock & Maxfield 1997: 217, quarry 8/9, no. 2.

938.	a. *RACLP;* b. *LII* ? (same face)	Mons Claudianus	Peacock & Maxfield 1997: 217, quarry 8/9, no. 3.
939.	a. *RACLP*; b. *LIX* (same face)	Mons Claudianus	Peacock & Maxfield 1997: 217, quarry 8/9, no. 4.
940.	a. *XL*; b. *LVIII*	Mons Claudianus	Peacock & Maxfield 1997: 217, quarry 8/9, no. 5.
941.	*n(umero) III*	Mons Claudianus	Peacock & Maxfield 1997: 217, quarry 8/9, no. 6.
942.	a. *RACLP*; b. *XX//* (same face); c. *XLVI* (another face)?	Mons Claudianus	Peacock & Maxfield 1997: 217, quarry 8/9, no. 7.
943.	*XLV*	Mons Claudianus	Peacock & Maxfield 1997: 217, quarry 8/9, no. 8.
944.	*VIII*	Mons Claudianus	Peacock & Maxfield 1997: 217, quarry 8/9, no. 9.
945.	a. *XLIX*; b. *XXII*; c. *RACLP* (all on same face)	Mons Claudianus	Peacock & Maxfield 1997: 217, quarry 8/9, no. 10.
946.	*XXX*	Mons Claudianus	Peacock & Maxfield 1997: 217, quarry 8/9, no. 11.
947.	*XXV*	Mons Claudianus	Peacock & Maxfield 1997: 217, quarry 8/9, no. 12.
948.	*XXIX*	Mons Claudianus	Peacock & Maxfield 1997: 217, quarry 8/9, no. 13.
949.	*XLIV*	Mons Claudianus	Peacock & Maxfield 1997: 217, quarry 8/9, no. 14.
950.	*XLVIII*	Mons Claudianus	Peacock & Maxfield 1997: 217, quarry 8/9, no. 15.
951.	*XXIII*	Mons Claudianus	Peacock & Maxfield 1997: 217, quarry 8/9, no. 16.
952.	a. *XLI*; b. *XXXXI*	Mons Claudianus	Peacock & Maxfield 1997: 217, quarry 8/9, no. 17.
953.	a. *XVIII*; b. *RACLP*; c. *XXXII*	Mons Claudianus	Peacock & Maxfield 1997: 217, quarry 8/9, no. 18.
954.	*XIIII*	Mons Claudianus	Peacock & Maxfield 1997: 217, quarry 8/9, no. 19.
955.	*XVI*	Mons Claudianus	Peacock & Maxfield 1997: 217, quarry 8/9, no. 20.
956.	*III*	Mons Claudianus	Peacock & Maxfield 1997: 217, quarry 8/9, no. 21.
957.	a. *XXVI;* b. *VIII XXXVI*	Mons Claudianus	Peacock & Maxfield 1997: 217, quarry 8/9, no. 22.
958.	a. *XXVII;* b. *RAQP XXXII;* c. *n(umero) XXXVI*	Mons Claudianus	Peacock & Maxfield 1997: 217, quarry 8/9, no. 23.

959.	a. *RACLP*; b. *XII*	Mons Claudianus	Peacock & Maxfield 1997: 217, quarry 8/9, no. 24.
960.	a. *Π̊ K L B;* b. *RACLP*; c. *XXIX*; d. *XIII*	Mons Claudianus	Peacock & Maxfield 1997: 217, quarry 8/9, no. 25.
961.	*n(umero) I*	Mons Claudianus	Peacock & Maxfield 1997: 217, quarry 10, no. 1.
962.	*n(umero) II* (adjacent block)	Mons Claudianus	Peacock & Maxfield 1997: 217, quarry 10, no. 2.
963.	*III* (adjacent block)	Mons Claudianus	Peacock & Maxfield 1997: 217, quarry 10, no. 3.
964.	*<u>EP</u> | LAT | ?OC*	Mons Claudianus	Peacock & Maxfield 1997: 217, quarry 11, no. 1.
965.	*C D T n(umero?) I*	Mons Claudianus	Peacock & Maxfield 1997: 217, quarry 17, no. 1.
966.	*n(umero) II?*	Mons Claudianus	Peacock & Maxfield 1997: 217, quarry 18, no. 1.
967.	*n(umero) III*	Mons Claudianus	Peacock & Maxfield 1997: 217, quarry 18, no. 2.
968.	*n(umero) VI*	Mons Claudianus	Peacock & Maxfield 1997: 217, quarry 18, no. 3.
969.	*n(umero) I Cb*	Mons Claudianus	Peacock & Maxfield 1997: 217, quarry 18, no. 4.
970.	a. *Cb III*; b. *n(umero) III*	Mons Claudianus	Peacock & Maxfield 1997: 217, quarry 19, no. 1.
971.	*Cb V*	Mons Claudianus	Peacock & Maxfield 1997: 217, quarry 19, no. 2.
972.	*Cb VI*	Mons Claudianus	Peacock & Maxfield 1997: 217, quarry 19, no. 3.
973.	*Cb VII*	Mons Claudianus	Peacock & Maxfield 1997: 217, quarry 19, no. 4.
974.	*Cb VIII*	Mons Claudianus	Peacock & Maxfield 1997: 217, quarry 19, no. 5.
975.	*Cb*	Mons Claudianus	Peacock & Maxfield 1997: 217, quarry 19, no. 6.
976.	*ΔAΔ*	Mons Claudianus	Peacock & Maxfield 1997: 217, quarry 19, no. 7.
977.	*O<u>EP</u>PELLIUS*	Mons Claudianus	Peacock & Maxfield 1997: 217, quarry 19, no. 8.
978.	*II*	Mons Claudianus	Peacock & Maxfield 1997: 217, quarry 20, no. 1.
979.	a. *ꞦACIP* (a–d all on same face of in situ block); b. *P D n(umero?) XIII*; c. *P R D* (in red paint); d. *XL*	Mons Claudianus	Peacock & Maxfield 1997: 218, quarry 22, no. 1.

980.	a. *P R D XIII*; b. *R̶ACLP*; c. *XLI*	Mons Claudianus	Peacock & Maxfield 1997: 218, quarry 22, no. 2.
981.	*P D XV*	Mons Claudianus	Peacock & Maxfield 1997: 218, quarry 22, no. 3.
982.	a. *P D n(umero?) III*; b. *R̶ACLP*	Mons Claudianus	Peacock & Maxfield 1997: 218, quarry 22, no. 4.
983.	*P D XII*	Mons Claudianus	Peacock & Maxfield 1997: 218, quarry 22, no. 5.
984.	*P D n(umero) II*	Mons Claudianus	Peacock & Maxfield 1997: 218, quarry 22, no. 6.
985.	*P D n(umero) V*	Mons Claudianus	Peacock & Maxfield 1997: 218, quarry 22, no. 7.
986.	*P D VI*	Mons Claudianus	Peacock & Maxfield 1997: 218, quarry 22, no. 8.
987.	*P D n(umero) VII*	Mons Claudianus	Peacock & Maxfield 1997: 218, quarry 22, no. 9.
988.	a. *R̶ACLP*; b. *P D XVII*	Mons Claudianus	Peacock & Maxfield 1997: 218, quarry 22, no. 10.
989.	a. *RA | CLP*; b. *P D XVI*; c. *XXXIV*	Mons Claudianus	Peacock & Maxfield 1997: 218, quarry 22, no. 11.
990.	a. *XXXIII;* b. *Π̊ I E*	Mons Claudianus	Peacock & Maxfield 1997: 218, quarry 22, no. 12.
991.	*P D C P n(umero) II*	Mons Claudianus	Peacock & Maxfield 1997: 218, quarry 22, no. 13.
992.	a. *RACLP*; b. *XLIII* (same face)	Mons Claudianus	Peacock & Maxfield 1997: 218, quarry 22, no. 14.
993.	*ΛΑΤΟΜΙΑC ΜΥΡΙCΜΟΥ | ΟΙΛΟΤΡΑΙΝΟC*	Mons Claudianus	Peacock & Maxfield 1997: 218, quarry 22, no. 15.
994.	*Cb n(umero) VI*	Mons Claudianus	Peacock & Maxfield 1997: 218, quarry 23, no. 1.
995.	*n(umero) IV*	Mons Claudianus	Peacock & Maxfield 1997: 218, quarry 24, no. 1.
996.	a. *P D n(umero) VI*; b. *P R D VI* (same face, red paint); c. *A CI P* (same face); d. *Π̊ OC* (same face).	Mons Claudianus	Peacock & Maxfield 1997: 218, quarry 24, no. 2.
997.	*XVIII*	Mons Claudianus	Peacock & Maxfield 1997: 218, quarry 24, no. 3.
998.	a. *RACLP*; b. *XXXIII*; c. *III* (all on same face)	Mons Claudianus	Peacock & Maxfield 1997: 218, quarry 25, no. 1.
999.	a. *R̶ACLP*; b. *II* (same face)	Mons Claudianus	Peacock & Maxfield 1997: 218, quarry 25, no. 2.

1000. *n(umero) VIII*	Mons Claudianus	Peacock & Maxfield 1997: 218, quarry 30, no. 1.
1001. *VII*	Mons Claudianus	Peacock & Maxfield 1997: 218, quarry 30, no. 2.
1002. *n(umero) II*	Mons Claudianus	Peacock & Maxfield 1997: 218, quarry 33, no. 1.
1003. *n(umero) V*	Mons Claudianus	Peacock & Maxfield 1997: 218, quarry 33, no. 2.
1004. *T n(umero?) I*	Mons Claudianus	Peacock & Maxfield 1997: 218, quarry 33, no. 3.
1005. *Cb n(umero) I*	Mons Claudianus	Peacock & Maxfield 1997: 218, quarry 36, no. 1.
1006. *EP II*	Mons Claudianus	Peacock & Maxfield 1997: 218, quarry 55, no. 1.
1007. a. *III EP*; b. *DENI*	Mons Claudianus	Peacock & Maxfield 1997: 218, quarry 55 no. 2.
1008. *EP V*	Mons Claudianus	Peacock & Maxfield 1997: 218, quarry 55, no. 3.
1009. a. *EP VI;* b. *VII*	Mons Claudianus	Peacock & Maxfield 1997: 218, quarry 55, no. 4.
1010. a. *EP VII;* b. *VII* (in red paint, prefix illegible)	Mons Claudianus	Peacock & Maxfield 1997: 218, quarry 55, no. 5.
1011. *VIII*	Mons Claudianus	Peacock & Maxfield 1997: 218, quarry 55, no. 6.
1012. *I?*	Mons Claudianus	Peacock & Maxfield 1997: 218, quarry 59, no. 1.
1013. a. *II*; b. *XXXIIX*	Mons Claudianus	Peacock & Maxfield 1997: 218, quarry 59, no. 2.
1014. a. *IV*; b. *XXXVII*	Mons Claudianus	Peacock & Maxfield 1997: 218, quarry 59, no. 3.
1015. a. *V*; b. *R̶A P*; c. *XXXIII*	Mons Claudianus	Peacock & Maxfield 1997: 218, quarry 59, no. 4.
1016. a. *VI*; b. *XXXIV*	Mons Claudianus	Peacock & Maxfield 1997: 218, quarry 59, no. 5.
1017. a. *VII*; b. *RACLI*; c. *XXXV* (same face as b.)	Mons Claudianus	Peacock & Maxfield 1997: 218, quarry 59, no. 6.
1018. a. *RACLP*; b. *XL IX*	Mons Claudianus	Peacock & Maxfield 1997: 218, quarry 59, no. 7.
1019. *X*	Mons Claudianus	Peacock & Maxfield 1997: 218, quarry 59, no. 8.
1020. *XI*	Mons Claudianus	Peacock & Maxfield 1997: 218, quarry 59, no. 9.
1021. *XII*	Mons Claudianus	Peacock & Maxfield 1997: 218, quarry 59, no. 10.
1022. a. *XIII*; b. *P R D* (in red paint on same face as a); c. *XXX*	Mons Claudianus	Peacock & Maxfield 1997: 218, quarry 59, no. 11.

1023. a. *XIV*; b. *RACLP*; c. *XXXI* (same face as b)	Mons Claudianus	Peacock & Maxfield 1997: 218, quarry 59, no. 12.
1024. *XV*	Mons Claudianus	Peacock & Maxfield 1997: 218, quarry 59, no. 13.
1025. a. *XVI*; b. *RACLP*; c. *XXXII* (same face as b)	Mons Claudianus	Peacock & Maxfield 1997: 219, quarry 59, no. 14.
1026. *XVII*	Mons Claudianus	Peacock & Maxfield 1997: 219, quarry 59, no. 15.
1027. *XVIII*	Mons Claudianus	Peacock & Maxfield 1997: 219, quarry 59, no. 16.
1028. *XIX*	Mons Claudianus	Peacock & Maxfield 1997: 219, quarry 59, no. 17.
1029. *XX*	Mons Claudianus	Peacock & Maxfield 1997: 219, quarry 59, no. 18.
1030. *XXI*	Mons Claudianus	Peacock & Maxfield 1997: 219, quarry 59, no. 19.
1031. *XXII*	Mons Claudianus	Peacock & Maxfield 1997: 219, quarry 59, no. 20.
1032. a. *XXIIII*; b. *RAC*	Mons Claudianus	Peacock & Maxfield 1997: 219, quarry 59, no. 21.
1033. *XXIIII* (another stone with the same number)	Mons Claudianus	Peacock & Maxfield 1997: 219, quarry 59, no. 22.
1034. *XXV*	Mons Claudianus	Peacock & Maxfield 1997: 219, quarry 59, no. 23.
1035. *XXV* (another stone with the same number)	Mons Claudianus	Peacock & Maxfield 1997: 219, quarry 59, no. 24.
1036. *XXVI*	Mons Claudianus	Peacock & Maxfield 1997: 219, quarry 59, no. 25.
1037. *XXVII*	Mons Claudianus	Peacock & Maxfield 1997: 219, quarry 59, no. 26.
1038. *XXIIX*	Mons Claudianus	Peacock & Maxfield 1997: 219, quarry 59, no. 27.
1039. *XXIX*	Mons Claudianus	Peacock & Maxfield 1997: 219, quarry 59, no. 28.
1040. a. *I* (east of quarry); b. *ΕΠ* (same face)	Mons Claudianus	Peacock & Maxfield 1997: 219, quarry 59, no. 29.
1041. *III*	Mons Claudianus	Peacock & Maxfield 1997: 219, quarry 60, no. 1.
1042. *IV*	Mons Claudianus	Peacock & Maxfield 1997: 219, quarry 60, no. 2.
1043. *ΦΛΑΔΙΔ*	Mons Claudianus	Peacock & Maxfield 1997: 219, quarry 60, no. 3.
1044. a. *XXIII*; b. *RACLP* (same face)	Mons Claudianus	Peacock & Maxfield 1997: 219, quarry 60, no. 4.

1045. a. *XVI;* b. *~~R~~ACLP* (same face)	Mons Claudianus	Peacock & Maxfield 1997: 219, quarry 60, no. 5.
1046. a. *XVII*; b. *~~R~~ACLP* (same face)	Mons Claudianus	Peacock & Maxfield 1997: 219, quarry 60, no. 6.
1047. a. *XIIX;* b. *RACLP* (same face)	Mons Claudianus	Peacock & Maxfield 1997: 219, quarry 60, no. 7.
1048. *RACLP XIX*	Mons Claudianus	Peacock & Maxfield 1997: 219, quarry 60, no. 8.
1049. *ACL* (in red paint = RACLP, the only legible part of many lines)	Mons Claudianus	Peacock & Maxfield 1997: 219, quarry 60, no. 9.
1050. a. *~~R~~ACIP*; b. *X* (same face)	Mons Claudianus	Peacock & Maxfield 1997: 219, quarry 61, no. 1.
1051. *RACIP*	Mons Claudianus	Peacock & Maxfield 1997: 219, quarry 61, no. 2.
1052. *ΛLIP*	Mons Claudianus	Peacock & Maxfield 1997: 219, quarry 61, no. 3.
1053. a. *RACLP*; b. *VIII*	Mons Claudianus	Peacock & Maxfield 1997: 219, quarry 61, no. 4.
1054. *KL«R*	Mons Claudianus	Peacock & Maxfield 1997: 219, quarry 61, no. 5.
1055. a. *~~R~~ACLP*; b. *VIV*	Mons Claudianus	Peacock & Maxfield 1997: 219, quarry 61, no. 6.
1056. *VII*	Mons Claudianus	Peacock & Maxfield 1997: 219, quarry 61, no. 7.
1057. a. *~~R~~ACIP*; b. *VI*	Mons Claudianus	Peacock & Maxfield 1997: 219, quarry 61, no. 8.
1058. *EP NII NoV*	Mons Claudianus	Peacock & Maxfield 1997: 219, quarry 62, no. 1.
1059. *E N C P ///* III	Mons Claudianus	Peacock & Maxfield 1997: 219, quarry 62, no. 2.
1060. *ΔΙΑΔΟΥ*	Mons Claudianus	Peacock & Maxfield 1997: 219, quarry 63, no. 1.
1061. a. *I*; b. *PRD I* (in red paint, same face)	Mons Claudianus	Peacock & Maxfield 1997: 219, quarry 65, no. 1.
1062. *II*	Mons Claudianus	Peacock & Maxfield 1997: 219, quarry 65, no. 2.
1063. *I ?*	Mons Claudianus	Peacock & Maxfield 1997: 219, quarry 67, no. 1.
1064. *III*	Mons Claudianus	Peacock & Maxfield 1997: 219, quarry 67, no. 2.
1065. *n(umero) VII*	Mons Claudianus	Peacock & Maxfield 1997: 219, quarry 67, no. 3.
1066. *RACLP*	Mons Claudianus	Peacock & Maxfield 1997: 219, quarry 68, no. 1.

1067. *CEPAΠIWNOC* | *EPΓOΔOTOY* | *NIKO-TYXAI*	Mons Claudianus	Peacock & Maxfield 1997: 219, quarry 75, no. 1.
1068. *C EP* (foot of slipway)	Mons Claudianus	Peacock & Maxfield 1997: 219, quarry 75, no. 2.
1069. *C PE n(umero) I* (foot of slipway)	Mons Claudianus	Peacock & Maxfield 1997: 219, quarry 75, no. 3.
1070. *C PE* (on small block in wadi)	Mons Claudianus	Peacock & Maxfield 1997: 219, quarry 79, no. 1.
1071. a. *IEPω* (on quarry face); b. *n(umero) I*	Mons Claudianus	Peacock & Maxfield 1997: 219, quarry 83, no. 1.
1072. a. *Π̊ ΠΔΔ′H′*; b. *LXXVI*; c. *PD XLVI*; d. *RACLP*; e. *R / / I* (all on same face)	Mons Claudianus	Peacock & Maxfield 1997: 219, quarry 84, no. 1.
1073. a. *PD* | *XLVIII*; b. *KACLP*	Mons Claudianus	Peacock & Maxfield 1997: 219, quarry 84, no. 2.
1074. *Π7*	Mons Claudianus	Peacock & Maxfield 1997: 219, quarry 84, no. 3.
1075. *ΠI*	Mons Claudianus	Peacock & Maxfield 1997: 219, quarry 84, no. 4.
1076. a. *RACLP*; b. *LXXIX*	Mons Claudianus	Peacock & Maxfield 1997: 219, quarry 84, no. 5.
1077. a. *PD XLIIII*; b. *LXXIIX*; c. *RAC /*	Mons Claudianus	Peacock & Maxfield 1997: 219, quarry 84, no. 6.
1078. *PD XXXII*	Mons Claudianus	Peacock & Maxfield 1997: 220, quarry 84, no. 7.
1079. a. *Π̊ OC*; b. *LXXIV*	Mons Claudianus	Peacock & Maxfield 1997: 220, quarry 84, no. 8.
1080. a. *PD XXXIIII*; b. *PACLP*; c. *LXXVI*	Mons Claudianus	Peacock & Maxfield 1997: 220, quarry 84, no. 9.
1081. *PACLP*	Mons Claudianus	Peacock & Maxfield 1997: 220, quarry 84, no. 10.
1082. *PD XXIII*	Mons Claudianus	Peacock & Maxfield 1997: 220, quarry 84, no. 11.
1083. *ex n(umero) XXVI*	Mons Claudianus	Peacock & Maxfield 1997: 220, quarry 84, no. 12.
1084. *PRD XXVI*	Mons Claudianus	Peacock & Maxfield 1997: 220, quarry 84, no. 13.
1085. *XIII RAD*	Mons Claudianus	Peacock & Maxfield 1997: 220, quarry 84, no. 14.
1086. *XVII*	Mons Claudianus	Peacock & Maxfield 1997: 220, quarry 84, no. 15.
1087. *PD XL /*	Mons Claudianus	Peacock & Maxfield 1997: 220, quarry 84, no. 16.
1088. a. *PRD*; b. *n(umero) III*; c. *n(umero) II*	Mons Claudianus	Peacock & Maxfield 1997: 220, quarry 84, no. 17.

1089. a. *PRD XVII*; b. *R LP*; c. *LVI*	Mons Claudianus	Peacock & Maxfield 1997: 220, quarry 84, no. 18.
1090. *PD XI IIII*	Mons Claudianus	Peacock & Maxfield 1997: 220, quarry 84, no. 19.
1091. *PD* | *LII*	Mons Claudianus	Peacock & Maxfield 1997: 220, quarry 84, no. 20.
1092. *P*	Mons Claudianus	Peacock & Maxfield 1997: 220, quarry 84, no. 21.
1093. *II*	Mons Claudianus	Peacock & Maxfield 1997: 220, quarry 84, no. 22.
1094. *P*	Mons Claudianus	Peacock & Maxfield 1997: 220, quarry 84, no. 23.
1095. *PD n(umero) XXXIII* (near necropolis)	Mons Claudianus	Peacock & Maxfield 1997: 220, quarry 84, no. 24.
1096. a. *ACLP*; b. *Π̊*	Mons Claudianus	Peacock & Maxfield 1997: 220, quarry 84, no. 25.
1097. a. *Π̊IA* ; b. *PRD n(umero) XVI*	Mons Claudianus	Peacock & Maxfield 1997: 220, quarry 84, no. 26.
1098. *LII*	Mons Claudianus	Peacock & Maxfield 1997: 220, quarry 84, no. 27.
1099. a. *PD n(umero) IIII* ; b. *Π̊IH*	Mons Claudianus	Peacock & Maxfield 1997: 220, quarry 84, no. 28.
1100. a. *RACLP*; b. *PD XXXVI*	Mons Claudianus	Peacock & Maxfield 1997: 220, quarry 84, no. 29.
1101. a. *P XVI*; b. *RACLP*	Mons Claudianus	Peacock & Maxfield 1997: 220, quarry 84, no. 30.
1102. a. *n(umero) VI*; b. *VIII*	Mons Claudianus	Peacock & Maxfield 1997: 220, quarry 85, no. 1.
1103. *C EP n(umero) IV*	Mons Claudianus	Peacock & Maxfield 1997: 220, quarry 85, no. 2.
1104. *C EP n(umero) IIII* (on block in front of fort)	Mons Claudianus	Peacock & Maxfield 1997: 220, quarry 85, no. 3.
1105. *C EP n(umero) III* (adjacent block)	Mons Claudianus	Peacock & Maxfield 1997: 220, quarry 85, no. 4.
1106. a. *XIX*; b. *LXXXVI*; c. *CEP n(umero) X*	Mons Claudianus	Peacock & Maxfield 1997: 220, quarry 90, no. 1.
1107. *XXXIX*	Mons Claudianus	Peacock & Maxfield 1997: 220, quarry 90, no. 2.
1108. *n(umero) XXII*	Mons Claudianus	Peacock & Maxfield 1997: 220, quarry 90, no. 3.
1109. *LXXXV*	Mons Claudianus	Peacock & Maxfield 1997: 220, quarry 90, no. 4.
1110. *P n(umero) XIII*	Mons Claudianus	Peacock & Maxfield 1997: 220, quarry 90, no. 5.
1111. *T n(umero) LXXXIII*	Mons Claudianus	Peacock & Maxfield 1997: 220, quarry 90, no. 6.

1112. Sundial mark	Mons Claudianus	Peacock & Maxfield 1997: 220, quarry 90, no. 7.
1113. *Cb*	Mons Claudianus	Peacock & Maxfield 1997: 220, quarry 93, no. 1.
1114. *Cb n(umero)III*	Mons Claudianus	Peacock & Maxfield 1997: 220, quarry 93, no. 2.
1115. *Cb n(umero)VI*	Mons Claudianus	Peacock & Maxfield 1997: 220, quarry 93, no. 3.
1116. *I III*	Mons Claudianus	Peacock & Maxfield 1997: 220, quarry 93, no. 4.
1117. *T n(umero) IIII*	Mons Claudianus	Peacock & Maxfield 1997: 220, quarry 94, no. 1.
1118. *n(umero) VII*	Mons Claudianus	Peacock & Maxfield 1997: 220, quarry 94, no. 2.
1119. *CEP I n(umero)III*	Mons Claudianus	Peacock & Maxfield 1997: 220, quarry 96, no. 1.
1120. *Π̊ ΛΓΛ′Δ′*	Mons Claudianus	Peacock & Maxfield 1997: 220, quarry 99, no. 1.
1121. *XXX*	Mons Claudianus	Peacock & Maxfield 1997: 220, quarry 99, no. 2.
1122. *XII*	Mons Claudianus	Peacock & Maxfield 1997: 220, quarry 99, no. 3.
1123. *n(umero)I*	Mons Claudianus	Peacock & Maxfield 1997: 220, quarry102, no. 1.
1124. a. *n(umero)I*; b. *T n(umero)I*	Mons Claudianus	Peacock & Maxfield 1997: 220, quarry 103, no. 1.
1125. *Tn(umero)II*	Mons Claudianus	Peacock & Maxfield 1997: 220, quarry 103, no. 2.
1126. *Tn(umero)IIII*	Mons Claudianus	Peacock & Maxfield 1997: 220, quarry 103, no. 3.
1127. *ΚΛΑΔΙ*	Mons Claudianus	Peacock & Maxfield 1997: 220, quarry 105, no. 1.
1128. *AM* (on rock face to west)	Mons Claudianus	Peacock & Maxfield 1997: 220, quarry 105, no. 2.
1129. *C EP EXLATARPOCR* (on block just west of loading ramp)	Mons Claudianus	Peacock & Maxfield 1997: 220, quarry 109, no. 1.
1130. *Cb*	Mons Claudianus	Peacock & Maxfield 1997: 220, quarry 113, no. 1.
1131. a. *n(umero) II*; b. *II* (same face)	Mons Claudianus	Peacock & Maxfield 1997: 220, quarry 117, no. 1.
1132. *III*	Mons Claudianus	Peacock & Maxfield 1997: 220, quarry 117, no. 2.
1133. *IIII*	Mons Claudianus	Peacock & Maxfield 1997: 220, quarry 117, no. 3.
1134. *VII V*	Mons Claudianus	Peacock & Maxfield 1997: 220, quarry 117, no. 4.

<table>
<tr><td>1135. a. VII; b. V</td><td>Mons Claudianus</td><td>Peacock & Maxfield 1997: 220, quarry 117, no. 5.</td></tr>
<tr><td>1136. a. LATQOPHERAYM PD; b. XXXII; c. RACLP (all on same face)</td><td>Mons Claudianus</td><td>Peacock & Maxfield 1997: 221, quarry 119, no. 1.</td></tr>
<tr><td>1137. LAT COCHLAX PD (on quarry face)</td><td>Mons Claudianus</td><td>Peacock & Maxfield 1997: 221, quarry 120, no. 1.</td></tr>
<tr><td>1138. ΚΑΙCΑΡΟC ΔΙΑ ΗΡΑΚΛΕΙΔΟΥ</td><td>Mons Claudianus</td><td>Peacock & Maxfield 1997: 221, quarry 129, no. 1.</td></tr>
<tr><td>1139. a. RACLP (on roughout of basin); b. A (palmbranch)</td><td>Mons Claudianus</td><td>Peacock & Maxfield 1997: 221, quarry 130, no. 1.</td></tr>
<tr><td>1140. CAESARIS n(ostri?) T n(umero?) VII</td><td>Mons Claudianus</td><td>Peacock & Maxfield 1997: 221, quarry unknown, no. 1.</td></tr>
<tr><td>1141. XXXIIII (near-finished stone near fort)</td><td>Mons Claudianus</td><td>Peacock & Maxfield 1997: 221, quarry unknown, no. 2.</td></tr>
<tr><td>1142. LXI (as above)</td><td>Mons Claudianus</td><td>Peacock & Maxfield 1997: 221, quarry unknown, no. 3.</td></tr>
<tr><td>1143 a. RACIP (on beam in fort); b. XLVIII</td><td>Mons Claudianus</td><td>Peacock & Maxfield 1997: 221, quarry unknown, no. 4.</td></tr>
<tr><td>1144. ΕΠ/ ΕΠΑΦ (on rock on west side of slipway up from fort)</td><td>Mons Claudianus</td><td>Peacock & Maxfield 1997: 221, quarry unknown, no. 5.</td></tr>
</table>

Carrara Marble: Luna

<table>
<tr><td>1145. M. Crasso c[os.] | n(umero) CXIIX</td><td>Carrara, 27</td><td>CIL XI 6723/16 = Dubois 1908: no. 30.</td></tr>
<tr><td>1146. [—]Tertull[o] c[o](n)s(ule) | [—]s event ƉLV</td><td>Rome /Palatino, 158?</td><td>Bruzza 1870: no. 326 = Dubois 1908: no. 49.</td></tr>
<tr><td>1147. a) Col(oniae?) X ↓ IX | FPS; b) Hilar(i)</td><td>Carrara / Cava di Gioia</td><td>CIL XI 6723/1a = Dubois 1908: no. 1.</td></tr>
<tr><td>1148. Col(oniae?) ↓ Hilar(i) f | P[.]AS</td><td>Carrara / Cava di Gioia</td><td>CIL XI 6723/1b = Dubois 1908: no. 2.</td></tr>
<tr><td>1149. Col(oniae?) ↓ XLC | ƆIO</td><td>Carrara / Cava di Gioia</td><td>CIL XI 6723/1c = Dubois 1908: no. 3.</td></tr>
<tr><td>1150. Col(oniae?) XCVII Hilar(i) a</td><td>Carrara / Cava di Gioia</td><td>CIL XI 6723/1d = Dubois 1908: no. 4.</td></tr>
<tr><td>1151. a) Col(oniae?) CCCXVI | PΛAΛE(?); b) Hilar(i) a</td><td>Carrara / Cava di Gioia</td><td>CIL XI 6723/1e = Dubois 1908: no. 5.</td></tr>
<tr><td>1152. n(umero) XXXV δ ς.</td><td>Luna</td><td>AE 1997: 506.</td></tr>
<tr><td>1153. a) Hilar(i); b) CQXXX HILAR</td><td>Luna</td><td>AE 1998: 436c.</td></tr>
</table>

1154. *Col(oniae?) CXXV SOL A*	Carrara / Fossacava	*CIL* XI 6723/5b = Dubois 1908: no. 17.
1155. *[C]ol(oniae) CXVIII⟨I*	Carrara / Campo la Piana	*CIL* XI 6723/15 = Dubois 1908: no. 29.
1156. *Col(oniae) CIX*	Rome / Porta Sa. Sebastiano	Dubois 1908: no. 40.
1157. a) *T(i)b(urtini?) lo(co) CXLIII*; b) *Er(otis) Cạẹṣ(aris servi?)*	Carrara / Cava di Gioia	*CIL* XI 6723/2a = Dubois 1908: no. 6.
1158. a) *T(i)b(urtini?) lo(co) CXLIV,* ; b) *Er(otis) Caes(aris servi?)*	Carrara / Cava di Gioia	*CIL* XI 6723/2b = Dubois 1908: no. 7.
1159. a) *T(i)b(urtini?) lo(co) CXLVII,* ; b) *Er(otis) Caes(aris servi?)*	Carrara / Cava di Gioia	*CIL* XI 6723/2c = Dubois 1908: no. 8.
1160. a) *T(i)b(urtini?) lo(co) CXLII,* ; b) *Er(otis) Caes(aris servi?)*	Carrara / Cava di Gioia	*CIL* XI 6723/2d = Dubois 1908: no. 9.
1161. *[—] VIIIIC | Er(otis) Caes(aris servi?)*	Carrara / Cava di Gioia	*CIL* XI 6723/2e = Dubois 1908: no. 10
1162. a) *Phil(onis) CCXXCIIII E INTER MANDR(?) VI | Phil(onis) C ↓ II F COR MVT(?) | AED*; b) *A XII COR Ioni[—] | S CAPIT(?)*	Carrara / Cava di Gioia	*CIL* XI 6723/3a,α.β. = Dubois 1908: no. 11.
1163. *Phil(onis) CX ↓ IIII D Teg. | AED VIIII(?)*	Carrara / Cava di Gioia	*CIL* XI 6723/3b = Dubois 1908: no. 12.
1164. *Phil(onis) ↓ Ị Ḍ Teg COL| II*	Carrara / Cava di Gioia	*CIL* XI 6723/3c = Dubois 1908: no. 13.
1165. *Phil(onis?) CCXXX SPIR(?) [O]PER(?) | S Cae*	Carrara / Fossacava	*CIL* XI 6723/5a = Dubois 1908: no. 16.
1166. *Abae(us) CCCXXIX b | PAVD*	Monte Strinato/Carrara	*AE* 1997: 510; Dolci 1997: 44, fig. 11.
1167. *Abae(us) X b | PAGA*	Monte Strinato/Carrara	*AE* 1997: 511; Dolci 1997: 43, fig. 10.
1168. *Abae(us) CCXVII A*	Figaia/Fantiscritti/Carrara	*AE* 1995: 492; Dolci 1995: 128.
1169. *Abae(us) CCLVII B | PVRB*	Figaia/Fantiscritti/Carrara	*AE* 1995: 493; Dolci 1995: 128.
1170. *Abae(us) H CXXD*	Rome / Porta del Popolo	Dubois 1908: no. 33.
1171. *Abae(us) H(?) CXXCI*	Ostia	Pensabene 1994: 204, no. 24.
1172. *Ero(tis) [l]oc(o) (?)*	Rome / Porta del Popolo	Dubois 1908: no. 42.
1173. *E[r]o(tis) C[aes(aris?)] | n(umero) CXXII*	Rome	Dubois 1908: no. 43.

1174. *Ero(tis) C[aes(aris?)]*	Tusculum	*CIL* XIV 2670 = Dubois 1908: no. 44.
1175. a) *Ephe(bi) Cae(saris?) IVNEM*; b) *Epheb(i) Aug(usti?)*	Carrara / Fossacava	*CIL* XI 6723/5d = Dubois 1908: no. 19.
1176. *Eph(ebi) Aug(usti) CCCCX* ↓ *II B*	Rome / Emporia?	Dubois 1908: no. 41.
1177. *Caes(aris) s(-) | CCCXLIIX*	Luna	*AE* 1997: 508.
1178. *Caes(aris) CCCXXCII | Ap(-)*	Luna	*AE* 1997: 509.
1179. *Caes I[—] | CCCXCIII*	Carrara / quarry near Torana?	*CIL* XI 6723/8 = Dubois 1908: no. 24.
1180. *Iul(ii?)Pr(imi?) X* ↓ *IIX BP*	Carrara / Fossacava	*CIL* XI 6723/5e = Dubois 1908: no. 20.
1181. *Pri(mi) CCXXXX*	Rome / Porta del Popolo	Dubois 1908: no. 31.
1182. *Eman(i) VD*	Rome / Porta del Popolo	Dubois 1908: no. 34.
1183. *X Ruf(i)*	Rome / Porta del Popolo	Dubois 1908: no. 35.
1184. *TI(berii?) IV(lii?) PRI(mi)*	Rome	Dubois 1908: no. 45.
1185. *A(—) T(iti) f(ilius) aed-(ilis) NL e?*	Luna	*AE* 1997: 504.
1186. *A(—) T(iti) (filius) aed-(ilis) m(etallorum?)*	Luna	*AE* 1997: 505.
1187. *TIH VII*	Luna	*AE* 1997: 512.
1188. *TH CDIIX*	Scalocchiella/Carrara	*AE* 1998: 436b; Dolci 1998: 128.
1189. *TH XXXVIEf*	Scalocchiella/Carrara	*AE* 1998: 436b; Dolci 1998: 128.
1190. *TH CCCVXX | VIIII*	Scalocchiella/Carrara ?	*AE* 1998: 436b; Dolci 1998: 128.
1191. CCXXIIX *ra(tio) m(armorum)*	Scalocchiella/Carrara	*AE* 1998: 436d; Dolci 1998: 130.
1192. *[—]c[—]X* ↓ *C[.]C*	Carrara / Cava di Gioia	*CIL* XI 6723/4a = Dubois 1908: no. 14.
1193. *TVΛXF(???)*	Carrara / Cava di Gioia	*CIL* XI 6723/4b = Dubois 1908: no. 15.
1194. *Pa (?) CC⊥XXIII*	Carrara / Fossacava	*CIL* XI 6723/5c = Dubois 1908: no. 18.
1195. *XI XXX*	Carrara / near villa Grassi	*CIL* XI 6723/9 = Dubois 1908: no. 25.
1196. *D A* ↓ *IX*	Carrara / Chiaruccia	*CIL* XI 6723/12 = Dubois 1908: no. 26.
1197. a) *MCXCLM;* b) CCCLM	Carrara / Sta Maria di Falleri	*CIL* XI 6723/13 = Dubois 1908: no. 27.

1198. *AR(?) CCXLIIII*	Carrara	*CIL* XI 6723/14 = Dubois 1908: no. 28.
1199. *la(?) l]o]co XXXVIIII*	Rome / Porta del Popolo	Dubois 1908: no. 32.
1200. *DXVI B PAS (?)*	Rome / Porta del Popolo	Dubois 1908: no. 36.
1201. *PA(S?) DCX↓IV*	Rome / Ponte Salaro	Dubois 1908: no. 38.
1202. *XXVCC PAS (?)*	Rome / S.Agnese cemetery	Dubois 1908: no. 37.
1203. *PA(S?) ↓XXVII*	Ostia/Ponte della Rifolta	Dubois 1908: no. 39.
1204. *loc(o) LVIIII δ ς.*	Luna	*AE* 1997: 507.
1205. *l(oco) CDXLI*	Rome / Via Appia	Dubois 1908: no. 51/3.
1206. *l(oco) DCCCX*	Rome / Presso la Magliana	Dubois 1908: no. 51/4.
1207. *n(umero) CIIII*	Rome / Terracina	*CIL* X 6337 = Dubois 1908: 16, no. 51/5.
1208. *n(umero) XXII δς*	Scalocchiella / Carrara	*AE* 1998: 436a; Dolci 1998: 126.
1209. *n(umero) VID*	Rome / Esquilin	Dubois 1908: no. 51/11.
1210. *n(umero) CCXXXXII*	Rome(?) ?	Dubois 1908: no. 51/14 = Bruzza 1870: no. 330.
1211. *PAR II de XX RC | OCCCES | SAROC ONT | CAEC. ST [—]*	Pisa	Dubois 1908: no. 54.
1212. *CCCXXIX N(?) | [—]*	Rome / Porta Salara	Dubois 1908: no. 46.
1213. *ACXX [—]*	Rome / Porta del Popolo	Dubois 1908: no. 47.
1214. a) *[—]EIME | [—]IIII;* b) *P*	Rome / Emporia	Dubois 1908: no. 48 = Bruzza 1870: no. 327.
1215. *CIX P*	Rome / Via Ostiense	Dubois 1908: no. 50.
1216. *D*	Rome / Porta del Popolo?	Dubois 1908: no. 51/1.
1217. *F*	Rome / Porta del Popolo	Dubois 1908: no. 51/2.
1218. *XXVIII*	Rome / Via Latina	Dubois 1908: no. 51/6.
1219. *↓XX | PO X*	Carrara / Fossacava	*CIL* XI 6723/5f = Dubois 1908: no. 21.
1220. *pa(?)XCIX*	Carrara / Fossa Piccola	*CIL* XI 6723/6 = Dubois 1908: no. 22.
1221. *III Mo PAGE (?)*	Carrara / Fossa Piccola	*CIL* XI 6723/7 = Dubois 1908: no. 23.
1222. *q ↓ XCIXO*	Rome / Chiesa di S. Antimo	Dubois 1908: no. 51/7.
1223. *R CCX*	Rome / Via Valeria	Dubois 1908: no. 51/8.

1224. *P*	Rome / Porta del Popolo	Dubois 1908: no. 51/9.
1225. *O XIX*	Rome / Concordia-temple	Dubois 1908: no. 51/10.
1226. *MCXCIXM*	Rome(?)	Dubois 1908: no. 51/12 = Bruzza 1870: no. 328.
1227. *[—]XXGLM[—]*	Rome(?)	Dubois 1908: no. 51/13 = Bruzza 1870: no. 329.
1228. *PE R XR* | *[—]*	Rome / SS. Cosma ?	Dubois 1908: no. 52.
1229. *Sev* | *II II de XXXIII*	Ostia	Dubois 1908: no. 53.

Parian and other White Marble: Paros

1230. *Imp(eratoris) Dom(itiani) Aug(usti)* | *Germ(anici) per* | *Chrez(imum) lib(ertum)* || *ӾC Σ* (rev.)	Rome, 83–96	Bruzza 1870: no. 277 = Dubois 1908: no. 267.
1231. *[I]mp(eratoris) Domitiani* | *Aug(usti) Germ(anici) loca* | *per Primum lib(ertum)*	Rome, 83–96	Bruzza 1870: no. 278 = Dubois 1908: no. 268.
1232. *Imp(eratoris) Domitiani Caesa(ris)* | *Aug(usti) Germ(anici) loco CLXXV [—?]*| *per Primum lib(ertum) [—?]*	Rome, 83–96	Hirschfeld 1905: 163 fn. 4 = Dubois 1908: no. 269 = *CIL* VI/8, 40849 = *ILS* 8713.
1233. *Ꞧ* | *Augurin(o) co(n)s-(ule)* | *⊥(oco) CĐXCVI*	Rome, 132	Bruzza 1870: no. 266 = Dubois 1908: no. 257.
1234. *Augurin(o)* | *co(n)s(ule)* | *LIIXLIIX* | *Ꞧ*	Ostia, 132	Baccini Leotardi 1989: no. 104.
1235. a) *M* | *NĐLXIII*; b) *Augurin(o) co(n)s(ule)* | *Ꞧ* | *L∞ĐCCCVI*	Portus, 132	Baccini Leotardi 1989: no. 118. See Fant 2001: 184, table 6.
1236. a) *n(umero) CLV* | *Praes(ente) et Ruf(ino) co(n)s(ulibus)* | *ex Ꞧ(atione) Aur(elii) (G)eorgi*; b) *l(oco) CCCLII* | *G*	Portus, 153	Baccini Leotardi 1989: no. 92. See Fant 2001: 184, table 6.
1237. *n(umero) CLIIII* | *Praes(ente) et Ruf(ino)* | *co(n)s(ulibus)* | *ex Ꞧ(atione) Aur(elii) Geor(gi)*	Portus, 153	Baccini Leotardi 1989: no. 93. See Fant 2001: 184, table 6.
1238. a) *n(umero) CXLVII* | *Pra(e)s(ente) et Ruf(ino) co(n)s(ulibus) ex Ꞧ(atione) A(urelii) Geor(gi)*; b) *l(oco) CCCCXIX* | *G*	Portus, 153	Baccini Leotardi 1989: no. 94. See Fant 2001: 184, table 6.

1239. a) *n(umero) CLXVII* | *Praes(ente) et Ruf(ino) co(n)s(ulibus)* | *ex R(atione) Aur(elii) Geor(gi)*; b) *l(oco) CCCLVII*	Portus, 153	Baccini Leotardi 1989: no. 95. See Fant 2001: 184, table 6.
1240. *n(umero) CLII* | *Praes(ente) et Rufi(no)* | *ex R(atione) Aur(eli) Geor(gi)* | *II C*	Portus, 153	Baccini Leotardi 1989: no. 96. See Fant 2001: 184, table 6.
1241. a) *Praes(ente) et Rufi(no)* | *ex R(atione) Aur(eli) Geor(gi)*; b) *l(oco)CCC[—]* | *LX* | *G*	Portus, 153	Baccini Leotardi 1989: no. 97. See Fant 2001: 184, table 6.
1242. a) *NKh* | *l(oco) LXII* | *G*; b) LL H | *Brad(ua) et Varo co(n)s(ulibus)* | *ex r(atione) Aur(eli) G(eorgi)*	Portus, 160	Baccini Leotardi 1989: no. 99 = Pensabene 1994: no. 58. cf. Fant 2001: 184, table 6.
1243. a) *loc(o) .XXXV*; b) *Hermo* | *loc(o) CDLXXV*; c) *Laelian(o)* | *Past(ore) co(n)s(ulibus)* |*(ex) r(atione) Sext(i) et* | *M[.]R*	Portus, 163	Baccini Leotardi 1989: no. 102 = Pensabene 1994: 121, no. 54. cf. Fant 2001: 184, table 6
1244. a) *loc(o)* | *XXI* | *Hermo* | *loc(o) DI* | ; b) *Macrino et Ce|lso co(n)s(ulibus)ex R(atione) Sext(i?) e|t Her(mae?)*	Portus, 164	Ostia inv. 39910 = Baccini Leotardi 1989: no. 103 = Pensabene 1994: no. 55. cf. Fant 2001: 184, table 6.
1245. *FLSTLC* | *RLVC*	Rome	Bruzza 1870: no. 267 = Dubois 1908: no. 258.
1246. *HERMOΛ* |*loc(o) CCXCV*	Paros	Bruzza 1870: no. 268 = Dubois 1908: no. 259/1 = *CIL* III 487a.
1247. *Hermo* | *loc(o)*	Paros	Bruzza 1870: no. 269 = Dubois 1908: no. 259/2 = *CIL* III 487b.
1248. *PIZ*	Rome	Bruzza 1870: no. 270 = Dubois 1908: no. 260.
1249. *CCCXV*	Paros	Bruzza 1870: no. 271 = Dubois 1908: no. 261 = *CIL* III 487f.
1250. *Hermo* | *loc(o) CCCLXXXVI*	Paros	Bruzza 1870: no. 272 = Dubois 1908: no. 262 = *CIL* III 487c.
1251. *Hermo* | *loc(o) CC|LXXXXVIII*	Paros	Bruzza 1870: no. 273 = Dubois 1908: no. 263 = *CIL* III 487d.
1252. *Hermo* | *loc[—]* | *[——]* | *n(umero) CCIIII*	Paros	Bruzza 1870: no. 274 = Dubois 1908: no. 264 = *CIL* III 487e.
1253. *R* | *CV* | *LXXX* | ∞ *DCCLXXVI*	Rome	Bruzza 1870: no. 275 = Dubois 1908: no. 265.
1254. ∞ *C[—]*	Rome	Bruzza 1870: no. 276 = Dubois 1908: no. 266.

1255. *Hermo* | *loc(o) CDLXXIV* | *loc(o) LXXV*	Ostia	Baccini Leotardi 1979: no. 87. See Fant 2001: 184, table 6.
1256. *l(oco)CCCLXI* | *G*	Ostia	Baccini Leotardi 1979: no. 88. See Fant 2001: 184, table 6.
1257. *l(oco) CCCLXVII* | *G*	Portus	Baccini Leotardi 1989: no. 98. See Fant 2001: 184, table 6.
1258. a) ꝶ | *l(oco)* ∞*CCCC* | *LXVII;* b) *M* | *n(umero) CLIII*	Portus	Baccini Leotardi 1989: no. 101. See Fant 2001: 184, table 6.
1259. L∞Đ	Portus	Baccini Leotardi 1989: no. 120. See Fant 2001: 184, table 6.

Parian and other White Marble: Pentelicos?

1260. *Serbilio Pudente* | *et Fufidio Pollione* | *co(n)s(ulibus) caesura Cla(udii) Hier Attici et Apollo*|*ni Lupi [—]*	Rome, 166	Bruzza 1870: no. 291 = Dubois 1908: no. 255 = Ameling 1983*b*: 216 no. 199.
1261. *l(oco)* ∞ *DCCCXXCVI* | ꝶ	Ostia	Baccini Leotardi 1989: no. 112.
1262. a) ∞ *DCCCXX* | *L[.] N II [.]* | *I;* b) *CCCXXCVIII;* c) *A*	Ostia	Baccini Leotardi 1989: no. 113.

Parian and other White Marble: Hymettos

1263. painted: *[—Eu]tyches ex ra-(tione)[—]* | *prob* | *[——]*	Rome	Bruzza 1870: no. 292 = Dubois 1908: no. 256.

Parian and other White marble: Greece?

1264. *n(umero) XXX* | *M(arco) Aur(e)lio Vero* | *III et L. V + + + C(aesare) II c(onsulibus)* | *ex r(atione) Aur(eli) Lis(—)*	Portus, 161	Baccini Leotardi 1989: no. 100.
1265. *ex r(atione) M(arci) Iulii Missi*	Rome	Bruzza 1870: no. 280 = Dubois 1908: no. 271.
1266. *CAE[—]* | *CER[—]*	Rome	Bruzza 1870: no. 281 = Dubois 1908: no. 272.
1267. *AGR AVG Do(mitiani) XX* | *III A NV ΛAC*	Rome	Bruzza 1870: no. 282 = Dubois 1908: no. 273.
1268. *AGR AVG Do(mitiani) DESA* | *NN XXXV ΛAC*	Rome	Bruzza 1870: no. 283 = Dubois 1908: no. 274.

1269. ┌┼┐ ↓ *XXXIII*	Rome	Bruzza 1870: no. 284 = Dubois 1908: no. 275.
1270. *[—]AB*	Rome	Bruzza 1870: no. 285 = Dubois 1908: no. 276.
1271. *l(oco) ∞ĐCCCLXXVI*	Ostia	Baccini Leotardi 1989: no. 105 = Pensabene 1994: 121, no. 57.
1272. a) *IIXXXIX*; b) *LXXIV*	Ostia	Baccini Leotardi 1989: no. 106
1273. a) *IR̶*; b) *III l(oco) DCCCXXXCVII*	Ostia	Pensabene 1994: 122, no. 59 = Baccini Leotardi 1989: no. 121.
1274. a) *n(umero) XXXXIIII* ; b) *l(oco) DXXXVII | CC*	Ostia	Pensabene 1994: 122, no. 63 = Baccini Leotardi 1989: no. 80.
1275. *[..] R V A / T*	Ostia	Pensabene 1994: 133, no. 87.
1276. *T I C V*	Ostia	Pensabene 1994: 180, no. 143.
1277. *DNGNF*	Ostia	Pensabene 1994: 181, no. 148.
1278. *Fl(avii) St(i)l(i)c(honis)*	Ostia	*CIL* XIV 165, Suppl. 2, p. 854 = Lugli & Filibeck 1935: 80 = Pensabene 1994: no. 181.
1279. *n(umero) XXVII co(n)s(ule?)*	Ostia	Pensabene 1994: 74, no. 23 = Baccini Leotardi 1989: 114.
1280. *CIXVIII | Mac(rino) et Cel-(so) co(n)s(ulibus)| ex r-(atione) Aur(eli) Suavis*	Ostia, 164	Fant 1989: no. 89[2].

Other Stones and inscribed seals on quarried marble

1281. *n(umero) XX [—] | Caes(aris) A[ug(usti)*	Rome, 2 BC	*CIL* VI/8, 40847 = Alföldy 1992.
1282. *Tr(aiani) Cae(saris)|L*	Rome, AD 98–117	*CIL* VI/8, 40850.
1283. *Imp(eratoribus) An[tonino et]|Vero n(umero?) [—]*	Rome, 161–9	*CIL* VI/8, 40851.

Bibliography

ABASCAL, J. M., and G. ALFÖLDY (1998), 'Zeus Theos Megistos en Segobriga', *Archivo español de arqueología* 71: 157–68.

ABASCAL PALAZÓN, J. M. (1994), *Los nombres personales en las inscripciones latinas de Hispania* (Anejos de Antigüedad y cristianismo II), Murcia.

ADAM, J.-P. (1999), *Roman Building. Materials and Techniques*, London & New York.

ADAMS, C. E. P. (2001), 'Who bore the burden? The Organization of Stone Transport in Roman Egypt', in D. J. Mattingly and J. Salmon (eds.), *Economies Beyond Agriculture in the Classical World*, London & New York, 171–92.

ALARCÃO, J., and R. ETIENNE (eds.) (1976), *Fouilles de Conimbriga II: épigraphie et sculpture*, Paris.

ALEKSANDROV, G. (1977), 'Latinski nadpisi ot Montana (dn. Mihailovgrad)', *Vekove* 5: 51–61.

ALFÖLDY, G. (1965*a*), *Bevölkerung und Gesellschaft der Römischen Provinz Dalmatien*, Budapest.

—— (1965*b*), 'Epigraphisches aus dem Rheinland I', *Bonner Jahrbücher* 165: 177–81.

—— (1969*a*), *Die Personennamen in der römischen Provinz Dalmatia*, Heidelberg.

—— (1969*b*), 'Epigraphica Norica', *Epigraphische Studien* 8: 1–34.

—— (1970), 'Patrimonium Regni Norici. Ein Beitrag zur Territorialgeschichte der römischen Provinz Noricum', *Bonner Jahrbücher* 170: 163–77.

—— (1974), *Noricum*, London & Boston.

—— (1975), *Römische Sozialgeschichte*, Wiesbaden.

—— (1987*a*), 'Epigraphisches aus dem Rheinland I', in G. Alföldy, *Römische Heeresgeschichte. Beiträge 1962–1985* (MAVORS III), Amsterdam, 177–81.

—— (1987*b*), 'Zum Kommando des römischen Heeres. J. B. Campbell: The Emperor and the Roman Army 31 BC – AD 235', in G. Alföldy, *Römische Heeresgeschichte. Beiträge 1962–85* (MAVORS III), Amsterdam, 19–25.

—— (1989), 'Die regionale Gliederung in der römischen Provinz Noricum', in G. Gottlieb (ed.), *Raumordnung im Römischen Reich. Zur regionalen Gliederung in den gallischen Provinzen, in Rätien, Noricum und Pannonien. Kolloquium an der Universität Augsburg anlässlich der 2000-Jahr-Feier der Stadt Augsburg vom 28. bis 29. Oktober 1985*, Munich, 37–55.

—— (1992), 'Un'avvertenza dell'architetto del tempio di Mars Ultor', in *Studi sull'epigrafia augustea e tiberiana di Roma* (Vetera 8), Rome, 33–4.

—— (1996), 'Spain', in A. K. Bowman, E. Champlin, and A. Lintott (eds.), *The Cambridge Ancient History. Second Edition. Volume X. The Augustan Empire, 43 BC – AD 69*, Cambridge, 449–63.

—— (2000*a*), *Provincia Hispania Superior*, Heidelberg.

—— (2000*b*), 'Das Neue Edikt des Augustus aus El Bierzo in Hispanien', *Zeitschrift für Papyrologie und Epigraphik* 131: 177–205.

—— (2003), 'Altar eines Benefiziariers', in A. Szabó and E. Tóth (eds.), *Bölcske. Römische Inschriften und Funde* (Libelli Archaeologici Ser. Nov. No. II), Budapest, 219–28.

—— (2004), 'Die "illyrischen" Provinzen Roms: von der Vielfalt zu der Einheit', in G. Urso (ed.), *Dall' Adriatico al Danubio. L'Illyrico nell' età greca e romana. Atti del convegno internazionale Cividale del Friuli, 25–27 settembre 2003*, Pisa, 207–20.

Alpers, M. (1995), *Das nachrepublikanische Finanzsystem. Fiscus und Fisci* (Untersuchungen zur antiken Literatur und Geschichte 45), Berlin.

Alston, R. (1995), *Soldier and Society in Roman Egypt. A Social History*, London & New York.

Ameling, W. (1983*a*), *Herodes Atticus. I. Biographie* (Subsidia Epigraphica. Quellen und Abhandlungen zur griechischen Epigraphik 11), Hildesheim & New York.

—— (1983*b*), *Herodes Atticus. II. Inschriftenkatalog* (Subsidia Epigraphica. Quellen und Abhandlungen zur griechischen Epigraphik 11), Hildesheim & New York.

Andreau, J. (1989), 'Recherches récentes sur les mines à l'époque romaine. I. Propriété et mode d'exploitation', *Revue Numismatique*, 86–112.

—— (1990), 'Recherches récentes sur les mines à l'époque romaine. II. Nature de la main d'oeuvre; Histoire des techniques et de la production', *Revue Numismatique*, 218–33.

Angeli Bertinelli, M. G. (1978), 'Culti e divinità della romana Luni nella testimonianza epigrafica', *Quaderni. Centro Studi Lunensi* 3: 3–32.

Ardaillon, E. (1904), 'Metalla', in C. Daremberg, E. Saglio, and E. Poittier (eds.), *Dictionnaire des Antiquités Grecques et Romaines d'après les Textes et les Monuments. Tome troisième, deuxième partie*, Paris, 1840–73.

Aubert, J.-J. (1994), *Business Managers in Ancient Rome. A Social and Economic Study of Institores, 200 BC – AD 250* (Columbia Studies in the Classical Tradition 21), Leiden, New York, & Cologne.

Austin, N. J. E., and N. B. Rankov (1995), *Exploratio. Military and Political Intelligence in the Roman World from the Second Punic War to the Battle of Adrianople*, London & New York.

Aymard, A. (1856), 'Les origines de la ville du Puy', in *Congrès scientifique de France, 22ᵉ session, Le Puy, 1855. II*, Paris, 323–403.

Baccini Leotardi, P. (1979), *Marmi di cava rinvenuti ad Ostia e considerazioni sul commercio dei Marmi in età romana* (Scavi di Ostia X), Rome.

—— (1989), *Nuove testimonianze sul commercio dei marmi in età imperiale* (Studi pubblicati dall'Istituo italiano per la storia antica. Fasc. XLIV), Rome.

Balboa de Paz, J. A. (1999), 'Un edicto Emperador Augusto hallado en El Bierzo', *Estudios Bercianos. Revista del Instituto de Estudios Bercianos* 25: 45–53.

Balil Illana, A., G. Pereira Menaut, and F. J. Sánchez-Palencia Ramos (1991), *Tabula Imperii Romani. Hoja K-29: Porto. Conimbriga – Bracara – Lucus – Asturica*, Madrid.

Banti, L. (1931), 'Antiche lavorazioni nelle cave Lunensi', *Studi Etruschi* 5: 475–97.

Barker, G. W. et al. (1997), 'The Wadi Faynan Project, Southern Jordan. A Preliminary Report on Geomorphology and Landscape Archaeology', *Levant* 29: 19–40.

BARKER, G. W. et al. (1998), 'Environment and Land Use in the Wadi Faynan, Southern Jordan. The Second Season of Geoarchaeoloy and Landscape Archaeology (1997)', *Levant* 30: 5–25.

—— (1999), 'Environment and Land Use in the Wadi Faynan, Southern Jordan. The Third Season of Geoarchaeology and Landscape Archaeology (1998)', *Levant* 29: 255–92.

—— (2000), 'Archaeology and Desertification in the Wadi Faynan. The Fourth (1999) Season of the Wadi Faynan Landscape Survey', *Levant* 32: 27–52.

BASTIANINI, G. (1986), 'Un ordine di scarcerazione: PBerol. inv 8997 (ChLA X 421)', in B. G. Mandilaras (ed.), *Proceedings of the XVIIIth International Congress of Papyrology, Athens, 25–31 May, 1986*, Athens, 351–6.

BAUCHHENSS, G. (1986), 'Hercules Saxanus, ein Gott der niedergermanischen Armee', in *Studien zu den Militärgrenzen Roms III. 13. Internationaler Limeskongress. Aalen 1983. Vorträge* (Forschungen und Berichte zur Vor- und Frühgeschichte in Baden-Württemberg, Band 20), Stuttgart, 90–5.

BELTRÁN MARTÍNEZ, A. (1939), *Las monedas latinas de Cartagena*, Murcia.

BENITO SERRA, P., and G. BACCO (1998), 'Forum Traiani: il contesto termale e l'indagine archeologica di scavo', *L'Africa Romana* 12: 1213–55.

BENNDORF, O., and R. HEBERDEY (1898), 'Vorläufige Berichte über die Ausgrabungen in Ephesus', *Jahreshefte des Österreichischen Ärchäologischen Institutes* 1: 54–82.

BENNETT, I. (1997), *Trajan Optimus Princeps. A Life and Times*, London & New York.

BENOIT, F. (1932), 'Notes et documents d'archéologie arlésienne. IX. Inscriptions inédites et nouvelles inscriptions du Musée Lapidaire', *Mémoires de L'Institut historique de Provence, tome IX*.

BÉRARD, F. (1992), 'Territorium legionis', *Cahiers Du Centre Gustave-Glotz* 3: 75–105.

BERNAND, A. (1969), *Les inscriptions grecques de Philae. Tome I. Époque ptolémaique*, Paris.

—— (1972), *De Koptos à Kosseir*, Leiden.

BERNAND, E. (1988), *Inscriptions grecques et latines d'Akôris* (IFAO/BE 103), Cairo.

—— (1992), 'A propos de l'autel dédié à Zeus Soleil, grand Sarapis, par l'architecte alexandrin Apollônios, fils d'Ammônios, au Mons Claudianus', *Zeitschrift für Papyrologie und Epigraphik* 91: 221–5.

—— (1999), *Inscriptions grecques d'Hermoupolis Magna et de sa nécropole* (IFAO/BE 123), Cairo.

BERNARD, H., and C. DOMERGUE (1991), 'Les lingots de plomb de l'épave romaine Sud Perduto (Bouches de Bonifacio, Corse)', *Bulletin de la Société des Sciences Historiques et Naturelles de la Corse* 111: 41–95.

BERTI, F. (1985), 'La nave romana di Valle Ponti', *Rivista Studi Liguri* 51: 553–70.

—— (1990), *Fortuna Maris. La nave romana di Comacchio*, Bologna.

BESNIER, M. (1920), 'Le commerce du plomb à l'époque romaine d'après les lingots estampillés', *Revue Archéologique* 13: 211–44.

—— (1921), 'Le commerce du plomb à l'époque romaine d'après les lingots estampillés (suite et fin)', *Revue Archéologique* 14: 98–130.

BILABEL, F. (1927), 'Librarius', in W. Kroll (ed.), *Paulys Real-Encyclopädie der Classischen Altertumswissenschaften*, Stuttgart, 137–9.

BINGEN, J. (1992), 'Les architektones du Mons Claudianus', in J. Bingen, et al. (1992), 39–55.

—— (1996), 'Dumping of Ostraka at Mons Claudianus', in D. M. Bailey (ed.), *Archaeological Research in Roman Egypt. The Proceedings of the Seventeenth Classical Colloquium of the Department of Greek and Roman Antiquities, British Museum, held on 1–4 December, 1993* (JRA Supplement 19), Ann Arbor, 29–38.

—— (1997), 'Lettres privées provenant de Raima', in J. Bingen, et al. (1997), 81–112.

—— A. Bülow-Jacobsen, W. E. H. Cockle, H. Cuvigny, L. Rubinstein, and W. V. Rengen (1992), *Mons Claudianus. Ostraca Graeca et Latina. I. O. Claud. 1 à 190* (FIFAO 29), Cairo.

—— —— —— —— F. Kayser, and W. V. Rengen (1997), *Mons Claudianus. Ostraca Graeca et Latina. II. O. Claud 191 à 416* (FIFAO 32), Cairo.

BIRD, D. G. (1972), 'The Roman Gold Mines of North-West Spain', *Bonner Jahrbücher* 172: 36–64.

BIRLEY, E. (1953), 'Marcus Cocceius Firmus. An Epigraphic Study', in E. Birley (ed.), *Roman Britain and the Roman Army. Collected Papers,* Kendal, 87–103.

—— (1963–4), 'Promotions and Transfers in the Roman Army II: The Centurionate', *Carnuntum Jahrbuch,* 21–33 (= E. Birley, *The Roman Army. Papers 1929–1986* (MAVORS IV), Amsterdam 1988: 206–20).

BIRLEY, A. R. (1987), *Marcus Aurelius. A Biography,* London & New York.

—— (1992), *Locus virtutibus patefactus? Zum Beförderungssystem in der Hohen Kaiserzeit* (Rheinisch-Westfälische Akademie der Wissenschaften. Geisteswissenschaften. Vorträge G 318), Opladen.

—— (1995), 'An Officer in Aurelian's Army', in R. Frei-Stolba and M. A. Speidel (eds.), *Römische Inschriften—Neufunde, Neulesungen und Neuinterpretationen. Festschrift für Hans Lieb* (AREA 2), Basel, 143–8.

—— (1999), *Septimius Severus. The African Emperor,* London & New York.

—— (2004), 'Britain 71–105: Advance and Retrenchment', in L. De Ligt et al. (eds.), *Roman Rule and Civic Life: Local and Regional Perspectives. Proceedings of the Fourth Workshop of the International Network Impact of Empire (Roman Empire, c. 200 B.C. – A.D. 476), Leiden, June 25–28: 2003,* Amsterdam, 97–112.

BISHOP, M. C., and J. C. N. COULSTON (2006), *Roman Military Equipment. From the Punic Wars to the Fall of Rome,* Oxford.

BLÁZQUEZ MARTÍNEZ, J. M. (1989), 'Administracion de las minas en epoca romana. Su evolucion', in C. Domergue (ed.), *Minería y metalurgia en las Antiguas Civilizaciones Mediterráneas y Europeas, Coloquio Internacional Asòciado. Madrid, 24–28 Octobre, 1985,* Madrid, 119–31.

BLECKMANN, B. (1997), 'Dioikesis. II. Rom', in *Der Neue Pauly. Enzyklopädie der Antike. Band 3,* Stuttgart & Weimar, 608.

BOATWRIGHT, M. T. (2000), *Hadrian and the Cities of the Roman Empire,* Princeton.

BOJANOVSKI, I. (1982), 'Anticko rudarstvo u unutrasnjosti provincije Dalmacije u svjetlu epigrafskih i numizmatickih izvora. L'Exploitation minière dans l'antiquité à l'intérieur de la province de Dalmatie, à la lumière des sources épigraphiques et numismatiques', *Arheoloski radovi i rasprave. Acta et dissertationes archaeologica* 8–9: 89–120.

Boulvert, G. (1970), *Esclaves et affranchis impériaux sous le Haut-Empire romain. Rôle politique et administratif*, Naples.

—— (1974), *Domestique et fonctionnaire sous le Haut-Empire romain. La condition de l'affranchi et de l'esclave du Prince*, Paris.

Bowman, A. K. (1996*a*), 'Provincial Administration and Taxation', in A. K. Bowman, E. Chapman, and A. Lintott (eds.), *The Cambridge Ancient History. Second Edition. Volume X. The Augustan Empire 43 BC – AD 69*, Cambridge, 344–70.

—— (1996*b*), *Egypt after the Pharaohs. 332 BC – AD 642: From Alexander to the Arab Conquest*, London.

Brand, C. E. (1968), *Roman Military Law*, Austin.

Breeze, D. J. (1974), 'The Organisation of the Career Structure of the Immunes and Principales of the Roman Army', *Bonner Jahrbücher* 174: 245–92.

—— (1987), 'The Roman Army in Cumbria', *Transactions of the Cumberland and Westmorland Antiquarian and Archaeological Society* 88: 9–22 (= Breeze and Dobson 1993, 317–30).

Breeze, D. J., and B. Dobson (1993), *Roman Officers and Frontiers* (MAVORS 10), Stuttgart.

Brodribb, G., and H. Cleere (1988), 'The Classis Britannica Bath-House at Beauport Park, East Sussex', *Britannia* 19: 217–74.

Brown, V. M., and J. A. Harrell (1995), 'Topographical and petrological survey of ancient roman quarries in the Eastern Desert of Egypt', in Y. Maniatis, N. Herz, and Y. Basiakos (eds.), *The Study of Marble and other Stones used in Antiquity. Asmosia III Athens: Transactions of the 3rd International Symposium of the Association for the Study of Marble and other Stones in Antiquity*, London.

Brun, J.-P. (2003), 'Le village dans les carrières du wâdî al-Hammâmât (Persou I)', in H. Cuvigny (ed.), *La route de Myos Hormos. L'armée romaine dans le désert Oriental d'Égypte. Praesidia du désert de Bérénice I* (FIFAO 48/1), 95–97.

—— H. Cuvigny, and M. Reddé (2003), 'Le mystère des tours et la question des skopeloi', in H. Cuvigny (ed.), *La route de Myos Hormos. L'armée romaine dans le désert Oriental d'Égypte. Praesidia du désert de Bérénice I* (FIFAO 48/1), 207–34.

Brunt, P. A. (1983), 'Princeps and equites', *Journal of Roman Studies* 73: 42–75.

—— (1990*a*), 'The "Fiscus" and its Development', in P. A. Brunt (ed.), *Roman Imperial Themes*, Oxford, 134–62.

—— (1990*b*), 'Procuratorial Jurisdiction', in P. A. Brunt (ed.), *Roman Imperial Themes*, Oxford, 163–87.

—— (1990*c*),'Publicans in the Principate', in P. A. Brunt (ed.), *Roman Imperial Themes*, Oxford, 354–432.

Bruun, C. (1991), *The Water Supply of Ancient Rome. A Study of Roman Imperial Administration* (Commentationes Humanarum Litterarum 93), Helsinki.

—— (2001), '*Adlectus amicus consiliarius* and a freedman *proc. metallorum et praediorum*. News on Roman Imperial Administration', *Phoenix* 55/3–4: 343–68.

Bruzza, L. (1870), 'Iscrizioni dei marmi grezzi', *Annali dell' Instituto di Corrispondenza Archaeologica* 42: 106–204.

Buckler, W. H., W. M. Calder, and C. W. M. Cox (1926), 'Monuments from Central Phrygia', *Journal of Roman Studies* 16: 53–94.

Bulat, M. (1989), 'Novi rimski natpisi is Osijeka. Neue römische Inschriften aus Osijek', *Osjecki Zbornik* 20: 31–51.

Bülow-Jacobsen, A. (1992*a*), 'The Excavation and Ostraca of Mons Claudianus', in *Proceedings of the XIXth Congress of Papyrology, Cairo 2–5 September 1989, Vol. I*, Cairo, 49–59.

—— (1992*b*), 'The Archive of Successus', in Bingen et al. 1992, 111–36.

—— (1996), *Mons Claudianus. Organisation, administration og teknik i et romersk stenbrud fra kejsertiden* (Studier fra Sprog- og Oldsidsforskningent), Copenhagen.

—— (1997*a*), 'The Correspondence of Dioscorus and others', in Bingen et al. 1997, 43–68.

—— (1997*b*), 'The Correspondence of Petenephotes', in Bingen et al. 1997, 69–80.

—— (1997*c*), 'Lists of Vigiles', in Bingen et al. 1997, 165–92.

—— (1997*d*), 'On Smiths and Quarries', in B. Kramer, W. Luppe, H. Maehler, and G. Poethke (eds.), *Akten des 21. Internationalen Papyrologenkongresses Berlin, 13.–19.8. 1995. Bd. I-II* (APF Beihefte 3), Stuttgart & Leipzig, 139–45.

—— (1998), 'Traffic on the roads between Coptos and the Red Sea', in O. E. Kaper (ed.), *Life on the Fringe. Living in the Southern Egyptian Deserts During the Roman and Early-Byzantine Periods. Proceedings of a Colloquium Held on the Occasion of the 25th Anniversary of the Netherlands Institute for Archaeology and Arabic Studies in Cairo 9–12 December 1996*, Leiden, 63–74.

—— (2001), 'Drinking and Cheating in the Desert', in T. Gagos and R. Bagnall (eds.), *Essays and Texts in Honor of J. David Thomas* (American Studies in Papyrology 42), Exeter, 119–23.

—— and H. Cuvigny (2007), 'Sulpicius Serenus, *procurator Augusti*, et la titulature des préfets de Bérénice', *Chiron* 37: 11–33.

—— —— and J. L. Fournet (1994), 'The identification of Myos Hormos. New papyrological evidence', *Bulletin de l'Institut Français d'Archéologie Orientale* 94: 27–42.

Buonocore, M. (1998), 'Le iscrizioni imperiali d'età Giulio-Claudia nella Regio IV. Nuove proposte di lettura', *Epigraphica* 60: 45–70.

Burnham, B. C. (1997), 'Roman mining at Dolaucothi. The implications of the 1991–3 excavations near the Carreg Pumsaint', *Britannia* 28: 325–36.

Burnham, B., and H. Burnham (2004), *Dolaucothi-Pumsaint. Survey and Excavations at a Roman Gold-Mining Complex 1987–1999*, Oxford.

Burton, G. P. (1976), 'The issuing of mandata to proconsuls and a new inscription from Cos', *Zeitschrift für Papyrologie und Epigraphik* 21: 63–8.

—— (1993), 'Provincial procurators and the public provinces', *Chiron* 23: 13–28.

—— (1998), 'Was there a Long-Term trend to centralisation of authority in the Roman Empire?', *Revue de Philologie* 72/1: 7–24.

Burton, G. P. (1999), 'Vectigal', in S. Hornblower and A. Spawforth (eds.), *The Oxford Classical Dictionary*, 3rd edn. Oxford, 1583.

—— (2000), 'The resolution of territorial disputes in the provinces of the Roman Empire', *Chiron* 30: 195–215.

Caamaño Gesto, J. M. (1984), 'Excavaciones en el campamento romano de Cidadela (Sobrado dos Monxes, Coruña). Memoria preliminar de la campaña 1981', *Noticiario Arqueologico Hispanico* 18: 235–54.

—— (1989), 'Estampillas de la cohors I Celtiberorum halladas en el campamento romano de Cidadela', *Callaecia* 11: 209–28.

Campbell, J. B. (1984), *The Emperor and the Roman Army 31 BC – AD 235*, Oxford.

—— (2000), *The Writings of the Roman Land Surveyors. Introduction, Text, Translation and Commentary* (Society for the Promotion of Roman Studies. Journal of Roman Studies Monograph 9), London.

Capenelli, D. (1989), 'Aspetti dell'amministrazione mineraria iberica nell'età del principato', in C. Domergue (ed.), *Minería y Metalurgia en las Antiguas Civilizaciones Mediterráneas y Europeas, Coloquio Internacional Asòciado. Madrid, 24–28 Octobre, 1985*, Madrid, 138–46.

Cauuet, B., C. Domergue, and M. Urteaga (2005), 'Mines et métallurgies en Aquitaine et en Hispanie septentrionale sous les Julio-Claudiens', in *L'Aquitaine et l'Hispanie septentrionale à l'époque julio-claudienne. Organisation et exploitation des espaces provinciaux. Colloque Aquitania Saintes, 11–13 septembre 2003* (Aquitania Supplément 13), Bordeaux, 423–60.

Cavalieri Manasse, G., G. Massari, and M. P. Rossignani (1982), *Guide archeologiche Laterza. Piemonte, Valle d'Aosta, Liguria, Lombardia*, Rome & Bari.

Čerškov, E. (1970), *Municipium DD kod Sočanica*, Pristina & Belgrade.

Christol, M. (1990), 'Ti. Claudius Proculus Cornelianus, procurateur de la région de Théveste', in *L'Africa romana, Atti del VII. Convegno di Studio (Sassari, 15–17 dicembre 1989)*, Sassari, 893–905.

—— (1999), 'Un aspect de l'administration impériale: le procurateur des mines de Vipasca', *Pallas* 50: 233–44.

—— and S. Demougin (1990), 'De Lugo à Pergame. La carrière de l'affranchi Saturninus dans l'administration impériale', *Mélanges d'Archéologie et d'Histoire de l'École française de Rome* 102/1: 159–211.

—— and T. Drew-Bear (1986), 'Documents latins du Phrygie', *Tyche* 1: 41–87.

—— (1987), 'Inscriptions de Dokimeion', in B. Rémy (ed.), *Anatolica Antiqua/Eski Anadolu. Receuil de travaux publiés par l'Institut Français d'Études Anatoliennes d'Istanbul*, Istanbul & Paris, 83–137.

—— —— (1991), 'Les carrières de Dokimeion à l'époque severienne', *Epigraphica* 53: 113–74.

Ciongradi, C., A. Timofan, and V. Bârcă (2008), 'Eine neue Erwähnung des *kastellum Starva* in einer Inschrift aus *Alburnus Maior*. Studium zu epigraphisch bezeugten *kastella* und *vici* im dakischen Goldbergwerksgebiet', *Zeitschrift für Papyrologie und Epigraphik* 165, 249–66.

Clauss, M. (1973), *Untersuchungen zu den principales des römischen Heeres von Augustus bis Diokletian. Cornicularii, speculatores, frumentarii*, Bochum.

Cleere, H., and C. David (1995), *The Iron Industry of the Weald*, 2nd edn. Cardiff.

Cockle, W. E. H. (1996), 'An Inscribed Architectural Fragment from Middle Egypt concerning the Roman Imperial Quarries', in D. M. Bailey (ed.), *Archaeological Research in Roman Egypt. The Proceedings of the Seventeenth Classical Colloquium of the Department of Greek and Roman Antiquities, British Museum, held on 1–4 December, 1993*, Ann Arbor, 23–8.

Colls, D., C. Domergue, F. Laubenheimer, and B. Liou (1975), 'Les lingots d'étain de l'épave Port-Vendres II', *Gallia* 33: 75–7.

—— R. Étienne, R. Lequément, B. Liou, and F. Mayet (1977), *L'épave Port-Vendres II et le commerce de la Bétique a l'époque de Claude* (Archaeonautica 1), Paris.

—— C. Domergue, and V. Guerrero Ayuso (1986), 'Les lingots de plomb de l'épave romaine Cabrera 5 (Ile de Cabrera, Baléares)', *Archaeonautica* 6: 31–80.

Corbier, M. (1974), *L'aerarium Saturni et l'aerarium militare. Administration et prosopographie sénatoriale.* (Collection de l'École Française de Rome XXIV), Rome.

Corsten, T. (1997), *Die Inschriften von Laodikeia am Lykos. Teil 1* (Inschriften griechischer Städte aus Kleinasien 49/1), Bonn.

Costabile, F., and O. Licandro (2000), 'Tessera Paeimeiobrigensis. Un nuovo editto di Augusto dalla "Transduriana" e l'imperium proconsulare del princeps', *Minima Epigraphica et Papyrologica* 3/4: 147–235.

Craddock, P. T. (1995), *Early Metal Mining and Production*, Edinburgh.

Crawford, D. J. (1980), 'The Emperor as Landowner (Review of 'G. M. Parassoglou, Imperial Estates in Roman Egypt, Amsterdam 1978')', *Classical Review* 30: 251–3.

Crawford, M. H. (1996), *Roman Statutes*, London.

Cuvigny, H. (1992*a*), 'Inscription inédite d'un ἐργοδότης dans une carrière du Mons Claudianus', in C. Décobert (ed.), *Itinéraires d'Égypte. Mélanges offerts au père Maurice Martin*, Cairo, 73–88.

—— (1992*b*), 'La mort et la maladie', in Bingen et al. 1992, 75–109.

—— (1996*a*), 'The Amount of Wages paid to the Quarry-Workers at Mons Claudianus', *Journal of Roman Studies* 86: 139–45.

—— (1996*b*), 'Arbejderne', in A. Bülow-Jacobsen (ed.), *Mons Claudianus. Organisation, administration og teknik i et romersk stenbrud fra kejsertiden*, Copenhagen, 14–32.

—— (1996*c*), 'Ulpius Himerus, procurateur impérial: I Pan 53', *Bulletin de l'Institut Français d'Archéologie Orientale* 96: 91–101.

—— (1997*a*), 'La mort et la maladie', in J. Bingen (ed.), *Mons Claudianus. Ostraca Graeca et Latina II (O. Claud. 191 à 416)*, Cairo, 19–41.

—— (1997*b*), 'Tableaux de Service', in J. Bingen (ed.), *Mons Claudianus. Ostraca Graeca et Latina II (O. Claud. 191 à 416)*, Cairo, 141–63.

—— (1998), 'Kainè, ville nouvelle: Une experience de regroupement familial au IIe s.è.chr.', in O. E. Kaper (ed.), *Life on the Fringe. Living in the Southern Egyptian Deserts During the Roman and Early-Byzantine Periods. Proceedings of a Colloquium held on the Occasion of the 25th Anniversary of the Netherlands Institute for Archaeology and Arabic Studies in Cairo 9–12 December 1996*, Leiden, 87–94.

Cuvigny, H. (2000*a*), *Mons Claudianus. Ostraca Graeca et Latina III. O. Claud. 417 à 631* (IFAO 38), Cairo.

—— (2000*b*), 'Procurator montis', in H. Melaerts (ed.), *Papyri in honorem Johannis Bingen octogenarii (P. Bingen)*, Leuven, 415–18.

—— (2001), 'Claudius Lucilianus, préfet d'aile et de Bérénice', in T. Gagos and R. Bagnall (eds.), *Essays and Texts in Honor of J. David Thomas* (American Studies in Papyrology 42), Exeter, 171–4.

—— (2002), 'Vibius Alexander, praefectus et épistratège de l'Heptanomie', *Chronique d'Égypte* 77: 238–48.

—— (2003*a*), 'Les documents écrits de la route de Myos Hormos à l'époque gréco-romaine', in H. Cuvigny (ed.), *La route de Myos Hormos. L'armée romaine dans le désert Oriental d'Égypte. Praesidia du désert de Bérénice I* (FIFAO 48/2), Cairo, 265–93.

—— (2003*b*), 'Le fonctionnement du réseau', in H. Cuvigny (ed.), *La route de Myos Hormos. L'armée romaine dans le désert Oriental d'Égypte. Praesidia du désert de Bérénice I* (FIFAO 48/2), Cairo, 295–360.

—— (2003*c*), 'La société civile des praesidia', in H. Cuvigny (ed.), *La route de Myos Hormos. L'armée romaine dans le désert Oriental d'Égypte. Praesidia du désert de Bérénice I* (FIFAO 48/2), Cairo, 361–97.

—— (2005*a*), *Ostraca de Krokodilô. La correspondence militaire et sa circulation. O. Krok. 1–151. Praesidia du désert de Bérénice* (FIFAO 51), Cairo.

—— (2005*b*), 'L'organigramme du personnel d'une carrière impériale d'après un ostracon du Mons Claudianus', *Chiron* 35: 309–53.

—— and G. Wagner (1986), 'Ostraca Grecs du Mons Claudianus', *Zeitschrift für Papyrologie und Epigraphik* 62: 63–73.

Daguet-Gagey, A. (1997), *Les opera publica à Rome (180–305 ap. J.-C.)* (Collection des Études Augustiniennes. Série Antiquité 156), Paris.

Damian, Paul (ed.) (2003), *Alburnus Maior I*, Bucharest.

Dardaine, S. (1983), 'La gens Argentaria en Hispania', *Mélanges de la Casa de Velázquez* 19: 5–15.

Daris, S. (2000), 'Legio XXII Deiotariana', in Y. Le Bohec and C. Wolff (eds.), *Les Légions de Rome sous le Haut-Empire. Actes du Congrès de Lyon, 17–19 septembre 1998* (Collection du Centre d'Études Romaines et Gallo-Romaines. NS 20), Lyon, 365–7.

Davies, O. (1935), *Roman Mines in Europe*, Cambridge.

Davies, R. W. (1989), *Service in the Roman Army*, Edinburgh.

Deininger, J. (1965), *Die Provinziallandtage der Römischen Kaiserzeit von Augustus bis zum Ende des dritten Jahrhunderts n. Chr.* (Vestigia 6), Munich.

De Laet, S. J. (1949), *Portorium. Étude sur l'organisation douanière chez les Romains, surtout à l'époque du Haut-Empire*, Brugge.

Desjardins, E. (1867–85), *Géographie historique et administrative de la Gaule romaine. I*, Paris.

Devijver, H. (1976–80), *Prosopographia militiarum equestrium quae fuerunt ab Augusto ad Gallienum*, Leuven.

Dise, R. L. (1991), *Cultural Change and Imperial Administration. The Middle Danube Provinces of the Roman Empire* (American University Studies. Series IX. History 99), New York, Bern, & Paris.

—— (1995), 'A reassessment of the functions of *beneficiarii consularis*', *Ancient History Bulletin* 9: 72–85.

Dixon, K. R., and P. Southern (1997), *The Roman Cavalry from the First to the Third Century AD*, London & New York.

Dobó, Á. (1968), *Die Verwaltung der römischen Provinz Pannonien von Augustus bis Diocletianus. Die provinziale Verwaltung*, Amsterdam.

Dobson, B. (1974), 'The significance of the centurion and "Primipilaris" in the Roman army and administration', in H. Temporini (ed.), *Aufstieg und Niedergang der Römischen Welt. Geschichte und Kultur Roms im Spiegel der Neueren Forschung II. Principat, Erster Band*, Berlin & New York, 393–434.

—— (1978), *Die Primipilares. Entwicklung und Bedeutung, Laufbahnen und Persönlichkeiten eines römischen Offiziersranges* (Beihefte der Bonner Jahrbücher 37), Cologne & Bonn.

Dodge, H. (1988), 'Decorative stones for architecture in the Roman Empire', *Oxford Journal of Archaeology* 7/1: 65–80.

—— (1991), 'Ancient Marble Studies. Recent research', *Journal of Roman Archaeology* 4: 28–50.

—— and B. Ward-Perkins (eds.) (1992), *Marble in Antiquity. Collected Papers of J. P. Ward-Perkins* (Archaeological Monographs of the British School at Rome 6), London.

Dolci, E. (1981–2), 'La localizzazione ed il rilevamento delle cave lunensi', *Quaderni. Centro Studi Lunensi* 6/7: 47–62.

—— (1983), 'Nuovi rilevamenti nelle cave lunensi. Notizie preliminari sui siti di Tarnone e Fossa Ficola', *Quaderni. Centro Studi Lunensi* 8: 53–88.

—— (1997), 'Un'officina imperiale nelle cave lunensi: il sito del monte Strinato a Carrara', *Quaderni. Centro Studi Lunensi.* ns 3: 27–46.

—— (1998), 'Una cava lunense scoperta di recente a Carrara: il sito della Scalocchiella', *Quaderni. Centro Studi Lunensi.* ns 4: 115–38.

Domergue, C. (1966), 'Les lingots de plomb romains du Musée Archéologique de Carthagène et du Musée Naval de Madrid', *Ausgrabungen und Funde* 39: 41–72.

—— (1971), 'El Cerro del Plomo, Mina "El Centenillo" (Jaen)', *Noticiario Arqueologico Hispanico* 16: 267–382.

—— (1983), *La mine antique d'Aljustrel (Portugal) et les tables de bronze de Vipasca*, Paris.

—— (1986), 'Dix-huit ans de recherche (1968–1986) sur les mines d'or romaines du Nord-Ouest de la Péninsule Ibérique', in *Actas I Congreso Internacional Astorga Romana 2*: 7–101.

—— (1987*a*), *Catalogue des mines et des fonderies antiques de la Péninsule Ibérique* (Publications de la Casa Velazquez. Serie Archeologie VIII), Madrid.

—— (1987*b*), 'Les lingots de plomb de l'époque romaine de Valle Ponti (Comacchio)', *Epigraphica* 49: 109–68.

Domergue, C. (1990), *Les mines de la Peninsule Iberique dans l'Antiquité Romaine* (Collection de l'École française de Rome 127), Rome.

—— (1998), 'A view of Baetica's external commerce in the 1st c. A.D. based on its Trade in Metals', in S. Keay (ed.), *The Archeology of Early Roman Baetica* (Journal of Roman Archaeology. Supplement Series 29), Portsmouth, RI, 201–15.

—— (2004), 'Le régime juridique des mines du domaine public à Rome. À propos d'un ouvrage récent', *Mélanges de la Casa Velázquez.* ns 34/2: 221–36.

—— and G. Herail (1978), *Mines d'or romaines d'Espagne. Le district de la Valduerna (León). Étude géomorphologique et archéologique,* Toulouse.

—— and M. Leroy (2000), 'L'état de la recherches sur les mines et les métallurgies en Gaule, de l'époque gauloise au haut Moyen Âge', *Gallia* 57: 3–10.

—— and T. Martin (1977), *Minas de Oro Romanas de la Provincia de León II. Hueña: Excavaciones 1972–1973* (Excavaciones arqueológicas en España 94), Madrid.

—— and C. Rico (2002), 'À propos de deux lingots de cuivre antiques trouvés en mer sur la côte languedocienne', in L. Rivet, M. Scialliano (eds.), *Vivre, produire et échanger: réflets méditeranéens. Mélanges offerts à Bernard Liou* (Archéologie et histoire romaine 8), Montagnac, 141–52.

—— and P. Sillieres (1977), *Minas de Oro Romanas de la Provincia de León I. La Corona de Quintanilla: Excavaciones 1971–1973. Las Coronas de Filiel, Boisan, Luyego 1 y 2: Exploraciones 1973.* (Excavaciones arqueológicas en España 94), Madrid.

—— P. Quarati, A. Nesta, and P. R. Trincherini (2006), 'Retour sur les lingots de plomb de Comacchio (Ferrara, Italie) en passant par l'archéométrie et l'épigraphie', in *Mineria Antigua: estudios regionales y temas de investigacion actual, Anejos de Archivo Espanol de Arqueologia,* Instituto de Historia, CSIC, Madrid (forthcoming), cf. http://eprintweb.org/S/ authors/All/ qu/Quarati/3 (August 11: 2007).

Dopico Caínzos, M. D. (1988), *La tabula Lougeiorum. Estudios sobre la implantacion romana en Hispania,* Vitoria/Gasteiz.

Dräger, M. (1993), *Die Städte der Provinz Asia in der Flavierzeit* (Europäische Hochschulschriften. Reihe III. Geschichte und Hilfswissenschaften 576), Frankfurt, Berlin, & Bern.

Drew-Bear, M. (1979), *Le nome Hermopolite: toponymes et sites* (American Studies in Papyrology 21), Ann Arbor.

Drew-Bear, T. (1994), 'Nouvelles Inscriptions de Dokimeion', *Mélanges d'Archéologie et d'Histoire de l'École française de Rome* 106: 747–844.

—— and W. Eck (1976), 'Kaiser-, Militär- und Steinbruchinschriften aus Phrygien', *Chiron* 6: 289–318.

—— and C. Naour (1990), 'Divinités de Phrygie', in W. Haase and H. Temporini (eds.), *Aufstieg und Niedergang der Römischen Welt. Geschichte und Kultur Roms im Spiegel der Neueren Forschung. Teil II: Principat, Band 18. Religion (Heidentum: Die Religiösen Verhältnisse in den Provinzen [Forts.]). 3. Teilband,* Berlin & New York, 1907–2044.

Drexhage, H.-J., H. Konen, and K. Ruffing (2002), *Die Wirtschaft der Römischen Reiches (1.–3. Jahrhundert). Eine Einführung*, Berlin.

Dubois, C. (1908), *Étude sur l'administration et l'exploitation des carrières marbres, porphyre, granit, etc. dans la monde romain*, Paris.

Duncan-Jones, R. (1994), *Money and Government in the Roman Empire*, Cambridge.

Dupuis, X. (2000), 'Les III publica Africae. Un exemple de personnel administratif subalterne en Afrique', *Cahiers du Centre Gustave-Glotz* 11: 277–94.

Dušanić, S. (1971*a*), 'The Antinous Inscription at Sočanica and the *metalla municipii Dardanorum*', *Živa Antika* 21: 241–61.

—— (1971*b*), 'The Heteroclite *metalli* on the Roman coins', *Živa Antika* 21: 535–54.

—— (1976), 'Le nord-ouest de la Mésie supérieure', in M. Mirković and S. Dušanić (eds.), *Inscriptions de la Mésie supérieure. Vol. I. Singidunum et le nord-ouest de la province*, Belgrade, 95–162.

—— (1977), 'Aspects of Roman mining in Noricum, Pannonia, Dalmatia and Moesia Superior', in H. Temporini and W. Haase (eds.), *Aufstieg und Niedergang der Römischen Welt. Geschichte und Kultur Roms im Spiegel der Neueren Forschung II. Principat, Sechster Band. Politische Geschichte (Provinzen und Randvölker: Lateinischer Donau-Balkanraum)*, Berlin & New York, 52–94.

—— (1978), 'Mounted cohorts in Moesia Superior', in *Akten des XI. Internationalen Limeskongresses, Székesfehérvár 1976*, Budapest, 237–46.

—— (1989), 'The Roman mines of Illyricum. Organization and impact on provincial life', in C. Domergue (ed.), *Minería y metalurgia en las Antiguas Civilizaciones Mediterráneas y Europeas, Coloquio Internacional Asòciado. Madrid, 24–28 Octobre, 1985*, Madrid, 148–56.

—— (1990), 'The legions and the fiscal estates in Moesia Superior: Some epigraphic notes', *Arheoloski vestnik* 41: 585–96.

—— (1994–5), 'Epigraphical notes on Roman mining in Dardania', *Starinar* 45–6: 27–34.

—— (2000), 'Army and mining in Moesia Superior', in G. Alföldy, B. Dobson, and W. Eck (eds.), *Kaiser, Heer und Gesellschaft in der Römischen Kaiserzeit. Gedenkschrift für Eric Birley* (HABES 31), Stuttgart, 343–64.

—— (2004*a*), 'The *princeps municipii Dardanorum* and the *metalla municipii Dardanorum*', *Živa Antika* 54: 5–32.

—— (2004*b*), 'Roman mining in Illyricum: Historical aspects', in G. Urso (ed.), *Dall' Adriatico al Danubio. L'Illyrico nell' età greca e romana. Atti del convegno internazionale Cividale del Friuli, 25–27 settembre 2003*, Pisa, 247–70.

Dworakowska, A. (1983), *Quarries in Roman Provinces*, Wroclaw.

Eck, W. (1979), *Die staatliche Organisation Italiens in der hohen Kaiserzeit*, Munich.

—— (1985), *Die Statthalter der germanischen Provinzen vom 1.–3. Jahrhundert* (Epigraphische Studien 14), Cologne & Graz.

—— (1994), 'M. Lucretius Iulianus, *procurator augustorum*. Zur Funktion und sozialen Wertschätzung von Provinzialprokuratoren', *Zeitschrift für Papyrologie und Epigraphik* 100: 559–76.

—— (1995*a*), 'Zur Durchsetzung von Anordnungen und Entscheidungen in der Hohen Kaiserzeit: die administrative Informationsstruktur', in W. Eck, *Die*

Verwaltung des Römischen Reiches in der Hohen Kaiserzeit. Ausgewählte und erweiterte Beiträge. 1. Band (AREA 1), Basel & Berlin, 55–79.

Eck, W. (1995*b*), 'Die Leitung und Verwaltung einer prokuratorischen Provinz', in W. Eck (ed.), *Die Verwaltung des Römischen Reiches in der Hohen Kaiserzeit. Ausgewählte und erweiterte Beiträge. 1. Band* (AREA 1), Basel & Berlin, 327–40.

—— (1995*c*), 'Prosopographica II', *Zeitschrift für Papyrologie und Epigraphik* 106: 249–54.

—— (1995*d*), 'Die staatliche Administration des römischen Reiches in der hohen Kaiserzeit. Ihre strukturellen Komponenten', in W. Eck (ed.), *Die Verwaltung des Römischen Reiches in der Hohen Kaiserzeit. Ausgewählte und erweiterte Beiträge. 1. Band* (AREA 1), Basel & Berlin, 1–28.

—— (1995*e*), 'Die Umgestaltung der politischen Führungsschicht—Senatorenstand und Ritterstand', in W. Eck (ed.), *Die Verwaltung des Römischen Reiches in der Hohen Kaiserzeit. Ausgewählte und erweiterte Beiträge. 1. Band* (AREA 1), Basel & Berlin, 103–78.

—— (1997*a*), 'Der Kaiser, die Führungsschichten und die Administration des Reiches (von Vespasian bis zum Ende des antoninischen Dynastie)', in W. Eck (ed.), *Die Verwaltung des Römischen Reiches in der Hohen Kaiserzeit. Ausgewählte und erweiterte Beiträge, 2. Band* (AREA 3), Basel & Berlin, 3–146.

—— (1997*b*), 'Prokonsuln und militärisches Kommando. Folgerungen aus Diplomen für prokonsulare Provinzen', in W. Eck (ed.), *Die Verwaltung des Römischen Reiches in der Hohen Kaiserzeit. Ausgewählte und erweiterte Beiträge, 2. Band* (AREA 3), Basel & Berlin, 187–202.

—— (1997*c*), 'Die Bedeutung der Claudischen Regierungszeit für die Administrative Entwicklung des römischen Reiches', in W. Eck (ed.), *Die Verwaltung des Römischen Reiches in der Hohen Kaiserzeit. Ausgewählte und erweiterte Beiträge, 2. Band* (AREA 3), Basel & Berlin.

—— (1999), '[II 3] Lucilius Capito', in H. Cancik and H. Schneider (eds.), *Der Neue Pauly. Enzyklopädie der Antike. Band 7*, Stuttgart & Weimar, 465.

—— (2000), 'The Growth of Administrative Posts', in A. K. Bowman, P. Garnsey, and D. Rathbone (eds.), *The Cambridge Ancient History. Second Edition. Volume XI. The High Empire, AD 70–192*, Cambridge, 238–65.

—— (2001*a*), 'Praefectus castrorum', in H. Cancik and H. Schneider (eds.), *Der Neue Pauly. Enzyklopädie der Antike. Band 10*, Stuttgart & Weimar, 243.

—— (2001*b*), 'Praefectus civitatium', in H. Cancik and H. Schneider (eds.), *Der Neue Pauly. Enzyklopädie der Antike. Band 10*, Stuttgart & Weimar, 244.

—— (2001*c*), 'Spezialisierung in der staatlichen Administration des Römischen Reiches in der Hohen Kaiserzeit', in L. DeBlois (ed.), *Administration, Prosopography and Appointment Policies in the Roman Empire. Proceedings of the First Workshop of the International Network. Impact of Empire (Roman Empire, 27 BC – AD 406). Leiden, June 28 – July 1, 2000*, Amsterdam, 1–23.

—— (2001*d*), 'Procurator', in H. Cancik and H. Schneider (eds.), *Der Neue Pauly. Encyclopädie der Antike. Band 10*, Stuttgart & Weimar, 366–9.

—— (2004*a*), *Köln in römischer Zeit. Geschichte einer Stadt im Rahmen des Imperium Romanum* (Geschichte der Stadt Köln. Band 1), Cologne.

—— (2004*b*), 'Augustus und die Grossprovinz Germanien', *Kölner Jahrbuch* 37: 11–22.

—— and F. Fernández (1991), 'Sex. Marius in einem Hospitiumvertrag aus der Baetica', *Zeitschrift für Papyrologie und Epigraphik* 85: 217–22.

—— and A. Pangerl (2005), 'Traians Heer im Partherkrieg. Zu einem Diplom aus dem Jahr 115', *Chiron* 35: 49–67.

—— A. Caballos, and F. Fernández (1996), *Das senatus consultum de Cn. Pisone patre* (Vestigia. Beiträge zur Alten Geschichte 48), Munich.

Eder, W. (1980), *Servitus Publica. Untersuchungen zur Entstehung, Entwicklung und Funktion der öffentlichen Sklaverei in Rom* (Forschungen zur Antiken Sklaverei), Wiesbaden.

Edmondson, J. C. (1989), 'Mining in the later Roman Empire and beyond. Continuity or disruption?', *Journal of Roman Studies* 79: 84–102.

Egger, R. (1961), *Die Stadt auf dem Magdalensberg. Ein Grosshandelsplatz. Die ältesten Aufzeichnungen über den Metallwarenhandel auf dem Boden Österreichs*, Graz, Vienna, & Cologne.

Eich, P. (2005), *Zur Metamorphose des politischen Systems in der römischen Kaiserzeit. Die Entstehung einer "personalen Bürokratie" im langen dritten Jahrhundert* (Klio Beihefte 9), Berlin.

Elkington, H. D. H. (1976), 'The Mendip lead industry', in K. Branigan and P. J. Fowler (eds.), *The Roman West Country. Classical Culture and Celtic Society*, Newton Abbot & London, 183–97.

Engelmann, H., and D. Knibbe (1989), *Das Zollgesetz der Provinz Asia* (Epigraphica Anatolica 14), Bonn.

Fabian, K.-D. (1979), 'Vilicus', in K. Ziegler, W. Sontheimer, and H. Gärtner (eds.), *Der Kleine Pauly. Lexikon der Antike in fünf Bänden. Band 5*, Munich, 1272–3.

Fant, J. C. (1985), 'Four unfinished sarcophagus lids at Docimium and the Roman imperial quarry system in Phrygia', *American Journal of Archaeology* 89: 655–62.

—— (1988). 'The Roman Emperor in the marble business. Capitalists, middlemen or philanthropists?', in N. Herz and M. Waelkens (eds.), *Classical Marble: Geochemistry, Technology, Trade. Proceedings of the NATO Advanced Research Workshop on Marble in Ancient Greece and Rome. Geology, Quarries, Commerce, Artifacts, Il Ciocco, Lucca, Italy, 9–13 May 1988*, Dordrecht.

—— (1989*a*), *Cavum Antrum Phrygiae. The Organization and Operations of the Roman Imperial Marble Quarries in Phrygia* (BAR International Series 482), Oxford.

—— (1989*b*), 'Poikiloi lithoi. The anomalous economics of the Roman imperial marble quarry at Teos', in *The Greek Renaissance in the Roman Empire. Papers from the Tenth British Museum Classical Colloquium, London 1986*, London, 206–18.

—— (1993*a*), 'Ideology, gift, and trade. A distribution model for Roman imperial marbles', in W. V. Harris (ed.), *The Inscribed Economy. Production and Distribution in the Roman Empire in the Light of the Instrumentum Domesticum*, Ann Arbor, 147–70.

Fant, J. C. (1993*b*), 'The Roman imperial marble trade. A distribution model', in R. Francovich (ed.), *Archeologia delle attività estrattive e metallurgiche. V Ciclo di Lezioni sulla Ricerca Applicata in Archeologia Certosa di Pontignano (SI)—Campiglia Marittima (LI), 9–21 settembre 1991*, Florence, 71–96.

—— (2001), 'Rome's marble yards', *Journal of Roman Archaeology* 14: 167–98.

Fernández Ochoa, C., M. Zarzalejos Prieto, P. Hevia Gómez, and G. Esteban Borrajo (1994), *Sisapo I. Excavaciones Arqueológicas en 'La Bienvenida', Almodóvar del Campo (Ciudad Real)*, Toledo.

Fernández-Posse, M. D. (1988), *La Corona y el Castro de Corporales II. Campaña de 1983 y prospecciones en La Valderia y La Cabrera (León)*, León.

—— and F. J. Sánchez-Palencia (1985), *La Corona y el Castro de Corporales I. Truchas (León). Campañas de 1978 a 1981*, León.

Fink, R. O. (1971), *Roman Military Records on Papyrus* (Philological Monographs 26), Cleveland.

Fitz, J. (1972), 'Verwaltung der pannonischen Bergwerke', *Klio* 54: 213–25.

—— (1990), 'Prosopographia Pannonica II', *Alba Regia* 24: 169–96.

—— (1993–5), 'Die Verwaltung Pannoniens in der Römerzeit', Budapest.

Fitzler, K. (1910), *Steinbrüche und Bergwerke im Ptolemäischen und Römischen Aegypten. Ein Beitrag zur antiken Wirtschaftsgeschichte*, Leipzig.

Flach, D. (1978), 'Inschriftenuntersuchungen zum römischen Kolonat in Nordafrika', *Chiron* 8: 441–92.

—— (1979), 'Die Bergwerksordnung von Vipasca', *Chiron* 9: 399–448.

—— (1985), 'Review of "Claude Domergue. La Mine antique d'Aljustrel. Portugal et les Tables de Bronze de Vipasca, Paris 1985"', *Gnomon* 57: 201–3.

—— (1989), 'Teilpacht im Römischen Bergbau', in C. Domergue (ed.), *Minería y Metalurgia en las Antiguas Civilizaciones Mediterráneas y Europeas, Coloquio Internacional Asòciado. Madrid, 24–28 Octobre, 1985*, Madrid, 133–7.

—— (1990), *Römische Agrargeschichte* (Handbuch der Altertumswissenschaft. 3. Abt., 9. Teil), Munich.

France, J. (2001), *Quadragesima Galliarum. L'organisation douanière des provinces alpestres, gauloises et germaniques de l'Empire romain* (Collection de l'École française de Rome 278), Rome.

Frank, T. (1959), *An Economic Survey of Ancient Rome. Vol. 1: Rome and Italy of the Republic*, 2nd edn. New Jersey.

Franke, T. (2000), 'Legio XXII Primigenia', in Y. Le Bohec and C. Wolff (eds.), *Les légions de Rome sous le Haut-Empire. Actes du Congrès de Lyon, 17–19 septembre 1998* (Collection du Centre d'Études Romaines et Gallo-Romaines. Nouvelle série 20), Lyon, 95–104.

Frend, W. H. C. (1956), 'A third-century inscription Relating to *angareia* in Phrygia', *Journal of Roman Studies* 46: 46–56.

Galsterer, H. (1971), 'Review of "András Mócsy, Gesellschaft und Romanisation in der römischen Provinz Moesia Superior, Amsterdam 1970"', *Bonner Jahrbücher* 171: 738–43.

—— (1988), 'Municipium Flavium Irnitanum. A Latin town in Spain', *Journal of Roman Studies* 78: 78–90.

García Bellido, A. (1966), 'Nuevos documentos militares de la Hispania Romana', *Archivo español de arqueología* 39: 24–40.

García-Bellido, M. P. (1994–5) 'Las torres-recinto y la explotación des plomo en Extremadura', *Anas* 7–8: 187–218.

—— (1998), 'Legionsstempel aus der Zeit des Agrippa auf hispanischen Bleibarren', *Bonner Jahrbücher* 198: 1–28.

García Merino, C. (1973), 'Las tierras del NO. de la Península Ibérica, foco de atracción para los emigrantes de la Meseta en época romana', *Hispanica Antiqua* 9–28.

Geerlings, W. (1983), 'Ins Bergwerk verurteilt—die "damnatio ad metalla". Bergwerksarbeit und Christenvergfolgung', *Der Anschnitt* 35: 130–136.

Gibson, E. (1978), 'The Rahmi Koç Collection. Inscriptions. Part I, Grave Monuments from the Plain of Altintas', *Zeitschrift für Papyrologie und Epigraphik* 28: 1–34.

González, J. (1986), 'The lex Irnitana. A new copy of the Flavian Municipal Law', *Journal of Roman Studies* 76: 147–243.

Greene, K. (2000), 'Industry and technology', in A. K. Bowman, P. Garnsey, and D. W. Rathbone (eds.), *The High Empire, AD 70–192. The Cambridge Ancient History. Second Edition. Volume XI*, Cambridge, 741–68.

Griepentrog, M. (1995), 'Bergbau im Umland von Munigua. Neue Ergebnisse zur Wirtschaftsgrundlage der Stadt', in *Provinzialrömische Forschungen. Festschrift für Günter Ulbert zum 65. Geburtstag*, Espelkamp, 237–52.

Gudea, N. (1977), 'Der Limes Dakiens und die Verteidigung der obermoesischen Donaulinie von Trajan bis Aurelian', in H. Temporini and W. Haase (eds.), *Aufstieg und Niedergang der Römischen Welt. Geschichte und Kultur Roms im Spiegel der Neueren Forschung II. Principat, Sechster Band. Politische Geschichte (Provinzen und Randvölker: Lateinischer Donau-Balkanraum)*, Berlin & New York, 848–87.

—— and Th. Lobüscher (2006), *Dacia. Eine römische Provinz zwischen Karpaten und Schwarzem Meer*, Mainz.

Gustafson, M. W. (1994), 'Condemnation to the mines in the later Roman Empire', *Harvard Theological Review* 87/4: 421–33.

—— (1997), 'Inscripta in fronte: Penal tattooing in late antiquity', *Classical Antiquity* 16/1: 79–105.

Habicht, C. (1969), *Die Inschriften des Asklepeions* (Altertümer von Pergamon VIII/3), Berlin.

Haensch, R. (1992), 'Das Statthalterarchiv', *Zeitschrift der Savigny-Stiftung für Rechtsgeschichte* 109: 209–317.

—— (1997), *Capita provinciarum. Statthaltersitze und Provinzverwaltung in der römischen Kaiserzeit* (Kölner Forschungen 7), Mainz.

—— (2000), 'Le rôle des officiales de l'administration provinciale dans le processus de décision', *Cahiers du Centre Gustave-Glotz* 9: 259–76.

Hagedorn, D. (1993), 'P. Hibeh II 273 + 217. Antrag auf Bezahlung von Säulen, Säulenbasen und Kapitellen', *Zeitschrift für Papyrologie und Epigraphik* 97: 97–101.

—— and J. C. Shelton (1975), 'Zur Entlassung von Strafgefangenen im Römischen Ägypten', *Zeitschrift für Papyrologie und Epigraphik* 18: 225–8.

Haley, E. W. (1991), *Migration and Economy in Roman Imperial Spain* (Aurea Saecula 5), Barcelona.

Halfmann, H. (2001), *Städtebau und Bauherren im Römischen Kleinasien. Ein Vergleich zwischen Pergamon und Ephesos* (Istanbuler Mitteilungen, Beiheft 43), Tübingen.

Hamilton-Dyer, S. (2001), 'The faunal remains', in V. A. Maxfield and D. P. S. Peacock (eds.), *Survey and Excavation. Mons Claudianus. 1987–1993. Volume II. Excavations: Part 1*, Cairo, 249–310.

Hanel, N., and P. Rothenhöfer (2005), 'Germanisches Blei für Rom. Zur Rolle des Römischen Bergbaus im rechtsrheinischen Germanien im frühen Prinzipat', *Germania* 83: 53–65.

Hankey, V. (1965), 'A marble quarry at Karystos', *Bulletin du Musée de Beyrouth* 18: 53–9.

Harris, W. V. (2000), 'Trade', in A. K. Bowman, P. Garnsey, and D. W. Rathbone (eds.), *The High Empire, AD 70–192. The Cambridge Ancient History. Second Edition. Volume XI*, Cambridge, 710–40.

Hartmann, M., and M. A. Speidel (1991), 'Die Hilfstruppen des Windischer Heeresverbandes', *Jahresbericht der Gesellschaft Pro Vindonissa* 3–33.

Hatch, M. J. (1997), *Organization Theory. Modern Symbolic and Postmodern Perspectives*, Oxford.

Hauken, T. (1998), *Petition and Response. An Epigraphic Study of Petitions to Roman Emperors, 181–249* (Monographs from the Norwegian Institute at Athens 2), Bergen.

Heberdey, R. (1912), 'Vorläufiger Bericht über die Grabungen in Ephesos', *Jahreshefte des Österreichischen Archäologischen Institutes* 12: 157–82.

Hengstl, J. (1981), 'Review of "Georges M. Parassoglou, imperial estates in Roman Egypt, Amsterdam 1978"', *Gnomon* 53: 37–41.

Hérail, G., and L. C. Pérez García (1989), 'Intérêt archéologique d'une étude géomorpho-litologique: Les gisements d'or alluvial du nord-ouest de l'Espagne', in C. Domergue (ed.), *Minería y metalurgia en las Antiguas Civilizaciones Mediterráneas y Europeas, Coloquio Internacional Asòciado. Madrid, 24–28 Octobre, 1985*, Madrid, 21–34.

Herrmann, P. (1988), 'Chresimus, procurator lapicidinarum. Zur Verwaltung der kaiserlichen Steinbrüche in der Provinz Asia', *Tyche* 3: 119–28.

—— (1990), *Hilferufe aus römischen Provinzen. Ein Aspekt der Krise des römischen Reiches im 3. Jhdt.n.Chr.* (Berichte aus den Sitzungen der Joachim Jungius-Gesellschaft der Wissenschaften E. V., Hamburg, Jahrgang 8, Heft 4), Hamburg.

—— (1998), *Inschriften von Milet. Teil 2. Inschriften n. 407–1019*, Berlin & New York.

Herz, P. (2005), 'Der römische Staat und die Wirtschaft. Staatliche Eingriffe in das Wirtschaftsleben (Kontrolle von Ressourcen)', in M. Polfer (ed.), *Artisanat et économie romaine: Italie et provinces occidentales de l'Empire. Acte du 3ᵉcolloque international d'Erpeldange (Luxembourg) sur l'artisanat romain—14–16 octobre 2004*, 17–30.

HILL, W., R. FEHLBAUM, and P. ULRICH (1994), *Organisationslehre 1. Ziele, Instrumente und Bedingungen der Organisation sozialer Systeme*, 5th edn. Bern, Stuttgart, & Vienna.

HIRSCHFELD, O. (1905), *Die kaiserlichen Verwaltungsbeamten bis auf Diokletian*, 2nd edn. Berlin.

—— (1913), 'Der Grundbesitz der römischen Kaiser in den ersten drei Jahrhunderten (1902)', in O. Hirschfeld (ed.), *Kleine Schriften*, Berlin, 516–75.

HIRT, A. M. (2004), 'Chresimus, *procurator a marmoribus*, und der Strassenbau. Überlegungen zu Logistik und Transportwesen kaiserlicher Steinbrüche', in R. Frei-Stolba (ed.), *Siedlungen und Verkehr im Römischen Reich. Römerstrassen zwischen Herrschaftssicherung und Landschaftsprägung. Akten des Internationalen Kolloquiums zu Ehren von Prof. Dr. H. E. Herzig, Universität Bern, 28. und 29. Juni 2000*, Basel, 115–33.

HORN, H. G. (ed.) (1987), *Die Römer in Nordrhein-Westfalen*, Stuttgart.

HORSTER, M. (2001), *Bauinschriften römischer Kaiser. Untersuchungen zu Inschriftenpraxis und Bautätigkeit in Städten des westlichen Imperium Romanum in der Zeit des Prinzipats* (Historia Einzelschriften 157), Stuttgart.

ISAAC, B. (1991), 'The Roman army in Judaea. Police duties and taxation', in V. A. Maxfield and M. J. Dobson (eds.), *Roman Frontier Studies 1989. Proceedings of the XVth International Congress of Roman Frontier Studies*, Exeter, 458–61.

JOHNSON, A. (1987), *Römische Kastelle des 1. und 2. Jahrhunderts n. Chr. in Britannien und in den germanischen Provinzen des Römerreiches*, Mainz.

JONES, A. H. M. (1960), 'The aerarium and the fiscus', in A. H. M. Jones (ed.), *Studies in Roman Government and Law*, Oxford, 102–14.

JONES, R. F. J. (1976), 'The Roman military occupation of north-west Spain', *Journal of Roman Studies* 176: 45–66.

JONES, G. D. B. (1980), 'The Roman Mines of Rio Tinto', *Journal of Roman Studies* 80: 146–65.

JONES, C. P. (1987), 'Stigma: Tattooing and branding in Graeco-Roman antiquity', *Journal of Roman Studies* 77: 139–55.

JONES, G. D. B. (1991), 'Dating and Dolaucothi', *Britannia* 22: 210–11.

—— and P. R. LEWIS (1971), 'The Dolaucothi gold-mines', *Bonner Jahrbücher* 171: 288–300.

—— and J. H. LITTLE (1973), 'Excavations of the Roman Fort at Pumpsaint, Carmarthenshire. Interim Report 1972', *Carmarthenshire Antiquary* 9: 3–28.

JONES, B., and D. J. MATTINGLY (1990), *An Atlas of Roman Britain*, Oxford.

KAMBITSIS, S. (1997), *Papyrus Graux III (P. Graux 30)* (École Pratique des Hautes Études, IVe Section. III Hautes études du monde gréco-romain 23), Geneva.

KARWIESE, S. (1995), *Gross ist die Artemis von Ephesos. Die Geschichte einer der grossen Städte der Antike*, Vienna.

Kaser, M. (1971), *Das Römische Privatrecht. Erster Abschnitt. Das Altrömische, das Vorklassische und Klassische Recht*, 2nd edn. (Handbuch der Altertumswissenschaft X.3.3.1), Munich.

—— and R. Knütel (2005), *Römisches Privatrecht*, 18th edn. Munich.

Kayser, F. (1993), 'Nouveaux textes grecs du Ouadi Hammamat', *Zeitschrift für Papyrologie und Epigraphik* 98: 111–56.

Khanoussi, M. (1991), 'Nouveaux documents sur la présence militaire dans la colonie julienne-augustéenne de Simitthus (Chemtou, Tunisie)', *Comptes Rendus de l'Académie des Inscriptions et Belles-Lettres* 824–39.

—— (1996), 'Les officiales marmorum Numidicorum', in M. Khanoussi, P. Ruggieri, and C. Vismara (eds.), *L'Africa Romana. Atti de XII convegno di studio Olbia, 12–15 dicembre 1996*, Sassari, 997–1016.

—— (1997), 'Le saltus Philomusianus et les carrières de marbre numidique', *Mitteilungen des Deutschen Archäologischen Instituts, Römische Abteilung* 104: 375–7.

Kienast, D. (1996), *Römische Kaisertabelle. Grundzüge einer römischen Kaiserchronologie*, 2nd edn. Darmstadt.

—— (1999), *Augustus. Prinzeps und Monarch*, Darmstadt.

Kierdorf, W. (1967), 'Hospitium', in K. Ziegler and W. Sontheimer (eds.), *Der Kleine Pauly. Lexikon der Antike in fünf Bänden. Band 2*, Stuttgart, 1234.

Klein, M. J. (1988), *Untersuchungen zu den kaiserlichen Steinbrüchen an Mons Porphyrites und Mons Claudianus in der östlichen Wüste Ägyptens*, Bonn.

Klemm, D. D. (1994), 'Chronologischer Abriss der antiken Goldgewinnung in der Ostwüste Ägyptens', *Mitteilungen des Deutschen Archäologischen Instituts, Abteilung Kairo* 50: 188–222.

—— and R. Klemm (1992), *Steine und Steinbrüche im Alten Ägypten*, Berlin.

Knapp, R. C. (1983), *Roman Córdoba* (University of California Publications. Classical Studies, Volume 30), Berkeley, Los Angeles, & London.

Koch, M. (1987), 'Neue Römische Inschriften aus Carthago Nova III', *Madrider Mitteilungen* 28: 127–34.

Kolb, A. (1993), *Die kaiserliche Bauverwaltung in der Stadt Rom. Geschichte und Aufbau der cura operum publicorum unter dem Prinzipat*, Stuttgart.

Kolb, F. (1995), *Rom. Die Geschichte der Stadt in der Antike*, Munich.

Kolb, A. (2000), *Transport und Nachrichtentransfer im Römischen Reich* (Klio. Beiträge zur Alten Geschichte. Beihefte Neue Folge 2), Berlin.

Kornemann, E. (1901), 'Collegium', in G. Wissowa (ed.), *Paulys Real-Encyclopädie der Classischen Altertumswissenschaft. Vierter Band*, Stuttgart, 380–480.

Korres, E. M. (1995), 'The ancient quarries at Mount Pentelikon', in Y. Maniatis, N. Herz, and Y. Basiakos (eds.), *The Study of Marble and Other Stones Used in Antiquity. Asmosia III Athens: Transactions of the 3rd International Symposium of the Association for the Study of Marble and Other Stones Used in Antiquity*, London.

Kozelj, T., and M. Wurch-Kozely (1995), 'Dragon-Houses of Southern Euboia. Military Guard-Places of the Quarries', in Y. Maniatis et al. (ed.), *The Study of Marble and Other Stones Used in Antiquity: Asmosia III Athens. Transactions of the*

3rd International Symposium of the Association for the Study of Marble and Other Stones Used in Antiquity, London, 17–31.

KRAUS, T. (1993*a*), 'Steinbruch- und Blockinschriften', in F. Rakob (ed.), *Simitthus I. Die Steinbrüche und die Antike Stadt*, Mainz, 55–64.

—— (1993*b*), 'Die Felsreliefs am Tempelberg', in F. Rakob (ed.), *Simitthus I. Die Steinbrüche und die Antike Stadt*, Mainz, 71–92.

—— J. RÖDER, and W. MÜLLER-WIENER (1967), 'Mons Claudianus—Mons Porphyrites. Bericht über die zweite Forschungsreise', *Mitteilungen des Deutschen Archäologischen Instituts, Abteilung Kairo* 22: 109–203.

LANG, A. (1911), *Beiträge zur Geschichte des Kaisers Tiberius*, Jena.

LAPORTE, J.-P. (1996), 'Notes sur l'aqueduc de Saldae (Bougie)', *L'Africa Romana* 11: 711–62.

LASSANDRO, D. (1995), 'I *damnati in metalla* in alcune testimonianze antiche', in M. Sordi (ed.), *Coercizione e mobilità umana nel mondo antico*, Milan, 271–7.

LASSÈRE, J. M. (1980), 'Le peuplement de la colonia Iulia Augusta Numidica Simitthus', *Antiquités Africaines* 16: 27–44.

LAUBENHEIMER-LEENHARDT, F. (1973), *Recherches sur les lingots de cuivre et de plomb d'époque romaine dans les régions de Languedoc-Roussillon et de Provence-Corse* (Revue Archéologique de Narbonnaise, Supp. 3), Paris.

LAZZARINI, S. (2001), *Lex metallis dicta. Studi sulla seconda tavola di Vipasca* (Minima epigraphica et papyrologica. Supplementa 2), Rome.

LE BOHEC, Y. (1989*a*), *La troisième légion Auguste*, Paris.

—— (1989*b*), *Les unités auxiliaires de l'armée romaine en Afrique Proconsulaire et Numidie sous le Haut Empire*, Paris.

—— (1992*a*), 'Notes sur les mines de Sardaigne à l'époque romaine', in M. Bonello Lai (ed.), *Sardinia antiqua. Studi in onore di Piero Meloni in occasione del suo settantesimo compleanno*, Cagliari, 255–64.

—— (1992*b*), 'Ti. Claudius Proculus Cornelianus, procurateur de la région de Theveste', *Zeitschrift für Papyrologie und Epigraphik* 93: 107–16.

—— (1994), *The Imperial Roman Army*, London.

LEGA, C. (2003), 'Ratio marmoraria. Iscrizioni estemporanee sulle pareti di un ambiente di età romana nel complesso del Barco Borghese a Monte Porzio Catone (RM)', *Mélanges d'Archéologie et d'Histoire de l'École française de Rome* 115/2: 563–92.

LEFEBVRE, S. (1998), 'Profils de carrière: douze procurateurs des Gaules et Germanies', *Cahiers du Centre Gustave-Glotz* 9: 247–64.

LEIST, G. A. (1896), 'Auction. Römisch', in G. Wissowa (ed.), *Paulys Real-Encyclopädie der Classischen Altertumswissenschaft. Zweiter Band*, Stuttgart, 2270–2.

LEONE, A. (1988), *Gli animali da trasporto nell' Egitto Greco, Romano e Bizantino*, Rome & Barcelona.

—— (1998), *Animali da trasporto nell' antico Egitto. Una rassegna papirologica dalla dinastia dei Lagidi ai Bizantini*, Naples.

LEROUX, P. (1977), 'L'armée de la péninsule ibérique et la vie économique', in *Armée et fiscalité dans le monde antique, Paris, 14–16 octobre 1976*, Paris, 341–71.

LeRoux, P. (1980), 'Review of "Inscriptions de la Mésie supérieure. Vol. I, Singidunum et nord-ouest de la province, par M. Mirkovic et S. Dusanic (...), Beograd 1976"', *Revue des Études Anciennes* 82: 374–5.

—— (1982), *L'Armée romaine et l'organisation des provinces Ibériques d'Auguste à l'invasion de 409*, Paris.

—— (1984), 'Autour de la notion d'inscription fausse', in *Épigraphie hispanique. Problèmes de méthode et d'édition*, Paris, 175–80.

—— (1985), 'Procurateur affranchi in Hispania: Saturninus et l'activité minière', *Madrider Mitteilungen* 26: 218–33.

—— (1989), 'Exploitations minières et armées romaines: essai d'interpretation', in C. Domergue (ed.), *Minería y metalurgia en las Antiguas Civilizaciones Mediterráneas y Europeas, Coloquio Internacional Asòciado. Madrid, 24–28 Octobre, 1985*, Madrid, 171–81.

—— (1992), 'L'armée romaine dans la Péninsule Ibérique sous l'empire: Bilan pour une décennie', *Revue des Études Anciennes* 94: 231–57.

—— (1995), *Romains d'Espagne. Cités et politique dans les provinces IIe siècle av. J.-C.–IIIe siècle ap. J.-C.*, Paris.

—— (1999), 'Vectigalia et revenus des cités en Hispanie au Haut-Empire', in *Il capitolo delle entrate nelle finanze municipali in occidente ed in oriente. Actes de la Xe Rencontre franco-italienne sur l'épigraphie du monde romain. Rome, 27–29 mai 1996*, Rome, 155–73.

Le Roy, C. (1961), 'Lakonika', *Bulletin de Correspondance Héllenique* 85: 206–35.

Lesquier, J. (1918), *L'armée romaine d'Égypte d'Auguste à Dioclétien* (MIFAO 41), Cairo.

Leveau, P. (1973), 'L'aile II de Thrace, la tribu des Mazices et les *praefecti gentis* en Afrique du Nord', *Antiquités Africaines* 7: 103–10.

Levick, B. (1999*a*), *Tiberius, the Politician* (rev. edn.), London & New York.

—— (1999*b*), *Vespasian*, London & New York.

—— (2000), *The Government of the Roman Empire. A Sourcebook*, London & New York.

Lewis, P. R. (1970), 'Roman Gold-Mining in North-West Spain', *Journal of Roman Studies* 60: 169–85.

Lewis, N. (1974), *Papyrus in Classical Antiquity*, Oxford.

—— (1989), *Papyrus in Classical Antiquity. A Supplement*, Brussels.

Lewis, M. J. T. (1997), *Millstone and Hammer. The Origins of Water Power*, Hull.

Lewis, P. R., and G. D. B. Jones (1969), 'The Dolaucothi Gold Mines I. The Surface Evidence', *Antiquaries Journal* 49: 244–72.

Lieb, H. (1998), 'Vindonissa und die römischen Lagerstädte', *Jahresbericht der Gesellschaft Pro Vindonissa*, 63–6.

Liebenam, W. (1900), *Städteverwaltung im römischen Kaiserreiche*, Leipzig.

Liermann, O. (1889), *Analecta epigraphica et agonistica* (Diss.), Halle.

Lintott, A. (1993), *Imperium Romanum. Politics and Administration*, London & New York.

Liou, B., and C. Domergue (1990), 'Le commerce de la Bétique au Ier siècle de nôtre ère. L'épave Sud-Lavezzi 2 (Bonifacio, Corse du Sud)', *Archaeonautica* 10: 56–94.

Lo Cascio, E. (2005), 'The Emperor and his Administration. B. The age of the Severans', in A. K. Bowman, P. Garnsey, and A. Cameron (eds.), *The Cambridge Ancient History. Second Edition. Volume XII. The Crisis of Empire A.D. 193–337*, 137–55.

Locher, J. (1999), *Topographie und Geschichte der Region am ersten Nilkatarakt in griechisch-römischer Zeit* (Archiv für Papyrusforschung, Beiheft 5), Stuttgart & Leipzig.

Long, L., and C. Domergue (1996), 'Le "véritable plomb de L. Flavius Verucla" et autres lingots. L'épave 1 des Saintes-Maries-de-la-Mer', *Mélanges d'Archéologie et d'Histoire de l'École française de Rome* 107/2: 800–67.

Lörincz, B. (1999), *Onomasticon Provinciarum Europae Latinarum (OPEL). Vol. II: Cabalicivs-Ixvs*, Vienna.

—— (2000), *Onomasticon Provinciarum Europae Latinarum (OPEL). Vol. III: Labareus-Pythea*, Vienna.

Mackensen, M. (2000), 'Erster Bericht über neue archäologische Untersuchungen im sog. Arbeits- und Steinbruchlager von Simitthus-Chemtou (Nordwest-Tunesien)', *Mitteilungen des Deutschen Archäologischen Instituts, Römische Abteilung* 107: 487–505.

—— (2005), *Militärlager oder Marmorwerkstätten. Neue Untersuchungen im Ostbereich des Arbeits- und Steinbruchlagers von Simitthus/Chemtou* (Simitthus III), Mainz.

Maischberger, M. (1997), *Marmor in Rom. Anlieferung, Lager- und Werkplätze in der Kaiserzeit* (Palilia 1), Wiesbaden.

Mangas, J. (1999), 'Los obligados por operae en las minas de Vipasca (Aljustrel, Portugal)', in J.-M. Pailler and P. Moret (eds.), *Mélanges C. Domergue 2* (Pallas 50), Toulouse, 245–51.

Marquardt, J. (1884), *Römische Staatsverwaltung. Zweiter Band*, 2nd edn. (Handbuch der Römischen Alterthümer 5), Leipzig.

Martin, S. D. (1989), *The Roman Jurists and the Organization of Private Building in the Late Republic and Early Empire* (Collection Latomus 204), Brussels.

Mason, H. J. (1974), *Greek Terms for Roman Institutions. A Lexicon and Analysis* (American Studies in Papyrology 13), Toronto.

Mateo, A. (1999), *Manceps, redemptor, publicanus. Contribución al estudio de los contratistas públicos en Roma*, Santander.

—— (2001), *Observaciones sobre el régimen jurídico de la minería en tierras públicas en época romana*, Santiago de Compostela.

—— (2003), 'Roman mining on public land: From the Republic to the Empire', in J.-J. Aubert (ed.), *Tâches publiques et entreprise privée dans le monde Romain*, Geneva, 123–33.

Maxfield, V. A. (1994–5), 'The Imperial Porphyry Quarries Project', *Journal of Egyptian Archaeology* 81: 17–21.

—— (1996), 'The Eastern Desert forts and the army in Egypt during the principate', in D. M. Bailey (ed.), *Archaeological Research in Roman Egypt. The Proceedings of the Seventeenth Classical Colloquium of the Department of Greek and Roman Antiquities, British Museum, held on 1–4 December, 1993* (JRS Supplementary Series 19), Ann Arbor, 9–19.

Maxfield, V. A. (2000), 'The deployment of the Roman auxilia in Upper Egypt and the Eastern Desert during the principate', in G. Alföldy, B. Dobson, and W. Eck (eds.), *Kaiser, Heer und Gesellschaft in der Römischen Kaiserzeit. Gedenkschrift für Eric Birley* (HABES 31), Stuttgart, 407–44.

—— (2001), 'Stone quarrying in the Eastern Egyptian Desert with particular reference to Mons Claudianus and Mons Porphyrites', in D. J. Mattingly and J. Salmon (eds.), *Economies beyond Agriculture in the Classical World*, London & New York, 143–70.

—— (2002), 'Armies and quarries. A preliminary look at the evidence from Roman Egypt', in P. R. Hill (ed.), *Polybius to Vegetius. Essays on the Roman Army and Hadrian's Wall Presented to Brian Dobson*, 69–81.

—— (1997), 'Mons Porphyrites', *Journal of Egyptian Archaeology* 83: 13.

—— (1998*a*), 'The archaeology of an industrial landscape. An interim report on the work of the Imperial Quarries (Mons Porphyrites) Project', in O. E. Kaper (ed.), *Life on the Fringe. Living in the Southern Egyptian Deserts During the Roman and Early-Byzantine Periods. Proceedings of a Colloquium held on the Occasion of the 25th Anniversary of the Netherlands Institute for Archaeology and Arabic Studies in Cairo 9–12 December 1996*, Leiden, 181–96.

—— (1998*b*), 'The Roman Imperial Porphyry Quarries Project, Gebel Dokhan', *Journal of Egyptian Archaeology* 84: 16–20.

—— (2001*a*), *The Roman Imperial Quarries. Survey and Excavation at Mons Porphyrites. Volume 1: Topography and Quarries*, London.

—— (2001*b*), *Survey and Excavation. Mons Claudianus. 1987–1993. Volume II. Excavations: Part 1*, Cairo.

—— and D. P. S. Peacock (1996), 'The Roman Imperial Porphyry Quarries Project, Gebel Dokhan', *Journal of Egyptian Archaeology* 82: 15–19.

McCabe, D. F. (1985), *Teos Inscriptions. Texts and List*, Princeton.

Meier, S. (1997), 'Blei-Fernhandel in römischer Zeit (2.Jh. v.Chr. - 4.Jh. n.Chr.)', *Helvetia Archaeologica* 109: 2–14.

Meiggs, R. (1973), *Roman Ostia*, 2nd edn. Oxford.

Meloni, P. (1990), *La Sardegna romana*, 3rd edn. Sassari.

Mendel, G. (1909), 'Catalogue des monuments grecs, romains et byzantins du Musée Impérial Ottoman de Brousse', *Bulletin de Correspondance Héllenique* 33: 245–435.

Mentella, G. (1990), 'L'imprenditoria privata nelle cave Lunensi alla Luce di CIL XI 6946', in F. Rebecchi (ed.), *Miscellanea di Studi Archeologici e di Antichità III*, Modena, 133–9.

Meredith, D. (1952), 'The Roman remains in the Eastern Desert of Egypt', *Journal of Egyptian Archaeology* 38: 94–111.

—— (1953), 'The Roman remains in the Eastern Egyptian Desert (continued)', *Journal of Egyptian Archaeology* 39: 95–106.

—— (1957), 'Inscriptions from the amethyst mines at Abu Diyeiba (Eastern Desert of Egypt)', *Eos. Commentarii Societatis Philologae Polonorum* 48/2: 117–19.

—— and A. Tregenza (1949), 'Notes on Roman Roads and Stations in the Eastern Desert I: (a) from Qena to Bab El-Mukheiniq, (b) the Roman Station in W. Abu Zawal', *Bulletin of the Faculty of Arts, Fouad I University* 11: 115–26.

Merkelbach, R., and J. Stauber (2001), *Steinepigramme aus dem griechischen Osten. Band 3. Der "Ferne Osten" und das Landesinnere bis zum Tauros*, Stuttgart & Leipzig.

Meyer, C. (1995), 'A Byzantine gold-mining town in the Eastern Desert of Egypt: Bir Umm Fawakhir, 1992–1993', *Journal of Roman Archaeology* 8: 192–224.

Millar, F. G. B. (1963*a*), 'The fiscus in the first two centuries', *Journal of Roman Studies* 53: 29–42 (= Millar 2004, 47–72).

—— (1963*b*), 'Review of 'H. G. Pflaum, Les carrières procuratoriennes équestres sous la Haut-Empire Romain I–III, Paris 1960–61', *Journal of Roman Studies* 53: 194–200 (= 'The equestrian Career under the Empire', in Millar 2004, 151–9).

—— (1964), 'Some evidence on the meaning of Tacitus Annals XII.60', *Historia* 13: 180–7.

—— (1965), 'The development of jurisdiction by Imperial Procurators. Further evidence', *Historia* 14: 362–7.

—— (1967), 'Emperors at work', *Journal of Roman Studies* 57: 9–19 (= Millar 2004, 3–22).

—— (1984), 'Condemnation to hard labour in the Roman Empire from the Julio-Claudians to Constantine', *Papers of the British School at Rome* 52: 124–47 (= Millar 2004, 120–50).

—— (1992), *The Emperor in the Roman World*, 2nd edn. London.

—— (1994), 'Latin in the epigraphy of the Roman Near East', in H. Solin et al. (eds.), *Acta Colloqui Epigraphici Latini. Helsinki 3–6 Sept. 1991* (Commentationes Humanarum Litterarum 104), XXX, 403–19 (= Millar 2006, 223–42).

—— (2004), *Rome, the Greek World, and the East. Volume 2. Government, Society, and Culture in the Roman Empire*, Chapel Hill & London.

—— (2006), *Rome, the Greek World, and the East. Volume 3. The Greek World, the Jews, and the East*, Chapel Hill & London.

Minnen, P. v. (2000), 'Prisoners of war and hostages in Graeco-Roman Egypt', *Journal of Juristic Papyrology* 30: 155–63.

Mirković, M. (1975), 'Iz istorije Polimlja u rimsko doba', *Godišnjak. Centar za balkanološa ispitivanja* 14/12: 95–108.

Mitchell, S. (1993), *Anatolia. Land, Men, and Gods in Asia Minor. Volume I*, Oxford.

—— (1999), 'The administration of Roman Asia from 133 BC to AD 250', in W. Eck (ed.), *Lokale Autonomie und römische Ordnungsmacht in den kaiserzeitlichen Provinzen vom 1. bis 3. Jahrhundert* (Schriften des Historischen Kollegs, Kolloquien 42), Munich, 17–46.

Mitthof, F. (1994), 'Quittung eines Soldaten an *πρεσβυτέροι κώμης* über verschiedene Lieferungen. Eine Neuedition von P. Alex.inv. 463', *Zeitschrift für Papyrologie und Epigraphik* 102: 207–12.

—— (2001), *Annona militaris. Die Heeresversorgung im spätantiken Ägypten. Ein Beitrag zur Verwaltungs- und Heeresgeschichte des Römischen Reiches im 3. bis 6. Jh. n. Chr.* (Pap Flor XXXII), Florence.

Mócsy, A. (1962), 'Pannonia', in *Paulys Real-Encyclopädie der Classischen Altertumswissenschaften. Band IX*, 516–776.

—— (1970), *Gesellschaft und Romanisation in der römischen Provinz Moesia Superior*, Amsterdam.

—— (1972), 'Das Problem der militärischen Territorien im Donauraum', *Acta Antiqua Academiae Scientiarum Hungaricae* 20: 133–68.

—— (1980), 'Zu den Auxiliarvici in Pannonien', in H. Galsterer and H. Wolff (eds.), *Studien zur antiken Sozialgeschichte. Festschrift Friedrich Vittinghoff*, Cologne & Vienna, 365–76.

Mommsen, Th. (1887), *Römisches Staatsrecht. Zweiter Band. 1. Abteilung* (Handbuch der Römischen Alterthümer), Leipzig.

—— (1899), *Römisches Strafrecht*, Leipzig.

Monthel, G., and P.-Y. Lambert (2002), 'La carrière gallo-romaine de Saint-Boil (Saône-et-Loire)', *Gallia* 52: 89–120.

Mrozek, S. (1968), 'Die kaiserlichen Bergwerksprokuratoren in der Zeit des Prinzipats', *Zeszyty Naukowe Uniwersytetu Mikolaja Kopernika w Toruniu. Nauki Humanistyczno-Spoleczne.* (Zeszyt 34. Historia IV), 45–65.

—— (1969), 'Die Arbeitsverhältnisse in den Goldbergwerken des römischen Dazien', in M. N. Andreev, J. Irmscher, E. Polay, and W. Warkallo (eds.), *Gesellschaft und Recht im Griechisch-Römischen Altertum. Eine Aufsatzsammlung. Teil 2*, Berlin, 139–55.

—— (1977), 'Die Goldbergwerke im römischen Dazien', in H. Temporini and W. Haase (eds.), *Aufstieg und Niedergang der Römischen Welt. Geschichte und Kultur Roms im Spiegel der Neueren Forschung II. Principat, Sechster Band. Politische Geschichte (Provinzen und Randvölker: Lateinischer Donau-Balkanraum)*, Berlin & New York, 95–109.

Nachtergael, G. (1996), 'Un proscynème d'Assouan-rive gauche', *Chronique d'Égypte* 70/1: 335–42.

Neesen, L. (1980), *Untersuchungen zu den direkten Staatsabgaben der Römischen Kaiserzeit (27 v.Chr.–284 n.Chr.)*, Bonn.

Negri, G. (1985), *Diritto minerario romano I. Studi esegetici sul regime delle cave private nel pensiero dei giuristi classici*, Milan.

Nelis-Clément, J. (2000), *Les beneficiarii: militaires et administratuers au service de l'Empire (1er s.a.C.–VIe s.p.C.)* (Ausonius Publications. Études 5), Bordeaux.

—— and F. Wiblé (1996), 'Aurelius Maximianus, bénéficiaire en poste à Martigny/ Forum Claudii Vallensium?', *Vallesia* 51: 271–309.

Nesselhauf, H. (1937), 'Neue Inschriften aus dem römischen Germanien und den angrenzenden Gebieten', *Bericht der Römisch-Germanischen Kommission* 27: 51–134.

Nicholas, B. (1962), *An Introduction to Roman Law*, Oxford.

Nicolet, C. (1993), 'Le *monumentum Ephesenum* et la délimitation du *portorium* d'Asie', *Mélanges d'Archéologie et d'Histoire de l'École française de Rome* 105/2: 929–59.

Noeske, H.-C. (1977), 'Studien zur Verwaltung und Bevölkerung der dakischen Goldbergwerke', *Bonner Jahrbücher* 177: 271–415.

Nonnis, D., and C. Ricci (1999), 'Vectigalia municipali ed epigrafia: un caso dall' Hirpinia', in *Il capitolo delle entrate nelle finanze municipali in occidente ed in oriente.*

Actes de la Xe Rencontre franco-italienne sur l'épigraphie du monde romain. Rome, 27–29 mai 1996, Rome, 41–59.

Nony, D. (1970), 'A propos des nouveaux procurateurs d'Astorga', *Archivo español de arqueología* 43: 195–201.

Nuber, H. U. (1977), 'Römische Steindenkmäler aus St. Ulrich und Afra in Augsburg', in J. Werner (ed.), *Die Ausgrabungen in St. Ulrich und Afra in Augsburg 1961–1968* (Münchner Beiträge zur Vor- und Frühgeschichte 23), Munich, 227–61.

—— (1985), 'Ein Bergwerkspächter in Rätien', in E. Keller, D. Reimann, and H. Stölzl (eds.), *Die Römer in Schwaben. Jubiläumsausstellung 2000 Jahre Augsburg veranstaltet vom Bayerischen Landesamt für Denkmalpflege und der Stadt Augsburg, Zeughaus, 23. Mai – 3. November 1985*, Munich, 130–1.

Oleson, J. P. (1984), *Greek and Roman Mechanical Water-lifting Devices. The History of a Technology*, Toronto & Buffalo.

Oliver, J. H. (1989), *Greek Constitutions of Early Roman Emperors from Inscriptions and Papyri* (Memoirs of the American Philosophical Society 178), Philadelphia.

Orejas, A. (1994), 'Les populations des zones minières du Nord-Ouest de la péninsule ibérique (Bassin NO du Douro, Leon–Espagne)', *Dialogues d'Histoire Ancienne* 20/1: 245–81.

—— (1996), *Estructura social y territorio. El impacto romano en la cuenca noroccidental del Duero*, Madrid.

—— and F. J. Sánchez-Palencia (2002), 'Mines, territorial organization, and social structure in Roman Iberia. Carthago Nova and the Peninsular Northwest', *American Journal of Archaeology* 106: 581–99.

—— and I. Sastre Prats (1999), 'Fiscalité et organisation du territoire dans le Nord-Ouest de la Péninsule Ibérique: civitates, tribut et ager mensura comprehensus', *Dialogues d'Histoire Ancienne* 25/1: 183–93.

Ørsted, P. (1985), *Roman Imperial Economy and Romanization. A Study in Roman Imperial Administration and the Public Lease System in the Danubian Provinces from the First to the Third Century A. D.*, Copenhagen.

Orth, F. (1924), 'Bergbau', in W. Kroll (ed.), *Paulys Real-Encyclopädie der classischen Altertumswissenschaften. Supplementband IV*, Stuttgart, 108–55.

Ott, J. (1995), *Die Beneficiarier. Untersuchungen zu ihrer Stellung innerhalb der Rangordnung des Römischen Heeres und zu ihrer Funktion* (Historia Einzelschriften 92), Stuttgart.

Panciera, S. (1957), *Vita economica di Aquileia in età romana*, Aquileia.

Papageorgakis, J. (1964), 'Die antiken Brueche des Karystischen Marmors', *Praktika tes Akadimias Athinon* 39: 262–84.

Parassoglou, G. M. (1978), *Imperial Estates in Roman Egypt* (American Studies in Papyrology 18), Amsterdam.

Parker, A. J. (1974), 'Lead ingots from a Roman ship at Ses Salines, Majorca', *International Journal of Nautical Archaeology* 3: 147–8.

—— (1992), *Ancient Shipwrecks of the Mediterranean and the Roman Provinces* (BAR International Series 580), Oxford.

Pašalić, E. (1967), 'Rolle und Bedeutung der römischen Eisenbergwerke in Westbosnien für den pannonischen Limes', in *Studien zu den Militärgrenzen Roms. Vorträge des 6. Internationalen Limeskongresses in Süddeutschland* (Beihefte der Bonner Jahrbücher, Band 19), Cologne & Graz, 127–30.

Patsch, C. (1899), 'Archäologisch-epigraphische Untersuchungen zur Geschichte der römischen Provinz Dalmatien. Dritter Theil', *Wissenschaftliche Mittheilungen aus Bosnien und der Herzegowina* 6: 154–273.

—— (1900), 'Archäologisch-epigraphische Untersuchungen zur Geschichte der römischen Provinz Dalmatien. Vierter Theil', *Wissenschaftliche Mittheilungen aus Bosnien und der Herzegowina* 7: 33–166.

Peacock, D. P. S. (1992), *Rome in the Desert. A Symbol of Power*, Southampton.

—— (1995), 'The *Passio Sanctorum Quattuor Coronatorum.* A petrological approach', *Antiquity* 69/263: 362–8.

—— (1996), 'A note on the distribution of porphyry and other rocks on Roman sites in the Eastern Desert', in D. M. Bailey (ed.), *Archaeological Research in Roman Egypt. The Proceedings of the Seventeenth Classical Colloquium of the Department of Greek and Roman Antiquities, British Museum, held on 1–4 December, 1993* (JRS Supplementary Series 19), Ann Arbor, 20–2.

—— and V. A. Maxfield (1997), *Survey and Excavation. Mons Claudianus 1987–1993. Volume I. Topography and Quarries* (FIFAO 37), Cairo.

Pékary, T. (1968), *Untersuchungen zu den Römischen Reichsstrassen*, Bonn.

—— (1979), 'Vectigal', in K. Ziegler, W. Sontheimer, and H. Gärtner (eds.), *Der Kleine Pauly. Lexikon der Antike in fünf Bänden. Band 5*, Munich, 1150.

Peña, J. T. (1989), 'P. Giss. 69: Evidence for the supplying of stone transport operations in Roman Egypt and the production of fifty-foot monolithic column shafts', *Journal of Roman Archaeology* 2: 126–32.

—— (1998), 'The mobilization of state olive oil in Roman Africa. The evidence of late 4th-c. Ostraca from Carthage', in *Carthage Papers. The Early Colony's Economy, Water Supply, a Public Bath, and the Mobilization of State Olive Oil*, Portsmouth, RI, 117–238.

Pensabene, P. (1978), 'A cargo of marble shipwrecked at Punta Scifo near Crotone (Italy)', *International Journal of Nautical Archaeology* 7/2: 105–18.

—— (1994), *Le vie del marmo. I blocchi di cava di Roma e di Ostia. Il fenomeno del Marmo nella Roma Antica* (Itinerari Ostiensi 7), Rome.

—— (2002), 'Il fenomeno del marmo nel mondo romano', in M. De Nucio and L. Ungaro (eds.), *I marmi colorati della Roma imperiale*, Venice, 3–67.

Petolescu, C. C. (1983), 'Numerus maurorum Hisp.', *Gerión* 1: 327–30.

—— (1988), 'Cronica Epigrafica a României', *Studii si Cercetari de Istorie Veche si Arheologie* 39: 403–8.

Petrović, P. (1979), *Inscriptions de la Mésie supérieure. Vol. IV. Naissus-Remesiana-Horreum Margi*, Belgrade.

—— (1995*a*), 'Der römische Bergbau in Ravna. Archäologische Skizzen', in B. Jovanović (ed.), *Ancient Mining and Metallurgy in Southeast Europe. International Symposium,*

Donji Milanovac. May 20–25: 1990 (Archaeological Institute Beograd Monographs 27), Bor, 195–202.

—— (1995*b*), *Inscriptions de la Mésie supérieure. Vol. III/2. Timacum Minus et la Vallée du Timok*, Belgrade.

Pflaum, H.-G. (1948), *Le marbre de Thorigny*, Paris.

—— (1950), *Les procurateurs équestres sous le Haut-Empire romain*, Paris.

—— (1955), 'Deux carrières équestres de Lambèse et de Zana (Diana Veteranorum)', *Libyca. Archéologie, Épigraphie* 3: 123–54.

—— (1960–1), *Les carrières procuratoriennes équestres sous le Haut-Empire romain* (Institut Français d'Archéologie de Beyrouth. Bibl. archéologique et historique 57), Paris.

—— (1970), 'La carrière de l'affranchi impérial Saturninus. Sous-procurateurs provinciaux équestres et procurateurs d'extraction affranchi', *Revue des Études Latines*, 297–310.

—— (1971), 'Une lettre de promotion de l'empereur Marc Aurèle pour un procurateur ducénaire de Gaule Narbonnaise', *Bonner Jahrbücher* 171: 349–66.

—— (1978), *Les Fastes de la province Narbonnaise* (XXXème supplément à Gallia), Paris.

—— (1981), *Scripta varia. La Gaule et l'Empire romain*, Paris.

—— (1982), *Les carrières procuratoriennes équestres sous le Haut-Empire romain. Supplément* (Institut Français d'Archéologie du Proche Orient Beyrouth-Damas-Amman. Bibliothèque archéologique et historique 112), Paris.

Piccottini, G. (1989), *Archäologischer Atlas von Kärnten*, Klagenfurt.

—— (1994*a*), 'Gold und Kristall am Magdalensberg', *Germania* 72/2: 467–77.

—— (1994*b*), *Mithrastempel in Virunum*, Klagenfurt.

—— H. Dolenz, F. Glaser, and R. Jernej (2002), 'Virunum', in M. Šašel Kos and P. Scherrer (eds.), *The Autonomous Towns of Noricum and Pannonia. Noricum*, Ljubljana, 103–34.

Piso, I. (1993), *Fasti Provinciae Dacicae I. Die senatorischen Amtsträger* (Antiquitas. Reihe 1. Abhandlungen zur Alten Geschichte 43), Bonn.

—— (2000), 'Les légions dans la province de Dacie', in Y. Le Bohec (ed.), *Les légions de Rome sous le Haut-Empire. Actes du Congrès de Lyon, 17–19 septembre, 1998* (Collection du Centre d'Études Romaines et Gallo-Romaines. Nouvelle série, no. 20), Lyon, 205–25.

—— (2004), 'Gli Illiri al Alburnus Maior', in G. Urso (ed.), *Dall' Adriatico al Danubio. L'Illyrico nell' età greca e romana. Atti del convegno internazionale Cividale del Friuli, 25–27 settembre 2003*, Pisa, 271–307.

—— (2005), 'La Mésie Supérieure et les Débuts de Sarmizegetusa', in M. Mirković (ed.), *Römische Städte und Festungen an der Donau. Akten der regionalen Konferenz. Organisiert von der Alexander von Humboldt-Stiftung, Beograd, 16–19 Oktober 2003*, Belgrade, 119–24.

Poveda Navarro, A. M. (2000), 'Societas Baliarica. Una nueva compañía minera romana de Hispania', *Gérion* 18: 293–313.

PREMERSTEIN, A. V. (1901), 'A commentariis', in G. Wissowa (ed.), *Paulys Real-Encyclopädie der Classischen Altertumswissenschaften*, Stuttgart, 759–68.

PRICE, S. R. F. (1984), *Rituals and Power. The Roman Imperial Cult in Asia Minor*, Cambridge.

QUILICI, L. (1974), *Collatia* (Forma Italiae. Regio I. Vol. X), Rome.

RAKOB, F. (1993), 'I. Zur Siedlungstopographie von Chemtou/Simitthus', in F. Rakob (ed.), *Simitthus I. Die Steinbrüche und die antike Stadt*, Mainz, 1–16.

—— (1994), 'II. Das Römische Steinbruchlager (Praesidium) in Simitthus', in F. Rakob (ed.), *Simitthus II. Der Tempelberg und das Römische Lager*, Mainz, 51–139.

—— (1997), 'Chemtou. Aus der römischen Arbeitswelt', *Antike Welt* 28: 1–20.

RANKOV, N. B. (1983), 'A contribution to the military and administrative history of Montana', in A. G. Poulter (ed.), *Ancient Bulgaria. Papers Presented to the International Symposium on the Ancient History and Archaeology of Bulgaria, University of Nottingham, 1981. Part II*, Nottingham, 40–73.

—— (1990), 'Frumentarii, the Castra Peregrina and the Provincial Officia', *Zeitschrift für Papyrologie und Epigraphik* 80: 176–82.

—— (1999), 'The governor's men: The Officium Consularis in provincial administration', in A. Goldsworthy and I. Haynes (eds.), *The Roman Army as a Community. Including Papers of a Conference held at Birkbeck College, University of London on 11–12 January, 1997* (Journal of Roman Archaeology. Supplementary Series 34), Portsmouth, RI, 15–34.

RATHBONE, D. (1993) 'Egypt, Augustus and Roman Taxation', *Cahiers du Centre Gustave Glotz* 4: 81–112.

—— (1996), 'The imperial finances', in A. K. Bowman, E. Champlin, and A. Lintott (eds.), *The Cambridge Ancient History. Second Edition. Volume X. The Augustan Empire, 43 BC – AD 69*, Cambridge, 309–23.

—— (2003), 'The control and exploitation of *ager publicus* in Italy under the Roman Republic', in J.-J. Aubert (ed.). *Tâches publiques et entreprise privée dans le monde romain. Actes du Diplôme d'Etudes Avancées, Universités de Neuchâtel et de Lausanne, 2000–2002*, Geneva, 135–78.

REINACH, S. (1890), 'Inscriptions inédites d'Asie Mineure & de Syrie. Recueillies par M. le capitaine Callier (1830–1834)', *Revue des Études Grecques* 3: 48–85.

RÉMY, B. (1995), *Inscriptions Latines d'Aquitaine (I.L.A.). Vellaves*, Bordeaux.

RENGEN, W. V. (1992), 'Les Laissez-Passer', in J. Bingen, A. Bülow-Jacobsen, W. E. H. Cockle, H. Cuvigny, L. Rubinstein, and W. V. Rengen (eds.), *Mons Claudianus. Ostraca Graeca et Latina. I. O. Claud. 1 à 190* (FIFAO 29), Cairo, 60–74.

—— (1995), 'A new Paneion at Mons Porphyrites', *Chronique d'Égypte* 70: 240–5.

—— (1997), 'La correspondance militaire', in Bingen et al. 1997, 193–227.

REYNOLDS, J. (1982), *Aphrodisias and Rome* (Journal of Roman Studies Monographs 1), London.

RICCARDI, E., and S. GENOVESI (2002), 'Un carico di piombo da Rena Maiore (Aglientu)', *L'Africa romana* 14: 1311–29.

RICHARDSON, J. S. (1976), 'The Spanish mines and the development of provincial taxation in the second century B.C.', *Journal of Roman Studies* 66: 139–52.

Ricl, M. (1997), *The Inscriptions of Alexandreia Troas* (Inschriften griechischer Städte aus Kleinasien 53), Bonn.

Ritterling, E. (1925), 'Legio', in W. Kroll (ed.), *Paulys Real-Encyclopädie der Classischen Altertumswissenschaft XII*, Stuttgart, 1186–830.

Röder, J. (1971), 'Marmor Phrygium. Die antiken Marmorbrüche von Iscehisar in Westanatolien', *Jahrbuch des Deutschen Archäologischen Instituts* 86: 253–312.

Röder, G. (1993), 'Die Steinbrüche des Numidischen Marmors von Chemtou', in F. Rakob (ed.), *Simitthus I. Die Steinbrüche und die Antike Stadt*, Mainz, 17–53.

Rodríguez Colmenero, A. (1997), *Aquae Flaviae, I. Fontes epigráficas da Gallaecia meridional interior*, Chaves.

Rostovtzeff, M. (1904), 'Geschichte der Staatspacht in römischer Kaiserzeit bis Diokletian', *Philologus Ergänzungsband* 9: 331–512.

—— (1957), *The Social and Economic History of the Roman Empire*, 2nd edn. Oxford.

Rosumek, P. (1982), *Technischer Fortschritt und Rationalisierung im Antiken Bergbau*, Bonn.

Rothenhöfer, P. (2003), 'Geschäfte in Germanien. Zur Ausbeutung von Erzlagerstätten unter Augustus in Germanien', *Zeitschrift für Papyrologie und Epigraphik* 143: 277–86.

—— (2005), *Die Wirtschaftsstrukturen im südlichen Niedergermanien. Untersuchungen zur Entwicklung eines Wirtschaftsraumes an der Peripherie des Imperium Romanum* (Kölner Studien zur Archäologie der Römischen Provinzen 7), Rahden.

Royden, H. L. (1988), *The Magistrates of the Roman Professional Collegia in Italy from the First to the Third Century A.D.* (Biblioteca di Studi Antichi 61), Pisa.

Sablayrolles, R. (1989), 'L'administration des mines de fer en Gaule Romaine', in C. Domergue (ed.), *Minería y metalurgia en las Antiguas Civilizaciones Mediterráneas y Europeas, Coloquio Internacional Asòciado. Madrid, 24–28 Octobre, 1985*, Madrid, 157–61.

—— (2005), 'Analyses d'économie antique: textes anciens et archéologie récente', in *L'Aquitaine et l'Hispanie septentrionale à l'époque julio-claudienne. Organisation et exploitation des espaces provinciaux. Colloque Aquitania Saintes, 11–13 septembre 2003* (Aquitania Supplément 13), Bordeaux, 415–22.

Sachers, E. (1932), 'Tabularius', in W. Kroll and K. Mittelhaus (eds.), *Paulys Real-Encyclopädie der Classischen Altertumswissenschaft. IVA*, Stuttgart, 1969–84.

Saguì, L. (1993), 'Crypta Balbi (Roma): Conclusione delle indagini archeologiche nell' eserda del monumento romano. Relazione preliminare', *Archaeologia Medievale* 20: 409–18.

Salama, P. (1951), *Les voies romaines de l'Afrique du Nord*, Alger.

Salerno, F. (2003), *Ad metalla. Aspetti giuridici del lavoro in miniera*, Naples.

Samsaris, D. (1985), 'Une inscription latine inédite trouvée près des frontières du territoire de la colonie de Philippes', *Klio* 67: 458–65.

—— (1987), 'Les mines et la metallurgie de fer et de cuivre dans la province romaine de Macédonie', *Klio* 69: 152–62.

SÁNCHEZ-PALENCIA, F. J. (1980), 'Prospecciones en las explotaciones auríferas del NO de España (Cuencas de los ríos Eria y Cabrera y Sierra del Teleno)', *Noticiario Arqueologico Hispanico* 8: 228–37.

—— (1986), 'El campamiento romano de Valdemeda, Manzaneda (León). Ocupacion militar y explotacion aurifera en el NW. peninsular', *Numantia* 2: 227–34.

—— and L. C. PEREZ GARCIA (1983), 'Las explotacionas auriferas y la ocupacion Romana del Noroeste de la Peninsula Iberica', in *Actas del II Seminario de Arqueologia del Noroeste, Santiago de Compostela 1980*, Madrid, 227–46.

—— M. D. FERNÁNDEZ-POSSE, J. FERNÁNDEZ MANZANO, Y. ÁLVAREZ GONZÁLEZ, and L. F. LÓPEZ GONZÁLEZ (1990), 'La Zona Arqueológica de Las Medulas (1988–89)', *Archivo español de arqueología* 63: 249–64.

SÁNCHEZ-PALENCIA RAMOS, F. J., M. D. FERNÁNDEZ-POSSE, J. FERNÁNDEZ MANZANO, and A. OREJAS (1996), *La zona arquéologica de Las Médulas, León. Guía Arqueológica*, León.

SANTOS YANGUAS, J. (1983), 'Cambios y pervivencias en las estructuras sociales indigenas. Sociedad indigena y sociedad romana en al area Astur', in *Indigenismo y Romanizacion en le Conventus Asturum*, Oviedo, 91–106.

ŠAŠEL, J. (1982), 'Zur Inschrift eines Zollbeamten aus dem Stadtgebiet des obermösischen Ulpiana', *Zeitschrift für Papyrologie und Epigraphik* 49: 211–13.

SASTRE PRATS, I. (1998), *Formas de dependencia social en el Noroeste peninsular (Transición del Mundo Prerromano al Romano y Época Altoimperial)*, León.

SAXER, R. (1967), *Untersuchungen zu den Vexillationen des römischen Kaiserheeres von Augustus bis Diokletian* (Epigraphische Studien 1), Cologne & Graz.

SCHÄFER, C. (1998), *Spitzenmanagement in Republik und Kaiserzeit. Die Prokuratoren von Privatpersonen im Imperium Romanum vom 2.Jh. v. Chr. bis zum 3.Jh. n. Chr.*, St. Katharinen.

SCHALLMAYER, E., K. EIBL, J. OTT, G. PREUSS, and E. WITTKOPF (1990), *Der römische Weihebezirk von Osterburken I. Corpus der griechischen und lateinischen Beneficiarier-Inschriften des Römischen Reiches* (Forschungen und Berichte zur Vor- und Frühgeschichte in Baden-Württemberg 40), Stuttgart.

SCHERRER, P. (1985), 'Zur Lage der statio Esc(ensis?) in Noricum', in M. Kandler, S. Karwiese, and R. Pillinger (eds.), *Lebendige Altertumswissenschaft. Festgabe zur Vollendung des 70. Lebensjahres von Hermann Vetters dargebracht von Freunden, Schülern und Kollegen*, Vienna, 255–7.

—— (1995), *Ephesos. Der neue Führer. 100 Jahre österreichische Ausgrabungen 1895–1995*, Vienna.

—— (2002), 'Vom regnum Noricum zur römischen Provinz: Grundlagen und Mechanismen der Urbanisierung', in M. Šašel Kos and P. Scherrer (eds.), *The Autonomous Towns of Noricum and Pannonia. Noricum*, Ljubljana, 12–70.

SCHEUERMANN, A. (1957), 'Diözese', in T. Klauser and E. Dassmann (eds.), *Reallexikon für Antike und Christentum*. Band 3, 1053–62.

SCHLEE, D. (2007), *Metal Mining in Upland Ceredigion. A Guide to the History and Protection of our Metal Mining Heritage*, Landeilo.

SCHNEIDER, K. (1958), 'Vilicus', in K. Ziegler and W. John (eds.), *Paulys Real-Encyclopädie der Classischen Altertumswissenschaft. VII A.2*, Stuttgart, 2136–41.

SCHNURBEIN, S. V. (1971), 'Ein Bleibarren der 19. Legion aus dem Hauptlager von Haltern', *Germania* 49: 132–6.

SCHÖNBAUER, E. (1929), *Beiträge zur Geschichte des Bergbaurechts* (Münchner Beiträge zur Papyrusforschung und antiken Rechtsgeschichte 12), Munich.

SHAW, I., and R. JAMESON (1993), 'Amethyst mining in the Eastern Desert: A preliminary survey at Wadi el-Hudi', *Journal of Egyptian Archaeology* 79: 81–97.

—— J. BUNBURY, and R. JAMESON (1999), 'Emerald mining in Roman and Byzantine Egypt', *Journal of Roman Archaeology* 12: 203–15.

SHEPHERD, R. (1993), *Ancient Mining*, London & New York.

SIDEBOTHAM, S. E. (1996), 'Newly discovered Sites in the Eastern Desert', *Journal of Egyptian Archaeology* 82: 181–92.

—— and R. E. ZITTERKOPF (1991), 'Survey of the 'Abu Sha'ar-Nile road', *American Journal of Archaeology* 95: 571–622.

—— H. BRANARD, J. A. HARRELL, and R. S. TOMBER (2001), 'The Roman quarry and installations in Wadi Umm Wikala and Wadi Semna', *Journal of Egyptian Archaeology* 87: 135–70.

SIJPESTEIJN, P. J. (1987), *Customs Duties in Graeco-Roman Egypt*, Zutphen.

—— (1993), 'Three small papyri from the Princeton collection', *Aegyptus* 73: 27–32.

SILLIERES, P. (1990), *Les voies de communication de l'Espagne méridionale*, Paris.

SIMIĆ, V. M., and M. R. VASIĆ (1977), 'La monnaie des mines romaines de l'Illyrie', *Revue Numismatique* 19: 48–61.

ŠKEGRO, A. (2000), 'Bergbau der römischen Provinz Dalmatien', *Godišnjak Jahrbuch* 31: 53–176.

SOLIN, H. (1996*a*), *Die stadtrömischen Sklavennamen. Ein Namenbuch. Drei Teile. I. Teil: Lateinische Namen* (Forschungen zur Antiken Sklaverei. Beiheft 2), Stuttgart.

—— (1996*b*), *Die stadtrömischen Sklavennamen. Ein Namenbuch. Drei Teile. II. Teil: Griechische Namen* (Forschungen zur Antiken Sklaverei. Beiheft 2), Stuttgart.

SORGE, A. (2002), 'Organization', in A. Sorge (ed.), *Organization*, London, 3–24.

SOTGIU, G. (1988), 'L'epigrafia latina in Sardegna dopo il CIL X e l'E.E. VIII', in W. Haase and H. Temporini (eds.), *Aufstieg und Niedergang der Römischen Welt. Geschichte und Kultur Roms im Spiegel der Neueren Forschung. Teil II: Principat, Band 11. Politische Geschichte (Provinzen und Randvölker: Sizilien und Sardinien; Italien und Rom; Allgemeines). 1.Teilband: Sizilien und Sardinien*, Berlin & New York, 552–739.

SPAUL, J. E. H. (1994), *Ala^2. The Auxiliary Cavalry Units of the pre-Diocletianic Imperial Roman Army*, Andover.

—— (2000), *Cohors^2. The Evidence for and a Short History of the Auxiliary Infantry Units of the Imperial Roman Army* (BAR International Series 841), Oxford.

SPEIDEL, M. P. (1977), 'The Roman army in Arabia', in H. Temporini and W. Haase (eds.), *Aufstieg und Niedergang der Römischen Welt. Geschichte und Kultur Roms im Spiegel der Neueren Forschung. II. Principat, Achter Band. Politische Geschichte*, Berlin & New York, 687–730 (= Roman Army Studies I, Amsterdam, 1984, 229–72).

Speidel, M. P. (1981–2), 'Noricum als Herkunftsgebiet der kaiserlichen Gardereiter', *Jahrbuch des Österreichischen Archäologischen Instituts* 53: 213–44.

—— (1984), *Regionarii in Lower Moesia*, ZPE 57: 185–8 (= Roman Army Studies II (MAVORS VIII), Stuttgart, 1992, 140–4).

Speidel, M. A. (1993), 'Miles ex cohorte. Zur Bedeutung der mit ex eingeleiteten Truppenangaben auf Soldateninschriften', *Zeitschrift für Papyrologie und Epigraphik* 95: 190–6.

—— (1994), 'Scribonius Proculus: *Curator aedium sacrarum et operum publicorum* in Rom oder in Luna? Überlegungen zu CIL XI 1340', *Zeitschrift für Papyrologie und Epigraphik* 103: 209–14.

—— (1996), *Die römischen Schreibtafeln von Vindonissa* (Veröffentlichungen der Gesellschaft Pro Vindonissa), Baden.

—— (1998), 'Legio IIII Scythica, its Movements and Men', in D. L. Kennedy (ed.), *The Twin Towns of Zeugma on the Euphrates. Rescue Work and Historical Studies*, Portsmouth, RI, 163–204.

—— (2000*a*), 'Geld und Macht. Die Neuordnung des staatlichen Finanzwesens unter Augustus', in *La révolution romaine après Ronald Syme* (Fondation Hardt. Entretiens sur l'antiquité classique. Tome XLVI), Geneva.

—— (2000*b*), 'Sold und Wirtschaftslage der römischen Soldaten', in G. Alföldy, B. Dobson, and W. Eck (eds.), *Kaiser, Heer und Gesellschaft in der Römischen Kaiserzeit. Gedenkschrift für Eric Birley* (HABES 31), Stuttgart, 65–94.

—— (2001), 'Legio operosa felix', *Electrum* 5: 153–6.

Steinby, M. (1978), 'Ziegelstempel von Rom und Umgebung', in H. Gärtner (ed.), *Paulys Real-Encyclopädie der Classischen Altertumswissenschaft. Suppl. XV*, Munich, 1489–531.

Steinwenter, A. (1916), 'Instrumentum', in W. Kroll (ed.), *Paulys Real-Encyclopädie der Classischen Altertumswissenschaften. IX*, Stuttgart, 1587–9.

Stevenson, G. H., and A. W. Lintott (1999), 'Clubs, Roman', in S. Hornblower and A. Spawforth (eds.), *The Oxford Classical Dictionary*, 3rd edn. Oxford, 352–3.

Stewart, N. S. (2002), *The Technology and Control of Mining in Roman Britain* (DPhil Thesis), Oxford.

Stoll, O. (1998), '"Silvanus im Steinbruch". Kulturtransfer durch Soldaten der legio IIII Scythica in Syrien?', in L. Schumacher (ed.), *Religion – Wirtschaft – Technik. Althistorische Beiträge zur Entstehung neuer kultureller Strukturmuster im historischen Raum Nordafrika/Kleinasien/Syrien*, St. Katharinen, 99–145.

Strack, P. L. (1933), *Untersuchungen zur römischen Reichsprägung des 2. Jahrhunderts*, Stuttgart.

Strobel, K. (1987), 'Einige Bemerkungen zu den historisch-archäologischen Grundlagen einer Neuformulierung der Sigillatenchronologie für Germanien und Rätien und zu wirtschaftsgeschichtlichen Aspekten der römischen Keramikindustrie', *Münstersche Beiträge zur Antiken Handelsgeschichte* 6/2: 75–115.

Strubbe, J. (1975), 'A group of imperial estates in Central Phrygia', *Ancient Society* 6: 229–50.

—— (1997), *ΑΡΑΙ ΕΠΙΤΥΜΒΟΙ. Imprecations Against Desecrators of the Grave in the Greek Epitaphs of Asia Minor. A Catalogue* (Inschriften griechischer Städte aus Kleinasien 52), Bonn.

Syme, R. (1969), 'Pliny, the Procurator', *Harvard Studies in Classical Philology* 73: 201–36.

Täckholm, U. (1937), *Studien über den Bergbau der Römischen Kaiserzeit*, Uppsala.

Talbert, R. J. A. (1984), *The Senate of Imperial Rome*, Princeton.

Taubenschlag, R. (1955), *The Law of Greco-Roman Egypt in the Light of the Papyri. 332 BC–640 AD*, 2nd edn. Warsaw.

Thomas, J. D. (1982), *The Epistrategos in Ptolemaic and Roman Egypt. Part II. The Roman Epistrategos* (Papyrologica Coloniensia 6/2), Opladen.

—— (1998), 'Communication between the Prefect of Egypt, the Procurators and the Nome Officials', in W. Eck (ed.), *Lokale Autonomie und römische Ordnungsmacht in den kaiserzeitlichen Provinzen vom 1. bis zum 3. Jahrhundert* (Schriften des Historischen Kollegs. Kolloquien 42), Munich, 181–95.

Thomasson, B. E. (1984), *Laterculi Praesidium. Volumen I*, Göteborg.

Todd, M. (1996), 'Pretia Victoria. Roman lead and silver mining on the Mendip Hills, Somerset, England', *Münstersche Beiträge zur Antiken Handelsgeschichte* 15/1: 1–18.

Tomber, R. S. (1996), 'Provisioning the desert. Pottery supply to Mons Claudianus', in D. M. Bailey (ed.), *Archaeological Research in Roman Egypt. The Proceedings of the Seventeenth Classical Colloquium of the Department of Greek and Roman Antiquities, British Museum, Held on 1–4 December, 1993* (Journal of Roman Archaeology, Supplement 19), Ann Arbor, 39–49.

Tomović, M. (1995), 'Roman mines and mining in the Mountain of Kosmaj', in B. Jovanović (ed.), *Ancient Mining and Metallurgy in Southeast Europe. International Symposium, Donji Milanovac, May 20–25: 1990* (Archaeological Institute Beograd Monographs 27), Bor, 203–12.

—— (2000), 'Kraku lu Jordan and Gold Mining and Metallurgy in Antiquity', *Starinar* 50: 155–83.

Tranoy, A. (1981), *La Galice romaine. Recherches sur le nord-ouest de la péninsule ibérique dans l'Antiquité* (Publications du Centre Pierre Paris 8), Paris.

Trincherini, P. R., P. Barbero, P. Quarati, C. Domergue, and L. Long (2001), 'Where do the lead ingots of the Saintes-Maries-de-la-Mer Wreck come from? Archaeology compared with physics', *Archaeometry* 43/3: 393–406.

Van der Veen, M. (1998), 'A life of luxury in the desert? The food and fodder supply to Mons Claudianus', *Journal of Roman Archaeology* 11: 101–16.

—— (2001), 'The botanical evidence', in V. A. Maxfield and D. P. S. Peacock (eds.), *Survey and Excavation. Mons Claudianus. 1987–1993. Volume II. Excavations: Part 1*, Cairo, 175–247.

Vanhove, D. (1996), 'Archaeological description', in D. Vanhove (ed.), *Roman Marble Quarries in Southern Euboea and the Associated Road Systems*, Leiden, New York, & Cologne, 16–44.

Velkov, V., and G. Aleksandrov (1994), *Epigrafski pametnici ot Montana i raiona*, Montana.

Velkov, V., and G. Aleksandrov (1988), '"Venatio Caesariana". Eine neue Inschrift aus Montana (Moesia Inferior)', *Chiron* 18: 271–6.

Veny, C. (1969–70), 'Diecisiete lingotes de plomo de una nave romana de Ses Salines (Mallorca)', *Ampurias* 31–2: 191–219.

—— (1979), 'Nuevos materiales de Moro Boti', *Trabajos de Prehistoria* 35: 465–79.

—— and D. Cerdá (1972), 'Materiales arqueológicos de dos pecios de la isla de Cabrera (Baleares)', *Trabajos de Prehistoria* 29: 299–310.

Vetters, H. (1977), 'Virunum', in H. Temporini and W. Haase (eds.), *Aufstieg und Niedergang der Römischen Welt. Geschichte und Kultur Roms im Spiegel der Neueren Forschung. II. Principat, Sechster Band. Politische Geschichte (Provinzen und Randvölker: Lateinischer Donau-Balkanraum)*, Berlin & New York, 302–54.

—— (1980), 'Review of "Geza Alföldy, Noricum, London-Boston 1974"', *Anzeiger für die Altertumswissenschaft* 30: 37–49.

Vittinghoff, F. (1953), 'Portorium', in *Paulys Real-Encyclopädie der Classischen Altertumswissenschaften XXII/1*, 346–400.

—— (1974), 'Das Problem des "Militärterritoriums" in der vorseverischen Kaiserzeit', in *Atti del convegno internazionale sul tema. I diritti locali nelle province romane con particolare riguardo alle condizioni giuridiche del suolo. Roma, 26–28 ottobre 1971*, Rome, 109–24.

Vives y Escudero, A. (1924), *La moneda hispánica*, Madrid.

Vulpe, R. (1976), 'Colonies et municipes de la Mesie inférieure', in R. Vulpe (ed.), *Studia Thracologica*, Bucharest, 289–314.

Waelkens, M. (1977), 'Phrygian votive and tombstones as sources of the social and economic life in Roman antiquity', *Ancient Society* 8: 277–315.

—— (1982), *Dokimeion. Die Werkstatt der repräsentativen kleinasiatischen Sarkophage. Chronologie und Typologie ihrer Produktion* (Archäologische Forschungen 11), Berlin.

—— (1985), 'From a Phrygian quarry. The provenance of the statues of the Dacian prisoners in Trajan's Forum at Rome', *American Journal of Archaeology* 89: 641–53.

—— (1986), *Die kleinasiatischen Türsteine. Typologische und epigraphische Untersuchungen der kleinasiatischen Grabreliefs mit Scheintür*, Mainz.

Wagner, W. (1938), *Die Dislokation der römischen Auxiliarfomationen*, Berlin.

Wagner, G. (1977), 'Legio IIII Scythica in Zeugma am Euphrat', in D. Haupt and H. G. Horn (eds.), *Studien zu den Militärgrenzen Roms II. Vorträge des 10. Internationalen Limeskongresses in der Germania inferior*, Cologne & Bonn, 517–39.

Wahl, J. (1993), 'Três Minas. Vorbericht über die archäologischen Ausgrabungen im Bereich der Römischen Goldbergwerke 1986/87', in H. Steuer and U. Zimmermann (eds.), *Montanarchäologie in Europa. Berichte zum Internationalen Kolloquium 'Frühe Erzgewinnung und Verhüttung in Europa' in Freiburg im Breisgau vom 4. bis 7. Oktober 1990*, Sigmaringen, 123–52.

Wallace, S. L. (1938), *Taxation in Egypt from Augustus to Diocletian*, New York.

Waltzing, J. P. (1895), *Étude historique sur les corporations professionnelles chez les Romains depuis les origines jusqu'à la chute de l'Empire d'Occident. Tome I*, Louvain.

—— (1899), *Étude historique sur les corporations professionnelles chez les Romains depuis les origines jusqu'à la chute de l'Empire d'Occident. Tome III*, Louvain.

Ward-Perkins, J. B. (1971), 'Quarrying in antiquity. Technology, tradition and social change', in *Proceedings of the British Academy. Volume LVII*, London, 137–58.

—— (1992*a*), 'Nicomedia and the marble trade (papers of the British School at Rome 48 (1980))', in H. Dodge and B. Ward-Perkins (eds.), *Marble in Antiquity. Collected Papers of J. B. Ward-Perkins* (Archaeological Monographs of the British School at Rome 6), Rome & London, 61–106.

—— (1992*b*), 'The roman system in operation (second shuffrey lecture)', in H. Dodge and B. Ward-Perkins (eds.), *Marble in Antiquity. Collected Papers of J. B. Ward-Perkins* (Archaeological Monographs of the British School at Rome 6), Rome & London, 23–30.

Watson, A. (1998), *The Digest of Justinian* (2 vols.), Philadelphia.

Wattenbach, W. (1870), *Passio Sanctorum Quattuor Coronatorum*, Leipzig.

Weaver, P. R. C. (1972), *Familia Caesaris. A Social Study of the Emperor's Freedmen and Slaves*, Cambridge.

Weber, E. (1995), 'Wieder einmal: die statio Esc-', in M. A. Speidel and R. Frei-Stolba (eds.), *Römische Inschriften—Neufunde, Neulesungen und Neuinterpretationen. Festschrift für Hans Lieb. Zum 65. Geburtstag dargebracht von seinen Freunden und Kollegen*, Basel, 171–8.

Wedenig, R. (1997), *Epigraphische Quellen zur städtischen Administration in Noricum* (Aus Forschung und Kunst 31), Klagenfurt.

Wells, C. M. (1972), *The German Policy of Augustus. An Examination of the Archaeological Evidence*, Oxford.

Werner, M. R. (1986), 'The Moesian limes and the imperial mining districts', in *Studien zu den Militärgrenzen Roms III. 13. Internationaler Limeskongress, Aalen 1983: Vorträge*, Stuttgart, 561–4.

Wheeler, E. L. (2000), 'Legio XV Apollinaris: From Carnuntum to Satala—and beyond', in Y. Le Bohec (ed.), *Les légions de Rome sous le Haut-Empire. Actes du Congrès de Lyon, 17–19 septembre, 1998* (Collection du Centre d'Études Romaines et Gallo-Romaines. Nouvelle série, no. 20), Lyon, 259–308.

Wickert, L. (1931), *Bericht über eine zweite Reise zur Vorbereitung von CIL II, Suppl. 2*, Berlin.

—— (1954), 'Princeps', in K. Ziegler (ed.), *Paulys Real-Encyclopädie der classischen Altertumswissenschaft. Band XXII/2*, Stuttgart, 1998–2296.

Wierschowski, L. (1984), *Heer und Wirtschaft. Das römische Heer der Prinzipatszeit als Wirtschaftsfaktor*, Bonn.

Wilkes, J. J. (1969), *Dalmatia*, London.

—— (2004), 'Cultural Identities in the Illyrian Provinces (2nd Century BC – 3rd Century AD): Some Old Problems Re-examined', in G. Urso (ed.), *Dall' Adriatico al Danubio. L'Illyrico nell' età greca e romana. Atti del convegno internazionale Cividale del Friuli, 25–27 settembre 2003*, Pisa, 309–18.

Willies, L. (1997), 'Roman mining at Río Tinto, Huelva, Spain', *Mining History* 13/3: 1–29.

WILMANNS, J. C. (1995), *Der Sanitätsdienst im Römischen Reich. Eine sozialgeschichtliche Studie zum römischen Militärsanitätswesen nebst einer Prosopographie des Sanitätspersonals* (Medizin der Antike. Beiträge zur antiken Medizin und zu deren Fortwirken 2), Hildesheim.

WILSON, D. R. (1971), 'Roman Britain in 1970. I. Sites explored', *Britannia* 2: 243–88.

WILSON, A. (2002), 'Machines, power and the ancient economy', *Journal of Roman Studies* 92: 1–32.

WINCKLER, A. (1895), 'Description de la voie romaine de Simittu-Colonia (Chemtou) à Tabraca (Tabarka)', *Revue Tunisienne*, 38–47.

WINKLER, G. (1969), *Die Reichsbeamten von Noricum und ihr Personal bis zum Ende der Römischen Herrschaft* (Österreichische Akademie der Wissenschaften. Philosophisch-Historische Klasse. Sitzungsbericht, 261. Band, 2. Abhandlung), Vienna.

—— (1977), 'Noricum und Rom', in H. Temporini and W. Haase (eds.), *Aufstieg und Niedergang der Römischen Welt. Geschichte und Kultur Roms im Spiegel der Neueren Forschung II. Principat, Sechster Band. Politische Geschichte* (Provinzen und Randvölker: Lateinischer Donau-Balkanraum), Berlin & New York, 183–262.

WOLLMANN, V. (1989), 'Nouvelles données concernant la structure socio-ethnique de la zone minière de la Dacia superior', in C. Domergue (ed.), *Minería y metalurgia en las Antiguas Civilizaciones Mediterrâneas y Europeas, Coloquio Internacional Asòciado. Madrid, 24–28 Octobre, 1985*, Madrid, 107–18.

—— (1996), *Minertul Metalifer, Extragerea Sarii si Carierele de Piatra în Dacia Romana. Der Erzbergbau, die Salzgewinnung und die Steinbrüche im römischen Dakien* (Bibliotheca Musei Napocensis XIII. Veröffentlichungen aus dem Deutschen Bergbau-Museum Bochum 63), Cluj-Napoca & Klausenburg.

WOLTERS, R. (1999), *Nummi Signati. Untersuchungen zur römischen Münzprägung und Geldwirtschaft* (Vestigia 49), Munich.

WOODS, A. (1987), 'Mining', in J. S. Wacher (ed.), *The Roman World*, 611–34.

WOYTEK, B. (2004*a*), 'Die Metalla-Prägungen des Kaisers Traian und seiner Nachfolger', *Numismatische Zeitschrift* 111/12: 35–68.

—— (2004*b*), 'Die Metalla-Prägungen des Kaisers Traian und seiner Nachfolger: Supplementum', *Mitteilungen der Österreichischen Numismatischen Gesellschaft* 44/4: 134–9.

WRIGHT, R. P., and M. W. C. HASSALL (1971), 'Roman Britain in 1970. II. Inscriptions', *Britannia* 2: 289–304.

WUILLEUMIER, P. (1948), *L'administration de la Lyonnaise sous le Haut-Empire*, Paris.

YALOURIS, E. (1986), 'Notes on the topography of Chios', in J. Boardman and C. E. Vaphopoulou-Richardson (eds.), *Chios. A Conference at the Homereion in Chios 1984*, Oxford, 141–68.

YOUTIE, H. C. (1978*a*), 'P. Mich. inv 974: *ΠΑΡΑΛΗΜΘΙΣ ΙΔΙΩΤΙΚΩΝ ΟΝΩΝ*', *Zeitschrift für Papyrologie und Epigraphik* 28: 245–8.

—— (1978*b*), 'Supplies for soldiers and stonecutters (P.Mich.inv. 6767)', *Zeitschrift für Papyrologie und Epigraphik* 28: 251–4.

Zaninović, M. (1977), 'The economy of Roman Dalmatia', in H. Temporini and W. Haase (eds.), *Aufstieg und Niedergang der Römischen Welt. Geschichte und Kultur Roms im Spiegel der Neueren Forschung II. Principat, Sechster Band. Politische Geschichte* (Provinzen und Randvölker: Lateinischer Donau-Balkanraum), Berlin & New York, 767–809.

Zawadzki, T. (1960), 'Sur une inscription de Phrygie relative au cursus publicus', *Revue des Études Anciennes* 62: 80–94.

Zucca, R. (1991), 'Le massae plumbae di Adriano in Sardegna', in A. Mastino (ed.), *L'Africa Romana. Atti dell' VIII convegno di studio. Cagliari, 14–16 dicembre 1990*, Sassari, 797–826.

—— (2001), 'Additamenta Epigraphica. All' amministratione della Sardegna da Augusto all' invasione Vandalica', in G. Angeli Bertinelli and A. Donati (eds.), *Varia Epigraphica. Atti del Colloquio Internazionale di Epigrafia. Bertinoro, 8–10 giugno 2000*, Faenza, 513–35.

Zwicky, H. (1944), *Zur Verwendung des Militärs in der Verwaltung der römischen Kaiserzeit*, Winterthur.

Index of Sources

A) Inscriptions

C) Literary Sources

D) Quarry Labels (Appendix nos.)

General Index

B

F

H

P

Q

W